Advances in Transportation Meteorology

Advances in Transportation Meteorology

Editors

Duanyang Liu
Hongbin Wang
Shoupeng Zhu

MDPI • Basel • Beijing • Wuhan • Barcelona • Belgrade • Manchester • Tokyo • Cluj • Tianjin

Editors

Duanyang Liu
Key Laboratory of Transportation Meteorology of China Meteorological Administration
Nanjing Joint Institute For Atmospheric Sciences
Nanjing
China

Hongbin Wang
Key Laboratory of Transportation Meteorology of China Meteorological Administration
Nanjing Joint Institute For Atmospheric Sciences
Nanjing
China

Shoupeng Zhu
Key Laboratory of Transportation Meteorology of China Meteorological Administration
Nanjing Joint Institute For Atmospheric Sciences
Nanjing
China

Editorial Office
MDPI
St. Alban-Anlage 66
4052 Basel, Switzerland

This is a reprint of articles from the Special Issue published online in the open access journal *Atmosphere* (ISSN 2073-4433) (available at: www.mdpi.com/journal/atmosphere/special_issues/Transportation_Meteorology).

For citation purposes, cite each article independently as indicated on the article page online and as indicated below:

LastName, A.A.; LastName, B.B.; LastName, C.C. Article Title. *Journal Name* **Year**, *Volume Number*, Page Range.

ISBN 978-3-0365-8461-4 (Hbk)
ISBN 978-3-0365-8460-7 (PDF)

Contents

atmosphere

Review

Observations and Forecasts of Urban Transportation Meteorology in China: A Review

Shoupeng Zhu [1], Huadong Yang [1,*], Duanyang Liu [1,*], Hongbin Wang [1], Linyi Zhou [1], Chengying Zhu [1], Fan Zu [1], Hong Wu [1], Yang Lyu [2], Yu Xia [3], Yanhe Zhu [2], Yi Fan [2], Ling Zhang [2] and Xiefei Zhi [2]

1 Key Laboratory of Transportation Meteorology of China Meteorological Administration, Nanjing Joint Institute for Atmospheric Sciences, Nanjing 210041, China
2 Collaborative Innovation Center on Forecast and Evaluation of Meteorological Disasters, Key Laboratory of Meteorological Disaster, Ministry of Education, Nanjing University of Information Science & Technology, Nanjing 210044, China
3 Institute of Urban Meteorology, China Meteorological Administration, Beijing 100089, China
* Correspondence: huadonyang@hotmail.com (H.Y.); liuduanyang@cma.gov.cn (D.L.)

Abstract: Against the backdrop of intensified global warming, extreme weather events such as dense fog, low visibility, heavy precipitation, and extreme temperatures have been increased and enhanced to a great extent. They are likely to pose severe threats to the operation of urban transportation and associated services, which has drawn much attention in recent decades. However, there are still plenty of issues to be resolved in improving the emergency meteorological services and developing targeted urban transportation meteorological services in modern cities. The present review briefly illustrates the current cutting-edge developments and trends in the field of urban transportation meteorology in China, including the establishment of observation networks and experiments and the development of early warning and prediction technologies, as well as the related meteorological commercial services. Meanwhile, reflections and discussions are provided in terms of the state-of-the-art observation channels and methods and the application of numerical model forecasts and artificial intelligence. With the advantages of various advanced technologies from multiple aspects, researchers could further expand explorations on urban transportation meteorological observations, forecasts, early warnings, and services. Associated theoretical studies and practical investigations are also to be carried out to provide solid scientific foundations for urban transportation disaster prevention and mitigation, for implementing the action of meteorological guarantees, and for the construction of a high-quality smart society.

Keywords: urban meteorology; transportation meteorology; observation; forecast; early warning; review; China

Citation: Zhu, S.; Yang, H.; Liu, D.; Wang, H.; Zhou, L.; Zhu, C.; Zu, F.; Wu, H.; Lyu, Y.; Xia, Y.; et al. Observations and Forecasts of Urban Transportation Meteorology in China: A Review. *Atmosphere* **2022**, *13*, 1823. https://doi.org/10.3390/atmos13111823

Academic Editor: Eduardo García-Ortega

Received: 30 August 2022
Accepted: 26 October 2022
Published: 2 November 2022

1. Introduction

In the context of the accelerated development of state-of-the-art information technologies, the smart city is a new urban development model that fully utilizes cloud computing, Internet of Things, and other innovative technological means to make intelligent responses to various demands for public services, social management, industrial operations, and other activities [1]. Transportation is one of the basic and strategic industries of the national economy in China, providing crucial support for cities' sustainable development [2]. Therefore, smart transportation is considered a priority area for the construction of smart cities and smart countries, of which the meteorological service is an inseparable part [3].

In fact, meteorology is one of the key factors affecting the safety and operation of urban transportation [4]. Taking China as an example, frequently occurring weather disasters tend to induce serious traffic accidents or blockages and to causes damage to traffic facilities. This is likely to generate severe threats to the safety and property of humans, as well as social and economic development. Meanwhile, changes in meteorological conditions would have an impact on the vehicle itself, the road conditions, the driver's judgment, and their responsiveness in the driving process, as well as the vehicle's interior environment. Moreover, different meteorological conditions may have different impacts on transportation. Compared with highways, although urban transportation is characterized by generally lower speeds of vehicles, it is confronted with a larger flow, more pedestrians, and more complex road conditions. Therefore, the impacts of meteorological conditions on urban roads and vehicle travel, as well as other traffic channels such as subways, are still significant and should not be ignored.

For instance, on 21 July 2012, Beijing suffered one of the most severe rainstorms in local history [5], causing the urban transportation system to be almost paralyzed. In total, 95 roads in the urban area were cut off due to the heavy precipitation. Severe waterlogging and circuit breakage brought huge losses to the citizens' lives and property, with at least 79 deaths and 1.84 billion dollars in economic losses [6]. More recently, in 2021, the Zhengzhou 7·20 (20 July) torrential rain event registered a record hourly precipitation rate of 201.9 mm and the 24-h precipitation reached 624.1 mm [7]. The disaster led to serious water accumulation in the Wulongkou Parking Lot of Zhengzhou Metro Line 5 and the surrounding area, causing one metro operating train to stop in the section ahead of this and resulting in the deaths of 14 passengers. Moreover, a number of facilities and equipment were damaged, and the entire network of the Zhengzhou subway was forced to shut down for more than 50 days. Therefore, it is certainly necessary to carry out real-time monitoring, early warning, and forecasting studies and focus on the services of urban transportation meteorology. Effective predictions of the meteorological impacts on urban transportation contribute to the development of effective measures of transportation management regarding safety and smoothness, disaster prevention, and impact mitigation in advance.

From the perspective of national policies, taking China as an example, the China Meteorological Administration (CMA) released the "Action Plan for the Development of Smart Meteorological Services (2019–2023)" in 2018 [8], making "the demonstration of smart traffic meteorological services" one of the key assignments to adapt to the high-quality construction of meteorological modernization and to improve the intelligence of meteorological services. In terms of the urban transportation meteorology, further integrations of road monitoring, intelligent grid forecasts, traffic management, map navigation, and other associated elements are required to establish impact-based urban transportation meteorological service models and indicators. On this basis, the business capability could be improved to identify the risks of road sections and transportation safety levels affected by severe weather.

Subsequently, in November 2021, the CMA, together with the China Ministry of Public Security, the China Ministry of Transport, the China State Railway Administration, and the China State Post Bureau, jointly formulated the "14th Five-Year Plan for Transportation Meteorological Safeguard" [9]. It points out the current state, where (1) the construction of a transportation meteorological monitoring station network has been taking shape; (2) the business services of transportation meteorological forecasting and early warning are developing rapidly, and (3) the cooperation mechanisms among multiple departments have been basically formed. However, there are still issues waiting to be resolved, such as (1) the pertinence of the transportation meteorological service needs to be strengthened and (2) the information fusion of transportation and meteorology has not yet been analyzed in depth. Correspondingly, the goals of (1) developing digital transportation meteorological forecast products based on intelligent grid forecasts and (2) developing technologies for transportation meteorological forecasting and early warning are proposed to promote transportation meteorology innovation and to create a high-quality transportation meteorological service system.

Moreover, in May 2022, to optimize the supply of high-quality meteorological services for the social economy, the China State Council issued the "Outlines for High-quality Meteorological Development (2022–2035)" [10], which specifically refines the important role of the meteorological service in the transportation industry, especially in the urban area. It requires the further implementation of a meteorological safeguard for the construction of a strong transportation system; the exploration of a modern, comprehensive transportation meteorological service platform; the strengthening of the transportation meteorological monitoring, forecasting, and early warning capabilities, and finally the construction of a smart system of urban transportation meteorological services.

In recent decades, with the rapid development of the social economy and urbanization, the rising demand for urban meteorological services has posed new challenges for professional meteorological fields. It is also expected to accelerate the deepening and expanding processes of these fields, such as urban transportation meteorology. In fact, in the context of intensified global warming, extreme meteorological events occur from time to time and are likely to become increasingly frequent, including dense fog, low visibility, heavy precipitation, extreme temperatures, and other analogous phenomena [11–14], which pose severe threats to the operation of urban transportation and associated services. Currently, there are still many issues to be resolved in improving the emergency meteorological services and developing targeted urban transportation meteorological services in modern cities. The study of urban transportation meteorological monitoring, early warning, and forecasting is of great scientific significance and application value in dealing with weather disasters, the rational planning of urban facilities, and the improvement of urban operating quality.

The present paper briefly reviews the current cutting-edge developments and trends in the field of urban transportation meteorology, especially in China, including the establishment of observation networks and experiments in Section 2 and the development of early warning and prediction technologies in Section 3. Meanwhile, Section 4 describes the related meteorological commercial services. Finally, reflections and discussions are provided in Section 5 in terms of the state-of-the-art observation channels and methods, the application of numerical model forecasts, and artificial intelligence. These are to contribute to a scientific basis and reference towards the operational urban transportation meteorological safeguard and urban transportation disaster prevention and mitigation in China.

2. Urban Transportation Meteorological Observation

2.1. Urban Meteorological Observation Network

Since the beginning of the 21st century, European and American countries, as well as Japan, have carried out meteorological observation experiments for urban areas. Among them, short-term observation experiments on various factors, such as near-surface turbulence, the vertical structure of the urban boundary layer, and the traceability and dispersion of air pollution are mostly implemented for periods of within one year. Selected examples are provided in Table 1 and are briefly described in the following paragraphs. On the other hand, there are also several long-term observation experiments lasting for longer than this (Table 2). Such observations are effectively utilized not only to reveal the characteristics of the urban atmosphere but also to validate and promote the development of numerical models. Various data based on these urban observation networks could be produced for research institutions and the public.

Table 1. Short-term urban meteorological observation experiments.

Name	Country	Time	Content
URBAN 2000 [15]	USA	October 2000	Tracer and meteorological field.
Joint Urban [16]	USA	July 2003	Tracer, dispersion, meteorological field, boundary layer structure, and urban energy balance.
Pentagon Shield [17]	USA	2004	Boundary layer thermal structure, wind field, tracer, and dispersion.

Table 1. *Cont.*

Name	Country	Time	Content
Madison Square Garden [18]	USA	2004	Wind field, tracer, and dispersion.
HEAT [19]	USA	2005	Air pollution, meteorological field, convective and mesoscale process.
ESCOMPTE [20]	France	July–September 2005	Dispersion, air pollution, meteorological field, boundary layer structure, and urban energy balance.
CAPITOUL [21]	France	June–July 2001	Tracer, dispersion, air pollution, meteorological field, boundary layer structure, and urban energy balance.
DAPPLE [22]	UK	February 2004	Tracer, dispersion, air pollution, and meteorological field.
REPARTEE [23]	UK	February 2005	Tracer, dispersion, air pollution, meteorological field, and boundary layer structure.
ClearfLo [24]	UK	May 2002	Air pollution, meteorology field, boundary layer structure, urban energy balance, and mesoscale process.
BUBBLE [25]	Switzerland	July 2006	Tracer, dispersion, meteorological field, boundary layer structure, and urban energy balance.
TOMACS [26]	Japan	October 2006	Meteorological field, convective and mesoscale process.

Table 2. Long-term urban meteorological observation experiments.

Name	Country	Time	Content
NYC mesonet [27]	USA	2003–present	Meteorological field, convective and mesoscale process.
DCNet [27]	USA	2003–present	Tracer and dispersion.
Helsinki Testbed [28]	Finland	2005–present	Dispersion, meteorological field, boundary layer structure, convective and mesoscale process.
METROS [29]	Japan	2002–2005	Boundary layer structure, convective and mesoscale process.
SUIMON [30]	China	2000–present	Air pollution, meteorological field, BL structure, urban energy balance, convective and mesoscale process.

In October 2000, the URBAN 2000 project [15] conducted several field experiments in Salt Lake City, which investigated the transportation and diffusion around a single downtown building, through the downtown area, and into the greater urban area. Moreover, meteorological measurements were conducted, including temperature and the 2D/3D wind field across the urban area. One mobile van was utilized to measure net radiation, sensible heat flux, and three levels of temperature to 18 m AGL. The project aims to evaluate and improve the hierarchy of atmospheric models being developed for simulating toxic agent dispersal from potential terrorist activities in urban environments. In the following Joint Urban 2003 project conducted in Oklahoma City [16], remote sensing instruments were used (radar profilers, lidars, sodars) to form a dense network. The network continuously measures the detailed wind and turbulence characteristics of the urban atmosphere from the ground through several kilometers above the ground, which lasts over one month. Another difference between URBAN 2000 and Joint Urban 2003 is that the former focuses on the urban nocturnal boundary layer (stable to neutral atmospheric conditions), while the latter focuses on the urban daytime boundary layer (neutral to unstable).

With a distinct perspective, the Pentagon Shield field program was implemented from 9 April to 16 May 2004 in Washington, D.C. [17]. It focuses on the effects of a single building (the Pentagon) on the flow field, chemical tracer transport, and dispersion. A unique aspect is the use of two higher-resolution scanning Doppler lidars with overlapping and synchronized scan patterns that work together, providing detailed data with 100-m resolution. Moreover, the Madison Square Garden field experiment [18] in New York City addresses its goal in cities with tall skyscrapers. They found that the mean wind speed and direction on the tops of tall downtown buildings are approximately equal to those near the surface at a nearby airport, but the mean wind speed is three times larger than that at street level. The HEAT program [19], conducted in Houston, Texas during the summer of 2005, mainly collected electrical data from the National Lightning Detection Network and a lightning-mapping system (LDAR II), and atmospheric variables such as temperature, moisture, winds, and aerosol from balloon-borne soundings, tethered atmospheric observation systems, and wind profilers, as well as mobile sounding units, airborne instruments, and three Doppler radar devices.

Similar field observation experiments were conducted in France, the UK, Switzerland, and Japan. The ESCOMPTE program [20], conducted in the Marseilles-Berre area in the south of France during Summer 2001, covered an area of 120 × 120 km. By utilizing surface measurement networks, remote sensing, and ship-borne, balloon-borne, and airplane measurements, the mean standard meteorological parameters and turbulent fluxes, ozone, ozone precursors, photochemically active trace gases, and aerosols were measured, and a database was therefore established. The CAPITOUL experiment [21], in the city of Toulouse in Southwest France, was conducted for one year from February 2004. Focusing on the urban climate, it revealed that the urban surface energy balance differs between summer and winter, while the city impacts the boundary layer when an urban breeze is observed. Aiming at improving the understanding of the physical processes affecting the street- and neighborhood-scale flow of air, traffic, and people, the DAPPLE project [22] was conducted at the intersection of Marylebone Road and Gloucester Place, London, in 2003. Data included the mean and turbulent winds at the intersection, carbon monoxide and nitrogen dioxide, traffic flow and composition, as well as personal exposure measurements of PM2.5, ultrafine particle counts, etc.

With respect to the urban meteorological observation networks in China, they have been developed and generally matured in the three major urban clusters, i.e., Beijing–Tianjin–Hebei, the Yangtze River Delta, and the Pearl River Delta. In addition to aiding scientific research, these observation networks also effectively support the forecasts of urban meteorology and transportation meteorology, as well as the decision making of associated departments [31,32].

In the area of Beijing–Tianjin–Hebei, scientific experiments in the Study of Urban Impacts on Rainfall and Fog/Haze (SURF) were conducted during 2014–2019, among which there were three main observational experiments [33]. Two of them focused on the summer thunderstorm processes (July–August 2015 and July–September 2016) to study the effects of urbanization on precipitation-triggering mechanisms, movement, and intensity. The other experiment focused on the winter haze (November 2015–January 2016) to study aerosol sources and processes of transportation and transformation. In this area, 1992 automatic weather stations were utilized, and the meteorological data were collected every 5 min. Radiosondes were launched twice-daily at 0000 and 1200 UTC, with additional soundings at 0600 UTC during flood seasons at the Nanjiao site. Based on existing operational instruments, the planetary boundary layer network was augmented with 5 wind profilers,

1 scanning Doppler lidar, 2 aerosol micropulse lidar, and 10 ceilometers. The turbulent flux and associated meteorological data were collected at the towers in Beijing and Tianjin. Based on these combined observations, the hockey-stick transition turbulence–wind relationships over the urban canopies [34,35] have been confirmed, which describes the roles of non-local large coherent turbulence eddies during turbulence intensity and mean wind speed under near-neutral conditions.

As for the Yangtze River Delta region, an urban meteorological and environmental meteorological observation network has been formed, consisting of transportation meteorology, urban environmental meteorology, ecological meteorology, agricultural meteorology, marine meteorology, climate resources, drought monitoring, lightning monitoring, hydrometeorology, and some other specialized observation systems. Aiming at extensive urban meteorological observations for megacities, Shanghai has built a combined ground-based and space-based system named Shanghai's Urban Integrated Meteorological Observation Network (SUIMON) [30]. By utilizing a dense observation network and various instrument types, the network collects wind, temperature, humidity, rain, and pressure every 1 min, and it has extended the observation to the vertical plane, thus providing a four-dimensional dataset of the area. A high frequency of severe convective precipitation events was found over the urban area and the mouth of the Yangtze River, which matches well with the spatial distribution of cloud-to-ground flash density. Suzhou has also constructed an urban heat island monitoring network for urban heat island-associated investigations. It was found that when the tree cover rate reached 40%, the daily average concentration of major air pollutants in the urban area decreases significantly [36]. In Hangzhou, a comprehensive detection system for compound atmospheric pollution has been established with a full range of detection items. Results show that the most recent decade of urban development in Hangzhou substantially reduced the atmospheric diffusion, and pollutant concentrations rose quickly in the urban area [37]. In Nanjing and Hefei, city-wide traffic visibility monitoring networks have been constructed for urban transportation meteorological services.

In the urban cluster of the Pearl River Delta, a comprehensive urban meteorological observation system has been developed, including a dense network of ground-based automatic stations, a variety of ground-based remote sensing equipment (e.g., wind profile radar, aerosol radar, and Doppler radar), a network of urban atmospheric composition monitoring stations, and a Global Positioning System/Meteorology (GPS/MET) water vapor monitoring network. Taking Shenzhen as an example, a generally complete urban meteorological disaster monitoring system has been formed since 1994, as well as a climate monitoring system [38]. Over the last few decades, the Shenzhen Urban Meteorological Observing Network (SUMON) has been developed comprehensively, with its spatial and temporal resolutions reaching 3.5 km and 1 min, respectively.

In addition, the meteorological observation towers also provide important support in the study of urban boundary layer physics and the atmospheric environment, as well as in the observation and monitoring of transportation meteorological-related elements. For instance, the 325-m flux tower (39°58′ N, 116°22′ E) at the Institute of Atmospheric Physics in Beijing, the 255-m flux tower (39°06′ N, 117°10′ E) at the Tianjin Meteorological Service in Tianjin, and the 356-m flux tower (22°40′ N, 113°54′ E) at the Shiyan Meteorological Observatory in Shenzhen provide diversified and solid foundations for local urban meteorological observations [39–41]. Databases of turbulent and gradient wind, temperature, and humidity are obtained at different layers, favoring more comprehensive studies of turbulent statistic characteristics and turbulent fluxes over the urban areas, as well as their synoptic and climatological features [42–44].

In recent years, via ground-based remote sensing techniques including wind profilers, microwave radiometers, and laser lidar, etc., three-dimensional observations of atmospheric temperature, humidity, wind field, water condensate, and aerosol have been well observed to enrich the local urban observation networks in several megacities and to further enhance the usage of these thermal and dynamic factors in numerical models [45]. Moreover, it is suggested that the effective combination of this equipment would help to obtain atmospheric profiles with higher spatial and temporal resolutions, which is also to be further developed in the future.

In general, during the construction of urban meteorological observation networks such as the abovementioned ones, the characteristics and development directions have been proposed to include five aspects [30]: multiple platforms, multiple variables, multiple scales, multiple linkages, and multiple functions. Detailed information is provided in Figure 1. In simple terms, the multiple platforms detecting multiple variables and considering multiple scales are to compose the comprehensive observation network via multiple linkages, which ultimately serve the users with multiple functions.

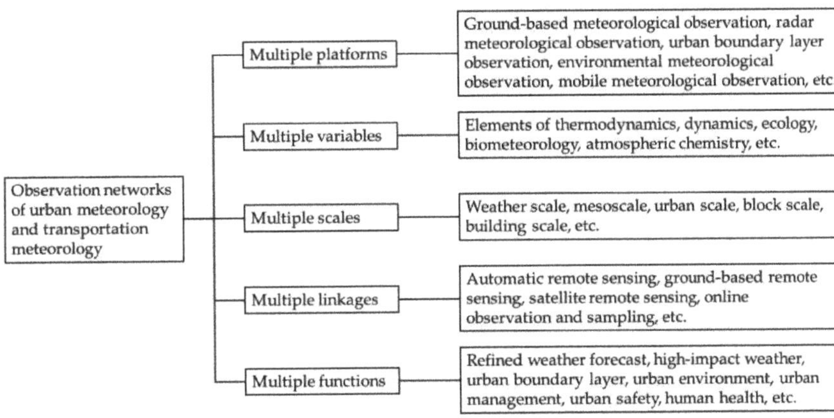

Figure 1. Characteristics and development directions of urban transportation meteorological observation network.

2.2. *Urban Meteorological Outfield Observation Experiments*

From the end of the last century to the beginning of this century, a series of large-scale outfield observation experiments have been successfully explored throughout the world, focusing on plenty of topics, among which the urban boundary layer meteorology and urban air pollution are two important and popular aspects [46]. In Beijing, China, the Beijing City Atmospheric Pollution Observation Field Experiment (BECAPEX, 2001–2003) and the Beijing Urban Boundary Layer Experiment (BUBLEX, 2004–2005) have been implemented to obtain three-dimensional structure integrated images of Beijing's atmospheric dynamics and chemical process [47]. More details have also been included, such as urban observations of the boundary layer, rainstorm adaptability, complex terrain and atmospheric circulation, flux and energy balance, as well as the urban thermal environment [48]. Associated studies have revealed that the air pollution in Beijing usually comes from local sources, but the sources in the vicinity also have a prominent influence [49]. Nanjing City has also carried out several experiments (e.g., periods in 2004 and 2006) with urban boundary layer observations to investigate the urban mixed layer, convective entrainment zone, and cloud feedback processes via LIDAR (Light Detection and Ranging), radiosonde, meteorological towers, turbulence measurements, and many other techniques [50–52], based on which the transition features of the mixed layer and the entrainment zone over Nanjing City have been revealed by comparison between urban and suburban areas. During the past decade, the urban meteorological observation experiments have been mainly focused on the research of high-impact weather mechanisms and their mitigation countermeasures, on the urban effects on weather and climate, and on the mutual feedback between urban aerosols and weather climate [53]. Meanwhile, the observation scope has been expanded from a single city in the past to multiple cities (city clusters and metropolitan areas) nowadays. These investigations could assist cities in facing, detecting, and discussing hazards such as storm surges, flooding, heatwaves, and air pollution episodes, especially in the changing climate.

There are also downscaling outfield experiments conducted. For instance, Sun Yat-sen University carried out an urban climate experiment in 2016–2017 in the suburbs of Guangzhou on a reduced scale [54]. The experimental site covered an area of 4800 m^2, which was far from surrounding buildings and had an impermeable surface. Two ideal street valley models were composed of roughly 2000 concrete building models, each with a height of 1.2 m, a width of 0.5 m, and a wall thickness of 1.5 cm. The effects of building heat capacities and street aspect ratios on the turbulence and temperature spatiotemporal characteristics of two-dimensional street valleys were investigated through outfield experiments under the typical non-stationary real meteorological conditions.

Moreover, through these outfield experiments on urban and transportation meteorology, the associated air pollution and source emission regarding $PM_{2.5}$, CO, SO_2, O_3, and even various heavy metals were also analyzed to investigate their spatial and temporal distributions and developing trends [55,56]. Studies have also revealed that, in winter, volatile organic compounds (VOCs) including benzene, toluene, ethylbenzene, and xylene (BTEX) could mainly be attributed to urban transportation, posing hazards for human health [57,58]. At the same time, corresponding assessments of health risks related to human exposure to urban transportation pollution and emissions are also currently under investigation using the databases from multiple experiments [59,60].

2.3. *Urban Transportation Meteorological Monitoring System*

Since the 1980s, the Standing International Road Weather Commission (SIRWEC) has been established worldwide by several countries to carry out studies on winter road weather predictions, winter road treatment methods, the Road Weather Information System (RWIS), and the Intelligent Transportation System (ITS). The frequent information exchange and technology sharing among members have promoted the development of urban transportation meteorology to a great extent in various countries [61].

In China, transportation meteorological monitoring mainly started in 2005, which is relatively late compared to developed countries [62]. At the early stage, it mostly focused on the highways outside of cities, while less attention has been paid to urban transportation [63,64]. At present, meteorological monitoring along urban transportation lines is being vigorously developed, and automatic meteorological monitoring stations have been deployed along the major roads. However, such monitoring equipment arrangements are still not sufficient; they are unevenly distributed, with little coverage in many provinces, especially in mountainous areas [65]. There is currently an urgent requirement to further increase and construct urban transportation meteorological stations [66].

In addition, remote sensing techniques have been widely used and have allowed effective progress in urban transportation meteorological monitoring applications due to the improvement in quantification precision, wide monitoring range, fast update time, and relatively low cost [67]. For instance, the unmanned aerial vehicle for remote sensing has such advantages as real-time measurement (wireless communication transmission), flexibility (fixed or mobile stations), high resolution (horizontally 3–5 m and vertically ~1 m), cost-effectiveness, etc., and can gather information in dangerous environments without any risk to flight crews, providing a powerful supplement for spaceborne remote sensing and airborne remote sensing [68]. Such technology has been increasingly used in obtaining spatial data, e.g., the conventional meteorological elements and urban transportation flow [69,70]. These improvements provide fundamental support for the implementation of the real-time monitoring of roads [71].

On the other hand, the usage of Internet of Things technology, which transmits data among multiple pieces of equipment via the internet in near-real time, provides an effective channel for the combination of meteorological information and urban transportation derived from multiple sensors [72–74]. By means of the automatic and timely communication between different equipment, it has generally realized the construction of urban transportation meteorological monitoring frameworks and plays a crucial role in the business operation of transportation management and emergency management [75,76]. Taking the Beijing Municipal Commission of Transportation as an example, based on the project titled "The internet of things application demonstration project of maintaining road traffic unblocked in extreme meteorological conditions", under the framework of the first batch of Internet of Things projects in Beijing, the meteorological information obtained by meteorological sensors is combined with transportation information, with schemes of prediction and early warning proposed regarding the depth of waterlogging, snow cover, ice formation, visibility, and others [77]. It has been successfully attempted and plays an important role in road transportation management under complex weather conditions [78]. Moreover, the Wuxi Meteorological Bureau of Jiangsu Province has also deployed Internet of Things-based urban waterlogging monitoring stations at some specific locations on main traffic roads [79]. Subsequently, a system of urban waterlogging monitoring and early warning has been constructed, realizing the automated, digitalized, and refined real-time online management of urban transportation meteorological conditions, especially the waterlogging risk [80].

With respect to rail-based urban transportation, based on sufficient experience in event prevention and emergency disposal, the Shanghai Urban Rail Transportation and Shanghai Meteorological Service Center jointly developed the Shanghai Rail Transportation Meteorological Assistant Decision-Making System in 2020, using the basic geographic data, observations, and monitoring information of local automatic weather stations within the Shanghai Meteorological Bureau. The real-time monitoring and risk warning of urban transportation meteorological disasters, as well as other functions, are therefore achieved [81]. In general, the system mainly consists of six main sections: a central map, risk warning, weather forecasts, impact, early warning information, and extreme weather statistics. By adopting technologically advanced monitoring methods and obtaining timely and effective early warning information, the meteorological risks of rail-based urban transportation could be "moved forward", building a solid foundation for improved operational urban rail transportation safety.

3. Urban Transportation Meteorological Early Warning and Forecast

The early warning and forecasting of urban transportation meteorology is closely related to the development of urban smart transportation, transportation planning, city management, and the improvement of citizen life. Previous studies [82] have pointed out that the existing issues needing to be addressed for the user mainly include street and channel wind speeds, precipitation and its phase state, road surface conditions, surface observation representativeness, refined forecasting (e.g., at the road scale), road surface temperature and visibility etc. In recent years, researchers have conducted diverse investigations on urban transportation meteorology in the field of smart cities based on a variety of statistical analysis techniques.

Silva et al. [83] pointed out that a precise weather forecast is one of the most crucial aspects in urban transportation and smart city big data analysis, which provides the underlying design basis for safe construction and production and stable system operation. Lu et al. [84] proposed to collect and analyze severe weather data reported on social media, and to use regression-based early warning models to estimate the urban transportation conditions. This could assist in urban transportation perception, forecasting, early warning, and decision-making, with intuitive visualization solutions. Wessel [85] statistically analyzes the impact of different weather phenomena on the cycling population, including not only the real-time weather conditions but also weather forecasts, especially regarding cloud, rainfall, snow, and thunderstorms, etc. They have a variety of leading or lagging impacts on the number of cyclists, which could be analyzed and obtained via statistical methodologies. The corresponding results could be beneficial for the formulation and promotion of policies in urban transportation planning and civilization construction. Taking the factors of seasons and weekday/non-working days into consideration, Simunek and Smutny [86] have established a transportation speed prediction conceptual model for lead times of within 1 week, combining meteorological elements such as air temperature, wind direction, wind speed, humidity, air pressure, and cloud cover, etc. It enriches the high-quality prediction of the urban transportation information, and it especially has a significant improvement effect on the estimation of arrival times and could aid the intelligent transportation system and urban transportation prediction.

Currently, artificial intelligence theories such as machine learning are flourishing, which brings not only new opportunities but also challenges to the field of urban transportation meteorology. Researchers are also trying to adopt relevant theories and techniques to explore the laws behind various meteorological phenomena [87]. In particular, deep learning models, with their high robustness and strong nonlinearity, have been widely used in the field of meteorological forecasting, including extrapolation forecasting of elements such as temperature and precipitation, multivariate statistical forecast modeling, and numerical model product applications, and also in urban transportation meteorological early warning and forecasting [88–90]. Some cases for nowcasting and predictions of short-term and even longer timescales are provided as follows.

For nowcasting and short-term urban transportation meteorological predictions, Jia et al. [91] introduced the precipitation factor into the deep neural network and Long Short-Term Memory (LSTM) methods, achieving a more accurate prediction of urban transportation flow than the raw deep learning networks. Nagy and Simon [92] have also demonstrated that both weather forecasts and their seasonal effects have important impacts on the prediction of urban transportation. Intelligent integrations of various temporal and spatial elements, such as weather, season, weekday/non-weekday, random events, and road conditions, could effectively improve the prediction of nonlinear transportation flows via deep learning frameworks such as the convolutional neural network (CNN) and LSTM. On this basis, Lee and Min [93] have constructed a Long Short-Term Memory Recurrent Neural Network (LSTM-RNN) model for urban transportation prediction using hourly observations and forecasts of temperature and precipitation, as well as the characteristics of urban activities affecting transportation operations. The model has been attempted and examined in Seoul, South Korea, and is demonstrated to be effective in improving the accuracy of urban transportation prediction. Moreover, Ali et al. [94] successfully established a deep hybrid neural network prediction model for urban transportation flow by using weather reality and forecasts as external factors affecting urban transportation flow, together with the spatiotemporal characteristics of daily urban transportation. Deb et al. [95] made full use of various deep learning methods to statistically analyze the time series correlations between weather status changes and transportation congestion magnitudes, and afterwards they conducted prediction experiments on transportation congestion by using regression analysis on the weather factors, which hence effectively improved the prediction ability regarding the urban transportation time. Based on the four situational factors of weather, season, weekday/non-weekday, and holiday/non-holiday, Ma et al. [96] constructed a daily urban transportation flow prediction model with a convolutional recurrent neural network and experimentally demonstrated the effectiveness and stability of the established scheme. Considering the nonlinear characteristics of traffic flow and the complex spatiotemporal correlations between transportation and weather, Nigam and Srivastava [97] defined a soft time threshold to capture the long-term impact of weather elements on transportation flow and proposed a hybrid CNN-LSTM model, which is capable of efficiently learning and predicting the transportation speed and flow issues in smart transportation operation and management. Tukymbekov et al. [98,99] designed an intelligent autonomous street light system based on the LSTM network using forecasts of weather and solar radiation, achieving the adaptive adjustment of the lighting system to effectively reduce energy consumption and serving urban transportation in a stable and safe manner. Moreover, with a weather-based transportation analysis method, Nasser and Simon [100] studied the relationship between transportation flow and weather factors at different frequencies and time intervals. It helps to reasonably determine and estimate the transportation flow under different meteorological conditions and to develop intelligent urban transportation systems in the construction of smart cities.

In addition, at longer timescales, based on urban transportation networks, social statistics, human flow data, and calendar data, as well as meteorological elements such as rainfall, snow, temperature, wind, etc., Zhou [101] simulated the correlation between the abovementioned factors and transportation events through deep learning algorithms, which helps to predict the frequency of possible accidents with cross-domain datasets and contributes to emergency management and decision-making in relevant departments. Furthermore, Ryu et al. [102] have constructed a time- and weather-aware deep learning neural network model with multiple modules using refined weather forecasts. It produces generally reliable and comprehensive long-term urban transportation condition predictions and effectively facilitates long-term urban transportation planning and management.

4. Commercial Services of Urban Transportation Meteorology

In the early 20th century, commercial meteorological services began to emerge worldwide and have been well developed, such as the numerical forecasts of the European Centre for Medium-Range Weather Forecasts [103,104], the marine meteorological navigation of the United States, the aviation meteorological services of Japan and New Zealand, and the transportation meteorological services of the United Kingdom and Finland [105,106]. At the beginning of the 21st century, several countries, including the United States, the United Kingdom, Germany, the Netherlands, Japan, and Singapore, etc., had already carried out the practice of smart transportation in the construction of smart cities and achieved practical success [107,108]. In October 2011, the United States National Weather Service of the National Oceanic and Atmospheric Administration installed multifunctional sensors on over 2000 passenger buses in New York. As the bus moves, data on temperature, humidity, and light levels can be collected every 10 s along the route, which are then immediately transmitted back to the National Weather Service center. Afterwards, the center integrates the collected meteorological, geographic, and many other types of data with official surveys. The natural disaster data network is also merged to capture the keys, to generate early warning system information, and to provide transportation meteorological services [109]. In 2014, Chicago launched the installation of street light sensors to collect urban pavement information and to detect meteorological data on temperature and wind. Using Internet of Things technology, all these sources are digitally connected, detected, analyzed, and integrated to allow the practical application of the intelligent transportation concept and to contribute to the construction and development of smart cities [110]. To date, a number of states in the United States have successively established regional transportation meteorological monitoring and forecasting systems, which provide real-time forecasts on transportation factors such as pavement temperature according to their respective road conditions [111,112].

As for China, in recent years, with the progress of meteorological monitoring, early warning and forecasting systems, as well as the rapid development of transportation-associated construction, the meteorological department has developed rapidly in terms of the research and application of transportation meteorological business [113]. In fact, according to the Chinese Ministry of Public Security's statistics, severe weather conditions account for nearly 40% of transportation accidents and roughly 65% of direct economic losses [114,115]. The CMA officially carried out a survey on the hidden risks of transportation meteorological hazards in 31 provinces during 2013–2015 by means of questionnaire distribution, on-the-spot investigation, and expert evaluation [116]. After quality control, a nationwide transportation meteorological disaster risk census database was then constructed, containing geographic information associated with hidden risks, meteorological information for observation, early warning, and prediction, accident information on transportation under certain disaster conditions, etc., favoring the establishment of a corresponding business system for early warning and forecasting [117]. Combining automation and human–computer interaction, transportation meteorological disaster early warning service products focusing on four high-impact meteorological elements, including low visibility, strong wind, heavy precipitation, and freezing rain and snow, would be generated for lead times of 3 days [118,119].

Moreover, meteorological departments and commercial companies in various provinces have also cooperated with transportation departments to jointly carry out research and development on a series of specific service products and platforms. For instance, the Key Laboratory of Transportation Meteorology of the China Meteorological Administration and the Nanjing Joint Institute for Atmospheric Sciences (formerly known as the Jiangsu Institute of Meteorological Sciences) have conducted investigations on the establishment of a transportation meteorological information service system and on the distributed transportation meteorological information sharing technology based on a web service, which is an early practice in smart city construction [120]. The Hebei Meteorological Service Center has also carried out research on data quality control methods for the monitoring of transportation conditions in Hebei Province, which provides accuracy assessments for the local transportation meteorological service business [121]. In Guangdong Province, Foshan has actively set up a large database of transportation meteorology, which integrates multi-source data to monitor the whole transportation flow and to optimize the route plan system [122]. From a broader perspective, the CMA Public Meteorological Service Center, the Huafeng Meteorological Media Group, and some other departments have also focused on transportation meteorology, including monitoring data fusion, road inversion algorithms, applied meteorological forecasting, as well as service system construction, and have achieved great progress [123]. A corresponding transportation meteorological service system platform combining information on transportation and meteorology is therefore established for decision-making and for specific users from multiple fields of traffic management, map navigation, logistics distribution, autopilot, Internet of Vehicles, vehicle–road collaboration, and many others [124,125].

Beijing, as the center of China from many perspectives, including culture, science, education, and international communication, has developed a generally complete framework for urban transportation meteorological services. The Beijing Municipal Commission of Transport and the Beijing Meteorological Bureau have cooperated closely to provide professional, refined, and targeted transportation meteorological services for associated departments, enterprises, and the public based on the intelligent grid forecast system [126]. They mainly focus on the safe operation of meteorological safeguard services through the development of professional meteorological monitoring and specific forecast products, as well as the construction of multifunctional service platforms. In addition, according to the demand of the Beijing Traffic Management Bureau for guiding and maintaining the transportation order on the city's roads, it is critical to determine and release the road closure standards when visibility levels of less than 50 m are observed [127]. However, the current refined monitoring techniques cannot reach this criterion; thus, they need to be further strengthened at the later stage of research and development, especially in the application of technologies such as real-time remote sensing and image recognition [128].

From a broader perspective, such as a national one, documents from multiple departments have proposed the task of developing smart transportation meteorological service demonstration [129,130]. In practical procedures, based on the vigorous development of artificial intelligence and computer technology, the role of deep learning algorithms such as deep neural networks is becoming increasingly prominent in urban transportation meteorological services, such as real-time monitoring and early warning systems [9]. In general, the implementation of deep learning in smart urban transportation meteorology could be separated mainly into three steps/subsystems: (1) data collection, storage, and query; (2) construction of an intelligent recognition algorithm; and (3) identification of the urban transportation meteorological risk. Corresponding measures are further described in detail in Figure 2.

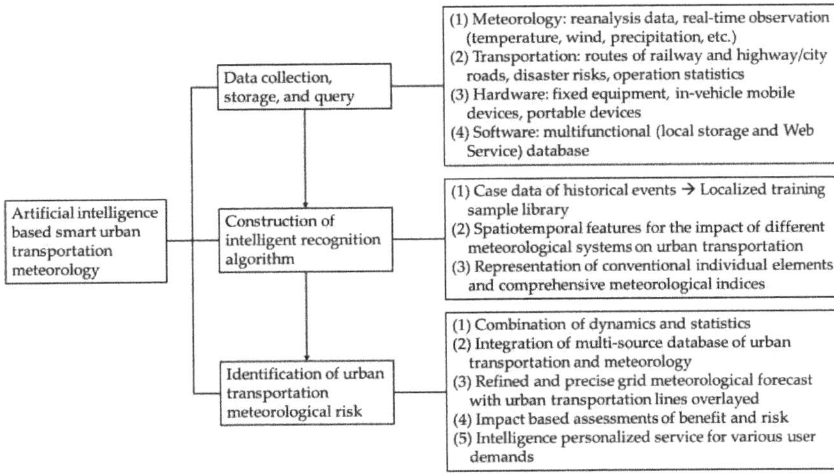

Figure 2. Detailed procedures of artificial intelligence-based smart urban transportation meteorology system.

5. Discussion

In the present era, a new round of scientific and technological revolution and industrial change is accelerating worldwide, in addition to Chinese social and economic development. It brings new strategic opportunities to the regional construction of urban transportation. With the construction progress of smart cities and modern transportation, targeted transportation meteorological monitoring, forecasting, and early warning systems are significantly ameliorating the problems of urban road resource shortages and traffic congestion, playing an important role in the construction of smart urban transportation. Nevertheless, in the context of intensified global warming, extreme meteorological events are becoming increasingly frequent, including dense fog, low visibility, heavy precipitation, extreme temperatures, and other analogous phenomena, which pose severe threats to the operation of urban transportation and associated services [131]. It has drawn much attention in recent decades, and there are still plenty of issues to be resolved in improving the emergency meteorological services and developing targeted urban transportation meteorological services in modern cities. These involve the research, development, and transformation of a new generation of meteorological service systems, but also cooperation among multiple departments and their own internal business adjustments.

In brief, the in-depth development of urban transportation meteorological services and the corresponding advancement of both theoretical and technological explorations are of great significance to improve the construction of smart cities and modern transportation. Hence, the following perceptions and discussions are to be proposed based on the above review and analyses.

(1) With the rapid development of observation facilities and methodologies, equipment such as radar, satellite, microwave radiometers, unmanned aerial vehicles, and mobile observations would further enrich the existing urban transportation meteorological observation system. The in-depth and effective integration of multi-source observations is favorable to establish a more comprehensive and more reliable urban transportation meteorological observation big data system with higher spatial and temporal resolutions. This would help to further reveal the spatiotemporal distribution and variation characteristics of urban transportation meteorology-associated factors and to provide solid support with a database for more accurate and effective forecasts and early warnings.

(2) Thus far, numerical weather prediction models have become the most important tool for meteorological forecasts around the world, which discretize the dynamical and physical equations of the atmosphere. Increasingly, operational business agencies have begun to develop a series of global numerical models with high spatial and temporal resolutions, generating more complete forecast systems. In this context, the quality of numerical forecast products has been continuously improved, with the product sources also being continuously expanded. However, their applications in the field of transportation meteorology, especially urban transportation meteorology, are still relatively lacking. The corresponding refinement and postprocessing of the model outputs are important scientific and technical issues that need to be investigated.

(3) Along with the recent advancement of machine learning, plenty of complex but efficient deep learning models (a branch of machine learning and artificial intelligence) are nowadays emerging in an endless stream and they have been considered as core technologies in many fields. However, many of them have not yet been timely and effectively applied in the field of meteorology, especially urban transportation meteorology, which needs extensive and in-depth experiments and analyses. At the same time, facing specific application scenarios such as urban transportation meteorology, it is always necessary to construct targeted, high-resolution meteorological observation datasets based on multi-source observation systems and collaborative observation experiments, which could give full play to the advantages of artificial intelligence's nonlinearity in data modeling and generate more reasonable and more accurate urban transportation meteorological forecast and early warning products.

(4) With regard to the different meteorological conditions and elements, they certainly tend to result in different impacts on urban transportation due to their different mechanisms of onset, development, and retreat. Respective research and development towards their optimal observation schemes, forecasts, and early warning technologies are necessary to predict the impacts of various meteorological conditions and complex weather events on all aspects of urban transportation in advance and to ultimately provide stable and reliable safeguarding services.

In summary, advances in meteorological observations and numerical models, as well as many other aspects, would bring great progress to urban transportation meteorology. Meanwhile, with the advantages of the vast range of data-driven statistical models and artificial intelligence frameworks, the urban transportation meteorological forecast and early warning system could also be continuously improved and optimized, providing a solid scientific foundation and high application value for the efficient prevention and mitigation of disasters and the high-quality construction of smart city transportation.

Author Contributions: Conceptualization, S.Z., H.Y. and D.L.; validation, H.W. (Hongbin Wang), L.Z. (Linyi Zhou) and C.Z.; formal analysis, S.Z. and H.Y.; investigation, Y.L., Y.Z. and Y.F.; resources, H.W. (Hongbin Wang), Y.X., L.Z. (Ling Zhang) and X.Z.; data curation, F.Z. and H.W. (Hong Wu); writing—original draft preparation, S.Z. and H.Y.; writing—review and editing, D.L. and H.W. (Hongbin Wang). All authors have read and agreed to the published version of the manuscript.

Funding: This research was jointly funded by the Basic Research Fund of the Chinese Academy of Meteorological Sciences (Grant Nos. 2022Y027 and 2021Y011), the research project of the Jiangsu Meteorological Bureau (Grant Nos. KQ202209 and KQ202114), the 333 Project of Jiangsu Province (Grant No. BRA2018420), the Beijing Foundation of NJIAS (Grant Nos. BJG201906, BJG202104 and BJG202209) and the General Program of Key Science and Technology in Transportation, the Ministry of Transport (Grant Nos. 2018-MS4-102 and ZL-2018-04).

Conflicts of Interest: The authors declare no conflict of interest.

References

1. Belli, L.; Cilfone, A.; Davoli, L.; Ferrari, G.; Adorni, P.; Di Nocera, F.; Dall'Olio, A.; Pellegrini, C.; Mordacci, M.; Bertolotti, E. IoT-enabled smart sustainable cities: Challenges and approaches. *Smart Cities* **2020**, *3*, 1039–1071. [CrossRef]
2. Xu, H.; Li, Y.; Zheng, Y.; Xu, X. Analysis of spatial associations in the energy–carbon emission efficiency of the transportation industry and its influencing factors: Evidence from China. *Environ. Impact Assess. Rev.* **2022**, *97*, 106905. [CrossRef]
3. Ding, H.; Li, X.; Cai, Y.; Lorenzo, B.; Fang, Y. Intelligent data transportation in smart cities: A spectrum-aware approach. *IEEE/ACM Trans. Netw.* **2018**, *26*, 2598–2611. [CrossRef]
4. Crevier, L.-P.; Delage, Y. METRo: A new model for road-condition forecasting in Canada. *J. Appl. Meteorol.* **2001**, *40*, 2026–2037. [CrossRef]
5. Yu, M.; Liu, Y. The possible impact of urbanization on a heavy rainfall event in Beijing. *J. Geophys. Res. Atmos.* **2015**, *120*, 8132–8143. [CrossRef]
6. Chen, Y.; Sun, J.; Xu, J.; Yang, S.; Zong, Z.; Chen, T.; Fang, C.; Sheng, J. Analysis and thinking on the extremes of the 21 July 2012 torrential rain in Beijing Part I: Observation and thinking. *Meteorol. Mon.* **2012**, *38*, 1255–1266.
7. Luo, Y.; Zhang, J.; Yu, M.; Liang, X.; Xia, R.; Gao, Y.; Gao, X.; Yin, J. On the influences of urbanization on the extreme rainfall over Zhengzhou on 20 July 2021: A convection-permitting ensemble modeling study. *Adv. Atmos. Sci.* **2022**, 1–17. [CrossRef]
8. China Meteorological Administration. *Action Plan for the Development of Smart Meteorological Services (2019–2023)*; China Meteorological Administration: Beijing, China, 2018.
9. China Meteorological Administration; China Ministry of Public Security; China Ministry of Transport; China State Railway Administration; China State Post Bureau. *The 14th Five-Year Plan for Transportation Meteorological Safeguard*; China Meteorological Administration: Beijing, China, 2021.
10. China State Council. *Outlines for High-quality Meteorological Development (2022–2035)*; China State Council: Beijing, China, 2022.
11. Ge, F.; Zhu, S.; Peng, T.; Zhao, Y.; Sielmann, F.; Fraedrich, K.; Zhi, X.; Liu, X.; Tang, W.; Ji, L. Risks of precipitation extremes over Southeast Asia: Does 1.5 °C or 2 °C global warming make a difference? *Environ. Res. Lett.* **2019**, *14*, 044015. [CrossRef]
12. Zhu, S.; Ge, F.; Fan, Y.; Zhang, L.; Sielmann, F.; Fraedrich, K.; Zhi, X. Conspicuous temperature extremes over Southeast Asia: Seasonal variations under 1.5 °C and 2 °C global warming. *Clim. Chang.* **2020**, *160*, 343–360. [CrossRef]
13. Zhu, S.; Ge, F.; Sielmann, F.; Pan, M.; Fraedrich, K.; Remedio, A.R.C.; Sein, D.V.; Jacob, D.; Wang, H.; Zhi, X. Seasonal temperature response over the Indochina Peninsula to a worst-case high emission forcing: A study with the regionally coupled model ROM. *Theor. Appl. Climatol.* **2020**, *142*, 613–622. [CrossRef]
14. Ge, F.; Zhu, S.; Luo, H.; Zhi, X.; Wang, H. Future changes in precipitation extremes over Southeast Asia: Insights from CMIP6 multi-model ensemble. *Environ. Res. Lett.* **2021**, *16*, 24013. [CrossRef]
15. Allwine, K.J.; Shinn, J.H.; Streit, G.E.; Clawson, K.L.; Brown, M. Overview of URBAN 2000. *Bull. Am. Meteorol. Soc.* **2002**, *83*, 521–536. [CrossRef]
16. Allwine, K.J.; Leach, M.J.; Stockham, L.W.; Shinn, J.S.; Hosker, R.P.; Bowers, J.F.; Pace, J.C. Overview of Joint Urban 2003: An atmospheric dispersion study in Oklahoma City. In Proceedings of the Symposium on Planning, Nowcasting, and Forecasting in the Urban Zone, Seattle, WA, USA, 10–15 January 2004.
17. Warner, T.; Benda, P.; Swerdlin, S.; Knievel, J.; Argenta, E.; Aronian, B.; Balsley, B.; Bowers, J.; Carter, R.; Clark, P.; et al. The Pentagon Shield Field Program: Toward Critical Infrastructure Protection. *Bull. Am. Meteorol. Soc.* **2007**, *88*, 167–176. [CrossRef]
18. Hanna, S.; White, J.; Zhou, Y.; Kosheleva, A. Analysis of Joint Urban 2003 (JU2003) and Madison Square Garden 2005 (MSG05) Meteorological and Tracer Data. In Proceedings of the 6th Symposium on the Urban Environment, Atlanta, GA, USA, 27 January–3 February 2006.
19. Orville, R.; Zhang, R.; Nielsen-Gammon, J.; Collins, D.; Ely, B.; Steiger, S. *Houston Environmental Aerosol Thunderstorm (HEAT) Project*; Texas A&M University Department of Atmospheric Sciences: College Station, TX, USA, 2004; p. 57.
20. Cros, B.; Durand, P.; Cachier, H.; Drobinski, P.; Fréjafon, E.; Kottmeier, C.; Perros, P.; Peuch, V.-H.; Ponche, J.-L.; Robin, D.; et al. The ESCOMPTE program: An overview. *Atmos. Res.* **2004**, *69*, 241–279. [CrossRef]
21. Masson, V.; Gomes, L.; Pigeon, G.; Liousse, C.; Pont, V.; Lagouarde, J.P.; Voogt, J.; Salmond, J.; Oke, T.R.; Hidalgo, J.; et al. The Canopy and Aerosol Particles Interactions in TOulouse Urban Layer (CAPITOUL) experiment. *Meteorol. Atmos. Phys.* **2008**, *102*, 135. [CrossRef]
22. Arnold, S.; ApSimon, H.; Barlow, J.; Belcher, S.; Bell, M.; Boddy, J.; Britter, R.; Cheng, H.; Clark, R.; Colvile, R.; et al. Introduction to the DAPPLE Air Pollution Project. *Sci. Total Environ.* **2004**, *332*, 139–153. [CrossRef]
23. Harrison, R.M.; Dall'Osto, M.; Beddows, D.C.S.; Thorpe, A.J.; Bloss, W.J.; Allan, J.D.; Coe, H.; Dorsey, J.R.; Gallagher, M.; Martin, C.; et al. Atmospheric chemistry and physics in the atmosphere of a developed megacity (London): An overview of the REPARTEE experiment and its conclusions. *Atmos. Chem. Phys.* **2012**, *12*, 3065–3114. [CrossRef]
24. Bohnenstengel, S.I.; Belcher, S.E.; Aiken, A.; Allan, J.D.; Allen, G.; Bacak, A.; Bannan, T.J.; Barlow, J.F.; Beddows, D.C.S.; Bloss, W.J.; et al. Meteorology, Air Quality, and Health in London: The ClearfLo Project. *Bull. Am. Meteorol. Soc.* **2015**, *96*, 779–804. [CrossRef]
25. Rotach, M.W.; Vogt, R.; Bernhofer, C.; Batchvarova, E.; Christen, A.; Clappier, A.; Feddersen, B.; Gryning, S.-E.; Martucci, G.; Mayer, H.; et al. BUBBLE—An urban boundary layer meteorology project. *Arch. Meteorol. Geophys. Bioclimatol. Ser. B* **2005**, *81*, 231–261. [CrossRef]

26. Nakatani, T.; Misumi, R.; Shoji, Y.; Saito, K.; Seko, H.; Seino, N.; Suzuki, S.-I.; Shusse, Y.; Maesaka, T.; Sugawara, H. Tokyo metropolitan area convection study for extreme weather resilient cities. *Bull. Am. Meteorol. Soc.* **2015**, *96*, ES123–ES126. [CrossRef]
27. Hicks, B.B.; Callahan, W.J.; Pendergrass, W.R., III; Dobosy, R.J.; Novakovskaia, E. Urban turbulence in space and in time. *J. Appl. Meteorol. Climatol.* **2012**, *51*, 205–218. [CrossRef]
28. Koskinen, J.T.; Poutiainen, J.; Schultz, D.M.; Joffre, S.; Koistinen, J.; Saltikoff, E.; Gregow, E.; Turtiainen, H.; Dabberdt, W.F.; Damski, J.; et al. The Helsinki Testbed: A mesoscale measurement, research, and service platform. *Bull. Am. Meteorol. Soc.* **2011**, *92*, 325–342. [CrossRef]
29. Takahashi, K.; Mikami, T.; Takahashi, H. Influence of the urban heat island phenomenon in Tokyo in land and sea breezes. In Proceedings of the Seventh International Conference on Urban Climate, Yokohama, Japan, 29 June–3 July 2009.
30. Tan, J.; Yang, L.; Grimmond, C.S.B.; Shi, J.; Gu, W.; Chang, Y.; Hu, P.; Sun, J.; Ao, X.; Han, Z. Urban integrated meteorological observations: Practice and experience in Shanghai, China. *Bull. Am. Meteorol. Soc.* **2015**, *96*, 85–102. [CrossRef]
31. Miao, S.; Jiang, W.; Liang, P.; Liu, H.; Wang, X.; Tan, J.; Zhang, N.; Li, J.; Du, W.; Pei, L. Advances in urban meteorological research in China. *J. Meteorol. Res.* **2020**, *34*, 218–242. [CrossRef]
32. Yu, M.; Dou, J.; Miao, S.; Chu, Y.; Sun, D. Advances of urban meteorological research: International conference on urban climate. *Adv. Meteorol. Sci. Technol.* **2019**, *9*, 16–22.
33. Liang, X.; Miao, S.; Li, J.; Bornstein, R.; Zhang, X.; Gao, Y.; Chen, F.; Cao, X.; Cheng, Z.; Clements, C.; et al. SURF: Understanding and Predicting Urban Convection and Haze. *Bull. Am. Meteorol. Soc.* **2018**, *99*, 1391–1413. [CrossRef]
34. Sun, J.; Mahrt, L.; Banta, R.M.; Pichugina, Y.L. Turbulence regimes and turbulence intermittency in the stable boundary layer during CASES-99. *J. Atmos. Sci.* **2012**, *69*, 338–351. [CrossRef]
35. Sun, J.; Lenschow, D.H.; LeMone, M.A.; Mahrt, L. The role of large-coherent-eddy transport in the atmospheric surface layer based on CASES-99 Observations. *Boundary-Layer Meteorol.* **2016**, *160*, 83–111. [CrossRef]
36. Zhu, Y.; Liu, H.; Shen, J.; Ji, Y. Influence of urban heat island on pollution diffusion in Suzhou. *Plateau Meteorol.* **2016**, *35*, 1584–1594.
37. Liu, H.; Ma, W.; Qian, J.; Cai, J.; Ye, X.; Li, J.; Wang, X. Effect of urbanization on the urban meteorology and air pollution in Hangzhou. *J. Meteorol. Res.* **2015**, *29*, 950–965. [CrossRef]
38. Mao, X.; Jiang, Y.; Zhuang, H.; Rao, H.; Tan, M. A brief introduction to Shenzhen urban meteorological observing network of networks. *Adv. Meteorol. Sci. Technol.* **2013**, *3*, 13–18.
39. Hu, F.; Li, X.; Chen, H.; Liu, G. Turbulence characteristics in the rough urban canopy layer. *Clim. Environ. Res.* **1999**, *3*, 252–258.
40. Miao, S.; Dou, J.; Chen, F.; Li, J.; Li, A. Analysis of observations on the urban surface energy balance in Beijing. *Sci. China Earth Sci.* **2012**, *55*, 1881–1890. [CrossRef]
41. Li, L.; Lu, C.; Chan, P.-W.; Zhang, X.; Yang, H.-L.; Lan, Z.-J.; Zhang, W.-H.; Liu, Y.-W.; Pan, L.; Zhang, L. Tower observed vertical distribution of PM2.5, O3 and NOx in the Pearl River Delta. *Atmos. Environ.* **2019**, *220*, 117083. [CrossRef]
42. Miao, S.; Chen, F. Enhanced modeling of latent heat flux from urban surfaces in the Noah/single-layer urban canopy coupled model. *Sci. China Earth Sci.* **2014**, *57*, 2408–2416. [CrossRef]
43. Huang, H.; Li, Y.; Han, S.; Wu, B.; Zhang, Y.; Li, C. Turbulent statistic characteristic of the urban boundary layer in Tianjin. *Plateau Meteorol.* **2011**, *30*, 1481–1487.
44. Liang, Y.; Wu, C.; Wu, D.; Liu, B.; Li, Y.J.; Sun, J.; Yang, H.; Mao, X.; Tan, J.; Xia, R.; et al. Vertical distributions of atmospheric black carbon in dry and wet seasons observed at a 356-m meteorological tower in Shenzhen, South China. *Sci. Total Environ.* **2022**, *853*, 158657. [CrossRef]
45. Wang, Z.; Zhang, X.; Mao, J.; Ji, C. Overview of ground-based remote sensing observation techniques for air temperature, humidity and wind profiles. *Meteorol. Hydrol. Mar. Instrum.* **2018**, *35*, 109–116.
46. Grimmond, C.S.B. Progress in measuring and observing the urban atmosphere. *Theor. Appl. Climatol.* **2006**, *84*, 3–22. [CrossRef]
47. Xu, X.; Ding, G.; Bian, L.; Xie, L. Characteristics of atmospheric environment of boundary layer structure of city community in BECAPEX and integrate influence. *Acta Meteorol. Sin.* **2004**, *5*, 663–671.
48. Li, J.; Shu, W. Observation and analysis of nocturnal low-level jet characteristics over Beijing in summer. *Chin. J. Geophys.* **2008**, *51*, 360–368.
49. Li, J.; Dou, J. Progress in urban meteorological experiments in Beijing. *Adv. Meteorol. Sci. Technol.* **2014**, *4*, 36–45.
50. Liu, H.; Jiang, W.; Sun, J.; Liu, G. An observation and analysis of the micrometeorological characteristics of the Nanjing urban boundary layer, eastern China. *J. Nanjing Univ. (Nat. Sci.)* **2008**, *1*, 99–106.
51. Mao, M.; Jiang, W.; Wu, X.; Qi, F.; Yuan, R.; Fang, H.; Liu, D.; Zhou, J. LIDAR exploring of the UBL in downtown of the Nanjing City. *Acta Sci. Circumstantiae* **2006**, *26*, 1723–1728.
52. Mao, M.; Jiang, W.; Gu, J.; Xie, C.; Zhou, J. Study on the mixed layer, entrainment zone, and cloud feedback based on lidar exploration of Nanjing city. *Geophys. Res. Lett.* **2009**, *36*. [CrossRef]
53. Baklanov, A.; Grimmond, C.; Carlson, D.; Terblanche, D.; Tang, X.; Bouchet, V.; Lee, B.; Langendijk, G.; Kolli, R.; Hovsepyan, A. From urban meteorology, climate and environment research to integrated city services. *Urban Clim.* **2018**, *23*, 330–341. [CrossRef]
54. Chen, G.; Wang, D.; Wang, Q.; Li, Y.; Wang, X.; Hang, J.; Gao, P.; Ou, C.; Wang, K. Scaled outdoor experimental studies of urban thermal environment in street canyon models with various aspect ratios and thermal storage. *Sci. Total Environ.* **2020**, *726*, 138147. [CrossRef]

55. Hajizadeh, Y.; Jafari, N.; Fanaei, F.; Ghanbari, R.; Mohammadi, A.; Behnami, A.; Jafari, A.; Aghababayi, M.; Abdolahnejad, A. Spatial patterns and temporal variations of traffic-related air pollutants and estimating its health effects in Isfahan city, Iran. *J. Environ. Health Sci. Eng.* **2021**, *19*, 781–791. [CrossRef]
56. Soleimani, A.; Toolabi, A.; Mansour, S.N.; Abdolahnejad, A.; Akther, T.; Fouladi-Fard, R.; Miri, M.; Mohammadi, A. Health risk assessment and spatial trend of metals in settled dust of surrounding areas of Lake Urmia, NW Iran. *Int. J. Environ. Anal. Chem.* **2022**, 1–14. [CrossRef]
57. Mojarrad, H.; Fard, R.F.; Rezaali, M.; Heidari, H.; Izanloo, H.; Mohammadbeigi, A.; Mohammadi, A.; Sorooshian, A. Spatial trends, health risk assessment and ozone formation potential linked to BTEX. *Hum. Ecol. Risk Assess. Int. J.* **2019**, *26*, 2836–2857. [CrossRef]
58. Wang, Y.; Bai, Y.; Zhi, X.; Wu, K.; Zhao, T.; Zhou, Y.; Xiong, J.; Zhu, S.; Zhou, W.; Hu, W.; et al. Two typical patterns of regional PM2.5 transport for heavy air pollution over Central China: Rapid transit transport and stationary accumulation transport. *Front. Environ. Sci.* **2022**. [CrossRef]
59. Mokhtari, M.; Jafari, N.; Mohammadi, A.; Hajizadeh, Y.; Ghanbari, R.; Nemati, S.; Abdolahnejad, A. Temporal and spatial trends of airborne asbestos fiber concentrations in the urban areas of Yazd, Iran. *Int. J. Environ. Sci. Technol.* **2018**, *16*, 2657–2666. [CrossRef]
60. Faraji, M.; Mohammadi, A.; Najmi, M.; Fallahnezhad, M.; Sabetkish, N.; Kazemnejad, A.; Shoormasti, R.S.; Fazlollahi, M.R.; Pourpak, Z.; Moin, M. Exposure to ambient air pollution and prevalence of asthma in adults. *Air Qual. Atmos. Health* **2021**, *14*, 1211–1219. [CrossRef]
61. Tian, H.; Zhang, N.; Zhang, K.; Yang, J.; Zhang, H. Advance on highway traffic meteorological research in foreign country. *J. Meteorol. Environ.* **2019**, *35*, 79–86.
62. Zhang, C.; Zhang, L.; Cheng, C.; Wang, B. Advances in road weather forecasting system and its future development. *J. Trop. Meteorol.* **2007**, *23*, 652–658.
63. Hu, J.; Lin, Z.; Cheng, T. Research progress on temperature prediction method for road icing. *Sci. Technol. Eng.* **2020**, *20*, 1–6.
64. Telang, S.; Chel, A.; Nemade, A.; Kaushik, G. Intelligent Transport System for a Smart City. In *Security and Privacy Applications for Smart City Development*; Tamane, S.C., Dey, N., Hassanien, A.E., Eds.; Studies in Systems, Decision and Control; Springer: Cham, Switzerland, 2021; Volume 308.
65. Lu, Z.; Liu, Z.; Li, Y.; Jia, A.; Ye, R. Research on highway traffic meteorological monitoring system. *Meteorol. Hydrol. Mar. Instrum.* **2021**, *38*, 16–19.
66. Chu, Y. On the system of road traffic's meteorological monitoring and early warning based on GIS. *J. Liaoning Police Coll.* **2017**, *19*, 58–62.
67. Wang, L.; Li, X.; Bao, Y.; Shao, Y. Research progress of remote sensing application on transportation meteorological disasters. *Remote Sens. Land Resour.* **2018**, *30*, 1–7.
68. Li, C.; Shen, L.; Wang, H.; Lei, T. The research on unmanned aerial vehicle remote sensing and its applications. In Proceedings of the 2nd International Conference on Advanced Computer Control, Shenyang, China, 27–29 March 2010; pp. 644–647.
69. Whitehead, K.; Hugenholtz, C.H.; Myshak, S.; Brown, O.; LeClair, A.; Tamminga, A.; Barchyn, T.E.; Moorman, B.; Eaton, B. Remote sensing of the environment with small unmanned aircraft systems (UASs), part 1: A review of progress and challenges. *J. Unmanned Veh. Syst.* **2014**, *2*, 69–85. [CrossRef]
70. Whitehead, K.; Hugenholtz, C.H.; Myshak, S.; Brown, O.; LeClair, A.; Tamminga, A.; Barchyn, T.E.; Moorman, B.; Eaton, B. Remote sensing of the environment with small unmanned aircraft systems (UASs), part 2: Scientific and commercial applications. *J. Unmanned Veh. Syst.* **2014**, *2*, 86–102. [CrossRef]
71. Barmpounakis, E.N.; Vlahogianni, E.I.; Golias, J.C. Unmanned Aerial Aircraft Systems for transportation engineering: Current practice and future challenges. *Int. J. Transp. Sci. Technol.* **2016**, *5*, 111–122. [CrossRef]
72. Atzori, L.; Iera, A.; Morabito, G. The Internet of Things: A survey. *Comput. Netw.* **2010**, *54*, 2787–2805. [CrossRef]
73. Chapman, L.; Muller, C.L.; Young, D.T.; Warren, E.; Grimmond, S.; Cai, X.-M.; Ferranti, E. The Birmingham urban climate laboratory: An open meteorological test bed and challenges of the smart city. *Bull. Am. Meteorol. Soc.* **2015**, *96*, 1545–1560. [CrossRef]
74. Zhang, N.; Chen, H.; Chen, X.; Chen, J. Semantic framework of Internet of Things for smart cities: Case studies. *Sensors* **2016**, *16*, 1501. [CrossRef]
75. Liu, C.; Ke, L. Cloud assisted Internet of things intelligent transportation system and the traffic control system in the smart city. *J. Control Decis.* **2022**, 1–14. [CrossRef]
76. Zhou, H.; Liu, B.; Wang, D. Design and research of urban intelligent transportation system based on the Internet of Things. In *Internet of Things. Communications in Computer and Information Science;* Wang, Y., Zhang, X., Eds.; Springer: Berlin/Heidelberg, Germany, 2012; Volume 312, pp. 572–580. [CrossRef]
77. The People's Government of Beijing Municipality. *Overall Plan for Internet of Things Application Construction in Beijing's Urban Safe Operation and Emergency Management*; Beijing Government General Office: Beijing, China, 2011; p. 14.
78. Beijing Municipal Commission of Transportation. *The Feasibility Study of "The Internet of Things Application Demonstration Project of Maintaining Road Traffic Unblocked in Extreme Meteorological Conditions"*; Beijing Municipal Commission of Transportation: Beijing, China, 2011; pp. 127–130.

79. Zhi, X.; Cui, B.; Ji, Y.; Zhu, S.; Ma, Z.; Zhang, X. Prediction of water level in urban waterlogging area based on deep learning approach. In Proceedings of the International Conference on Electrocommunication, Intelligent Computing and Systems, Xi'an, China, 18–19 June 2022.
80. Zhang, X.; Sun, H.; Zhang, L. Discussion on locations of road weather information system station on smart expressway. *Highway* **2022**, *67*, 248–254.
81. Geng, K. Strategy for refined control of meteorological risks in Shanghai urban rail transit. *Urban Mass Transit.* **2022**, *25*, 86–90.
82. Miao, S.; Wang, Y. Advances and prospects of urban meteorology research: Meeting users' needs. *Adv. Meteorol. Sci. Technol.* **2014**, *4*, 6–14.
83. Silva, B.N.; Khan, M.; Han, K. Big Data Analytics Embedded Smart City Architecture for Performance Enhancement through Real-Time Data Processing and Decision-Making. *Wirel. Commun. Mob. Comput.* **2017**, *2017*, 9429676. [CrossRef]
84. Lu, H.; Zhu, Y.; Shi, K.; Lv, Y.; Shi, P.; Niu, Z. Using Adverse Weather Data in Social Media to Assist with City-Level Traffic Situation Awareness and Alerting. *Appl. Sci.* **2018**, *8*, 1193. [CrossRef]
85. Wessel, J. Using weather forecasts to forecast whether bikes are used. *Transp. Res. Part A Policy Pract.* **2020**, *138*, 537–559. [CrossRef]
86. Simunek, M.; Smutny, Z. Traffic Information Enrichment: Creating Long-Term Traffic Speed Prediction Ensemble Model for Better Navigation through Waypoints. *Appl. Sci.* **2020**, *11*, 315. [CrossRef]
87. Reichstein, M.; Camps-Valls, G.; Stevens, B.; Jung, M.; Denzler, J.; Carvalhais, N.; Prabhat. Deep learning and process understanding for data-driven Earth system science. *Nature* **2019**, *566*, 195–204. [CrossRef] [PubMed]
88. Taillardat, M.; Mestre, O.; Zamo, M.; Naveau, P. Calibrated ensemble forecasts using quantile regression forests and ensemble model output statistics. *Mon. Weather Rev.* **2016**, *144*, 2375–2393. [CrossRef]
89. Handler, S.L.; Reeves, H.D.; McGovern, A. Development of a probabilistic subfreezing road temperature nowcast and forecast using Machine Learning. *Weather Forecast.* **2020**, *35*, 1845–1863. [CrossRef]
90. Han, S.; Xu, J.; Yan, M.; Liu, Z. Using multiple linear regression and BP neural network to predict critical meteorological conditions of expressway bridge pavement icing. *PLoS ONE* **2022**, *17*, e0263539. [CrossRef]
91. Jia, Y.; Wu, J.; Xu, M. Traffic flow prediction with rainfall impact using a deep learning method. *J. Adv. Transp.* **2017**, *2017*, 6575947. [CrossRef]
92. Nagy, A.M.; Simon, V. Survey on traffic prediction in smart cities. *Pervasive Mob. Comput.* **2018**, *50*, 148–163. [CrossRef]
93. Lee, Y.-J.; Min, O. Long short-term memory recurrent neural network for urban traffic prediction: A case study of Seoul. In Proceedings of the 21st International Conference on Intelligent Transportation Systems (ITSC), Maui, HI, USA, 4–7 November 2018; pp. 1279–1284. [CrossRef]
94. Ali, A.; Zhu, Y.; Chen, Q.; Yu, J.; Cai, H. Leveraging spatio-temporal patterns for predicting citywide traffic crowd flows using deep hybrid neural networks. In Proceedings of the IEEE 25th International Conference on Parallel and Distributed Systems (ICPADS), Tianjin, China, 4–6 December 2019; pp. 125–132. [CrossRef]
95. Deb, B.; Khan, S.R.; Tanvir Hasan, K.; Khan, A.H.; Alam, M.A. Travel Time Prediction using Machine Learning and Weather Impact on Traffic Conditions. In Proceedings of the 2019 IEEE 5th International Conference for Convergence in Technology (I2CT), Pune, India, 29–31 March 2019. [CrossRef]
96. Ma, D.; Song, X.; Li, P. Daily traffic flow forecasting through a contextual convolutional recurrent neural network modeling inter- and intra-day traffic patterns. *IEEE Trans. Intell. Transp. Syst.* **2020**, *22*, 2627–2636. [CrossRef]
97. Nigam, A.; Srivastava, S. Macroscopic traffic stream variables prediction with weather impact using hybrid CNN-LSTM model. In *Adjunct Proceedings of the 2021 International Conference on Distributed Computing and Networking (ICDCN '21)*; Association for Computing Machinery: New York, NY, USA, 2021; pp. 1–6.
98. Tukymbekov, D.; Saymbetov, A.; Nurgaliyev, M.; Kuttybay, N.; Nalibayev, Y.; Dosymbetova, G. Intelligent energy efficient street lighting system with predictive energy consumption. In Proceedings of the International Conference on Smart Energy Systems and Technologies (SEST), Porto, Portugal, 9–11 September 2019.
99. Tukymbekov, D.; Saymbetov, A.; Nurgaliyev, M.; Kuttybay, N.; Dosymbetova, G.; Svanbayev, Y. Intelligent autonomous street lighting system based on weather forecast using LSTM. *Energy* **2021**, *231*, 120902. [CrossRef]
100. Nasser, A.; Simon, V. A novel method for analyzing weather effect on smart City traffic. In Proceedings of the IEEE 22nd International Symposium on a World of Wireless, Mobile and Multimedia Networks (WoWMoM), Pisa, Italy, 7–11 June 2021; pp. 335–340. [CrossRef]
101. Zhou, Z. Attention based stack ResNet for citywide traffic accident prediction. In Proceedings of the 20th IEEE International Conference on Mobile Data Management (MDM), Hong Kong, China, 10–13 June 2019; pp. 369–370. [CrossRef]
102. Ryu, S.; Kim, D.; Kim, J. Weather-Aware Long-Range Traffic Forecast Using Multi-Module Deep Neural Network. *Appl. Sci.* **2020**, *10*, 1938. [CrossRef]
103. Lyu, Y.; Zhi, X.; Zhu, S.; Fan, Y.; Pan, M. Statistical calibrations of surface air temperature forecasts over East Asia using pattern projection methods. *Weather Forecast.* **2021**, *36*, 1661–1674. [CrossRef]
104. Zhu, S.; Zhi, X.; Ge, F.; Fan, Y.; Zhang, L.; Gao, J. Subseasonal forecast of surface air temperature using superensemble approaches: Experiments over Northeast Asia for 2018. *Weather Forecast.* **2021**, *36*, 39–51. [CrossRef]
105. Haltiner, G.J.; Williams, R.T. Some recent advances in numerical weather prediction. *Mon. Weather Rev.* **1975**, *103*, 571–590. [CrossRef]

106. Bauer, P.; Thorpe, A.; Brunet, G. The quiet revolution of numerical weather prediction. *Nature* **2015**, *525*, 47–55. [CrossRef]
107. Espinosa, J.M.; Holm, E.; White, M. Part 3: Technology: Creating Intelligent, Coordinated Transit: Moving New Mexico the Smart Way. *Transp. Res. Rec. J. Transp. Res. Board* **2005**, *1927*, 137–148. [CrossRef]
108. Kueper, D. The smart transportation guidebook: Planning and designing highways and streets that support sustainable and livable communities. *ITE J.* **2008**, *78*, 38–42.
109. Zhang, C. Big data will change several major thinking categories. *J. Huazhong Univ. Sci. Technol. (Soc. Sci. Ed.)* **2015**, *29*, 120–125.
110. English, N.; Zhao, C.; Brown, K.L.; Catlett, C.; Cagney, K. Making Sense of Sensor Data: How Local Environmental Conditions Add Value to Social Science Research. *Soc. Sci. Comput. Rev.* **2020**. [CrossRef]
111. Minhoto, M.J.C.; Pais, J.C.; Pereira, P.A.A.; Picado-Santos, L.G. Predicting asphalt pavement temperature with a three-dimensional finite element method. *Transp. Res. Rec.* **2005**, *1919*, 96–110. [CrossRef]
112. Hosseini, F.; Hossain, S.M.K.; Fu, L.; Johnson, M.; Fei, Y. Prediction of pavement surface temperature using meteorological data for optimal winter operations in parking lots. In Proceedings of the 16th International Conference on Cold Regions Engineering, Salt Lake City, UT, USA, 19–22 July 2015; pp. 440–451.
113. Zhai, Y.; Li, X. Advances in traffic meteorological service under the influence of disastrous weather. *J. Catastrophol.* **2015**, *30*, 144–147.
114. Kang, Y.; Wang, S.; Yang, X.; Li, J.; Xu, W.; Shang, K. Progress of traffic meteorological researches about monitoring and forecasting services on express highways. *J. Arid. Meteorol.* **2016**, *34*, 591–603.
115. China Intelligent Transportation Association. *Framework and Technical Specification for Expressway Traffic Meteorological System*; China Intelligent Transportation Association: Beijing, China, 2022; p. 11.
116. Cui, B.; Liu, Y.; Wang, X.; Zhang, X. Analysis and research on the characteristics of hidden risks of highway meteorological disasters. In Proceedings of the 32nd Annual Conference of the Chinese Meteorological Society, Tianjin, China, 14–16 October 2015; pp. 29–33.
117. Yang, F.; Xia, W.; Chen, X. Investigation of highway system during survey on disaster bearing body. *City Disaster Reduct.* **2021**, *137*, 35–38.
118. Meng, C.; Zhang, C. Development and verification of a numerical forecast model for road meteorological services. *J. Appl. Meteorol. Sci.* **2012**, *23*, 451–458.
119. Zhao, N.; Meng, X.; Ma, C.; Qu, X.; Zhang, J. Weather forecast indexes of dense fogs based on traffic weather monitoring data. *Meteorol. Sci. Technol.* **2015**, *43*, 145–150.
120. Feng, D.; Tang, W.; Liu, Y.; Wang, M.; Li, A.; Qu, H. Design and development of an early risk warning system of highway traffic meteorological disasters. *Meteorol. Sci. Technol.* **2018**, *46*, 822–828.
121. Qu, X.; Ma, C.; Liu, J.; Yang, G.; Zhou, Y. Methods for controlling quality of meteorological monitoring data on expressway surface state. *Meteorol. Sci. Technol.* **2012**, *40*, 203–206.
122. Sun, N.; Shi, H.; Han, G.; Wang, B.; Shu, L. Dynamic path planning algorithms with load balancing based on data prediction for smart transportation systems. *IEEE Access* **2020**, *8*, 15907–15922. [CrossRef]
123. Xiao, Y.; Xi, S.; Wang, L. Research on refined forecast test and correction of maximum wind speed along expressway in Henan province. *Meteorol. Environ. Sci.* **2022**, *45*, 29–35.
124. Wang, Z.; Han, Y.; Li, A. Advance in research and operation in traffic meteorological service in China. *Adv. Meteorol. Sci. Technol.* **2017**, *7*, 85–89.
125. Chen, Y.; Song, J.; Zhang, X.; Wang, Y.; Zhang, L. Analysis of the current status and innovative models of "immersive" specialized meteorological services in meteorological departments. *Bull. Sci. Technol.* **2021**, *37*, 18–23.
126. Wu, J.; Liao, H. Weather, travel mode choice, and impacts on subway ridership in Beijing. *Transp. Res. Part A Policy Pract.* **2020**, *135*, 264–279. [CrossRef]
127. Li, A.; Wu, H.; Liu, Y.; Yang, J.; Tian, H.; Pan, J. Risk assessment and region partition of low visibility disasters on highway in China. *Meteorol. Mon.* **2018**, *44*, 676–683.
128. Tian, H.; Wu, H.; Yang, J.; Gao, J.; Zhang, N.; Zhang, K. Demand analysis of meteorological decision-making service for highway traffic. *Meteorol. Environ. Sci.* **2018**, *41*, 70–76.
129. China Meteorological Administration; China National Development and Reform Commission. *The 14th Five-Year Plan for National Meteorological Development*; China Meteorological Administration: Beijing, China, 2021; p. 52.
130. China Meteorological Administration; Ministry of Science and Technology of the People´s Republic of China; Chinese Academy of Sciences. *Development Strategy of Meteorological Science and Technology in China (2021–2035)*; China Meteorological Administration: Beijing, China, 2022; p. 46.
131. Lu, H.; Chen, M.; Kuang, W. The impacts of abnormal weather and natural disasters on transport and strategies for enhancing ability for disaster prevention and mitigation. *Transp. Policy* **2019**, *98*, 2–9. [CrossRef]

atmosphere

MDPI

Article

Analysis of Spatio-Temporal Characteristics of Visibility in the Yellow and Bohai Seas Based on Observational Data

Lei Zhang [1,2], Mei Xu [1,2], Xiaobin Qiu [3], Dongbin Zhang [4], Rongwei Liao [5,*], Xiaoyi Fang [5], Bingui Wu [2,3] and Fanchao Meng [6]

1 Tianjin Meteorological Information Centre, Tianjin 300074, China
2 Tianjin Key Laboratory for Oceanic Meteorology, Tianjin 300074, China
3 Tianjin Institute of Meteorological Science, Tianjin 300074, China
4 National Meteorological Information Centre, Beijing 100081, China
5 State Key Laboratory of Severe Weather, Chinese Academy of Meteorological Sciences, Beijing 100081, China
6 Tianjin Climate Center, Tianjin 300074, China
* Correspondence: liaorw@cma.gov.cn

Abstract: In the Yellow and Bohai Seas, the detailed characteristics of visibility are analyzed based on automatic hourly observation data of marine visibility between 2019 and 2021. The results show that the annual average visibility in the Yellow and Bohai Seas is 13.346 km. The average visibility at high latitudes is higher than that at low latitudes in the Yellow and Bohai Seas. The low visibility area is mainly distributed in the southwest of the Yellow Sea. There are obvious seasonal differences in visibility in the Yellow and Bohai Seas. Visibility is high from September to November, with maximum values in October. Visibility is lowest in July when the maximum visibility is low and the minimum visibility is high. The visibility in spring is overall relatively low, and the areas of low visibility appear in the southwest of the Yellow Sea. The visibility in autumn is overall relatively high, and the areas of high visibility occur in the northern part of the Bohai and Yellow Seas. The visibility has significant intraday variation. The visibility around sunset is significantly higher than that around sunrise. The hourly visibility is low between 4:00 and 9:00, with the lowest visibility most likely around 7:00. The hourly visibility is high between 16:00 and 21:00, with the highest visibility most likely around 18:00. Low visibility occurs frequently between November and April, most of all in March. Low visibility most often occurs between 4:00 and 7:00. Low visibility may occur at any time between November and April, and also in mornings between May and August. It occurs less often at other times.

Keywords: climatology; visibility; Yellow Sea and Bohai Sea; observation data

Citation: Zhang, L.; Xu, M.; Qiu, X.; Zhang, D.; Liao, R.; Fang, X.; Wu, B.; Meng, F. Analysis of Spatio-Temporal Characteristics of Visibility in the Yellow and Bohai Seas Based on Observational Data. *Atmosphere* **2023**, *14*, 1101. https://doi.org/10.3390/atmos14071101

Academic Editor: Albert Gabric

Received: 5 June 2023
Revised: 26 June 2023
Accepted: 28 June 2023
Published: 30 June 2023

1. Introduction

The Yellow and Bohai Seas form an important maritime area in the north of China, and many of its ports are busy with shipping. Atmospheric visibility is an important meteorological factor affecting marine transportation. Low visibility will make it difficult to observe and position, which can easily cause marine traffic accidents such as collisions, resulting in casualties, property losses, and environmental pollution. The monitoring and forecasting of visibility can be improved by studying the changes in visibility more precisely [1–4].

Many researchers have studied maritime visibility. Visibility research methods can be roughly divided into three categories: field observation [5–7], satellite remote sensing [8,9] and numerical simulation [10–13]. Research areas have been mostly concentrated in the Newfoundland Sea area on the east coast of Canada, the sea area south of the Kamchatka Peninsula, the California sea area on the west coast of the United States, the sea area off the northeast coast of Scotland in the United Kingdom and the Yellow Sea of China [14–18].

Fog is the most common and severe low–visibility weather, receiving much attention. Visibility is an indicator used to distinguish different intensities of fog. The weather processes that lead to low visibility and thus impact maritime navigation mainly occur in the atmospheric boundary layer at sea. However, most of our understanding of the atmospheric boundary layer comes from studies carried out on land. A series of observation plans of maritime visibility have been implemented such as CALSPAN, CEWCOM and Project Haar [19–21]. By such means, detailed information about the maritime atmospheric boundary layer formed by low visibility has been obtained. At the same time, the boundary layer structure of sea fog with low visibility is analyzed. These field observation plans have raised awareness of low visibility and related concerns [22–25]. The research provides an important reference for the accurate observation and judgment of low visibility weather. However, these observations are mostly experiments, and real-time and continuous observation data cannot be obtained.

Changes in sea visibility in the Yellow and Bohai Seas have long been a matter of concern. Due to a lack of data obtained by direct observation at sea [26–28], the temporal and spatial characteristics of visibility have few studies over the whole Yellow and Bohai Seas. However, detailed observational data are essential if sea visibility in this region is to be effectively monitored and forecasted. For this reason, the observation of marine visibility needs to be improved in the Yellow and Bohai Seas [2,29,30].

In recent years, through the deployment of automatic stations and buoys and the accumulation of ship observation data, the visibility observation capability of the Yellow Sea and the Bohai Sea has been gradually enhanced. In this study, climatic characteristics of maritime visibility and the frequency of low-visibility conditions in the Yellow and Bohai Seas are analyzed in detail, based on high-temporal-resolution automatic observation data of marine visibility obtained.

2. Materials and Methods

This study uses the automatic observation data of hourly marine visibility between 2019 and 2021, which is sourced from the National Meteorological Information Center. The hourly automatic observation data set can monitor changes in meteorological parameters, as well as hydrological information, in key sea areas with high time resolution in a continuous and long-term manner. Automatic visibility observation devices are carried on observation platforms such as buoys and oil platforms. The observation instruments of visibility include forward scattering instruments and transmission instruments. Two instruments obtain visibility, respectively, by measuring the scattering coefficient and transmission coefficient of air [31]. The average observation error of the instruments is less than 20%, which meets the range of visibility measurement uncertainty specified by WMO [32,33]. The National Meteorological Information Center has integrated and controlled the quality of data from automatic observation device which have been added to the Yellow and Bohai Seas in recent years, thereby enriching the ocean observation data of this area and providing data support for the development of marine meteorological research. The 2 m temperature, sea surface temperature, 10 m u-component of wind, and 10 m v-component of wind of ERA5 monthly averaged reanalysis data was used to analyze the climate characteristics in the Yellow and Bohai seas from 1991 to 2020 (https://cds.climate.copernicus.eu/cdsapp#!/dataset/reanalysis-era5-single-levels-monthly-means?tab=form, accessed on 30 June 2023) [34].

Automatic hourly observation data of marine visibility are obtained as follows: If less than 3 h are missing in a day, the data for that day are retained. If more than 3 h are missing in a day, the observation data for that day are considered absent. There are 12 missing times, with missing times accounting for less than 0.04% of the total times. The annual average value of visibility is the average value of visibility from January to December in a calendar year. The seasonal division method is as follows: winter extends from December in one year through to February of the next; spring from March to May; summer from June to August; and autumn from September to November. The standard deviation of visibility is calculated based on the hourly average visibility data in the Yellow and Bohai Sea.

According to Koschmieder's law (Formula (1)), the visibility is inversely proportional to the atmospheric extinction coefficient.

$$VIS = \frac{-ln(\varepsilon)}{\sigma} \quad (1)$$

where *VIS* is visibility, and σ is atmospheric extinction coefficient, and ε is contrast threshold.

The atmospheric extinction coefficient can be considered as spatially continuous and can be Linear interpolation. For the visible light interpolation process, visibility is first converted to atmospheric extinction coefficient, and then the extinction coefficient is interpolated, and then converted back to visibility.

The spatial distribution of visibility is interpolated using natural interpolation method and extrapolated using linear method [35]. The spatiotemporal characteristics of monthly visibility in the Yellow and Bohai Sea from 2019 to 2021 were analyzed by the EOF method. The seasonal probability distribution function of visibility, at intervals of 2 km, is determined by the frequency of visibility. The probability distribution function of visibility has been normalized. The cumulative probability distribution function is used to analyze the distribution of visibility data in February and July. The cumulative probability distribution function of visibility has also been normalized. According to the "Convention on the International Regulations for Preventing Collisions at Sea" [36], special safety measures are required for sea navigation with visibility below 2 nautical miles. The frequency characteristics of visibility from 2 km to 5 km were analyzed by classification in this study. Average visibility of less than 5 km is defined as low visibility. The frequency of low visibility is the times accumulated when the average visibility is less than 5 km. The study area (117~127.5° E, 35~41.5° N) is the Yellow Sea and the Bohai Sea (Figure 1a). The selected period of visibility observation data covers three years from 2019 to 2021. The spatial distribution of observation stations covers the same sea area (Figure 1b).

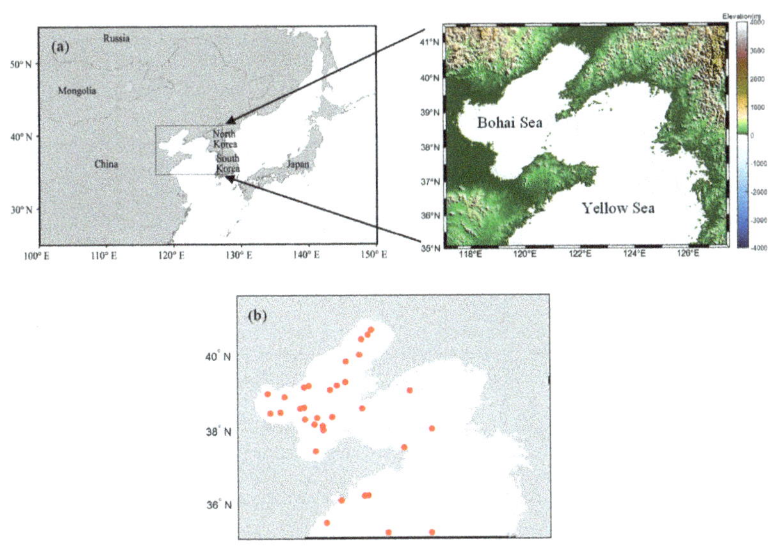

Figure 1. The study region (**a**) and Spatial distribution of observation stations (**b**) (The red dots denote the observation stations).

3. Results

3.1. Annual and Seasonal Characteristics of Visibility

The climate characteristics of the Yellow and Bohai Seas were analyzed using ERA5 reanalysis data on 2 m air temperature, sea surface temperature, and wind from 1991 to 2020 (Figure 2). From the spatial distribution, it can be seen that both the 2 m air temperature and sea surface temperature in the Yellow and Bohai Sea have the characteristics of low in the north and high in the south. In the Yellow and Bohai Seas, the climate is obviously a monsoonal climate, and southern winds prevail in summer, and northwestern winds prevail in winter.

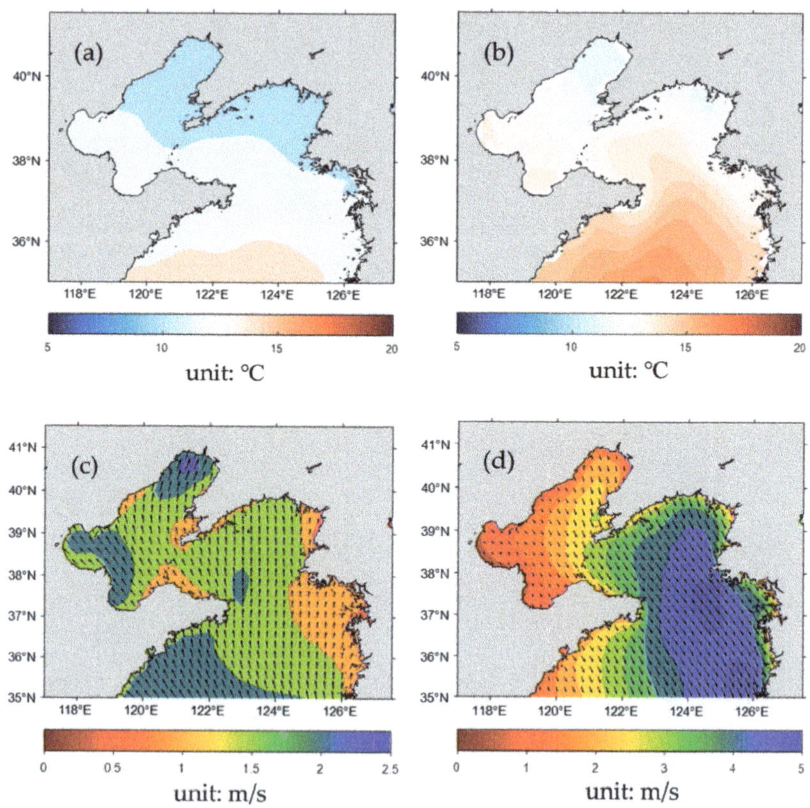

Figure 2. The climate characteristics in the Yellow and Bohai Seas from 1991 to 2020 ((**a**) 2 m air temperature, (**b**) sea surface temperature, (**c**) wind in summer, (**d**) wind in winter).

In the Yellow and Bohai Seas, the average visibility at high latitudes is higher than that at low latitudes (Figure 3a). The low visibility area is mainly distributed in the southwest of the Yellow Sea. From the seasonal distribution of visibility (Figure 3b–e), the visibility in spring is overall relatively low, and the areas of low visibility appear in the southwest of the Yellow Sea. The visibility in autumn is overall relatively high, and the areas of high visibility occur in the northern part of the Bohai and Yellow Seas. Throughout the four seasons, the visibility in the southwest of the Yellow Sea has been consistently low.

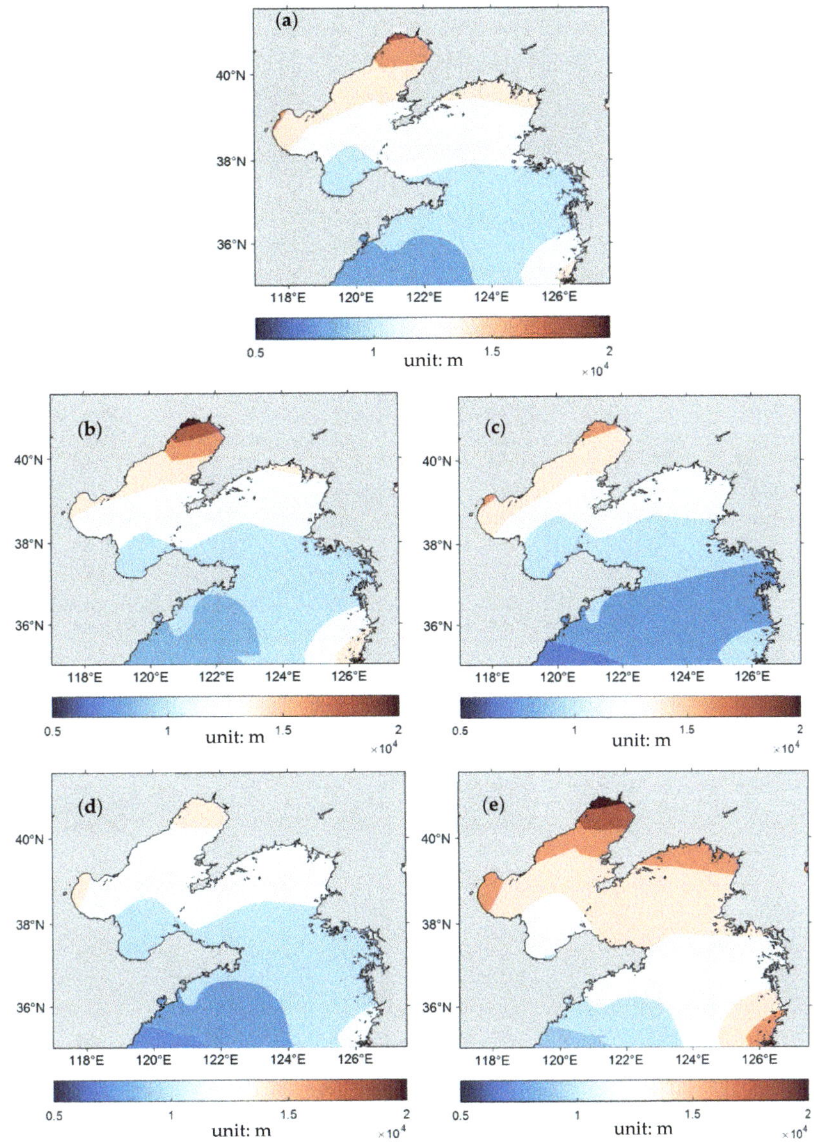

Figure 3. The distribution of annual (**a**) and seasonal ((**b**) winter, (**c**) spring, (**d**) summer, (**e**) autumn) average visibility in the Yellow and Bohai Seas from 2019 to 2021.

Between 2019 and 2021, annual average visibility in the Yellow and Bohai Seas was 13.346 km, and the standard deviation of the annual average visibility was 4.351 km (Table 1). As shown by the values in Table 1, the maximum average and median of visibility both occurred in autumn and were 15.514 km and 15.696 km, respectively. The minimum average and median of visibility both occurred in summer and were 12.529 km and 12.312 km (Table 1), respectively. The average and median visibility shows significant differences between that in autumn and in summer. The average and median visibility in winter and spring were similar.

Table 1. Seasonal average and standard deviations of visibility (unit/km) in the Yellow and Bohai Seas from 2019 to 2021.

	Winter	Spring	Summer	Autumn	Annual
Average	12.757	12.599	12.529	15.514	13.346
Median	12.489	12.692	12.312	15.696	13.286
Standard Deviation	4.749	4.007	3.612	4.231	4.351

The maximum standard deviation of visibility was 4.749 km, in winter, and the minimum standard deviation of visibility was 3.612 km, in summer. Although the average visibility in winter and summer is similar, the standard deviation of visibility in winter is significantly greater than that in summer, indicating that there is a significant difference in the distribution of visibility in winter and summer.

Analysis of the data reveals obvious seasonal differences in the distribution of visibility in the Yellow and Bohai Seas (Figure 4). Maximum values of PDF (Probability Density Function) in summer and autumn are greater than 0.2, while maximum values of PDF in winter and spring are lower than 0.2. The maximum value of visibility distribution in winter and spring is obtained at 12–14 km, while the maximum values of visibility distribution in summer and autumn are obtained at 10–12 km and 16–18 km, respectively. The area of statistical distribution of low visibility is greatest in winter, and the area of high visibility is greatest in autumn.

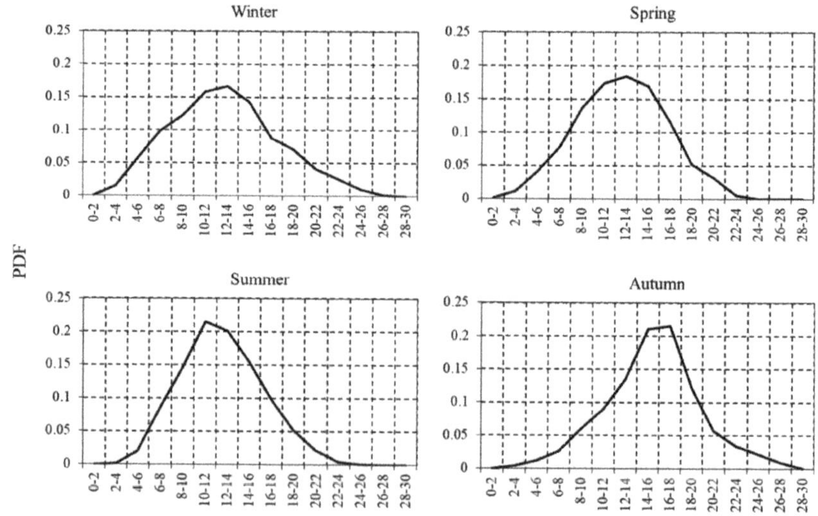

Figure 4. Seasonal distribution of visibility in the Yellow and Bohai Seas from 2019 to 2021.

3.2. *Monthly Characteristics of Visibility*

The average visibility was high and was more than 15 km from September to November (Table 2). The average visibility was smaller in February and July, and the visibility was less than 12 km. The maximum average visibility appeared in October and was 15.659 km. The minimum average visibility appeared in July and was 11.677 km.

Table 2. Monthly average and standard deviations of visibility (unit/km) in the Yellow and Bohai Seas from 2019 to 2021.

	Jan	Feb	Mar	Apr	May	Jun	Jul	Aug	Sept	Oct	Nov	Dec
Average	12.494	11.794	12.101	12.728	12.972	12.268	11.677	13.632	15.635	15.659	15.242	13.902
Median	12.439	11.098	12.131	12.436	13.174	12.062	11.527	13.782	16.009	15.548	15.444	13.908
Standard Deviation	4.386	4.705	4.601	3.841	3.439	3.633	2.906	3.938	3.419	4.533	4.614	4.904

The median visibility was high from September to November, which also exceeded 15 km (Table 2). The median visibility was low in February and July, and was less than 12 km. The maximum median visibility appeared in September and was 16.009 km. The minimum median visibility appeared in February and was 11.098 km. The maximum average visibility and the maximum median visibility appeared in October and September, respectively. The occurrence time of the minimum average visibility and the minimum median visibility appeared in July and February, respectively.

From the monthly data distribution (Figure 5), it can also be observed that changes in monthly median visibility are similar to changes in the average value. Notably, the maximum visibility is low and the minimum visibility is high in July.

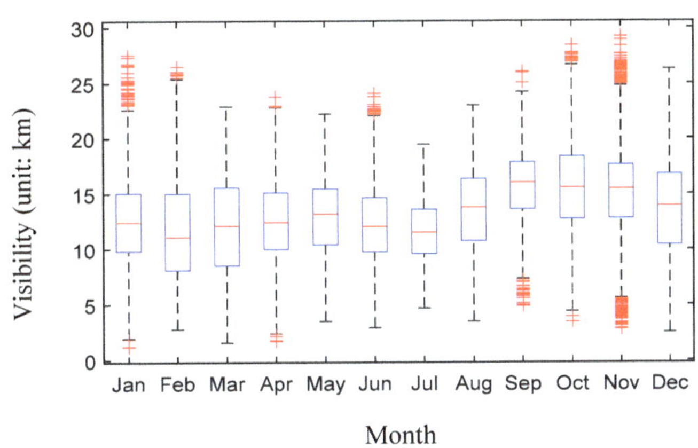

Figure 5. Monthly distributions of visibility in the Yellow and Bohai Seas from 2019 to 2021.

The minimum value of the standard deviation of visibility appeared in July and was 2.906 km (Table 2). The maximum value of standard deviation of visibility appeared in December, which was 4.904 km. The standard deviation of visibility in July was small, indicating that the distribution of visibility in July was concentrated, and the fluctuation of visibility in July was small. The standard deviation of visibility in December is large (Figure 5).

The average and median visibility in February and July are very close, but the standard deviation in February is significantly greater than that in July, resulting in a large difference in the distribution of visibility between February and July. It can be clearly seen from the comparison of the cumulative distribution function (CDF) in February and July (Figure 6) that the occurrence probability of low visibility of less than 6 km in February is significantly larger than that in July, and the occurrence probability of a high visibility of more than 16 km in February is also significantly larger than that in July. The frequency of low visibility weather and high visibility weather in February is greater than that in July. Although the average and median visibility in February and July are very close, there is a huge difference in terms of low visibility between February and July, which has attracted much attention.

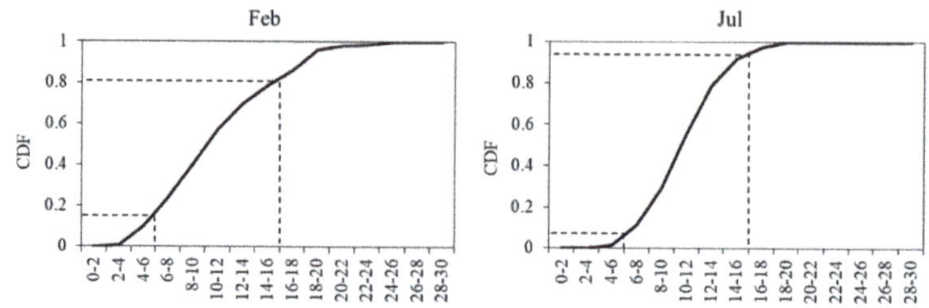

Figure 6. Cumulative distribution function of visibility in February and July in the Yellow and Bohai Seas from 2019 to 2021.

The spatiotemporal characteristics of visibility in the Yellow and Bohai Sea from 2019 to 2021 were analyzed using the EOF method (Figure 7). The variance contribution of the first and second modes are 64.4% and 17.4%, respectively, and the total variance contribution of the two modes is 81.8%. The first mode is positive in the northern Bohai Sea and negative in the eastern Yellow Sea. The second mode is negative in the west of Bohai Sea and positive in the north and southeast of Yellow Sea. The time coefficients of the first and second modes have obvious changes with time, and the frequency of positive–negative alternation of the first mode time coefficient is higher.

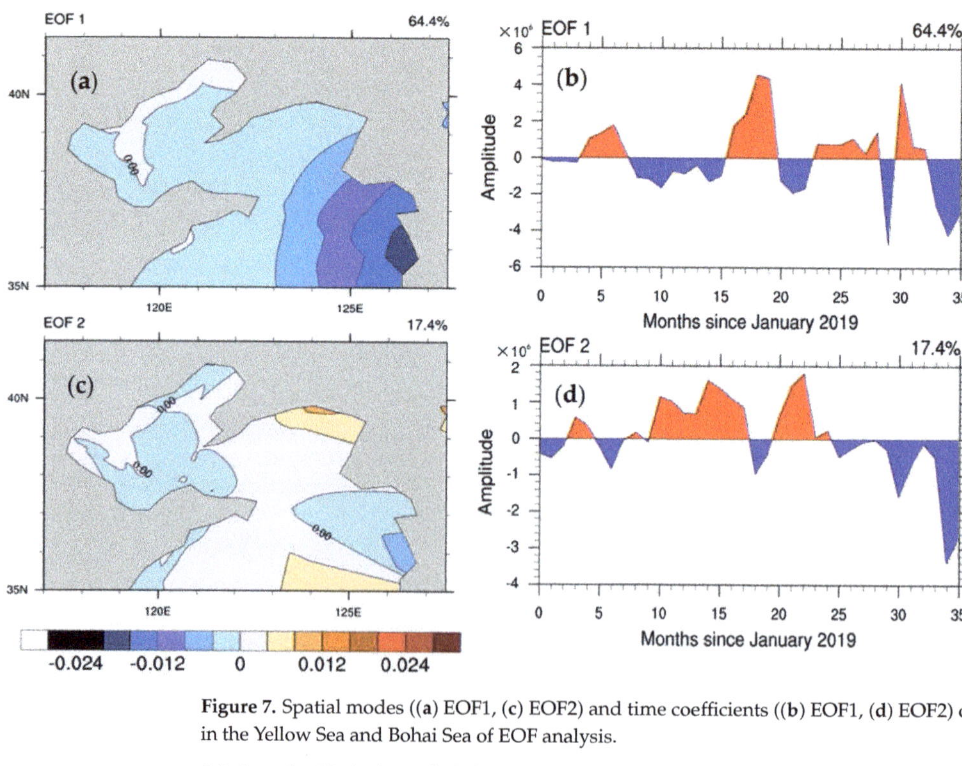

Figure 7. Spatial modes ((**a**) EOF1, (**c**) EOF2) and time coefficients ((**b**) EOF1, (**d**) EOF2) of visibility in the Yellow Sea and Bohai Sea of EOF analysis.

3.3. Intraday Variations of Visibility

It can be seen from the curve of hourly visibility in the Yellow and Bohai Seas (Figure 8) that the hourly visibility is low between 4:00 and 9:00, and the lowest visibility is most likely around 7:00. The hourly visibility is high between 16:00 and 21:00, and highest visibility most likely around 18:00. The visibility has obvious intraday variation characteristics. The visibility around sunset is significantly higher than that around sunrise.

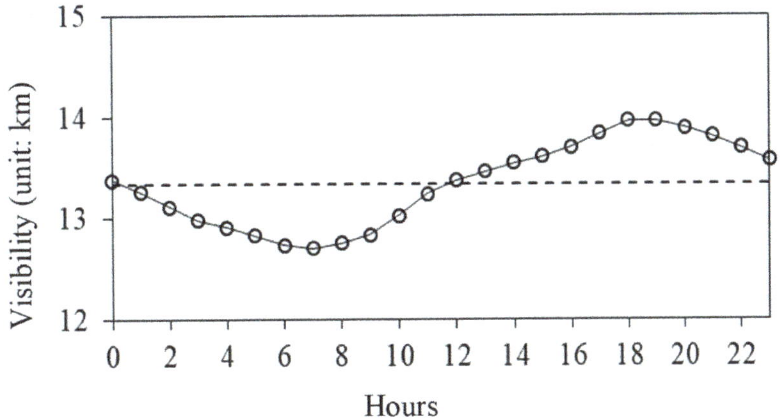

Figure 8. Hourly visibility in the Yellow and Bohai Seas from 2019 to 2021.

Considering the distribution of visibility in the Yellow and Bohai Seas on an hourly and monthly basis (Figure 9), it can be seen that visibility between September and November is significantly higher than in other months. There is an obvious low-visibility period between 5:00 and 9:00 in the June–September period, with the lowest visibility occurring around 7:00 in July. There is also a high-visibility period between 18:00 and 21:00 in September, with the highest visibility occurring around 19:00.

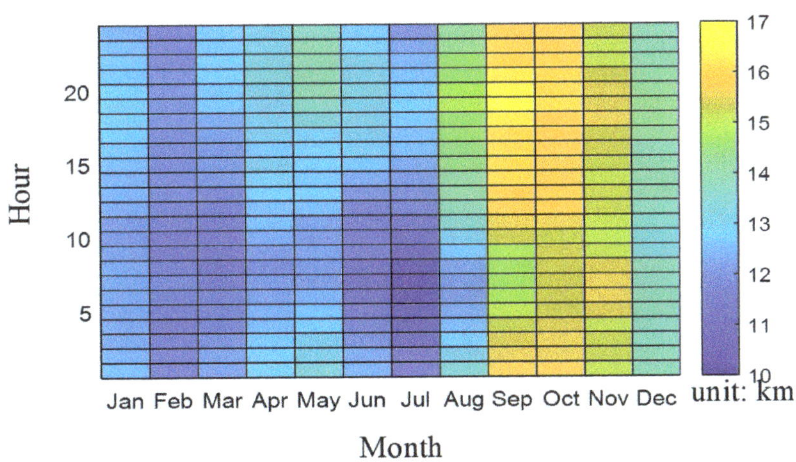

Figure 9. Monthly and hourly distributions of visibility in the Yellow and Bohai Seas from 2019 to 2021.

3.4. Characteristics of Low Visibility

Because low visibility has attracted more attention in previous studies, we focus on low visibility in our analysis. From the monthly statistical chart of low-visibility frequency in the Yellow and Bohai Seas (Figure 10), it can be seen that low visibility occurs frequently between November and April, and most frequently of all in March. The frequency of low visibility from July to October is relatively low.

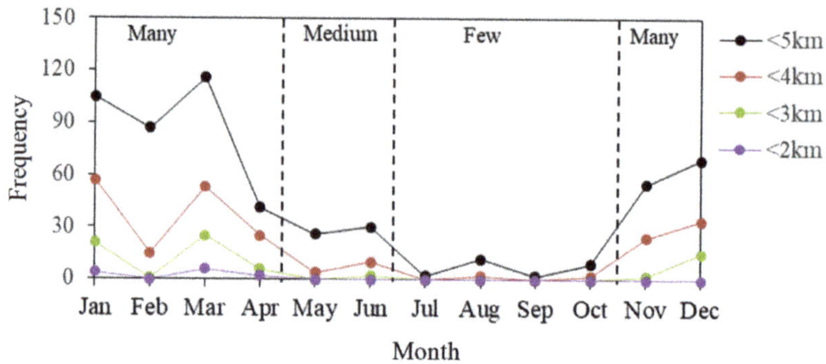

Figure 10. Monthly frequencies of low visibility in the Yellow and Bohai Seas from 2019 to 2021.

The change of low visibility in the Yellow and Bohai Seas during the years from 2019 to 2021 can be divided into three stages: many, medium and few. The many stage of low visibility is from November to April of the next year. The medium stage is May and June. The few stage is from July to October (Figure 10). In the many stage, the frequency of visibility of less than 5 km is higher and is more than 30 times. In the few stage, the frequency of visibility of less than 5 km is less than 15 times. In the medium stage, the frequency of visibility of less than 5 km is between 15 and 30 times. The variation characteristics of visibility of less than 4 km are similar as those of visibility of less than 5 km. The frequency of low visibility in the many stage is significantly higher than that in the few stage.

It can be seen from the hourly frequency of low visibility in the Yellow and Bohai Seas (Figure 11) that there are many instances of visibility less than 5 km between 4:00 and 7:00. Low visibility occurs less frequently between 13:00 and 18:00. The frequency of low visibility around sunrise is significantly higher than that around sunset. The curve of hourly low-visibility frequency is similar to the curve of hourly visibility, showing obvious intraday variation characteristics, but the two curves are not completely corresponding.

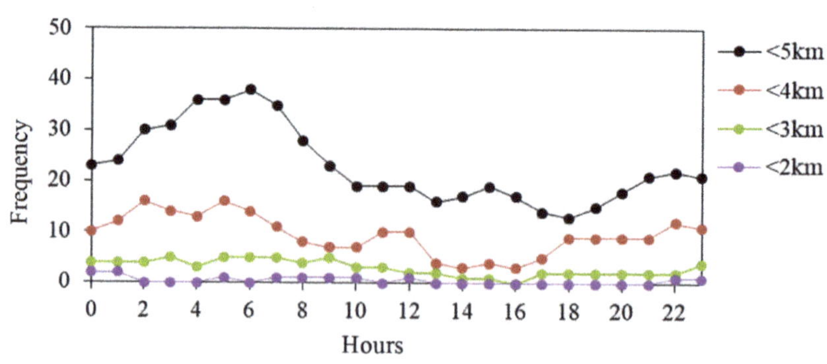

Figure 11. Hourly frequencies of low visibility in the Yellow and Bohai Seas from 2019 to 2021.

From the days–hours statistical chart of low visibility in the Yellow and Bohai Seas (Figure 12), it can be seen that low visibility is most likely between November and April, when it may occur at any time of the day. Low visibility is also widespread during mornings between May and August. It occurs less frequently at other times.

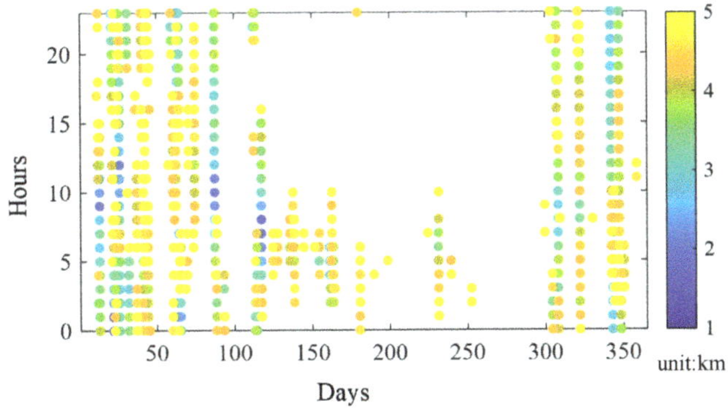

Figure 12. Daily and hourly distributions of low visibility values in the Yellow and Bohai Seas from 2019 to 2021 (Average visibility of less than 5 km is defined as low visibility).

4. Discussion

At present, the visibility monitoring of the Yellow and Bohai Seas is mainly based on field observation data and satellite data, both of which have advantages and disadvantages. The field observation data offer high accuracy, but limited coverage. Satellite data have advantages in spatial distribution, but their observation accuracy is low. Recently, the number of automatic observation devices of visibility in the Yellow and Bohai Seas has been increased, and their observation mode has been changed from manual observation three times a day to automatic observation with high time resolution. Using automatic observation data of hourly marine visibility, the precise characteristics of visibility in the region were analyzed in this study.

We found that low visibility in the Yellow and Bohai Seas occurs most frequently between November and April, and most frequently of all in March. The frequency of low visibility from July to October is relatively low. Using satellite data, Wu et al. [37] found that sea fog in the Bohai Sea is most prevalent in December, followed by April, but is less likely to occur between August and October. In the Yellow Sea, fog is most prevalent between March to June, and most of all in April, with less fog between August and November, least of all in October. The study region of this paper covered both the Yellow Sea and the Bohai Sea. The periods of low visibility identified in this study are basically consistent with those found by Wu et al. [37]. The reported periods of high visibility indicate the common time characteristics of the Yellow and Bohai Seas.

This study used high-time-resolution observation data, and because observation modes have recently changed from manual observation three times a day to a higher-frequency automatic observation, more detailed data can now be obtained. The hourly visibility is low between 4:00 and 9:00, with low visibility most likely around 7:00. The hourly visibility is high between 16:00 and 21:00, and high visibility most likely around 18:00. The visibility has obvious intraday variation characteristics. The visibility around sunset is significantly higher than that around sunrise. From hourly marine visibility data, we found that low visibility is most likely between 4:00 and 7:00, and most likely of all at 6:00. Low visibility is least likely between 17:00 and 19:00, and least likely of all around 18:00. The curve of hourly low-visibility frequency is similar to the curve of hourly visibility, showing obvious intraday change characteristics, but the two curves are not consistent. Zheng et al. [38] studied daily variations in sea fog in the Bohai Sea using artificial observation data obtained three times a day. They found that fog was most likely to occur at 8:00, and less likely at 14:00 and at 20:00. The results of this study are supported by those obtained by Zheng et al. [38]. We give the hourly variation of low visibility, which is more detailed than the characteristics reflected by Zheng et al. [38].

Although the average visibility in July is small, the low visibility is very few; the two are not contradictory. The maximum visibility is low and the minimum visibility is high in July, resulting in the overall average visibility is low, while the frequency of low visibility weather is very few. The average and median of visibility in February and July are very close, but the standard deviation of February is significantly greater than that of July, resulting in a large difference in the distribution of visibility between February and July. There is a huge difference in terms of low visibility between February and July. The frequency of low-visibility weather and high-visibility weather in February are greater than that in July.

In spring, the areas of low visibility appear in the southwest of the Yellow Sea. In autumn, the areas of high visibility occur in the northern part of the Bohai and Yellow Seas. Using satellite data, Wu et al. [37] found that the frequency of sea fog in the western Yellow Sea is significantly higher than that in the Bohai Sea in spring, and the frequency of sea fog in autumn is the lowest among the four seasons. The results of this study are supported by those obtained by Wu et al. [37].

According to Koschmieder's law, visibility is inversely proportional to the extinction coefficient. The attenuation of light intensity caused by the absorption and scattering of light by substances such as gas molecules, aerosol particles, and water droplets when visible light passes through the atmosphere [39]. Therefore, the magnitude of this coefficient is influenced by the composition of the atmosphere. The weather phenomena that affect the extinction coefficient include fog, haze, precipitation, snowfall, and dust. The corresponding meteorological elements include humidity, aerosol concentration, precipitation and snowfall [40,41].

The data in this study covers the period from 2019 to 2021. There is a great deal of scientific evidence to support that the global pollutant levels dropped during the pandemic of COVID-19. Some pollutants have the ability to act as fog condensation nuclei, and reduced pollutants may lead to increased visibility [42,43]. The visibility during 2020–2021 in the Yellow Sea and Bohai Sea may be better than the average. The data used in this study covers a short time period. There is, therefore, a need for longer-term high-resolution visibility observation data, so that variations in visibility in the Yellow and Bohai Sea area can be better understood.

At present, although the extent of sea visibility observation covers the whole of the Yellow and Bohai Sea region, there is an uneven spatial distribution of observation stations. The density of observation stations close to the coast is higher than that of those located far offshore. In terms of spatial representation, there is still a significant difference between offshore and the open sea. The observations of marine visibility in the Yellow and Bohai Seas needs to be increased continuously.

5. Conclusions

Based on the hourly observation data of visibility which we obtained for the Yellow and Bohai Seas during the years 2019–2021, and our analysis of climatic characteristics of sea visibility in this region during this time, the following conclusions can be stated.

(1) Between 2019 and 2021, the annual average visibility of the Yellow Sea and Bohai Sea was 13.346 km, and the standard deviation of the annual average visibility was 4.351 km. There were obvious differences in the distribution of visibility across the seasons. The maximum average visibility was 15.514 km, in autumn. The minimum average visibility of 12.529 km occurred in summer. The maximum standard deviation of visibility of 4.749 km was recorded in winter and the minimum standard deviation of 3.612 km was recorded in summer. The highest values of visibility distribution in winter and spring were in the range of 12–14 km, while the highest values of visibility distribution in summer and autumn were in the range of 10–12 km and 16–18 km, respectively. The areas of low visibility appear in the southwest of the Yellow Sea in spring. The areas of high visibility occur in the northern part of the Bohai and Yellow Seas in autumn. The low visibility area is mainly distributed in the southwest of the Yellow Sea.

(2) Changes in monthly median visibility were similar to changes in average value. The monthly visibility was high from September to November. The average value of visibility was highest in October and lowest in July. In July, the maximum visibility is low, and the minimum visibility is high. The visibility has obvious intraday variation characteristics. The visibility around sunset is significantly higher than that around sunrise. The hourly visibility is low between 4:00 and 9:00, with low visibility most likely around 7:00. The hourly visibility is high between 16:00 and 21:00, and high visibility most likely around 18:00.
(3) Between November and April, low visibility occurs frequently, most frequently in March. Between July and October, the frequency of low visibility was relatively low. Low visibility occurred frequently between 4:00 and 7:00. The change of low visibility in the Yellow and Bohai Seas can be divided into three stages: many, medium and few. The many stage of low visibility is from November to April of the next year. The medium stage is May and June. The few stage is from July to October. In the Yellow and Bohai seas, low visibility mainly occurs at any time between November and April, and also during mornings between May and August. It occurs less often at other times.

Author Contributions: L.Z. analyzed the data, made the figures and wrote the paper. M.X., X.Q., D.Z., R.L., X.F. and B.W. gave scientific advice and helped improve the paper. F.M. revised the language and characters. All the authors contributed to the discussion and final version of the paper. All authors have read and agreed to the published version of the manuscript.

Funding: This study was funded by Innovation and Development Project of China Meteorological Administration (CXFZ2022J028, CXFZ2023J004), the National Key Research and Development Program of China (2022YFC3090600), Bohai Rim Regional Science and Technology Collaborative Innovation Fund (QYXM202101, QYXM202202), The Natural Science Foundation of Tianjin (22JCQNJC00370), Basic Research Fund of Chinese Academy of Meteorological Sciences (Grant No. 2023Z016) and the Transverse Item-Research Project of the Chinese Academy of Meteorological Sciences (Grant No. IN_JS_2022031).

Institutional Review Board Statement: Not applicable.

Informed Consent Statement: Not applicable.

Data Availability Statement: The data used in this study are publicly available via the National Meteorological Information Centre http://www.nmic.cn/ (Recently accessed date: 30 May 2022).

Acknowledgments: The authors acknowledge, with deep gratitude, the contribution of the anonymous reviewers whose time and effort in providing comments has improved this paper substantially.

Conflicts of Interest: The authors declare that there are no conflicts of interest regarding the publication of this paper.

References

1. Wu, D.; Wu, X.J.; Zhu, X.X. *Fog and Haze*; China Meteorological Press: Beijing, China, 2009.
2. Fu, G.; Li, P.Y.; Zhang, S.P.; Gao, S.H. A Brief Overview of the Sea Fog Study in China. *Adv. Meteor. Sci. Technol.* **2016**, *2*, 20–28.
3. Niu, S.J.; Lu, C.S.; Lü, J.J.; Xu, F.; Zhao, L.J.; Liu, D.Y.; Yue, Y.Y.; Zhou, Y.; Yu, H.Y.; Wang, T.S. Advances in Fog Research in China. *Adv. Meteor. Sci. Technol.* **2016**, *2*, 6–14.
4. Zhang, G.C. The Progress of Fog Forecast Operation in China. *Adv. Meteor. Sci. Technol.* **2016**, *2*, 42–48.
5. Tanimoto, Y.; Xie, S.P.; Kai, K.; Okajima, H.; Tokinaga, H.; Murayama, T.; Nonaka, M.; Nakamura, H. Observations of marine atmospheric boundary layer transitions across the summer Kuroshio Extension. *J. Clim.* **2009**, *6*, 1360–1374. [CrossRef]
6. Zhang, S.P.; Zhang, X.; Shi, X.M. Observational analysis of a fog event in the Oyashio Extension area. *J. Mar. Meteorol.* **2022**, *1*, 1–11.
7. *WMO: Guide to Meteorological Instruments and Methods of Observation*, 7th ed.; WMO-No. 8; World Meteorological Organization: Geneva, Switzerland, 2008.
8. Wu, D.; Lu, B.; Zhang, T.C.; Yan, F.Q. A method of detecting sea fogs using CALIOP data and its application to improve MODIS-based sea fog detection. *J. Quant. Spectrosc. Radiat. Transf.* **2015**, *153*, 88–94. [CrossRef]
9. Wu, X.J.; Li, Y.; Huang, B.; Wang, X.; Song, W. Application of dynamic threshold method to sea fog detection with FY-2 satellite. *J. Mar. Meteor.* **2017**, *2*, 31–41.

10. Fu, G.; Xu, J.; Zhang, S.Q. Comparison of Modeling Atmospheric Visibility with Visible Satellite Imagery. *Period. Ocean Univ. China* **2011**, *4*, 1–10.
11. Wang, Y.M.; Gao, S.H.; Fu, G.; Sun, J.L.; Zhang, S.P. Assimilating MTSAT-derived humidity in nowcasting sea fog over the Yellow Sea. *Weather. Forecast.* **2014**, *29*, 205–225. [CrossRef]
12. Jiang, Y.X.; Zhang, S.P.; Xie, S.P.; Chen, Y.; Liu, H.K. Effects of a cold ocean eddy on local atmospheric boundary layer near the Kuroshio Extension: In situ observations and model experiments. *J. Geophys. Res.* **2019**, *11*, 5779–5790. [CrossRef]
13. Tian, M.; Wu, B.G.; Huang, H.; Zhang, H.S.; Zhang, W.Y.; Wang, Z.Y. Impact of water vapor transfer on a Circum-Bohai-Sea heavy fog: Observation and numerical simulation. *Atmos. Res.* **2019**, *229*, 1–22. [CrossRef]
14. Koracin, D.; Dorman, C.E. *Marine Fog: Challenges and Advancements in Observations, Modeling, and Forecasting*; Springer International Publishing: Cham, Switzerland, 2017.
15. Fu, G.; Song, Y.J. Climatology Characteristics of Sea Fog Frequency over the Northern Pacific. *Period. Ocean Univ. China* **2014**, *10*, 35–41.
16. Philip, A.; Bergot, T.; Bouteloup, Y.; Bouyssel, F. The impact of vertical resolution on fog forecasting in the kilometric-scale model AROME: A case study and statistics. *Weather. Forecast.* **2016**, *5*, 1655–1671. [CrossRef]
17. Koracin, D.; Dorman, C.E.; Lewis, J.M.; Hudson, J.G.; Wilcox, E.M.; Torregrosa, A. Marine fog: A review. *Atmos. Res.* **2014**, *143*, 142–175. [CrossRef]
18. Zhang, S.P.; Xie, S.P.; Liu, Q.Y.; Yang, Y.Q.; Wang, X.G.; Ren, Z.P. Seasonal variations of Yellow Sea fog: Observations and mechanisms. *J. Clim.* **2009**, *24*, 6758–6772. [CrossRef]
19. Goodman, J. The microstructure of California coastal fog and stratus. *J. Appl. Meteorol. Clim.* **1977**, *10*, 1056–1067. [CrossRef]
20. Pilie, R.; Mack, E.; Rogers, C.; Katz, U.; Kocmond, W. The formation of marine fog and the development of fog-stratus systems along the California Coast. *J. Appl. Meteorol. Clim.* **1979**, *18*, 1275–1286. [CrossRef]
21. Findlater, J.; Roach, W.; McHugh, B. The haar of North-East Scotland. *Q. J. Roy. Meteor. Soc.* **1989**, *115*, 581–608. [CrossRef]
22. Moores, J.E.; Komguem, L.; Whiteway, J.A.; Lemmon, M.T.; Dickinson, C.; Daerden, F. Observations of near-surface fog at the Phoenix Mars landing site. *Geophys. Res. Lett.* **2011**, *38*, L04203. [CrossRef]
23. Lewis, J.M.; Koracin, D.; Redmond, K.T. Sea fog research in the United Kingdom and United States: A historical essay including outlook. *B. Am. Meteorol. Soc.* **2004**, *3*, 395–408. [CrossRef]
24. Lewis, J.M.; Koracin, D.; Rabin, R.; Businger, J. Sea fog off the California Coast: Viewed in the Context of Transient Weather Systems. *J. Geophys. Res.* **2003**, *108*, 4457. [CrossRef]
25. Oliver, D.A.; Lewellen, W.S.; Williamson, G.G. The interaction between turbulent and radiative transport in the evelopment of fog and low level stratus. *J. Atmos. Sci.* **1978**, *35*, 301–316.
26. Fan, G.F.; Ma, H.; Zhang, X.W.; Liu, Y. Impacts of relative humidity and PM2.5 concentration on atmospheric visibility: A comparative study of hourly observations of multiple stations. *Acta. Meteor. Sin.* **2016**, *6*, 959–973.
27. Xiao, S.R.; Zhou, J.; Wu, Q.Y.; Xu, M.; Shang, G.Q.; Shi, L.F. Improved method for nighttime visibility measurement based on digital photography. *Appl. Opt.* **2014**, *6*, 1016–1022.
28. Zhao, L.J.; Niu, S.J.; Zhang, Y.; Xu, F. Microphysical characteristics of sea fog over the east coast of Leizhou Peninsula, China. *Adv. Atmos. Sci.* **2013**, *4*, 1154–1172. [CrossRef]
29. Qu, P.; Xie, Y.Y.; Liu, L.L.; Lin, Y.; He, N.G. Character Analysis of Sea Fog in Bohai Bay from 1988 to 2010. *Plateau Meteor.* **2014**, *1*, 285–293.
30. Yang, M.; Guo, X.Y.; Zheng, J.Y.; Sun, Q. Long-Term Trend and Inter-Annual Variation of Ocean Heat Content in the Bohai, Yellow, and East China Seas. *Water* **2022**, *17*, 2763. [CrossRef]
31. China Meteorological Administration. *Specification for Automatic Ground Meteorological Observation*; China Meteorological Press: Beijing, China, 2020.
32. *GB/T 12763.3-2020*; National Standards of the People's Republic of China—Specifications for Marine Survey—Part 3: Marine Meteorological Observations. Ministry of Natural Resources: Beijing, China, 2020.
33. Organization, W.M. WMO Guide to Meteorological Instruments and Methods of Observation. World Meteorological Organization (WMO): Geneva, Switzerland, 2008.
34. Hersbach, H.; Bell, B.; Berrisford, P.; Hirahara, S.; Horányi, A.; Muñoz-Sabater, J.; Nicolas, J.; Peubey, C.; Radu, R.; Schepers, D.; et al. The ERA5 global reanalysis. *Q. J. R. Meteorol. Soc.* **2020**, *146*, 1999–2049. [CrossRef]
35. Amidror, I. Scattered data interpolation methods for electronic imaging systems: A survey. *J. Electron. Imaging.* **2002**, *2*, 157–176. [CrossRef]
36. IMO. Convention on the International Regulations for Preventing Collisions at Sea, 1972 (COLREGs). Available online: https://www.imo.org/en/About/Conventions/Pages/COLREG.aspx (accessed on 30 June 2023).
37. Wu, X.J.; Li, S.M.; Liao, M.; Cao, Z.Q.; Wang, L.; Zhu, J. Analyses of seasonal feature of sea fog over the Yellow Sea and Bohai Sea based on the recent 20 years of satellite remote sensing data. *Acta. Oceanol. Sin.* **2015**, *1*, 63–72.
38. Zheng, Y.; Li, R.; Shi, D.D.; Wang, Y.N.; Sun, M.N. Characteristics of offshore and coastal sea fog in the mid-west Bohai Sea. *Mar. Forecasts* **2016**, *6*, 74–80.
39. Griffing, G.W. Relations between the prevailing visibility, nephelometer scattering coefficient and sunphotometer turbidity coefficient. *Atmos. Environ.* **1980**, *14*, 577–584. [CrossRef]

40. Chen, J.; Zhao, C.S. A Review of Influence Factors and Calculation of Atmospheric Low Visibility. *Adv. Meteor. Sci. Technol.* **2014**, *4*, 44–51.
41. Clark, P.A.; Harcourt, S.A.; Macpherson, B.; Mathison, C.T.; Cusack, S.; Naylor, M. Prediction of visibility and aerosol within the operational Met Office unified model. I: Model formulation and variational assimilation. *Q. J. R. Meteorol. Soc.* **2008**, *134*, 1801–1816. [CrossRef]
42. Liu, D.Y.; Yan, W.L.; Qian, J.L.; Liu, M.; Wang, Z.D.; Cheng, M.N.; Peng, H.Q. A movable fog-haze boundary layer conceptual model over Jianghuai area, China. *Front. Environ. Sci.* **2021**, *9*, 802316. [CrossRef]
43. Qian, J.L.; Liu, D.Y.; Yan, S.Q.; Cheng, M.N.; Liao, R.W.; Niu, S.J.; Yan, W.L.; Zha, S.Y.; Wang, L.L.; Chen, X.X. Fog scavenging of particulate matters in air pollution events: Observation and simulation in the Yangtze River Delta, China. *Sci. Total. Environ.* **2023**, *876*, 162728. [CrossRef]

atmosphere

Article

Spatiotemporal Distributions and Vulnerability Assessment of Highway Blockage under Low-Visibility Weather in Eastern China Based on the FAHP and CRITIC Methods

Tian Jing [1,2], Duanyang Liu [2,*], Yunxuan Bao [1,2,3,*], Hongbin Wang [2], Mingyue Yan [2,4] and Fan Zu [2]

1 Collaborative Innovation Center on Forecast and Evaluation of Meteorological Disasters, Nanjing University of Information Science & Technology, Nanjing 210041, China; jingtian819216471@163.com
2 Key Laboratory of Transportation Meteorology of China Meteorological Administration, Nanjing Joint Institute for Atmospheric Sciences, Nanjing 210017, China
3 Engineering Research Center for Internet of Things Equipment Super-Fusion Application and Security in Jiangsu Province, Wuxi University, Wuxi 214105, China
4 Highway Monitoring & Response Center, Ministry of Transport of the People Republic China, Beijing 100029, China
* Correspondence: liuduanyang@cma.cn (D.L.); baoyunxuan@163.com (Y.B.)

Abstract: In this study, the spatiotemporal distributions of highway blockage and the low-visibility weather events in eastern China are studied by taking Jiangsu Province as an example. Based on the record table data of highway-blocking events, a vulnerability evaluation model for the highway network in Jiangsu Province is established using the weight assignment methods of the fuzzy analytic hierarchy process (FAHP) and criteria importance though intercriteria correlation (CRITIC). By using the geographic information system, the vulnerability evaluation map of road network in low-visibility weather in Jiangsu Province is finally drawn. The results show that the monthly blockage events on Jiangsu highways are more frequent in the north than in the south and are more frequent along the coast than inland, with the highest occurrence number in winter and a second peak in May. There are basically no blockage events from July to October. Traffic blockage on Jiangsu highways mainly occurs between 22:00 and 08:00 Beijing time. In the afternoon, there are almost no highway-blocking events caused by low-visibility weather. The vulnerability of highway blockage in Jiangsu Province is high in the north and low in the south and high in coastal areas and relatively low in inland. The section K6-K99 of the G30 Lianhuo Highway is the most sensitive.

Keywords: highways; road blockage; fuzzy analytic hierarchy process; CRITIC weight assignment method; road network vulnerability; spatiotemporal distribution

Citation: Jing, T.; Liu, D.; Bao, Y.; Wang, H.; Yan, M.; Zu, F. Spatiotemporal Distributions and Vulnerability Assessment of Highway Blockage under Low-Visibility Weather in Eastern China Based on the FAHP and CRITIC Methods. *Atmosphere* **2023**, *14*, 756. https://doi.org/10.3390/atmos14040756

Academic Editor: Leonardo Primavera

Received: 14 March 2023
Revised: 17 April 2023
Accepted: 18 April 2023
Published: 21 April 2023

1. Introduction

Traffic blockage is a prolonged obstruction or blockage of road access due to sudden traffic-related events, which may be caused by issues such as severe weather, geological disasters, traffic accidents, and planned traffic blockage (road and bridge maintenance, major social events, etc.) [1–3]. Concurrent with the rapid development of China's economy, the mileage and traffic flow of highways have also experienced rapid growth. Prolonged traffic blockage affects the flow, increases the risk of major accidents, threatens the lives of drivers, and has negative impacts on social and economic development [4]. Research has shown that rainfall, snow, fog, and typhoons can all lead to highway blockages [5]. Researchers have found that among all the adverse weather conditions, low visibility causes the greatest hazards during vehicle operation [6]. Many studies have demonstrated that approximately a quarter of all traffic blockages are caused by low-visibility weather, such as dense fog, and the rate of highway accidents in dense fog is 10 times higher than that in normal weather [7].

Low-visibility weather is a catastrophic weather phenomenon in which horizontal visibility is reduced to less than 1000 m by heavy precipitation (e.g., water droplets or ice crystals), fog, etc. [8]. On 15 November 2017, the Chu-Xin highway in Anhui Province was affected by dense fog. More than 30 vehicles collided in succession, and the road was severely blocked. On the morning of 28 January 2021, a series of car crashes occurred on section 1026 of the Hu-Chong highway in Qianjiang, Hubei Province, which was also caused by dense fog. Approximately 20 vehicles were involved in a pileup, which caused a prolonged highway-blocking event [9].

There are many studies regarding the incidence, severity, and risk of highway crashes caused by low-visibility weather [10,11]. Hamilton et al. [12] and Abdel-Aty et al. [13] found that drivers are more likely to be involved in fatal crashes while driving in low-visibility weather. Alghamd et al. [14] used Poisson regression to analyze the 30-year crash data and found that the severity of crashes on foggy days is over 2.55 times higher than on other days. Perry et al. [15] found more severe crashes related to low visibility on highways than that on other roads. Wu et al. [16] applied a binary logistic regression model to actual traffic flow and weather data from two areas in Florida to compare the traffic patterns during persistent fog events to those during sunny periods. Their results show that the risk of accident increases under foggy conditions. Feng et al. found there is a high correlation between the variations of fog-related traffic accidents and low-visibility weather on four highway sections in China [17]. By analyzing the traffic accidents and associated weather conditions in England and Wales, Edwards et al. found that there is an obvious seasonal variation of traffic accidents [18].

Many researchers conducted studies on the traffic vulnerability of roads. Berdica [19] proposed a definition of vulnerability for the road traffic network for the first time, which refers to the vulnerability factors of the road network to adverse external influences. Husdal et al. [20] considered the road network vulnerability as the non-functionality of the road transport network under certain circumstances, emphasizing the loss or impact of an event on the network. Sohn [21] and Scott [22] proposed a scenario-based approach which can identify the key locations in the road network and investigate the vulnerability of the network. Huang et al. [23] explored the factors influencing the degree of injury and death in traffic accidents through the aspect of accident severity. Christopher et al. [24] studied the impact of environment on the safe driving of electric bicycles based on the traffic statistics from the National Statistical Yearbook. Wang et al. [25] used questionnaires to obtain the information on the driving safety of e-bike drivers. Yu et al. [26] analyzed the impact of traffic speed on traffic safety by acquiring vehicle speed information based on coil detectors. There are also many methods of machine learning [27] and computational intelligence [28,29] that have been applied to the analysis and identification of traffic accident risk.

The whole vulnerability assessment mainly refers to the definition of vulnerability to the catastrophe, including three parts: indicator screening, indicator empowerment, and evaluation methods [30]. Among them, the combination of subjective and objective methods was adopted in the index weighting. The subjective method is the fuzzy analytic hierarchy process (FAHP), which optimizes the weight calculation process and can achieve the consistency of judgment matrix and the unity of the thoughts of decision makers [31]. The method of criteria importance though intercriteria correlation (CRITIC) is adopted for objective weighting, which has more adaptability in the weighting for indicators with stronger relevance [32]. The combination of subjective and objective weighting methods can make up for the shortcomings of a single method and solve the defects of a single evaluation method. This combination model has been widely used in the assessments of natural disaster risk, natural resources and carrying capacity [33].

Much of the research on the impact of low-visibility weather on highways is focused on the traffic accident rate or traffic flow, while the research on traffic blockage is mainly focused on the macroscopic level, lacking detailed analysis on the specific types of catastrophic weather. There is no good method for evaluating the vulnerability of motorway networks and no specific study on the vulnerability of road network under specific weather hazard. There is also no specific spatial matching method for the low-visibility weather and motorway-blocking events.

To address the abovementioned problems, this study uses highway control needs as guidance and takes the highway-blocking events under low-visibility weather conditions in Jiangsu Province as the main research object. Instead of lumping all weather factors into one category, we select low-visibility weather—which accounts for a relatively high proportion of highway weather hazards in eastern China—as the target weather type, filter out all the low-visibility weather events that cause highway blockage, and propose a corresponding model to evaluate the vulnerability of the road network to specific weather hazards. Different weather events require specific response measures, which can effectively improve the efficiency of disaster prevention and mitigation and reduce the cost of emergency management. The aim of this work is to improve the existing theory of highway traffic blockage and provide technical support to ensure the smooth and safe operation of regional highway traffic.

The rest of this paper is organized as follows. The materials and methods are described in Section 2. The highway blockage characteristics, distribution of low-visibility weather, and blockage vulnerability are analyzed in Section 3. Finally, the discussions and main conclusions are given in Sections 4 and 5, respectively.

2. Materials and Methods

2.1. Study Area

The study area is Jiangsu Province (166°18′–121°57′ E and 30°45′–35°20′ N), which is located in the eastern part of the Jianghuai Plain along the coast of mainland China and in the mid-latitude zone along the east coast of the Asian continent. This region is in the climate transition zone between the subtropical and warm temperate zones and is characterized by the East Asian monsoon climate. However, it is influenced both by the westerlies in the middle latitudes and the easterlies in subtropical and low latitudes, thus causing a variety of meteorological disasters. There is a network of more than 50 highways (approximately 5000 km in total) in Jiangsu Province, as shown in Figure 1, which suffers from low-visibility weather more and more frequently in recent years. Traffic blockage due to fog and other low-visibility conditions has occurred repeatedly. Therefore, the road safety problems caused by low visibility have received close attention from all sectors of society [34].

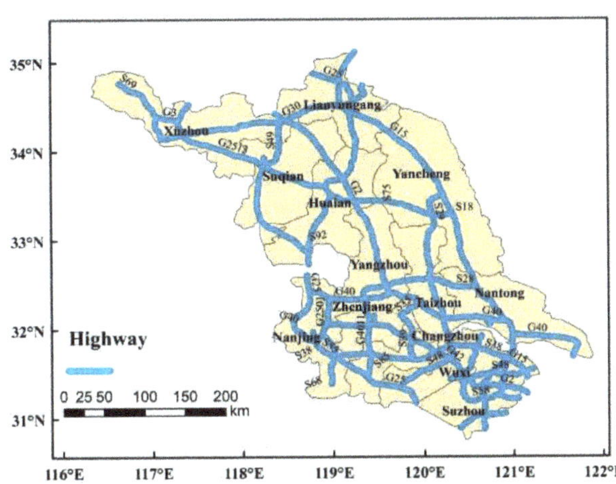

Figure 1. Distribution of highway networks in Jiangsu Province.

2.2. Traffic Blockage Data

The traffic blockage data used in this study are the records of traffic disruptions on the highway network in Jiangsu Province in 2020 obtained from the Road Network Monitoring and Emergency Response Centre of the Ministry of Transport. The records follow the standards established by the Ministry of Transport Highway Traffic Blockage Information Reporting System (Transportation Highway Development (2006) No. 451). There are 16 items in the records, including the province, reporting unit, route name, route number, starting and ending 100 m distance marker, reason for the blockage, length (mileage) of the blockage, status, blockage type, information event classification, site description, disposal measures, blockage discovery time, reporting time, and expected recovery time. Considering that highway blockage information is reported manually, there may be statistical errors in the records. The time series of data are corrected in advance, and then data quality control is performed based on the revised correlation of blockage causes and site descriptions.

2.3. Research Methodology

2.3.1. Determination of Blockage Events Caused by Low-Visibility Weather

In addition to low-visibility weather, the causes of highway blockages include rainfall (water), snowfall (snow), icy roads, and other factors. However, this study focuses on highway blockages caused by low-visibility factors. In order to pinpoint the location of the section where the blockage event occurs, geographic information systems (GIS) technology is used to analyze each blockage event, and the maximum segmentation unit selected in this study is 1 km.

The specific analysis steps are shown in Figure 2. In Part 1, the SPLIT function module in Python is used to slice the data in the record table, and the matching function is used to calibrate and match the road network data with the highway blockage events data. In Part 2, GIS technology is used to spatially match the blockage events with the highway network and visualize the spatial distribution.

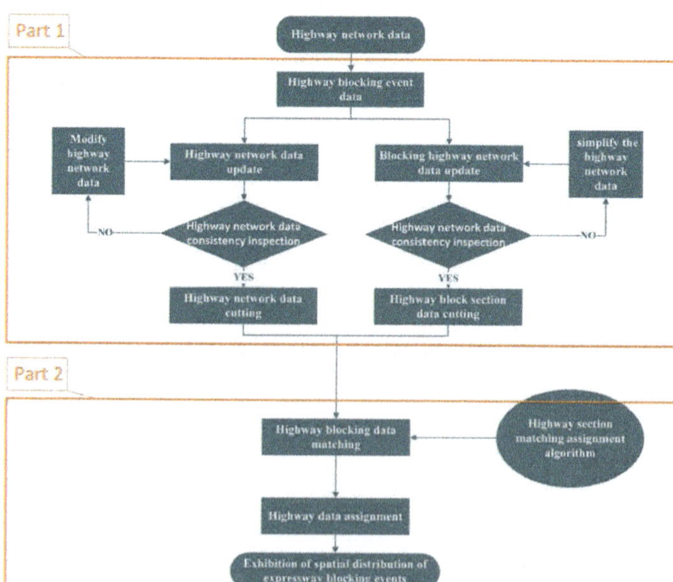

Figure 2. Flow chart of matching assignment of highway block sections.

Data pre-processing is performed first. New highway data are generated through the integration of national and provincial highway route codes. To correct the manually recorded highway blockage data, each highway blockage event in the spatial data of the highway network is analyzed by using the correlation between the route codes and pile numbers, and the incorrect blockage events are removed. The POINT_REMOVE simplification algorithm of GIS platform is then used to simplify the highway network data.

The SPLIT function is used as a processing tool for each blockage event on each highway section with a cell length of 1 km. First, each highway blockage is matched with the corresponding highway data since the starting and ending points of each blockage event are known. Then, the corresponding event is plotted spatially via GIS and attributed with the appropriate information from the nearest highway marker. Traffic events are counted as multiple events if the starting and ending markers that demarcate the extent of the event include more than one section. Each section between the markers is therefore counted as one event.

Based on the occurrence frequency of traffic blockage events under low-visibility weather in 2020, this study classifies the highway sections into six levels: slight risk (<8), less risk (8–14), medium risk (15–21), severe risk (22–28), more severe risk (29–35), and extreme risk (>35).

2.3.2. FAHP Weight Assignment Method

In order to evaluate the vulnerability of the highway network in Jiangsu Province under low-visibility weather, the FAHP method is used to obtain the subjective weights for the analysis of vulnerability weight. The results can be used as one of the reference indexes for the final analysis of vulnerability [35,36].

Based on the hierarchical analysis (AHP), the problem and influencing factors are characterized into the three layers of target, criterion, and indicator. Secondly, the indicator factors u_1–u_n with the same affiliation and hierarchy are compared in pairs to measure their importance, and the scale of importance is represented by 1 to 9 and its reciprocal. The judgment matrix $H_{n\times n}$ is established though Equation (1):

$$H_{n\times n} = \left(a_{ij}\right)_{n\times n} \tag{1}$$

In Equation (1), a_{ij} is the AHP importance scale of u_i and u_j relative to the upper factors, in which a_{ii} is equal to 1 and a_{ji} is the reciprocal of a_{ij}. The meanings of the importance scales are shown in Table 1.

Table 1. Meanings of importance scales.

Scale	Description
1	Equally important
3	Slightly important
5	Obviously important
7	Strongly important
9	Extremely important
2, 4, 6,8	Median value

According to the scale conversion formula, the judgment matrix is transformed into a fuzzy complementary judgment matrix $W_{n\times n}$ though Equations (2) and (3):

$$w_{ij} = log_{\alpha}a_{ij} + 0.5 \tag{2}$$

$$W_{n\times n} = \left(w_{ij}\right)_{n\times n} \tag{3}$$

where a_{ij} is the same variable as that in Equation (1). α can be used to adjust the difference between the weights of the final indicators, which is determined by the decision maker according to the actual situation. In this study, α is set to be equal to or greater than 81 to ensure that the w_{ij} is between 0 and 1. It can be seen from Equation (2) that w_{ij} represents the relative importance of u_i and u_j and that $w_{ij} + w_{ji}$ is equal to 1. When w_{ij} is equal to 0.5, u_i and u_j are of equal importance; when w_{ij} is greater (less) than 0.5, u_i (u_j) is more important than u_j (u_i).

The fuzzy complementary judgment citation is further transformed into the fuzzy consistent judgment matrix $R_{n\times n}$ through the following equations:

$$r_i = \sum_{j=1}^{n} w_{ij} \tag{4}$$

$$r_{ij} = \frac{r_i - r_j}{2(n-1)} + 0.5 \tag{5}$$

$$R_{n\times n} = \left(r_{ij}\right)_{n\times n} \tag{6}$$

where r_{ij} is the FAHP importance scale of u_i and u_j corresponding to the upper factors, and it satisfies the conditions of $r_{ii} = 0.5$, $r_{ij} + r_{ji} = 1$ and $r_{ij} = r_{ik} - r_{jk} + 0.5$, $(i, j, k = 1, 2, \ldots, n)$.

The fuzzy judgement matrix is consistent with no need for consistency testing, and it can reflect the subjective thought of the decision maker. The subjective weight W_1 corresponding to each indicator can be obtained via the characteristic root method.

2.3.3. CRITIC Weight Assignment Method

Considering that there is a strong correlation between the selected indicators, the CRITIC method was adopted in this study to objectively evaluate the vulnerability weight, which was regarded as one of the final vulnerability reference indicators [37].

The CRITIC method is an objective weight assignment method which is commonly used for the analysis of data with strong correlations between indicators while considering the variability among indicators concurrently. By objectively calculating the indicators of data, each indicator was assigned a different weight, and the calculation steps are as follows [38].

The standard deviation can be used to measure the contrast intensity and dispersion degree of indicators. A larger standard deviation represents a greater dispersion degree, which indicates larger differences between samples and larger assigned corresponding weights. The standard deviation can be calculated by Equation (7).

$$S_j = \sqrt{\frac{\sum_{i=1}^{n}\left(x_{ij} - \frac{1}{n}\sum_{i=1}^{n} x_{ij}\right)^2}{n-1}} \tag{7}$$

where x_{ij} denotes the ith sample for the jth indicator, S_j is the standard deviation of the jth indicator, and n is the total number of samples for the jth indicator.

Correlation is expressed as the correlation coefficient between indicators. The stronger the correlation between indicators is, the more conflicting the indicators are and the higher the repetition rate of information expression. Therefore, the corresponding weights of the indicators can be reduced to a certain extent. The correlation coefficient R can be calculated in Equation (8).

$$R_j = \sum_{i=1}^{p} (1 - r_{ij}) \tag{8}$$

where R_j indicates the correlation coefficient of the jth indicator with the other indicators, r_{ij} denotes the correlation coefficient between the ith and jth indicators, and p is the total number of indicators.

The weight of the jth indicator W_{2j} can be obtained from Equation (9).

$$W_{2j} = \frac{S_j \times R_j}{\sum_{j=1}^{p} S_j \times R_j} \tag{9}$$

2.3.4. Portfolio Empowerment Method

The CRITIC method considers the correlation between indicators more and pays attention to the information content of the data itself, but it is easily affected by the two-level value. The subjective weight assignment will ignore the information brought by the data itself. Therefore, the combination of subjective and objective weight assignment methods can be used for more reliable and accurate evaluation.

After the subjective and objective weighting methods are used to determine the weight of the assessment indicators, the proportion of subjective and objective weights in the overall weight should be clarified to better reflect the difference in the importance between multiple assessment indicators. W_1 and W_2 are the weights derived from the FAHP and CRITIC methods. The combined weight was obtained using the linear combination, as shown in Equation (10).

$$W = \alpha W_1 + \beta W_2 \tag{10}$$

where α and β denote the weight allocation coefficients.

In order to find the best combined weight, the optimal weight assignment coefficients which minimized the standard deviation of W were obtained though the following equations:

$$x_i = \alpha_i W_{1i} + \beta_i W_{2i} \tag{11}$$

$$\delta = \sqrt{\frac{\sum_{i=1}^{j} (x_i - \sum_{i=1}^{j} \frac{x_i}{j})}{j}} \tag{12}$$

where j is the maximum of sample numbers and both α and β are between 0 and 1 and their sum is equal to 1.

3. Results

3.1. Characteristics of Highway Blockage

3.1.1. Annual Variation of Highway Blockage

Based on the causes, site descriptions, and treatment measures, a total of 1340 highway blockages due to the low visibility were extracted from highway blockage records in Jiangsu Province in 2020, with a total annual cumulative blockage mileage of 69,466.3 km. In addition, the cumulative monthly blockage mileage and monthly blockage frequency were calculated to analyze the spatial and temporal distributions of highway blockages in Jiangsu Province.

As shown in Figure 3, the proportions of annual highway blockage events that led to blockage mileage less than 100 km, between 100–200 km, and 200–300 km are 85.5, 9.8, and 4.7%, respectively. The blockage events with maximum cumulative mileage and frequency on the highways of Jiangsu Province in 2020 basically occurred in January and February, while the minimum values were observed in August and September, when there was 0 km of cumulative blockage and 0 events were reported. The frequency of blockage was higher in January, February, May, November, and December, when the mileage of individual blockage was also higher. The number of blockage events in May shows a second peak, which is different from the decreasing trend observed in March–April.

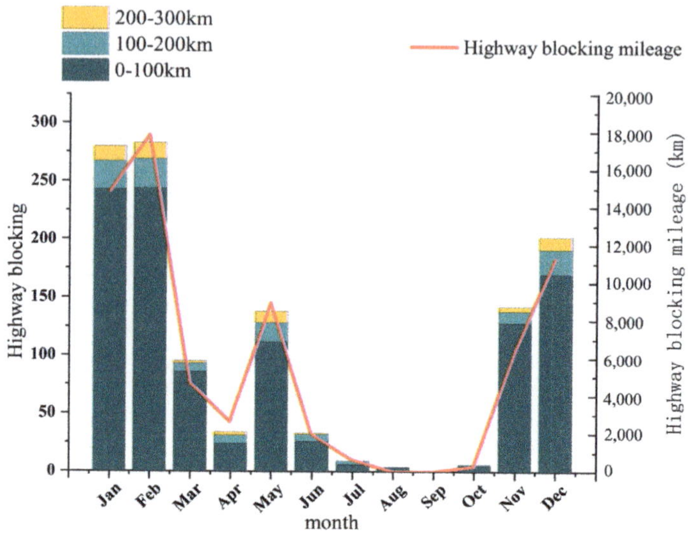

Figure 3. Annual variations of frequency (bars) and mileage (line) of highway-blockage events in Jiangsu Province in 2020.

The spatial distribution of highway blockage events in Jiangsu Province in 2020 (Figure 4) shows high occurrence in the north and low occurrence in the south. For the northern Jiangsu Province, the number of highway-blocking events in the eastern part is also substantially higher than that in the western part. It can be seen form Figure 4 that, as on the major north–south highways along the eastern coast and in the central region of Jiangsu Province, there are severe highway-blocking events in the north sections of the G15 Shenhai Highway and the G2 Beijing–Shanghai Highway. Among them, there are more than 37 blockage events in the K760–K1163 section of the G15 Highway, making this section one of the most severely blocked sections.

On the G25 Changshen Highway, the peak number of blockage events reached 64 in the K1762 section. The number of highway blockages in southern Suzhou was relatively small, with the cities of Wuxi (except Yixing) and Suzhou having four or fewer blockage events. There was no highway blockage on the S9 Sushao Highway and the G50 Shanghai–Chongqing Highway in Suzhou. There were at least 11 blockage events in Nanjing, Changzhou, and Yixing on the G25 Changshen Highway, sections K2060–K2190, in the western part of southern Jiangsu, which is a high value area within the entire southern Jiangsu region. Highway blockages in the central region of Suzhou were mainly concentrated in the K710–K1038 section of the G2 Highway. Niu et al. [39] conducted a GIS-based study on blocked highways in Jiangsu Province as a function of low visibility and found that the G15 Shenhai Highway is the key highway blocked by fog, which is basically consistent with the findings in this study.

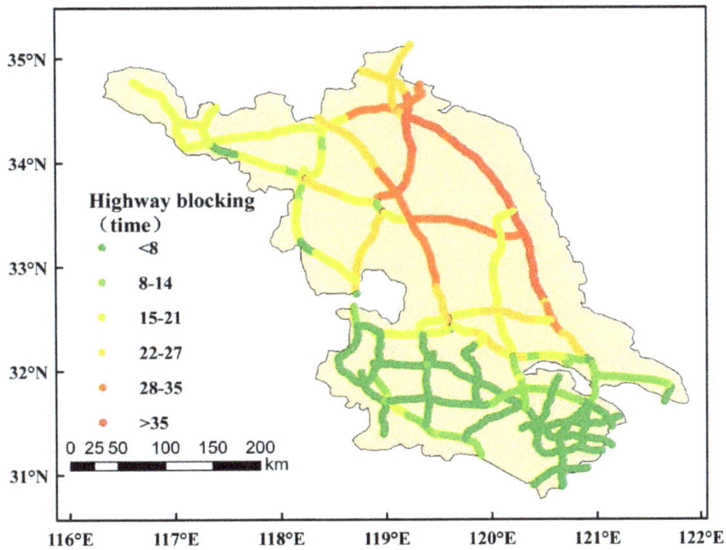

Figure 4. Spatial distribution of highway-blocking events in Jiangsu Province in 2020.

The annual variation of highway blockage spatial distribution in Jiangsu Province in 2020 (Figure 5) shows that there are more highway-blocking events in November, December, and January than that in other months, with 6–8 blockage events in January in the northern section of the G15 Shenhai Highway (K850–K989) and the central section of the G2 Beijing–Shanghai Highway (K712–K958). The southern section of the G15 Shenhai Highway K1090–K1173 was blocked less than the northern section, with six events. There were fewer than six blockage events in the southern part of Jiangsu Province, with most blockage occurring in the K43–K76 section of the S38 Changhe Highway in Wuxi (4–6 events). In February, the highway blockage is the most severe, with the highest number of 12 times on the G15 Shenhai Highway in Lianyungang and 10 blockage events occurring on the S18 Yanhuai Highway in Huaiyin and Yancheng. The blockage events in the K710–K971 section of the G2 Beijing–Shanghai Highway and the K126–K261 section of the S28 Qiyang Highway reached a moderate level of six events in the central region of Suzhou.

There were fewer than two blockage events in the entirety of southern Jiangsu. The spatial distribution of highway blockages in Jiangsu Province in February showed a decreasing trend from north to south and from east to west, while the blockage events in March were mainly concentrated in the K819–K1736 section of the G25 Changshen Highway in the city of Huaiyin and in the K38–K136 section of the S18 Yanhuai Highway in the city of Yancheng in April. There was a sudden increase in the number of blockage events in May, mainly in the K890–K1093 section of the G15 Shenhai Highway along the eastern coast. Some blockage events occurred in the K854–K1087 section of the G15 Shenhai Highway in June but were substantially fewer than that in May. The number of highway-blocking events increased further in December. The highway-blocking events were still mainly concentrated in the coastal area and in the K854–K1087 section of the G15 Shenhai Highway, with 5–8 blockage events recorded in the K854–K1094 section. The K0–K206 section of the G30 Lianhuo Highway also had relatively high risk, with five to six recorded blockage events, while the number of highway blockages remained below two in southern Jiangsu.

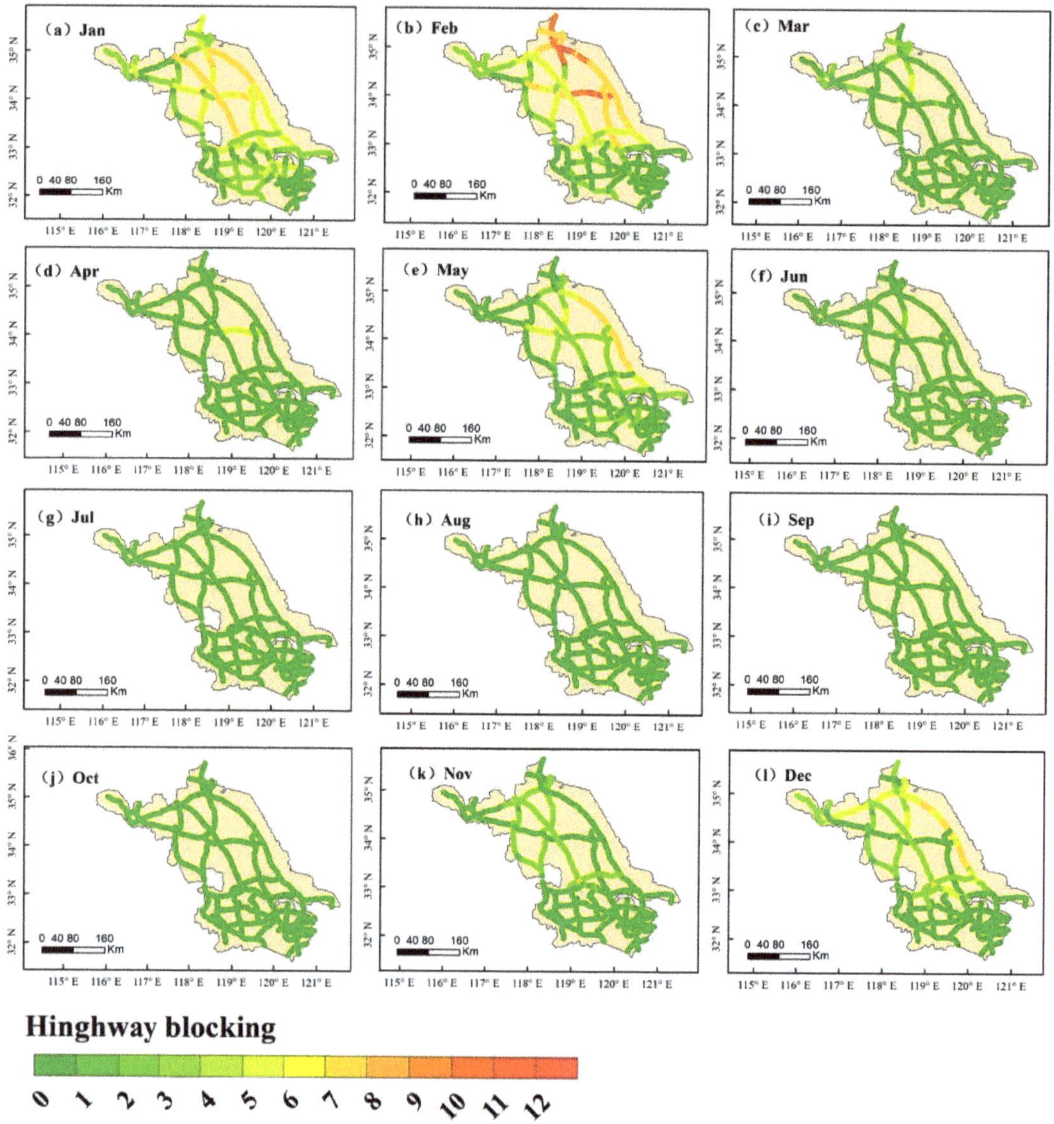

Figure 5. Annual variation of highway blockage distribution in Jiangsu Province in 2020.

3.1.2. Diurnal Variation of Highway-Blocking Events

As seen from the diurnal variation of the highway blockage frequency in Jiangsu Province in 2020 (Figure 6), highway-blocking events are mainly concentrated in the late night and early morning hours. The number of blockages reached the peak value between 00:00 and 01:00 Beijing time (BT) with the value of 158. From 00:00 to 04:00 BT, there was a decreasing trend in the frequency of highway blockages, and increased between 04:00 and 05:00 BT, reaching the second highest value of 135. The trend then decreased until 09:00 BT, with almost no blockage events from 10:00–21:00 BT. The number of blockage events started to increase after 21:00 BT and reached 130 at 23:00 BT. During the late night and early morning hours, there was a greater number of events with blockage mileage over 200 km, as longer and more severe blockage events were common during this period.

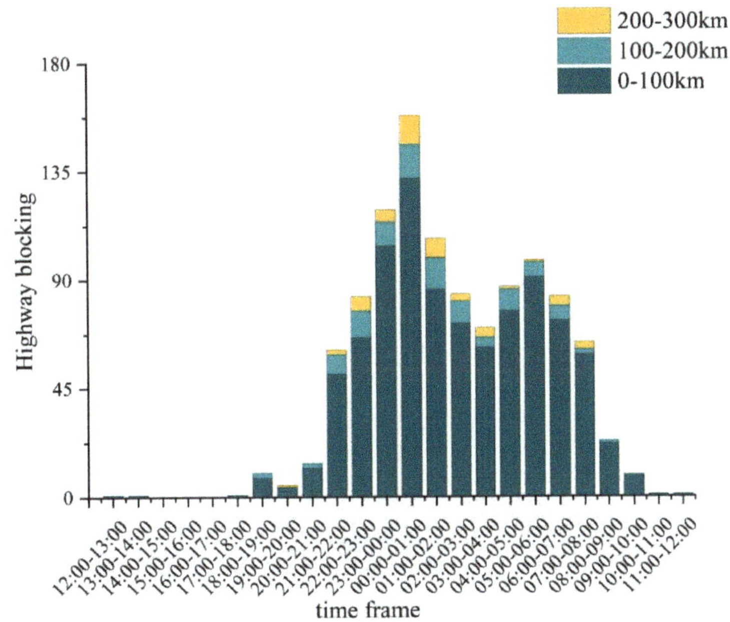

Figure 6. Diurnal variation of the annual average frequency of highway-blocking events in Jiangsu Province in 2020.

The spatial distributions of highway-blocking events during different time periods in Jiangsu Province in 2020 are shown in Figure 7. It shows that the highway-blocking events in Jiangsu Province are mainly concentrated between 00:00 and 08:00 BT. In general, the number of events in the north is higher than that in the south, and more events occur along the coast than inland. The overall trend shows the characteristic of decreasing from the eastern coast to the west, with the most highway-blocking events on the eastern coast and in the K888–K1004 section of the G15 Shenhai Highway. The section K744–K955 in the G2 Beijing–Shanghai Highway also has more highway-blocking events than average. During the period of 09:00–17:00 BT, the number of highway blockages in the entire province was relatively low; only the K765–K1098 section of the G15 Shenhai Highway and the K40–K140 section of the S18 Yanhuai Highway in the eastern coastal area had more than eight blockage events. There was an obvious increase in the number of highway blockages in the K0–K90 section of the G30 Lianhuo Highway in Lianyungang and the K1840–K1933 section of the G25 Changshen Highway in Huaiyin from 18:00–23:00 BT compared to the those during the period of 09:00–17:00 BT, showing an overall increasing trend from the inland to the eastern coastal area from 18:00–23:00 BT.

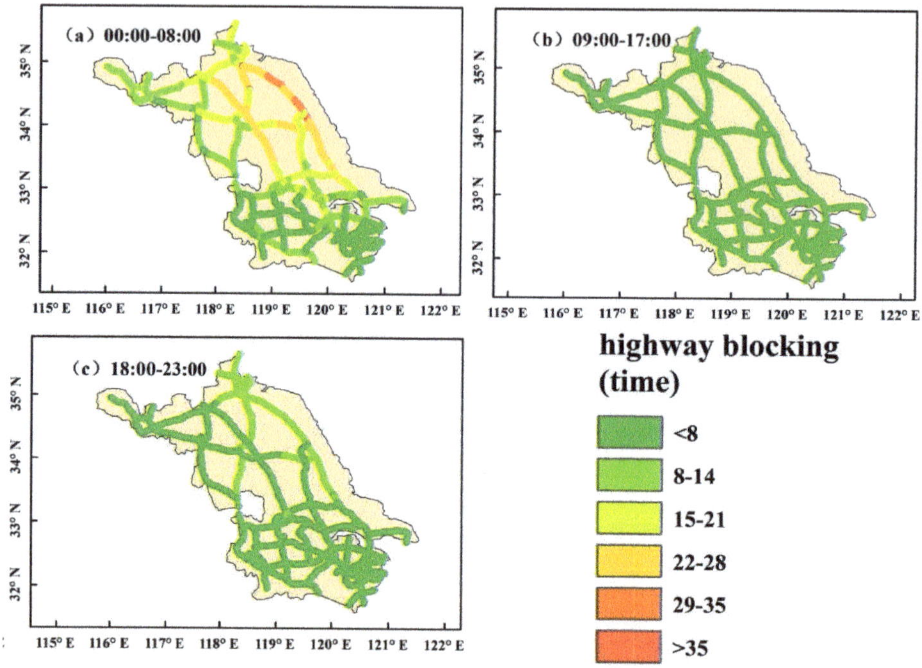

Figure 7. Spatial distributions of highway-blocking events during different time periods in Jiangsu Province, 2020.

3.2. Vulnerability of Highways to Low-Visibility Weather in Jiangsu Province

This study determines the vulnerability of each road unit in the Jiangsu provincial highway network to low-visibility weather by selecting data from the records of highway disruption events. The blockage frequency, the cumulative duration of blockage, the duration of blockage response, the duration of blockage rescue, and the duration of highway blockage with false alarm were used as the vulnerability indicators. The specific indicators are shown in Table 2.

Table 2. Indicators of highway network vulnerability assessment.

Target Level	Guideline Level	Number	Program Level	Number
Vulnerability	Sensitivity	A	Blockage frequency	A1
			Cumulative duration of blockage	A2
			Blockage severity	A3
	Emergency response capability	B	Duration of blockage response	B1
			Duration of blockage rescue	B2
			Duration of highway blockage with false alarm	B3

According to the indicator system in Table 2, the relative weights of the indicators corresponding to each fuzzy consistent judgment matrix were calculated using the characteristic root method to obtain the subjective weight. By unitizing the highway network data, the weight was calculated using the CRITIC method. The optimal weight allocation coefficients were determined to be those with the minimum standard deviation calculated using Equations (11) and (12). The results are shown in Figure 8, and the final weight W assignment was obtained using Equation (13). The final weights are shown in Table 3.

$$W = 0.5583W_1 + 0.4417W_2 \tag{13}$$

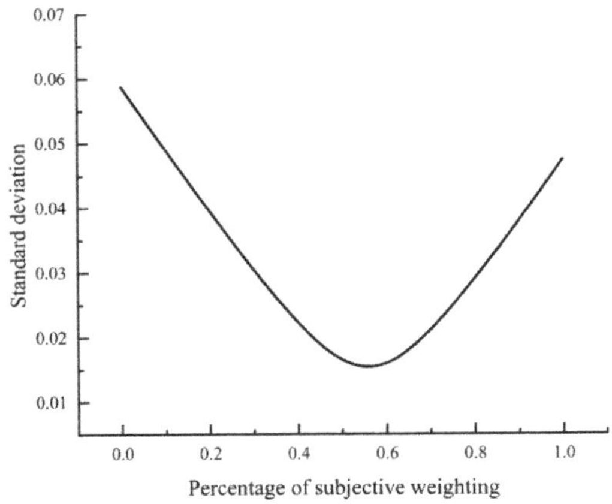

Figure 8. Variation of standard deviation with weight coefficient of W_1.

Table 3. Weights of indicators for the vulnerability assessment.

Target Level	Guideline Level	Program Level	W_1	W_2	W
Vulnerability	Sensitivity	frequency of blockage	6.75%	28.34%	16.29%
		cumulative duration of blockage	17.57%	10.06%	14.25%
		Blockage severity	19.14%	11.65%	15.83%
	Emergency response capabilities	the duration of blockage response	21.99%	15.66%	19.19%
		Blocked rescue duration	17.87%	16.59%	17.30%
		the duration of highway blockages with false alarms	19.14%	17.71%	18.51%

First, each vulnerability index in Table 3 was calculated in a standardized way. Then, each vulnerability index of each road network unit was calculated. The final evaluation value was obtained according to W, and the evaluation results was matched to the highway network through GIS technology, as shown in Figure 9. It is obvious that the sensitivity of highway network in Jiangsu Province to low-visibility weather presents a decreasing trend from north to south. The vulnerability of highway network is higher in coastal areas and lower in southern Jiangsu. It is worth noting that the frequency of highway blockage in Nantong City is low, but the vulnerability is high, indicating poor capacity of disaster relief in this section. The S49 Suyang highway section in central Jiangsu is highly sensitive. Combined with the analysis of blockage events, the rescue response time of this section is relatively high. The high vulnerability of G30 Lianhuo Highway in Lianyungang City is mainly due to the high severity of a single highway-blocking event. For the highways in coastal areas, although the frequency of highway-blocking events is higher, the cumulative severity of highway blockage is lower, and the misreporting of highway blockage time is less. It indicates that the abilities of disaster resistance and emergency response are good in this area, which leads to a medium level of vulnerability. The density of the road network in the whole of southern Jiangsu is relatively high, but it is kept at a low vulnerability due to the low frequency of highway blockage in low-visibility weather.

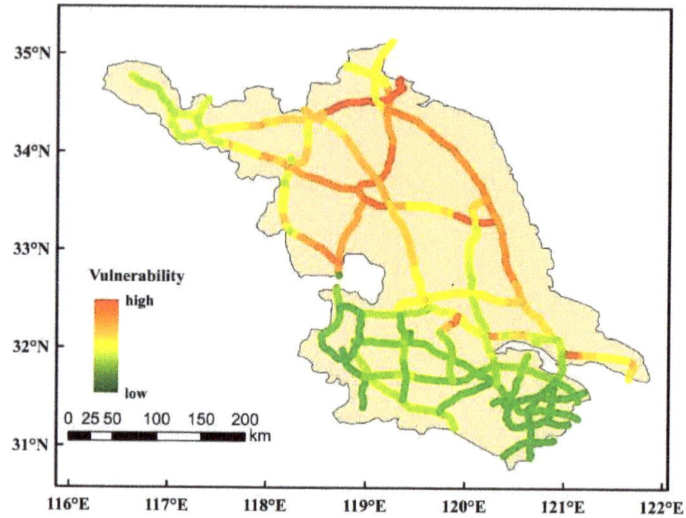

Figure 9. Distribution map of highway vulnerability in Jiangsu Province.

4. Discussion

4.1. Cause Analysis of the Blockage Distribution Due to Low-Visibility Weather

The reasons for the spatiotemporal distribution of highway-blocking events in Jiangsu Province were analyzed in relation to existing studies. Consistent with previous research, it has been observed that the hours between 00:00 and 04:00 BT are conducive to the occurrence of fog in Jiangsu Province; there is favorable environment of high air humidity, low wind speed, and cooling at nighttime.

Fog tends to last until 08:00 BT in the morning and begins to dissipate after sunrise, when the solar radiation increases, the temperature near the ground rises, and the relative humidity of the air decreases. This is one of the reasons that contribute to the sharp decline in the blockage events related to low-visibility weather starting at 08:00 BT [40].

Generally, after 10:00 BT, the wind speed at ground level gradually increases until approximately 14:00 BT, when it reaches the maximum value for the day. Aerosol particles in the air are also more likely to disperse during this period. Therefore, the probability of low-visibility weather is lower, and the number of highway-blocking events begins to decrease accordingly [41]. When the lighting condition is better in the afternoon, the visibility is less likely to be affected and to drop low enough to cause highway blockage, which makes the occurrence of blockage events less frequent.

In the coastal area of northern Jiangsu, the topography is more complex, and surface water is more widely available. Coastal geography also provides abundant water vapor and appropriate conditions for the formation of fog [42,43]. As a result, the frequency of highway blockage due to low visibility in the coastal area of northern Jiangsu is higher year-round.

In summary, the seasonal distribution of highway-blocking events in Jiangsu obviously shows fewer events in summer and more events in winter which is consistent with the seasonal distribution of low-visibility weather. The diurnal variation of highway-blocking events shows that more events occur in the nighttime and that fewer occur in the daytime.

4.2. Vulnerability Assessment and Analysis of Highway Network in Low-Visibility Weather

In this study, the highway network was segmented into units by GIS technology, and the highway-blocking events from the highway blockage data record table were matched with the road network information by matching function. Highway-blocking events can be quickly positioned on specific units, and the spatial distribution map of blockage events can be effectively obtained.

The blockage events that have happened on each road network unit were sorted out and analyzed, and the comprehensive weight assignment method combining the FTHP and CRITIC methods was adopted to calculate the blockage event information of each road unit, and the final vulnerability assessment result of the entire highway network in Jiangsu Province to low-visibility weather was obtained. The result can be used to effectively avoid traffic blockage in practice and provide reference for road safety operation management.

The distribution of road network vulnerability to low-visibility weather in Jiangsu Province is higher in the north and lower in the south. The coastal and central regions are more vulnerable than the southern inland regions. Among them, the highways in coastal areas have higher frequencies of blockage events, but their overall vulnerability is at a medium level due to the high capability of disaster resistance and emergency response. In the whole network, the vulnerability is higher at the intersections of highways.

In the whole highway network, the vulnerability of east–west highways—such as highways G30 and S18—which have poor performance in disaster preparedness is higher. The northern part of G30 is the worst in terms of disaster resilience. In addition, highway G25 in the north is highly sensitive, which leads to higher vulnerability. It is worth noting that although the annual blockage frequency of G40 in Nantong City is low, the overall vulnerability also reaches a medium level due to the long false alarm blockage time in this section.

Studies on motorway disruptions in low-visibility weather have focused more on the traffic flow and travel mode chosen by drivers [44,45]. Chaudhuri et al. [46] used the 10-year motorway accident data in Barcelona to build a risk model for the urban motorway network through Bayesian networks. However, accidents are contingent, and a single high damage accident can affect the risk assessment results of the whole network. In contrast, the traffic-blocking events used in this study have covered a wider range, taking traffic congestion in low-visibility weather into account, and can reduce the impact of a single event to a certain extent.

Miomir et al. [47] used the fuzzy MARCOS method to analyze accident risk at 38 points along a 7.3 km stretch of road, which is better for shorter roads but is not applicable to the emergency preparedness planning and design for the whole road network. Some scholars used accident data in their studies of road traffic risk and vulnerability, which are inextricably linked to drivers, and some have also studied the characteristics of drivers in traffic accidents.

It is therefore difficult to define whether the specific factors that contribute to each accident are external environmental factors or the internal factors of drivers in the study of highway accidents. In contrast, the traffic-blocking events are based on field record data, which allows for a clearer identification of the weather factors that led to the blockage.

It is worth noting that the geographical environment of the specific highway was not taken into account in this study for the vulnerability modelling process. However, in the analysis of road traffic risk, the presence of bridges and tunnels, as well as the road gradient and road surface material, are also important reference indicators which will be considered in the analysis of key blocked sections in the future study.

4.3. Prevention and Control Measures in the Sections with High Blockage Frequency

In this study, we have selected the three most severely blocked highways, G15, G2, and G25, which need the focus of attention of meteorological and traffic authorities. The blocked sections in these highways were classified into three levels according to the number of highway-blocking events: lighter, medium, and serious (Table 4). The meteorological department should prepare forecasts for low-visibility weather, and the transport department should prepare in advance for low-visibility weather. Additional warning signs should be erected on the blocked roads to remind drivers to pay attention to current road conditions and slow down. On the other hand, light-emitting diode (LED) fog lights and monitoring equipment should be installed in the sections prone to severe low-visibility blockages, and reserve ramps should be built in excessively long sections prone to blockage to clear the traffic flow in the low-visibility weather. Staff and duty points should be arranged to be on duty in advance in the blockage-prone periods and sections. Additional weather stations are also needed to enhance monitoring capability. In the lighter-level sections, the density of monitoring stations should be improved to 10 km per station, while the density for the sections of medium and serious levels needs to be increased to 5 km per station. Appropriate emergency plans and rescue measures for different levels of traffic-blocking events also need to be improved. It is important to warn drivers before the occurrence of blockage, and contact drivers in time to strengthen the diversion of traffic flow after the end of blockage.

Table 4. Actions suggested by the traffic department.

Route Number	Highway Blockage	Highway Blockage Section	Fog Lights Suggestions	Road Proposal
G15	lighter	K1164–K1184	Unidirectional mounting	Place warning signs
		K1217–K1251		
		K823–K835		
	medium	K1182–K1216	Bidirectional installation	Diversion road
		K760–K822		Diversion road
		K836–K844		Place warning signs
	serious	K846–K1163		Diversion road
		K2122–K2190		Place warning signs
G2	lighter	K1089–K1190	Unidirectional mounting	Diversion road
		K972–K991		Place warning signs
		K1039–K1088		
	medium	K710–734	Bidirectional installation	Place warning signs
		K979–K1061		Diversion road
	serious	K735–K970		Place warning signs
G25	lighter	K1644–K1693	Unidirectional mounting	Place warning signs
	medium	K1827–K1932	Bidirectional installation	Diversion road
		K1763–1825		Place warning signs
	serious	K1695–K1760		

5. Conclusions

This study used the traffic blockage data of highways in Jiangsu Province in 2020 to identify and locate highway-blocking events for individual highway marker. By using the Python programming language, GIS technology and mathematical statistics, the spatiotemporal distribution characteristics of highway-blocking events in Jiangsu Province were analyzed, and the vulnerability of Jiangsu highway network to low-visibility weather was assessed through the FAHP and CRITIC weight assignment methods. The main conclusions are as follows.

Jiangsu Province has a distinct seasonal variation of highway blockage with a bimodal pattern of annual variation. Blockage events are more likely to occur in winter, followed by spring and autumn, and they are less likely to occur in summer. Most highway-blocking events occur during late night to early morning, followed by early evening, with minimal occurrence in the afternoon. The peak time of blockage occurrence is during 00:00–01:00 BT, with no blockage occurring during 09:00–21:00 BT.

More highway-blocking events occur in the north of Jiangsu Province than in the south, and more blockage events occur in the eastern coast than in the western hinterland, which indicates an overall decreasing trend from the coastal area to inland. Lianyungang and Yancheng are the cities with the highest incidence of highway blockage, whereas the most blockage events occur on the G15 Shenhai highway in the coastal area.

The main distribution characteristic of vulnerability for the highways in Jiangsu Province is higher in the north and lower in the south. The vulnerability of highways in coastal areas is higher than in inland areas. Among them, the K2122–K2190 section of the G25 Shenhai Highway, the K735–K970 section of the G2 Beijing–Shanghai highway, the K1595–K1760 section of the G25 Changshen Highway, and the K6–K99 section of the G30 Lianhuo Highway are highly sensitive to low-visibility weather. The primary factors of high vulnerability are different in different regions. Northern Jiangsu is dominated by high sensitivity and low capabilities of disaster resistance and emergency response, while central Jiangsu is dominated by the capabilities of disaster resistance and emergency response. The vulnerability of highways in southern Jiangsu is at low level due to the few blockage events.

It is important to note that the time series of the highway blockage data used in the study is relatively short, with only 1-year data available (2020), which imposes certain limitations on the reliability of our results. It is expected that a longer time series will be available in the future to verify the spatiotemporal distributions obtained from the analysis presented here. Furthermore, the quantitative analyses of the spatial and temporal variabilities of highway-blocking events cannot be conducted due to the complexity of road environment factors, while only qualitative analyses can be conducted through literature research, practical investigation, and professional knowledge. Currently, it is not possible to quantitatively analyze the characteristics and causes of highway-blocking events by combining various meteorological elements and geographic environmental factors, which must be continuously improved in the future.

Author Contributions: Conceptualization, T.J. and D.L.; methodology, D.L. and Y.B.; investigation, H.W. and M.Y.; resources, M.Y.; data curation, F.Z.; writing—original draft preparation, T.J.; writing—review and editing, D.L. and Y.B. All authors have read and agreed to the published version of the manuscript.

Funding: This research was jointly funded by the Wuxi Social Development Science and Technology Demonstration project (N20201012) and the Beijige Open Foundation (BJG202201, BJG202209).

Institutional Review Board Statement: Not applicable.

Informed Consent Statement: Not applicable.

Data Availability Statement: Data available on request due to restrictions, e.g., privacy or ethical; the data presented in this study are available on request from the corresponding author. The data are not publicly available due to confidentiality.

Conflicts of Interest: The authors declare no conflict of interest.

References

1. Huang, C.F. The basic principle of integrated risk assessment. In Proceedings of the First International Conference on Risk Analysis and Crisis Response, Shanghai, China, 25–26 September 2007; pp. 22–27.
2. Ministry of Transport of the People's Republic of China. *Traffic Blocking Information Reporting System of Ministry of Transport (Trial Operation, Ministry of Transport)*; Ministry of Transport of the People's Republic of China: Beijing, China, 2006.
3. The Ministry of Transport of the People's Republic of China. *Traffic Blocking Information Reporting System of the Ministry of Transport*; The Ministry of Transport of the People's Republic of China: Beijing, China, 2011.
4. Song, J.; Li, Y.; Zhang, H.; Tian, H.; Cheng, Z.; Yi, X. Analysis of the Influence of Severe Weather on Highway Traffic Blocking. *Highway* **2021**, *66*, 248–256.
5. Yan, K.; Dong, L. Data Statistics and Analysis of Blocked Road Information in China. *J. Highw. Transp. Res. Dev.* **2009**, *16*, 121–124.
6. Zhang, Y.; Ou, B.; Sun, X. Cause analysis and solution of highway traffic accident in foggy weather. *China Sci. Technol. Inf.* **2008**, *20*, 294.
7. Yan, M. The Charaeteristies and Numerical Simulations of Dense Fog Along Nanjing To Shanghai highway. Master's Thesis, Nanjing University of Information Science & Tcchnology, Nanjing, China, 2011.
8. Huang, B. Study on the Traffic Risk Analysis and Forewarning Management of Freeway Under Disaster Weather. Master's Thesis, Shanghai Jiao Tong University, Shanghai, China, 2012.

9. Gao, Y.; Gong, J. Characteristics and mechanism of road traffic accidents. *J. Saf. Environ.* **2022**, *6*, 12–15. [CrossRef]
10. Zhang, J.; Tan, G.; Wu, H.; Zhang, Y.; Zhang, D. Characteristics on Highway Traffic Accidents Caused by the Disastrous Weather in Hebei and the Early-warning Indexes System of Highway Traffic Under Foggy Weather Condition. *J. Arid Meteorol.* **2016**, *34*, 370–375.
11. Wang, Y.; Liang, L.; Evans, L. Fatal crashes involving large numbers of vehicles and weather. *J. Safety Res.* **2017**, *63*, 1–7. [CrossRef]
12. Hamilton, B.; Tefft, B.; Arnold, L.; Grabowski, J. Hidden highways: Fog and traffic crashes on America's roads. 2014. Available online: http://aaafoundation.org/wp-content/uploads/2017/12/FogAndCrashesReport.pdf (accessed on 13 March 2023).
13. Abdel-Aty, M.; Ekram, A.A.; Huang, H.; Choi, K. A study on crashes related to visibility obstruction due to fog and smoke. *Accid. Anal. Prev.* **2011**, *43*, 1730–1737. [CrossRef]
14. Al-Ghamdi, A.S. Experimental evaluation of fog warning system. *Accid. Anal. Prev.* **2007**, *39*, 1065–1072. [CrossRef]
15. Perry, A.H. *Environmental Hazards in the British Isles*; George Allen & Unwin: London, UK, 1982; p. 73.
16. Wu, Y.; Abdel-Aty, M.; Lee, J. Crash risk analysis during fog conditions using real-time traffic data. *Accid. Anal. Prev.* **2017**, *114*, 4–11. [CrossRef]
17. Feng, L.; Deng, Y.; Li, A.; Li, Y.; Song, J. The characteristics of traffic accidents in highway in fog days and their relationship with visibility. *Sci. Technol. Rev.* **2020**, *38*, 160–168.
18. Edwards, J.B. The temporal distribution of road accidents in adverse weather. *Meteorol. Appl.* **1999**, *6*, 59–68. [CrossRef]
19. Berdica, K. An introduction to road vulnerability: What has been done, is done and should be done. *Transp. Policy* **2002**, *9*, 117–127. [CrossRef]
20. Husdal, J. The vulnerability of road networks in a cost-benefit perspective. In Proceedings of the Transportation Research Board Annual Meeting (TRB 2005), Washington, DC, USA, 9–13 January 2005.
21. Sohn, J. Evaluating the significance of highway network links under the flood damage: An accessibility approach. *Transp. Res. Part A Policy Pract.* **2006**, *40*, 491–506. [CrossRef]
22. Scott, D.M.; Novak, D.C.; Aultman-Hall, L.; Guo, F. Network robustness index: A new method for identifying critical links and evaluating the performance of transportation networks. *J. Transp. Geogr.* **2006**, *14*, 215–227. [CrossRef]
23. Huang, H.; Chin, H.C.; Haque, M.M. Severity of driver injury and vehicle damage in traffic crashes at intersections: A Bayesian hierarchical analysis. *Accid. Anal. Prev.* **2008**, *40*, 45–54. [CrossRef] [PubMed]
24. Cherry, C.R.; Weinert, J.X.; Xinmiao, Y. Comparative environmental impacts of electric bikes in China. *Transp. Part D Transp. Environ.* **2009**, *14*, 281–290. [CrossRef]
25. Williams, A.F.; Shabanova, V.I. Responsibility of drivers, by age and gender, for motor-vehicle crash deaths. *J. Saf. Res.* **2003**, *34*, 527–531. [CrossRef]
26. Yu, R.; Quddus, M.; Wang, X.; Yang, K. Impact of data aggregation approaches on the relationships between operating speed and traffic safety. *Accid. Anal. Prev.* **2018**, *120*, 304–310. [CrossRef]
27. De Ona, J.; López, G.; Mujalli, R.; Calvo, F.J. Analysis of traffic accidents on rural highways using Latent Class Clustering and Bayesian Networks. *Accid. Anal. Prev.* **2013**, *51*, 1–10. [CrossRef]
28. Kashani, A.T.; Mohaymany, A.S. Analysis of the traffic injury severity on two-lane, two-way rural roads based on classification tree models. *Saf. Sci.* **2011**, *49*, 1314–1320. [CrossRef]
29. Harb, R.; Yan, X.; Radwan, E.; Su, X. Exploring precrash maneuvers using classification trees and random forests. *Accid. Anal. Prev.* **2009**, *41*, 98–107. [CrossRef] [PubMed]
30. Cannon, T.; Twigg, J.; Rowell, J. Social Vulnerability, Sustainable Livelihoods and Disasters. *Benfield Hazard Res. Cent.* **2003**, *93*, 22. Available online: https://www.researchgate.net/profile/J-Twigg/publication/254398816_Social_Vulnerability_Sustainable_Livelihoods_and_Disasters/links/56656a1308ae4931cd621574/Social-Vulnerability-Sustainable-Livelihoods-and-Disasters.pdf (accessed on 13 March 2023).
31. Li, X.; Xiao, G.R.; Cai, S.H. Evaluation of water environmental sensitivity based on fuzzy analytic hierarchy process combined with web text. *J. Geo-Inf. Sci.* **2019**, *21*, 1832–1844. Available online: http://www.dqxxkx.cn/CN/10.12082/dqxxkx.2019.190165 (accessed on 13 March 2023).
32. Qi, Q.S.; Yang, Z.L. GRA and CRITIC Method for Intuitionistic Fuzzy Multiattribute Group Decision Making and Application to Development Potentiality Evaluation of Cultural and Creative Garden. *Math. Probl. Eng.* **2021**, *2021*, 9957505. [CrossRef]
33. Kumar, A.; Sharma, M.P.; Rai, S.P. A novel approach for river health assessment of Chambal using fuzzy modeling, India. *Desalination Water Treat.* **2017**, *58*, 72–79. [CrossRef]
34. Gong, Q.; Chen, B.Z.; Zhi, P.P.; Li, Y.H. Comprehensive Evaluation for Design for Design Scheme of Metro Auxiliary Power supply System Based on FAHP. *IOP Conf. Ser. Mater. Sci. Eng.* **2021**, *2*, 1043. [CrossRef]
35. Wu, Z.R. The Application of FAHP in Decisions of Pavement Maintenance. *IOP Conf. Ser. Earth Environ. Sci.* **2017**, *1*, 61. [CrossRef]
36. Jakiel, P.; Fabianowski, D. FAHP model used for assessment of highway RC bridge structural and technological arrangements. *Expert Syst. Appl.* **2015**, *42*, 4054–4061. [CrossRef]
37. Zhong, S.H.; Chen, Y.; Miao, Y.J. Using improved CRITIC method to evaluate thermal coal suppliers. *Sci. Rep.* **2023**, *13*, 195. [CrossRef]
38. Sun, B.; Yu, H.X.; Liu, B.J. Initial Water Rights Allocation in Hanjiang District of Meizhou City Based on Improved CRITIC Method. *Water Resour. Power* **2022**, *40*, 61–65. [CrossRef]

39. Niu, S.; Liu, J.; Li, Y.; Sha, H.; Zhang, J. Data analysis method of blocked road information based on GIS. *J. Chang. Univ.* **2015**, *35*, 211–215. [CrossRef]
40. Zong, C.; Qian, W.; Bao, Y.; Yuan, C.; Zhou, L.; Cui, C.; Wang, H. Temporal-Spatial Variations of Summerdense fog and Its Meteorological Influence Factors in Jiangsu Province. *Meteorol. Mon.* **2019**, *45*, 968–977.
41. Cao, J.; Guo, P. Feature analysis of fog concentration and climatic tendency of Jiangsu province. *J. Meteorol. Sci.* **2016**, *36*, 632–638.
42. Wang, B.; Pu, M.; Tian, L.; Zhang, Z.; Wu, J. Climate Characteristics and Impact Factors of Low-Visibilitydense fog on Jiangsu Coast highway. *Meteorol. Mon.* **2016**, *42*, 192–202.
43. Yu, G.; Wang, B.; Chen, P.; Huang, L.; Xie, X. Analysis of Characteristics of a Long-Lasting Fog-Haze Event in Jiangsu 2013. *Meteorol. Mon.* **2015**, *41*, 622–629.
44. Liu, Y.; Huang, X.; Duan, J.; Zhang, H. The assessment of traffic accident risk based on grey relational analysis and fuzzy comprehensive evaluation method. *Nat. Hazards* **2017**, *88*, 1409–1422. [CrossRef]
45. Zhu, C. The Method of Road Traffic Risk Assessment. *IOP Conf. Ser. Mater. Sci. Eng.* **2020**, *768*, 032001. [CrossRef]
46. Chaudhuri, S.; Saez, M.; Varga, D.; Juan, P. Spatiotemporal modeling of traffic risk mapping: A study of urban road networks in Barcelona, Spain. *Spat. Stat.* **2023**, *53*, 100722. [CrossRef]
47. Stanković, M.; Stević, Ž.; Das, D.K.; Subotić, M.; Pamučar, D. A New Fuzzy MARCOS Method for Road Traffic Risk Analysis. *Mathematics* **2020**, *8*, 457. [CrossRef]

atmosphere

Article

Study on Risk Prediction Model of Expressway Agglomerate Fog-Related Accidents

Jianyang Song [1], Hua Tian [1,2,*], Xiaoyu Yuan [1], Jingjing Gao [1], Xihui Yin [3], Zhi Wang [1,2], Meichao Qian [1] and Hengtong Zhang [3]

1 Public Meteorological Service Center of China Meteorological Administration, Beijing 100081, China; songjy@cma.gov.cn (J.S.)
2 Key Laboratory of Transportation Meteorology of China Meteorological Administration, Nanjing 210041, China
3 Road Network Monitoring and Emergency Response Center of the Ministry of Transport, Beijing 100029, China
* Correspondence: tianh1@cma.gov.cn

Abstract: Based on meteorological observations, traffic flow data and information of traffic accidents caused by fog or agglomerate fog along the expressways in Jiangsu Province and Anhui Province in China from 2012 to 2021, key impact factors including meteorological conditions, road hidden dangers and traffic flow conditions are integrated to establish the prediction model for risk levels of expressway agglomerate fog-related accidents. This model takes the discrimination of the occurrence conditions of agglomerate fog as the starting term, and determines the hazard levels of agglomerate fog-related accidents by introducing the probability prediction value of meteorological conditions for fog-related accident as the disaster-causing factor. On this basis, the hourly road traffic flow and the location of road sections with a hidden danger of agglomerate fog are taken as traffic and road factors to construct the correction scheme for the hazard levels, and the final predicted risk level of agglomerate fog-related accident is obtained. The results show that for the criteria of disaster-causing factor classification threshold, 72.3% of fog-related accidents correspond to a hazard of a medium level or above, and 86.2% of the road traffic flow conditions are consistent with the levels of the traffic factor defined based on parametric indexes. For risk level prediction, six out of the seven agglomerate fog-related accidents correspond to the level of higher risk or above, which can help provide meteorological support for traffic safety under severe weather conditions. Moreover, the model takes into account the impacts of traffic flow and the road environment, which is conducive to further improving the reliability of the risk assessment results.

Keywords: expressway; agglomerate fog; risk level prediction of fog-related accidents; meteorological conditions; road hidden dangers; traffic flow conditions

Citation: Song, J.; Tian, H.; Yuan, X.; Gao, J.; Yin, X.; Wang, Z.; Qian, M.; Zhang, H. Study on Risk Prediction Model of Expressway Agglomerate Fog-Related Accidents. *Atmosphere* **2023**, *14*, 960. https://doi.org/10.3390/atmos14060960

Academic Editor: Alexander V. Chernokulsky

Received: 10 April 2023
Revised: 15 May 2023
Accepted: 25 May 2023
Published: 31 May 2023

1. Introduction

Fog is one of the most common disastrous weather events on expressways [1,2]. With ab increasing road network density and the changing climate environment, the impacts of fog on expressway traffic safety and traffic efficiency are becoming increasingly serious. In China, the accumulated mileage blocked by fog is 1.78 times the total national expressway mileage per year on average [3,4].

The occurrence, development and dissipation of fog are caused by multiple processes (thermodynamical, radiative, dynamical and microphysical), and these processes interact nonlinearly with each other. The micro-physical characteristics of fog can impact the duration, radiation and visibility of fog. Many studies [5–10] on the micro-physical characteristics of fog are conducted, and present the variation characteristics number, concentration and size of fog droplets, which can provide some reference for the improvement of parameterization schemes in numerical models via a better understanding of the mechanism of fog occurrence. For example, Haeffelin et al. [11] used the ParisFog dataset to investigate the effect of hydrated aerosols on visibility, the role of aerosols' microphysical and chemical properties on supersaturation and droplet activation, and the role of turbulence and sedimentation on fog life cycles. Guo et al. [12] used the data collected in the project Low-Visibility Weather Monitoring and Forecasting in the Beijing–Tianjin region to study the microphysical properties of aerosol, cloud condensation nuclei, the fog droplet spectrum and liquid water content for an unusual fog-haze event that lasted for one week in North China. They presented the physical characteristics of aerosol accumulation, as well as the transition and mixture of aerosol and fog. Using a ground-based counterflow virtual impactor, Duplessis et al. [13] measured the size distributions of fog droplet and aerosol near Halifax on the eastern coast of Canada, as well as the fog droplet residuals. In addition, many studies analyzed the macro characteristics related to the formation, development and dissipation of fog, such as the synoptic pattern and meteorological factors (wind speed, relative humidity and moisture) [14–16].

Since the 21st century, many scholars have gradually applied multi-source traffic monitoring data to propose various real-time accident risk prevention and control techniques by considering the comprehensive effects of road traffic flow, weather conditions and road features [17–20]. Xu et al. [21] took into account the meteorological elements of precipitation and visibility when using logistic regression to assess the impacts of environmental factors and real-time traffic conditions on expressway crash risks, thus improving the prediction accuracy of expressway accident occurrence by 6.8%. Based on the real-time traffic flow data on foggy days, Wu et al. [22] estimated the influences of traffic and weather variables on rear-end collisions using the random logistic regression and negative binomial distribution models.

In China, systematic studies have been conducted on various aspects including the disaster-causing mechanism in the foggy section of expressways, dense fog or visibility monitoring and forecasting as well as road traffic safety and security measures [23–27]. Specifically, the quantitative impact assessment of foggy weather on expressway traffic safety is the key to defending against fog damage. Based on machine learning algorithms, using traffic accident information and meteorological observations, some important accident-related variables are selected, such as time, geolocation and the meteorological environment. Then, the mathematical models of the accident probabilities are built, which can be used to assess the real-time traffic safety state on expressways during foggy days [28,29]. Additionally, the factors indicating accidents under low visibility conditions are selected from the observed or simulated traffic parameters including upstream and downstream traffic volume, speed and occupancy rate, and the road traffic safety status under foggy conditions is quantitatively evaluated by detecting the number of traffic conflicts or safety distance [30]. The occurrence of accidents is linked to drivers, vehicles and roads (environment), but only a few scholars have integrated multi-source information (such as traffic, weather and visual information) into risk prediction due to the complexity and data availability of road traffic systems. For example, Qu et al. [31] introduced the single traffic flow and road environment to establish a risk level prediction model of fog disasters on expressways in Hebei Province. Tian et al. [32] established a weather risk warning index system for expressway traffic safety control by introducing the traffic flow, road alignment and location type. However, in general, the spatio-temporal resolutions of these forecast models are low, and the timeliness is poor. Moreover, the input data of non-meteorological factors in the model are static, and thus the dynamic driving capability of the models is obviously limited.

Agglomerate fog is a low-visibility weather phenomenon with locality, abruptness and spatio-temporal inhomogeneity, and it is also a difficult problem during road traffic weather monitoring, forecasting and early warning services. In China, the rate of traffic accidents caused by agglomerate fog is found to be 2.5 times that caused by other severe weather events, and the number of casualties in agglomerate fog-related accidents accounts for

29.5% of the total number of casualties in traffic accidents [33]. Up to now, many scholars have carried out studies on expressway agglomerate fog, with their focus on fog formation and dissipation [34–36], simulation and diagnosis [37–39], distribution law [40–42] and disaster-causing mechanisms [43–45], while there are few studies on the impact forecasting or risk early warning of agglomerate fog traffic accidents. To this end, taking the Jiangsu and Anhui area (hereinafter referred to as the "test area", as shown in Figure 1) where agglomerate fog accidents occur frequently as an example, this study establishes a risk level prediction model for expressway agglomerate fog accidents by integrating the key impact factors (meteorological environment, road hidden dangers and traffic flow conditions) and proposing the factor classification threshold determination method. This model provides a new approach to predict the agglomerate fog-related accident risk level. It is noteworthy to mention that the introduction of dynamic traffic parameters and the determination of factor classification thresholds in this study is more objective than that in past studies. We hope the results of this study can help improve the fine-resolution meteorological impact prediction and disaster prevention capability for expressway traffic safety under severe weather conditions.

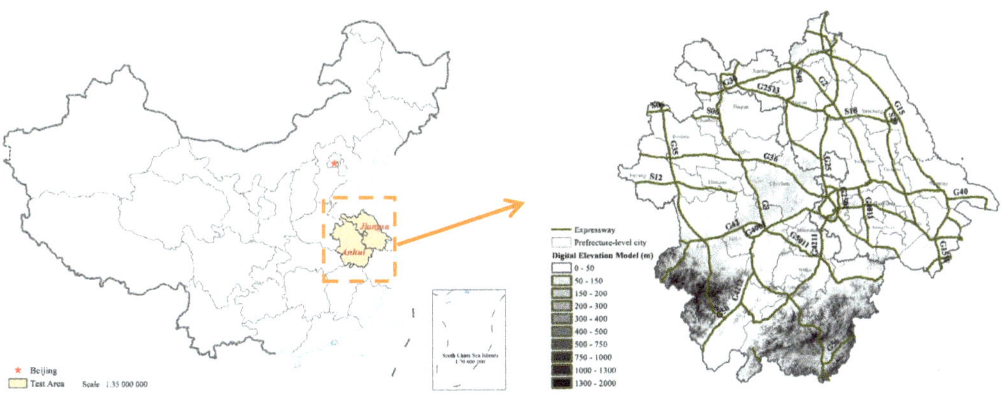

Figure 1. Basic information of the test area.

The remainder of this paper is organized as follows. Section 2 describes the data sources. The modeling method is provided in Section 3. The values and calculation procedures of the disaster-causing factor, traffic factor and road factor are given in Section 4. Section 5 presents the application and validation of the risk level prediction model. Finally, Section 6 gives the conclusions and discussion.

2. Data

In this study, the meteorological observation data are obtained from 616 traffic meteorological stations, 317 regional meteorological stations and 30 national meteorological stations along the expressways in the test area, which are provided by the National Meteorological Information Center of the China Meteorological Administration. The data quality control method refers to the "Quality Control of Meteorological Observation Data-Surface" (QX/T 118-2020) of the meteorological industry standard of the People's Republic of China. The traffic accident data in foggy or agglomerate foggy days are from the traffic control departments and news reports of media. The two kinds of data cover the period from 2012 to 2021. The traffic flow data such as vehicle flow rate and congestion index are calculated via road section estimation and road matching based on the mobile location information from internet navigation and national heavy-load freight, which covers the period from 2018 to 2021. These kinds of data are derived from the National Intelligent Road Network Monitoring Platform, and the website is http://hmrc.palmgo.cn/lwzx2/a1c64c3e6c9b76efcbccb8effd58fcad.html (accessed on 15 May 2023). In terms of the information of road sections with hidden dangers due to agglomerate fog, this study uses the information of road sections with frequent agglomerate fog released by the Traffic Administration Bureau of the Ministry of Public Security of the People's Republic of China in recent years, and the results of expressway traffic meteorological disaster risk survey by the China Meteorological Administration.

3. Modeling Method

3.1. Index Selection

Road traffic accident risk is jointly determined by the driver, vehicle and road environment. Considering the predictability of traffic accident systems, three types of indexes (weather, traffic and road) are selected to construct a risk level prediction model for expressway agglomerate fog-related accidents.

The traffic accidents in foggy days are closely related to the synoptic background. The hazard of meteorological conditions for agglomerate fog-related accidents is selected as the disaster-causing factor and the core index to construct the risk level prediction model. In addition to its low visibility, fog can often cause the reduction in the road friction coefficient through the interaction between fog droplets and dust, or through forming a thin layer of ice on cold road surfaces. By using the random forest and support vector machine algorithms, Song et al. [28,29] established a model depicting the relationship of the probability of fog-related accidents within an hour with the meteorological elements (visibility, relative humidity, wind, air temperature, etc.) and related derived variables, where recursive feature elimination and principal component analysis were used for feature selection. By using the results of the two models, the probability prediction value P of the meteorological conditions for the occurrence of fog-related accidents is obtained by weighting, which is used as the disaster-causing factor. The formula is as follows:

$$P = \sum_{i=1}^{2} p_i \times \alpha_i \quad (1)$$

where p_1 is the probability prediction value output by random forest model, p_2 is the probability prediction value output by the support vector machine model, and α_i is the weight coefficient. Considering risk prevention and control, it is hoped that the events are not missed. Hence, the ratio between the recall rates of the two models in the training set is used as the criterion for weight assignment. For the training sample consisting of the same accident group and control group, the recall rate of the random forest model and support vector machine model is 75.4% and 81.4%, respectively. Therefore, α_1 and α_2 are 0.48 and 0.52, respectively.

The traffic factor is a dynamic correction index of the risk level prediction model for expressway agglomerate fog-related accidents. The traffic operation of road network is closely related to traffic meteorological disasters, and the traffic flow situation should be considered when studying unfavorable weather effects [46]. Taking the sections of the Beijing–Shanghai Expressway, Beijing–Taipei Expressway and Nanjing–Luoyang Expressway in the test area where fog-related traffic accidents frequently occur as an example, the accumulated number of fog-related accidents (Figure 2a) is generally consistent with the annual average foggy days (visibility < 1 km) along the expressways in terms of spatial distribution (Figure 2b). The determination coefficient of the power function fitting curve is 0.106, which passes the confidence test at a 95% confidence level. However, it is also influenced by the operation status of expressway traffic (Figure 2c), and the determination coefficient of the power function fitting curve is 0.078, which passes the confidence test at a 95% confidence level. From the perspective of temporal distribution (figure omitted), dense fog occurs frequently during 03:00–08:00 BST (Beijing standard time, the same below) and peaks during 05:00–07:00 BST, while fog-related accidents occur mainly during 05:00–10:00 BST and peak during 07:00–08:00 BST when the traffic flow increases rapidly. Hence, this study chooses hourly traffic flow prediction as the traffic factor for the dynamic correction of the risk level of the occurrence of expressway agglomerate fog-related accidents.

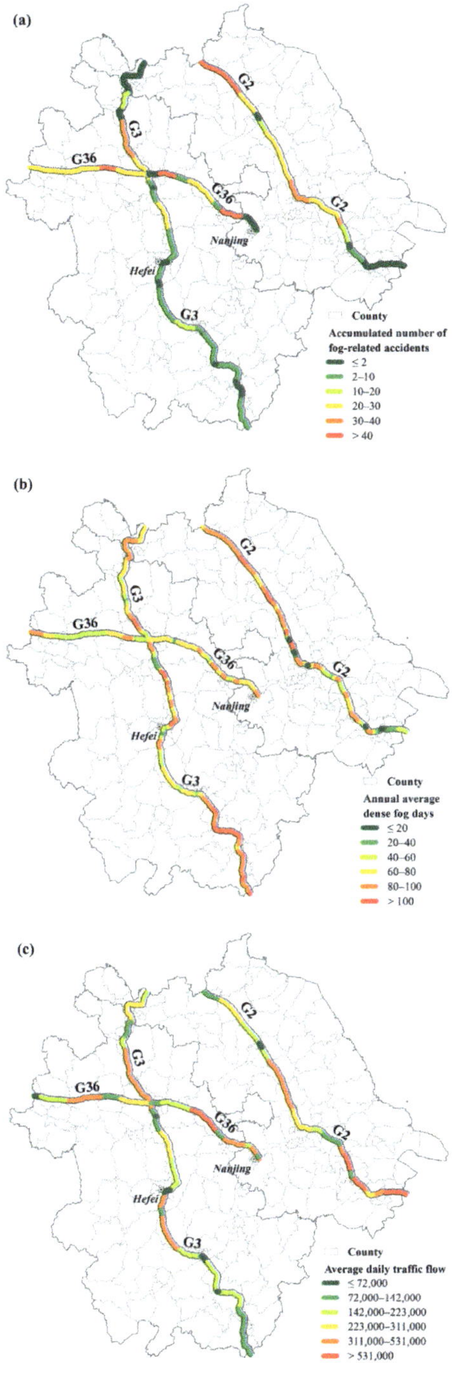

Figure 2. Spatial distributions of (**a**) accumulated number of fog-related accidents from 2013 to 2018, (**b**) annual average dense fog days from 2013 to 2018, and (**c**) average daily traffic flow from 2018 to 2021.

The road factor is a static correction index of the risk level prediction model for expressway agglomerate fog-related accidents. Agglomerate fog is usually formed under the background of meso–micro-scale circulation systems over mountainous areas, river valleys and areas with dense river networks [45], exhibiting specificity in terms of the geographical environment of roads. In this study, the location information of segmented roads in the test area is collected as the road factor, which is used for the static correction of the risk level by identifying the special form of a disaster-pregnant environment with hidden dangers in the risk level prediction model.

3.2. Assessment Procedure

As shown in Figure 3, the assessment procedure consists of three key steps: discrimination of the occurrence of agglomerate fog, risk level initial prediction of agglomerate fog-related accidents based on the disaster-causing factor, and risk level correction of agglomerate fog-related accidents based on traffic and road factors. To achieve the operationalization and visualization of this prediction model, the hierarchical threshold determination method is used to quantify the factors in the model.

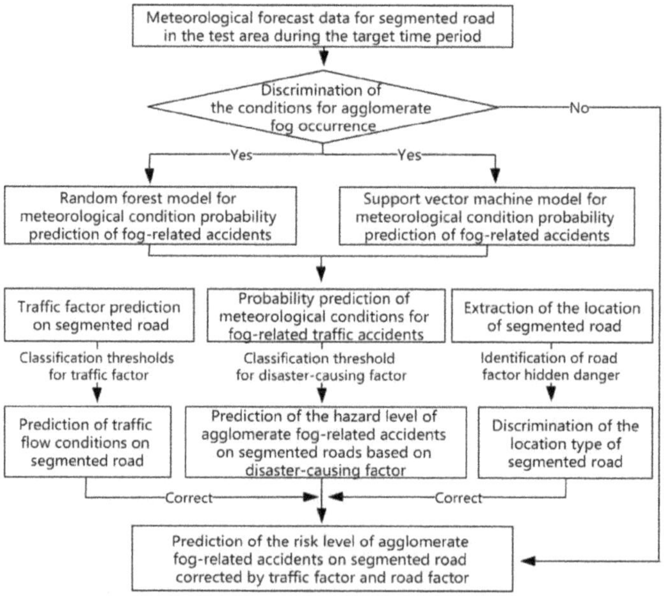

Figure 3. Flow chart for risk assessment of agglomerate fog-related accidents.

In the first step, the occurrence of expressway agglomerate fog is taken as the starting term of the risk level prediction model for expressway agglomerate fog-related accidents. If the meteorological forecast data on segmented road in the test area meet the predetermined conditions for the occurrence of agglomerate fog, the risk level of traffic accident is further calculated; otherwise, the risk is directly determined to be low. The test area consists of two parts: Jiangsu Province and Anhui Province. According to the data of agglomerate fog-related traffic accidents recorded by the traffic department, the variation characteristics of meteorological factors (visibility, relative humidity, temperature and wind) around agglomerate fog occurrence are analyzed to establish the meteorological forecast indexes for agglomerate fog in the two provinces separately (Table 1). Specific details can be found in Tian et al. [35] and Gao et al. [36].

Table 1. Meteorological forecast indexes for agglomerate fog in the test area.

Meteorological Characteristics of Agglomerate Fog	Jiangsu Province	Anhui Province
Background conditions	Fog weather background	Fog weather background
Relative humidity	>92%	>86%
Daily temperature decrease	>7 °C	>8 °C
Wind speed	<2 m s^{-1}	<1 m s^{-1}

In the second step, the pre-trained meteorological probability prediction model for fog-related accidents is utilized to obtain the probability prediction value of meteorological conditions for fog traffic accidents on corresponding road sections. According to the mapping relationship between the configured ranges of the disaster-causing factor at different hazard levels and the risk levels of agglomerate fog-related traffic accidents, five levels are initially determined, which are in the order of the extremely high level (Level 5), high level (Level 4), medium level (Level 3), low level (Level 2) and extremely low level (Level 1).

In the third step, the defined thresholds for grading the traffic factor and road factor are utilized to classify the traffic flow conditions (peak and off-peak periods) and road locations (special and ordinary types). Combined with the emergency handling experience of public security traffic administration departments, the hazard levels of meteorological conditions for agglomerate fog-related accidents are adjusted. On this basis, four risk levels are obtained (Table 2), where Level I (severe risk), Level II (very high risk), Level III (high risk) and Level IV (general risk) indicate the extremely high, very high, high and general possibilities of the occurrence of traffic accidents induced by expressway agglomerate fog, respectively.

Table 2. Classification of the risk levels for the occurrence of expressway agglomerate fog-related accidents.

Hazard Level of Disaster-Causing Factor	Ordinary Location		Special Location	
	Off-Peak	Peak	Off-Peak	Peak
Extremely high (Level 5)	I	I	I	I
High (Level 4)	II	I	I	I
Medium (Level 3)	III	II	II	II
Low (Level 2)	IV	III	III	III
Extremely low (Level 1)	No	IV	IV	IV

4. Factor Values and Calculation

4.1. Classification of Disaster-Causing Factor

According to Equation (1), the probability prediction values of meteorological conditions for fog-related accidents corresponding to 418 fog events [28,29] in the training set are calculated, and then the frequency of disaster occurrence at a probability interval of 0.05 is calculated by using the statistical method of histogram. Figure 4 reveals that a significant negative skewness appears in the distribution of disaster frequency corresponding to the probability of meteorological conditions, with the skewness and kurtosis being −1.36 and 1.21, respectively, and the left side of the peak shows a monotonically increasing trend. Thus, we count the frequency of fog-related events in the left range of the peak at intervals of 0.01 probabilities. Then, the first occurrence of three consecutive intervals with a frequency of more than or equal to 2 is defined as the change point where the accident frequency begins to increase significantly. The average of the meteorological condition probability prediction value corresponding to the continuous interval is calculated and is used to determine the initial probability value of the meteorological conditions that induce traffic accidents on foggy days. It is found that the probability value of disaster-causing meteorological conditions is 0.19, which is taken as the critical threshold for disaster-causing factor at Levels 1–2.

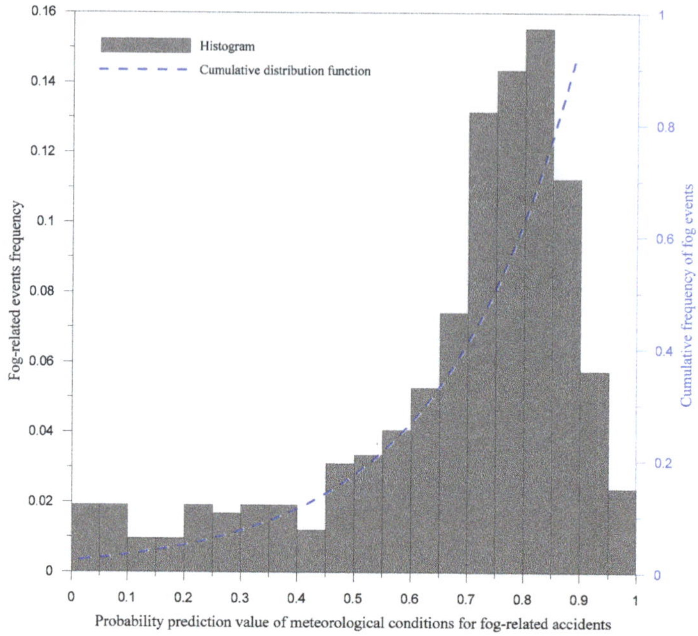

Figure 4. Histogram of disaster frequency of fog events corresponding to disaster-causing factor.

The thresholds for disaster-causing factors at Levels 2–5 are further determined based on the cumulative distribution function. The fitting equation is determined according to the features of the cumulative distribution functions of the probability of meteorological conditions for fog-related accidents and the frequency of fog events, which conforms to the exponential characteristics. On this basis, the predicted values of the probability of meteorological conditions corresponding to the cumulative frequency of 25%, 50% and 75% are used as the critical thresholds for disaster-causing factor at Levels 2–3, Levels 3–4 and Levels 4–5, respectively. Using the samples of 47 fog events and 141 non-fog events from the validation set [28,29], Table 3 validates the rationality of the of hazard of a disaster-causing factor. The results show that the frequency of disasters at the five hazard levels is consistent with the criteria for index classification. Approximately 72.3% of fog-related accidents correspond to a hazard of a medium level or above, while the false alarm rate is about 7.1%.

Table 3. Defined hazard levels for disaster-causing factor and corresponding effect validation.

Hazard	Extremely Low	Low	Medium	High	Extremely High
	[0, 0.19)	[0.19, 0.60)	[0.60, 0.75)	[0.75, 0.84)	[0.84, 1]
Number of accidents	2	11	14	12	8
Number of non-accidents	105	26	7	3	0

4.2. Classification of Traffic Factor

With the increasing traffic flow, the car following distance on the expressway becomes smaller, which makes it prone to causing traffic accidents due to low visibility, slippery road conditions or improper operation. The traffic risk under foggy weather conditions is basically proportional to the traffic volume, and the traffic flow can be divided into off-peak (normal) and peak (risk) periods according to the variations in hourly traffic volume [32]. Considering that each province has different management standards for expressways within its jurisdiction, the classification thresholds for traffic factor are determined in each province.

The hourly traffic flow of the expressways in the test area during 2018 to 2020 is extracted in sections based on the county level, which is further divided into several sections at 5th-percentile interval. Furthermore, the average value of the congestion index in each section is calculated in two provinces. The congestion index is a comprehensive parameter characterizing the operation state of road traffic and the change in traffic flow, which is expressed as follows:

$$I_{A,T} = \sum_{i=1}^{N} l_i \times \beta_{i,T}, i \in A \quad (2)$$

where $I_{A,T}$ is the congestion index in the analysis area, A, during the period, T (unit: km h), l_i is the length of road section i (unit: km), $\beta_{i,T}$ is the cumulative congestion (speed less than 40 km h^{-1}) duration on road section i during the period, T (unit: h), i is the road section number, and N is the total number of road sections within the analysis area.

With the increasing road utilization rate, the mutual interference between vehicles is aggravated, and the growth characteristics of the congestion index with traffic volume also changes significantly. The variations in the average congestion index in the unit percentile section of hourly traffic flow are shown in Figure 5. It is found that the line type usually changes from near-linear growth to near-exponential growth. In this study, split points are set from 15% to 85% at an interval of 5%, and the linear fitting formula and the exponential fitting formula between the mean congestion index and the corresponding percentile before and after split points are calculated separately; the goodness-of-fit values on the two fitting curves are recorded separately. The split point corresponding to the maximum value of the average goodness of fit is determined as the position where the congestion index abruptly changes, and the corresponding percentile value of traffic flow is adopted as the classification threshold for the traffic factor. It is found that the largest value of the average goodness of fit appears at the split point that adopts the 55th percentile of the historical hourly traffic flow dataset, which can be regarded as the cut-off point when the traffic flow becomes saturated with conflict from the free and stable state. Accordingly, the 55th percentile value of the above historical hourly traffic flow dataset is defined as the classification threshold for traffic the factor (9871 vehicles h^{-1} in Jiangsu Province and 5405 vehicles h^{-1} in Anhui Province). If the hourly traffic flow on the segmented road during the target time period is higher than the threshold, it is considered the peak (risk) traffic flow condition; otherwise, it is regarded as the off-peak (normal) condition.

Considering the difficulty of obtaining real-time traffic flow data, this study constructs the parametric index of the traffic factor by calculating the average hourly traffic flow during 2018–2020 based on the spatio-temporal distribution characteristics of traffic flow with the county-level sections, with months and hours as basic statistical units, which is used to simulate the traffic flow conditions on corresponding road sections in similar periods. Additionally, to characterize the distinct features of the sharp increase in traffic flow and peak hours on holidays (such as the New Year's Day and the Spring Festival), the parametric indexes for the traffic factor during holiday periods are constructed differentially. Table 4 validates the rationality of the classification of the traffic factor using the observed traffic flow in 2021. The results show that the parametric index of the traffic factor constructed from historical data has a strong positive correlation with that constructed from the observed data (statistically significant at the 99% confidence level), with which can well-simulate the trend variations of hourly road traffic flow. Furthermore, about 86.2% of the traffic factor levels are consistent with the conditions of road traffic flow defined based on the observed data.

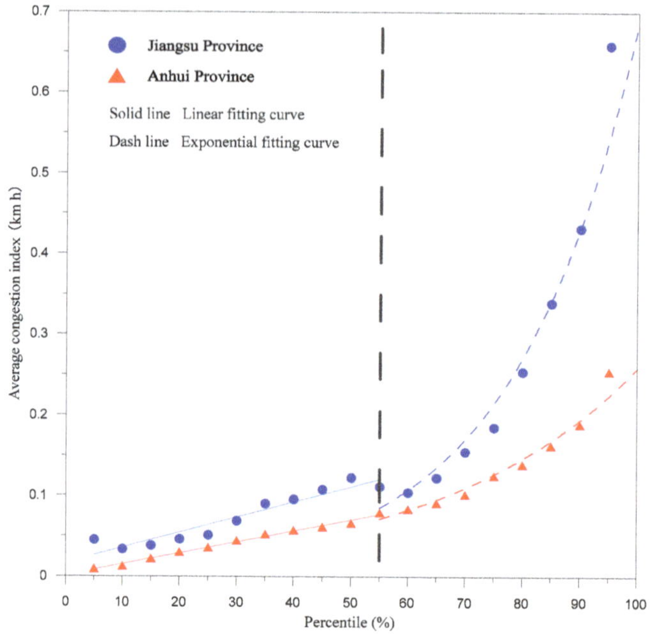

Figure 5. The variation curve of congestion index corresponding to the percentile of hourly traffic flow (taking the split point of 55% as an example).

Table 4. Effect validation of the classification of traffic factor levels.

Validation Scope	Pearson Correlation Coefficient between Parametric Index of Hourly Traffic Flow and Observed Data	Consistency of Traffic Factor Levels Classified Based on Parametric Index and Observed Data	
		Consistent with Traffic Flow Conditions	Inconsistent with Traffic Flow Conditions
Jiangsu Province	0.850	85.6%	14.4%
Anhui Province	0.867	87.0%	13.0%
Test area	0.860	86.2%	13.8%

4.3. Identification of Road Factor Hidden Danger

Using the spatial analysis technique based on the geographic information system, the road within a range of 1 km around the road section with frequent agglomerate fog in the test area are marked as special location, and the rest are marked as ordinary locations. Under similar weather conditions, the topographical features around the special road section are more conducive to the formation and maintenance of agglomerate fog, which help increase the occurrence probability of agglomerate fog-related accidents.

5. Application and Validation

5.1. Overall Situation

Seven agglomerate fog-related accidents in the test area from 2015 to 2021 are selected as the test samples to assess the application of the risk prediction model for expressway agglomerate fog-related accidents. The hindcasts give the risk level of test samples and the classification of each factor, as shown in Table 5. Overall, six out of seven agglomerate fog-related accidents correspond to risk level III or above, where three correspond to the level of severe risk and three are at the level of higher risk. For the No. 4 traffic accident on the Huaibei section of the Sixu Expressway (S06), the risk of agglomerated fog-caused accidents is predicted to be low as the daily temperature decrease fails to reach the conditions for agglomerate fog formation, while the meteorological, traffic and road factors are all conducive to the occurrence of traffic accidents. The introduction of traffic factor and road factor has appropriately raised the risk levels of agglomerate fog-related accidents on local road sections, especially for cases in which a low-visibility condition is not evident around the location of the traffic accident. For example, the visibility at the adjacent traffic weather station I5814 in accident No. 1 is approximately 2.6 km, and the disaster-causing factor corresponds to the level of low hazard. However, considering that it is a special location with frequent agglomerate fog events, the model adjusts the risk of the occurrence of agglomerate fog-related accidents on this road section from level IV to level III. In accident No. 2, the visibility at Station I2858 near the accident location is higher than 3 km before and after the accident. However, affected by the increase in traffic flow on the National Day, the traffic operation on this road section is in a peak condition. Therefore, the model adjusts the risk of agglomerate fog-related accident from level IV to level III. It can be seen that the model is of good indicative significance for the risk of agglomerate fog-related accidents, especially for the identification and warning of road sections and periods of risks under atypical disaster-causing meteorological conditions.

Table 5. Validation of the model application based on agglomerate fog-related accidents.

Number	Accident Occurrence Period		Location	Situation	Distance of Traffic Station from the Accident Location and Corresponding Average/Minimum Visibility	Agglomerate Fog Index	Hazard Factor	Traffic Factor	Road Factor	Risk Level
1	13 February 2021	07:00–08:00 BST	Tongling, Anhui, Shanghai–Chongqi-ng Expressway (G50)	7 accidents of several vehicles scraping each other and rear-end collision	1 km (I5814) 2666/1630 m	matches the conditions	Level 2	Off-peak	Special	III
2	3 October 2019	06:00–07:00 BST	Bengbu, Anhui, Nanjing–Luoyang Expressway (G26)	10 people dead and 7 injured in 4 accidents	8 km (I2858) 3768/3432 m	matches the conditions	Level 2	Peak	Ordinary	III
3	15 November 2017	07:00–08:00 BST	Fuyang, Anhui, Chuzhou–Xincai Expressway (S12)	18 people dead and 21 injured in multi-point and multi-vehicle collisions	1 km (I2754) 80/57 m 71/68 m	matches the conditions	Level 5	Peak	Special	I
		08:00–09:00 BST				matches the conditions	Level 5	Peak	Special	
4	5 February 2017	08:00–09:00 BST	Huaibei, Anhui, Sixian–Xuchang Expressway (S06)	16 vehicles damaged and 6 people injured	3 km (I1358) 226/165 m	mismatch with the conditions	Level 5	Peak	Special	No
5	2 April 2016	12:00–13:00 BST	Changzhou, Jiangsu, Shanghai–Chengdu Expressway (G42)	51 vehicles damaged, 3 people dead and 31 injured	5 km (M9112) 1058/846 m	matches the conditions	Level 2	Peak	Ordinary	III
6	7 December 2015	00:00–01:00 BST	Yancheng, Jiangsu, Shenyang–Haikou Expressway (G15)	3 people dead and 3 injured in multi-vehicle collisions	4 km (M9437) 87/75 m	matches the conditions	Level 5	Peak	Ordinary	I
7	23 May 2015	06:00–07:00 BST	Lianyungang, Jiangsu, Shenyang–Haikou Expressway (G15)	4 people dead and 8 injured in dozens of rear-end collisions	3 km (M9433) 197/115 m	matches the conditions	Level 4	Peak	Ordinary	I

5.2. Typical Cases

From 07:35 BST to 08:57 BST on 15 November 2017, a multi-point and multi-vehicle rear-end collision occurred on the road section from 191 km to 194 km along the downward direction of the Chuzhou–Xincai Expressway (S12) due to sudden agglomerate fog, resulting in 18 deaths, 21 injuries and 70 vehicles damaged.

Figure 6 provides the output of the risk prediction model of the expressway agglomerate fog-related accidents. It can be seen that the risk level in northwestern Anhui is obviously higher than that in other road networks in the test area before and after the occurrence of accidents. Since the early morning of November 15, the coverage of higher-risk or above of agglomerate fog-related accidents has gradually expanded from the northwest to southeast, and rapidly weakened from southeast to northwest after reaching its peak during 06:00–07:00 BST. From 08:00 BST to 09:00 BST, there was generally no risk of agglomerate fog-related accidents along the expressway in the test area, but the accident section still showed the severe risk level, indicating that the simulation results are reasonable and can provide targeted tips for determining the risk of local road traffic safety.

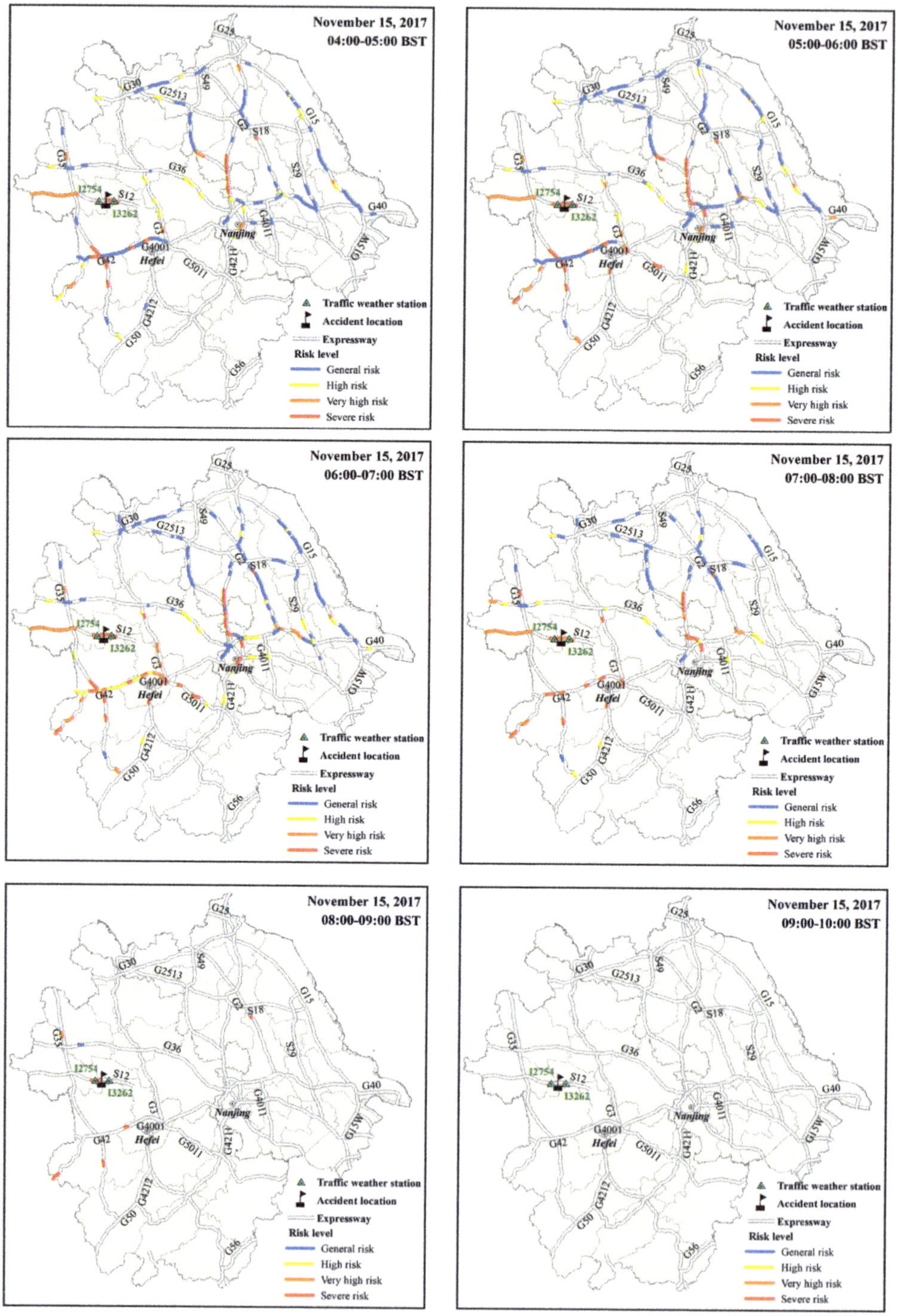

Figure 6. Assessment results of the risk level prediction model of expressway agglomerate fog-related accidents (from 04:00 BST to 10:00 BST on 15 November 2017).

From the evolution of road traffic risks at the accident location (Figure 7), the entire accident section (191–194 km) showed an extremely high risk level of traffic accidents induced by agglomerate fog two hours before the first traffic accident. After 09:00 BST, the entire accident section returned to a no-risk situation, coinciding with the end time of this series of traffic accidents. Particularly, the section east of the accident location (Station I3262) is the section with the earliest occurrence time of severe risk (Level I), and the closer section west of the accident location (Station I2754) is the section where the risk of an event at Level I finally disappears.

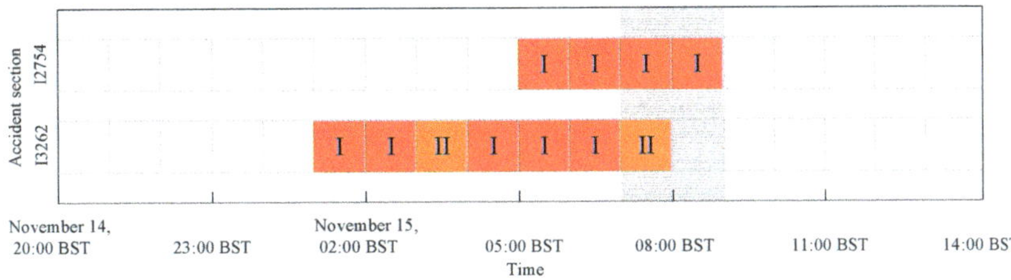

Figure 7. Evolution process of the risk levels of agglomerate fog-related accidents in the accident section (the shading indicates the period when the accident occurred).

6. Conclusions and Discussion

For severe weather-related traffic accidents, the key impact factors including meteorological conditions, road hidden dangers and traffic flow conditions are integrated to establish the risk assessment procedure and risk level prediction model for expressway agglomerate fog-related accidents, which consists of three core steps—discrimination of the conditions for agglomerate fog occurrence, risk level initial prediction of agglomerate fog-related accidents based on disaster-causing factors and risk level correction of agglomerate fog-related accidents based on traffic and road factors.

The probability prediction value of meteorological conditions for fog-related accidents is taken as the disaster-causing factor. The thresholds for five levels of disaster-causing factor are determined according to the statistical relationship between the frequency of historical fog-related events and the probability of meteorological conditions in the corresponding periods. The validation reveals that approximately 72.3% of fog-related accidents correspond to a hazard of the medium level or above.

The predicted value of hourly road traffic flow is taken as the traffic factor, and the thresholds of traffic factor levels are determined in each province based on the variation characteristics of the congestion index increasing with the traffic volume. There is a good consistency between the traffic factor levels defined based on the parametric index of traffic flow and the observed traffic data in 2021, where the traffic flow conditions with the same type account for about 86.2%.

Based on the analysis and validation of seven cases of agglomerate fog-related accidents from 2015 to 2021, it is found that three cases correspond to the level of higher risk and three correspond to the level of severe risk, indicating that the prediction results can support the demand for meteorological support for traffic safety under severe weather conditions. In addition, the comprehensive consideration of traffic flow and road environment impacts can help in the accurate identification of key prevention areas on foggy or agglomerate foggy days and the timely research and judgment of the risk periods, which can improve the quality of prediction of the risk of agglomerate fog-related accidents.

This study proposes a new research idea and methodological exploration for the risk prediction of agglomerate fog-related accidents, especially for the dynamic consideration of the impact of road traffic flow conditions and the objective calculation of the factor classification thresholds. However, the prediction accuracy is restricted by the limited road condition data. In the model prediction, the real-time-measured information of traffic flow parameters is not introduced, and some other factors such as road shape and vehicle type are not considered. In the future, we need to use more fog-accident data and more detailed traffic and meteorological data to conduct studies on the influencing mechanism of unfavorable weather conditions and the associated relationship. On this basis, by introducing the real-time traffic flow parameters and more impact factors such as road characteristics and vehicle types, we may continuously modify and improve the risk prediction model of agglomerate fog-related accidents, which is beneficial to further enhancing the reliability of the assessment results.

Author Contributions: Methodology, J.S.; validation, J.S. and X.Y. (Xihui Yin); formal analysis, J.S. and H.T.; data curation, J.G. and H.Z.; writing—original draft preparation, J.S.; writing—review and editing, H.T. and X.Y. (Xiaoyu Yuan); visualization, J.G. and M.Q.; project administration, H.T. and Z.W. All authors have read and agreed to the published version of the manuscript.

Funding: This research was funded by the Meteorological Technology Innovation Platform Project of China Meteorological Service Association (grant number CMSA2023MB023), the Special Innovation and Development Program of China Meteorological Administration (grant number CXFZ2023J072) and the Innovation Fund Project of Public Meteorological Service Center of China Meteorological Administration (grant number K2021002).

Institutional Review Board Statement: Not applicable.

Informed Consent Statement: Not applicable.

Data Availability Statement: The used data in this study are obtained from the data platform of China Meteorological Administration and the National Intelligent Road Network Monitoring Platform.

Conflicts of Interest: The authors declare no conflict of interest.

References

1. Li, X.; Tang, B.; Zeng, C. Research and development of expressways in foggy area safety and security technology. *J. Sci. Technol. Eng.* **2018**, *18*, 196–206.
2. Cornejo-Bueno, S.; Casillas-Pérez, D.; Cornejo-Bueno, L.; Chidean, M.I.; Caamaño, A.J.; Cerro-Prada, E.; Casanova-Mateo, C.; Salcedo-Sanz, S. Statistical analysis and machine learning prediction of fog-caused low-visibility events at A-8 motor-road in Spain. *Atmosphere* **2021**, *12*, 679. [CrossRef]
3. Wang, B.; Xu, F.; Tian, X.; Yuan, C. A review of fog weather research methods in China in recent years. *Meteorol. Sci. Technol.* **2014**, *42*, 23–30.
4. Jiang, Y.; Zhang, N.; Li, A.; Wu, H.; Li, R.; Yang, J. Effects of weather on highway traffic capacity in China: Characteristics and causes of roadblocks due to fog events. *J. Pure Appl. Geophys.* **2020**, *177*, 5027–5040. [CrossRef]
5. Lu, C.; Liu, Y.; Niu, S.; Zhao, L.; Yu, H.; Cheng, M. Examination of microphysical relationships and corresponding microphysical processes in warm fogs. *Acta Meteorol. Sin.* **2013**, *27*, 832–848. [CrossRef]
6. Mazoyer, M.; Burnet, F.; Denjean, C.; Roberts, G.; Haeffelin, M.; Dupont, J.; Elias, T. Experimental study of the aerosol impact on fog microphysics. *Atmos. Chem. Phys.* **2019**, *19*, 4323–4344. [CrossRef]
7. Liu, Q.; Wu, B.; Wang, Z.; Hao, T. Fog droplet size distribution and the interaction between fog droplets and fine particles during dense fog in Tianjin, China. *Atmosphere* **2020**, *11*, 258. [CrossRef]
8. Wang, Y.; Niu, S.; Lu, C.; Lv, J.; Zhang, J.; Zhang, H.; Zhang, S.; Shao, N.; Sun, W.; Jin, Y.; et al. Observational study of the physical and chemical characteristics of the winter radiation fog in the tropical rainforest in Xishuangbanna, China. *Sci. China Earth Sci.* **2021**, *64*, 1982–1995. [CrossRef]
9. Weston, M.; Francis, D.; Nelli, N.; Fonseca, R.; Temimi, M.; Addad, Y. The first characterization of fog microphysics in the United Arab Emirates, an arid region on the Arabian Peninsula. *Earth Space Sci.* **2022**, *9*, e2021EA002032. [CrossRef]
10. Wang, Y.; Lu, C.; Niu, S.; Lv, J.; Jia, X.; Xu, X.; Xue, Y.; Zhu, L.; Yan, S. Diverse dispersion effects and parameterization of relative dispersion in urban fog in eastern China. *J. Geophys. Res. Atmos.* **2023**, *128*, e2022JD037514. [CrossRef]
11. Haeffelin, M.; Bergot, T.; Elias, T.; Tardif, R.; Carrer, D.; Chazette, P.; Colomb, M.; Drobinski, P.; Dupont, E.; Dupont, J.; et al. Parisfog: Shedding new light on fog physical processes. *Bull. Am. Meteorol. Soc.* **2010**, *91*, 767–783. [CrossRef]
12. Guo, L.; Guo, X.; Fang, C.; Zhu, S. Observation analysis on characteristics of formation, evolution and transition of a long-lasting severe fog and haze episode in North China. *Sci. China Earth Sci.* **2015**, *58*, 329–344. [CrossRef]
13. Duplessis, P.; Bhatia, S.; Hartery, S.; Wheeler, M.J.; Chang, R.Y.-W. Microphysics of aerosol, fog and droplet residuals on the Canadian Atlantic coast. *Atmos. Res.* **2021**, *264*, 105859. [CrossRef]
14. Gultepe, I.; Tardif, R.; Michaelides, S.C.; Cermak, J.; Bott, A.; Bendix, J.; Müller, M.D.; Pagowski, M.; Hansen, B.; Ellrod, G.; et al. Fog research: A review of past achievements and future perspectives. *Pure Appl. Geophys.* **2007**, *164*, 1121–1159. [CrossRef]

15. Pu, M.; Zhang, G.; Yan, W.; Li, Z. Features of a rare advection-radiation fog event. *Sci. China Ser. D Earth Sci.* **2008**, *51*, 1044–1052. [CrossRef]
16. Penov, N.; Stoycheva, A.; Guerova, G. Fog in Sofia 2010–2019: Objective circulation classification and fog indices. *Atmosphere* **2023**, *14*, 773. [CrossRef]
17. Snæbjörnsson, J.T.; Baker, C.J.; Sigbjörnsson, R. Probabilistic assessment of road vehicle safety in windy environments. *J. Wind. Eng. Ind. Aerodyn.* **2007**, *95*, 1445–1462. [CrossRef]
18. Abdel-Aty, A.M.; Hassan, H.M.; Ahmed, M.; Al-Ghamdi, S.A. Real-time prediction of visibility related crashes. *Transp. Res. Part C Emerg. Technol.* **2012**, *24*, 288–298. [CrossRef]
19. Ahmed, M.; Abdel-Aty, A.M.; Yu, R. Assessment of interaction of crash occurrence, mountainous freeway geometry, real-time weather, and traffic data. *Transp. Res. Rec.* **2012**, *2280*, 51–59. [CrossRef]
20. Yuan, J.; Abdel-Aty, A.M.; Wang, L.; Lee, J.; Wang, X.; Yu, R. Real-time crash risk analysis of urban arterials incorporating bluetooth, weather, and adaptive signal control data. In Proceedings of the Transportation Research Board 97th Annual Meeting, Washington, DC, USA, 7–11 January 2018.
21. Xu, C.; Wang, C.; Liu, P. Evaluating the combined effects of weather and real-time traffic conditions on freeway crash risks. *Weather Clim. Soc.* **2018**, *10*, 837–850. [CrossRef]
22. Wu, Y.; Abdel-Aty, A.M.; Cai, Q.; Lee, J.; Park, J. Developing an algorithm to assess the rear-end collision risk under fog conditions using real-time data. *Transp. Res. Part C Emerg. Technol.* **2018**, *87*, 11–25. [CrossRef]
23. Zhang, W.; He, Y.; Liu, H. *Traffic Safety Enhancement Technique of Expressways in Fog Areas*; China Communications Press: Beijing, China, 2008.
24. Tian, X.; Wu, J.; Yan, M.; Yuan, C.; Jiao, S. Some new developments of the heavy fogs' observation and forecast on the expressway. *J. Meteor. Sci.* **2009**, *29*, 414–420.
25. Zhao, H.; Wang, W.; Li, Z.; Wang, Y. Effects of fog on traffic and transportation and the counterm easures in China. *J. Meteor. Environ.* **2010**, *26*, 58–62.
26. Wang, L. Study on Speed Limit of Freeway in Foggy Day Based on Traffic Flow Condition and 85%th Speed. Master's Thesis, Beijing Jiaotong University, Beijing, China, 2015.
27. Zhang, Y.; Wang, Y.; Zhu, Y.; Yang, L.; Ge, L.; Luo, C. Visibility prediction based on machine learning algorithms. *Atmosphere* **2022**, *13*, 1125. [CrossRef]
28. Song, J.; Wang, Z.; Li, A.; Tian, H.; Gao, J.; Wang, Y. Research on meteorological risk model for traffic safety on highway under foggy weather condition. *J. Nat. Disaster.* **2022**, *31*, 60–68.
29. Song, J.; Tian, H.; Gao, J.; Wang, Z.; Li, A.; Chen, Y. An hourly meteorological model for fog accident discriminant based on machine learning technology. *J. Meteor. Sci. Technol.* **2023**, *51*, 149–156.
30. Zang, C.; Peng, H.; Zhang, S.; Lyu, C. Real-time traffic safety evaluation method for freeway in fog. *China Saf. Sci. J.* **2017**, *27*, 110–115.
31. Qu, X.; Zhang, D.; Guo, R.; Qi, Y.; Zhao, Z.; Wu, D. Forecasting method on integrated risk level of traffic condition based on weather conditions for highway of Hebei Province. *J. Arid Meteorol.* **2019**, *37*, 345–350.
32. Tian, H.; Wu, X.; Cai, G.; Song, J.; Kong, C.; Feng, L. *Weather Risk Warning and Countermeasures for Traffic Control*; China Meteorological Press: Beijing, China, 2019.
33. Guo, P. Research on the Early Warning and Safety Assurance Technology of Expressways Air-Mass Fog. Master's Thesis, Chang'an University, Xi'an, China, 2016.
34. Lestringant, R.; Bergot, T. Analysis of small-scale spatial variability of fog at Paris-Charles de Gaulle airport. *Atmosphere* **2021**, *12*, 1406. [CrossRef]
35. Tian, H.; Gao, J.; Li, A.; Wu, Y.; Zhu, C.; Zhao, L.; Zhang, H. Meteorological characteristics of local agglomerate fog on Jiangsu highway. *Desert Oasis Meteorol.* **2019**, *13*, 39–46.
36. Gao, J.; Tian, H.; Li, A.; Song, J.; Zhu, X. Analysis of agglomerate fog meteorological characteristics in Anhui Province based on traffic accident data. *Pure Appl. Geophys.* **2022**, *180*, 313–333. [CrossRef]
37. Wan, X.; Bao, Y.; Yan, M.; Yuan, C.; Qian, W. Simulation of the dumpling fog on the Nanjing-Shanghai Expressway based on different land-surface schemes. *J. Meteor. Sci.* **2010**, *30*, 487–494.
38. Shi, D.; Li, C.; Shi, Y.; Zhang, Y. Study on the localization diagnosis of extra heavy fog on the background of the fog weather based on machine learning algorithms. *J. Catastrophol.* **2018**, *33*, 193–199.
39. Pagowski, M.; Gultepe, I.; King, P. Analysis and modeling of an extremely dense fog event in southern Ontario. *J. Appl. Meteorol.* **2004**, *43*, 3–16. [CrossRef]
40. Ding, Q.; Bao, Y.; Yuan, C.; Yan, M. Occurrence pattern and local characteristics of dumpling fog events on Shanghai = Nanjing expressway. *J. Meteor. Sci.* **2013**, *33*, 634–642.
41. Cui, C.; Bao, Y.; Yuan, C.; Jiao, S.; Wang, L.; Xie, X. Study on temporal and spatial occurrence pattern of fog on the coastal expressway of Jiangsu province. *J. Sci. Technol. Eng.* **2015**, *15*, 6–20.
42. Yamamoto, A. Climatology of the traffic accident in Japan on the expressway with dense fog and a case study. In Proceedings of the 11th International Road Weather Conference, Sapporo, Japan, 26–28 January 2002.
43. Hu, S.; Zhu, Y. Study on the mechanism of impact of agglomerate fog weather on expressway traffic safety. *J. J. China Foreign Highw.* **2013**, *2*, 290–292.

44. Li, X.; Tang, B.; Zeng, C. Transient analysis of the driving behavior in the dumpling fog sections of expressway based on catastrophe theory. *China J. Highw. Transp.* **2018**, *31*, 78–88.
45. Lu, Z.; Yang, C.; Zheng, Z.; Cui, G.; An, Y.; Gu, F. Study on warning level of expressway dense fog. *J. Meteor. Environ. Sci.* **2022**, *45*, 103–111.
46. Bao, G.; Zhou, D.; Zheng, L.; Dai, Q.; Ma, S.; Liu, W. A study of severe snowfall disaster risk zoning along highways in Qinghai Province. *Desert Oasis Meteorol.* **2019**, *13*, 109–116.

atmosphere

Article

Variation Characteristics of Pavement Temperature in Winter and Its Nowcasting for Xianyang Airport Expressway, China

Lei Feng [1,2], Hua Tian [1,2,*], Xiaoyu Yuan [1,*], Lei Miao [1] and Mingyu Lin [1]

1 Public Meteorological Service Center of China Meteorological Administration, Beijing 100081, China
2 Key Laboratory of Transportation Meteorology of China Meteorological Administration, Nanjing Joint Institute for Atmospheric Sciences, Nanjing 210041, China
* Correspondence: tianh1@cma.gov.cn (H.T.); yuanxy@cma.gov.cn (X.Y.)

Abstract: Based on the pavement temperature observation data of the transportation meteorological stations along the Xianyang Airport Expressway, China, as well as the datasets of precipitation and sunshine hours obtained from the nearby weather stations, the variation characteristics of local pavement temperatures are investigated for winter in this study. Results indicate that during the daytime, the pavement temperatures are always higher on sunny and cloudy days than those on rainy and snowy days, while during the nighttime, the temperatures on sunny and cloudy days are higher than those on the days with freezing rain and snow, and with the temperatures on rainy and snowy days without icing being further higher. In general, the pavement temperatures in winter features significant periodic oscillations with cycles of roughly 24 h, 12 h, 8 h, 6 h, 5 h and 4 h, which differ slightly at different times for different stations. Moreover, the nowcasting experiments on the local pavement temperatures are also carried out using a regression model via extracting the corresponding periodic features. It shows the mean absolute errors of about 0.6 °C, 1.2 °C, and 1.5 °C for lead times of 1 h, 2 h, and 3 h, respectively. The nowcasting skills are higher on rainy and snowy days, while are inferior on sunny days. For nowcasting cases initialized at nighttime (daytime), the mean absolute errors are 0.4 °C (0.7 °C) and 0.9 °C (1.4 °C) for lead times of 1 h and 2 h. Examinations suggest that the nowcasting system could be well utilized in plain areas of China, whereas it shows relatively larger biases in plateau areas with complex terrain.

Keywords: pavement temperature; nowcasting; variation characteristics; forecast validation

Citation: Feng, L.; Tian, H.; Yuan, X.; Miao, L.; Lin, M. Variation Characteristics of Pavement Temperature in Winter and Its Nowcasting for Xianyang Airport Expressway, China. *Atmosphere* **2023**, *14*, 361. https://doi.org/10.3390/atmos14020361

Academic Editor: Graziano Coppa

Received: 13 January 2023
Revised: 28 January 2023
Accepted: 8 February 2023
Published: 11 February 2023

1. Introduction

In winter, the low temperature combined with the rain and snow tend to trigger freezing events and lead the road to be slippery. It increases the braking distance and reduces the anti-sideslip ability of vehicles, which poses severe threats to transportation safety. Previous statistics have revealed that the collision and scraping accidents of vehicles in snowy weathers are roughly 14 times of those in sunny weathers [1]. In China, the traffic accidents in snowy days account for 6.93% of the total, thus being the second most crucial meteorological factor for inducing traffic accidents. On 7 December 2001, slight snowfall in Beijing caused a severe traffic jam in the city due to its coincidence with the off-duty traffic peak [2,3].

The pavement temperature forecast is one of the most important parameters for predicting road icing and snow. It directly affects the effectiveness of road icing and snow state identification in the near future. The current study focuses on the Xi'an Xianyang Airport Expressway in China, which is the traffic artery leading to the airport in Xi'an with large traffic flows. However, its safe and smooth operation is certainly threatened by weathers of low temperature, rain, snow and freezing. Therefore, it is of great significance to investigate the temporal variation characteristics of local pavement temperature in winter and its nowcasting, which is favorable for effective forecasts of snow and icing on the airport expressway. At the same time, the nowcasting method established for this region is helpful to improve the pavement temperature prediction for other areas [4].

With respect to the prediction techniques of pavement temperatures, the physical models have already been developed based on the surface energy balance earlier in the United States and several countries in Europe [5–9]. Additionally, these models have also been continuously improved and optimized by revealing the interactions between meteorological and pavement conditions [10,11]. Generally, the mechanisms of pavement temperature predictions have been studied well, and multiple numerical models have been used worldwide [12]. Meanwhile, on this basis, combined with the multi-source data fusion analysis and forecasting system, several methods have been developed on refined predictions of pavement temperatures, which could be scrolled at high frequencies and could better meet the requirements of transportation users [13,14].

In China, investigations on pavement transportation meteorology started relatively late. In recent decades, some provinces such as Jiangsu, Beijing and Hebei have organized transportation meteorological operational businesses to deal with the severe effects of low temperature, rainy, snowy and freezing weathers on pavement traffic [15–17]. Multiple physical and statistical models have been established to automatically predict the pavement temperatures, which are examined and demonstrated to be effective in local forecast businesses [18]. These models are generally combined with the applications of numerical prediction products and the lead times mainly range from 0 to 24–48 h [19,20]. It is of great significance for the identification of low temperature, rainy, snowy and freezing events in the short term [21,22]. The predictions of pavement conditions are affected by plenty of factors including not only the accuracy of meteorological forecasts but also the initial observation of transportation road surfaces, the shading effect of surrounding obstructions, the emission of human activities, and the pavement heat conductions, etc. Therefore, the forecast accuracy of pavement conditions such as temperatures is generally limited and less skillful than that of the conventional meteorological elements [23,24].

The accurate perception and prediction of pavement status play crucial roles in the realization of vehicle–road coordination and establishment of the smart highway. The refined meteorological services on hundred-meter-level and minute-level resolutions are becoming an important guarantee for smart transportation capacities at all weather conditions. In recent years, the departments of public security, transportation and meteorology have cooperated on the joint prevention and mitigation of severe weather impacts, which brings about the increasing demands for forecasts of low temperature and icing pavements with higher resolution and accuracy. In general, the forecast skills tend to decrease (increase) with longer (shorter) lead times. Tang and Guo [25] used the autoregressive summation moving average method to explore the fluctuation pattern of winter pavement temperature in the near future under the impacts of external factors, and constructed a short-term forecast model of winter pavement temperature with lead times of 3 h based on the transportation meteorological observations at minute-by-minute intervals. Further, Wang et al. [26] carried out the nowcasting experiments using the random forest regression method for transportation meteorological stations alongside the Ning-Su-Xu Expressway in Jiangsu Province, detecting the influences of different input schemes and parameters of different observation stations on the pavement temperature forecasts. On the basis of the physical METRo model with surface energy balance principles, Qu et al. [27] conducted studies on the 0–6 h forecasts of pavement temperatures for the Beijing–Zhangjiakou Olympic Winter Games-associated expressway, revealing that the forecast during nighttime is characterized by root-mean-square errors of ~1 °C and is superior to that during daytime. Comparisons indicate that the statistical methods constructed on the observation analyses show generally higher forecast skills than the physical models based on the surface energy balance for the nowcasting of pavement temperatures [25–27].

In the current paper, the variation characteristics of observed pavement temperature in winter are to be investigated for two transportation meteorological stations along the Xianyang Airport Expressway in Xi'an, China. On this basis, a pavement temperature nowcasting method is proposed via extracting the periodic features of the temperature temporal series and constructing the associated regression model, which is examined for pavement temperature nowcasting experiments at a 10 min rolling update and a 10 min interval. Meanwhile, the impacts of different sample amounts, different weather conditions, and different initialization times on the nowcasting effects are also investigated in detail. In addition, similar examinations on pavement temperature nowcasting are carried out for multiple expressways to further verify the applicability of the proposed method for different regions.

2. Dataset and Methodology

2.1. Dataset

The hourly observation datasets of the two transportation meteorological stations (V0001 (108.92° E, 34.37° N) and V0002 (108.90° E, 34.43° N), as shown in Figure 1) along the Xianyang Airport Expressway in Xi'an, China were obtained for the two winters of 2018/2019 and 2019/2020 from the data platform of China Meteorological Administration. The two stations are located at the two ends of the No. 2 Grand Bridge over the Weihe River. Station V0001 is in the suburban area. There are many building facilities around, and the traffic flow is relatively complex. Station V0002 is located in the outer suburbs. It is surrounded mainly by farmland, and there are also a few buildings. The altitudes of Stations V0001 and V0002 are 37.1 m and 40.7 m, respectively. The observation elements include pavement temperature, surface air temperature, wind speed, wind direction and relative humidity. The pavement temperature was measured by using embedded thermal sensor with platinum rod-shaped probe. It was noted that most transportation meteorological stations have not yet been equipped with weighing precipitation sensors due to the maintenance difficulties and the solid precipitation such as snowfall could hardly be monitored. Therefore, the hourly precipitations over the two transportation meteorological stations were approximately substituted by the observations of the neighboring national weather station (Station 57131), which is 10.1 km and 7.3 km far from Stations V0001 and V0002, respectively. In addition, comprehensive observations such as sunshine hours from Station 57131 were also employed to determine weather conditions (sunny, cloudy, rainy, and snowy, etc.) at the transportation meteorological stations. Based on these hourly observation datasets, the basic characteristics of pavement temperature variations on the expressways were analyzed for the wintertime.

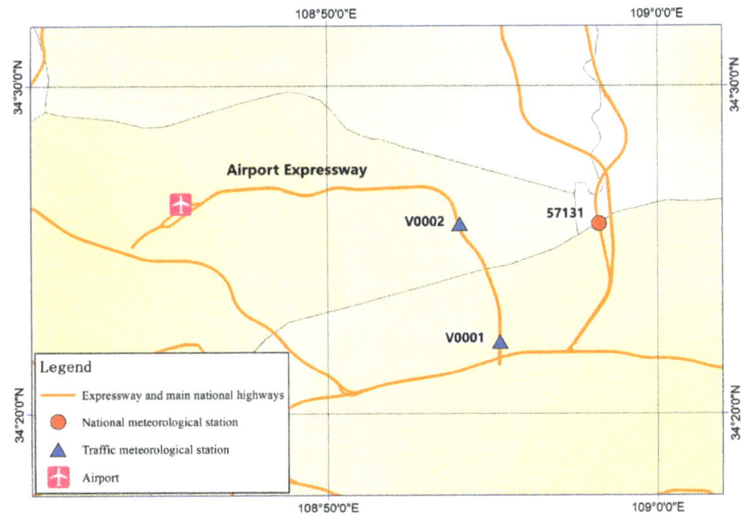

Figure 1. Locations of the transportation meteorological stations along Xianyang Airport Expressway in Xi'an, China, and the neighboring national meteorological station.

In order to verify the rationality of taking the national station observation (e.g., precipitation, sunshine hours, etc.) as substitutions for the two transportation meteorological stations, Table 1 shows the statistics of observed temperature and wind among the three stations for the total of 181 days in the two winters of 2018/2019 and 2019/2020. It is indicated that the temperatures at Station V0001 (V0002) are slightly higher (lower) than those at Station 57131, with their differences in daily mean, maximum and minimum temperatures being 0.65 °C (1.19 °C), 0.76 °C (1.03 °C) and 0.58 °C (1.35 °C), respectively. Such spatial differences might be associated with the urban heat island effects and the local environment of these stations. Besides, the correlation coefficients between the daily mean temperature series of the two transportation meteorological stations and the national observation station are both greater than 0.97, with that of Station V0001 being higher than V0002 and reaching up to 0.99. As for the wind speed, the daily averages and maximums at the two transportation meteorological stations are all lower than those at the national observation station. The daily average and maximum wind speeds at Station V0001 (V0002) are characterized by values of 0.77 m/s (1.13 m/s) and 1.00 m/s (1.41 m/s) lower than those of Station 57131, respectively, and show correlation coefficients of 0.95 (0.80) and 0.95 (0.68) with the observations from the national station. The comparison analyses demonstrate that the meteorological conditions at the two transportation meteorological stations are generally similar to those of the neighboring national meteorological station. Therefore, the observations of precipitation and sunshine hours at Station 57131 could be reasonably taken as the corresponding status of the two transportation meteorological stations.

Table 1. Statistics of observed temperature and wind speed for Station 57131 and Stations V0001 and V0002 in the two winters, including the respective temporal averages (TAs, units: °C for temperature and m/s for wind speed) and the correlation coefficients (CCs) of the two transportation meteorological stations V0001 and V0002 to the national station 57131.

Station ID	Daily Average Temperature		Daily Maximum Temperature		Daily Minimum Temperature		Daily Average Wind Speed		Daily Maximum Wind Speed	
	TA	CC	TA	CC	TA	CC	TA	CC	TA	CC
57131	3.22	-	3.61	-	2.83	-	2.05	-	2.62	-
V0001	3.87	0.99	4.37	0.99	3.41	0.99	1.28	0.95	1.62	0.95
V0002	2.03	0.98	2.58	0.98	1.48	0.97	0.92	0.80	1.21	0.68

Observations with high frequency and high accuracy are the most basic fundamentals to carry out effective early warnings and predictions of pavement low temperature and road icing. In this study, the pavement temperature observations at the two transportation meteorological stations were obtained every 10 min during the winter in 2018/2019 (90 days in total, composing 12,960 samples for each station) for investigations of pavement temperature nowcasting. In general, they had relatively complete datasets, except the period of 7:00 to 13:00 on 30 December 2018 and several other missing observations, which were further supplemented with linear interpolations. Calculations on the observations show that the mean pavement temperature and the corresponding mean square deviation are 3.31 °C (3.17 °C) and 4.54 °C (3.73 °C) for Station V0001 (V0002), respectively. Significant differences (at the 95% confidence level according to the Student's *t*-test) are observed between the two stations.

Based on the nowcasting experiments on pavement temperatures for the two transportation meteorological stations along the Xianyang Airport Expressway, the proposed method was further promoted and examined towards three stations in three different-climate regions (Beijing, Hubei and Tibet). The method applicability was verified for the three different places via pavement temperature nowcasting experiments at a 10 min rolling update and a 10 min interval for the winter of 2021/2022. The geographical information of the three transportation meteorological stations is displayed in Table 2.

Table 2. Geographical information of the three transportation meteorological stations in Beijing, Hubei and Tibet, respectively.

Station ID	Region	Expressway	Longitude	Latitude	Altitude
A1412	Changping, Beijing	West Sixth Ring Road (G4501)	116.11° E	40.13° N	75 m
Q0007	Enshi, Hubei	Shanghai-Chongqing Expressway (G50)	109.37° E	30.24° N	708 m
U1801	Lhasa, Tibet	Ya'an–Yecheng Expressway (G4218)	90.95° E	29.45° N	3610 m

2.2. Methodology

2.2.1. Classification of Weather Types

Using the precipitation and sunshine hour observations from Station 57131, the weather conditions were classified into several categories, which represented the weather of the two transportation meteorological stations along the airport expressway. The variation characteristics and nowcasting skills of pavement temperatures were subsequently analyzed towards different weather conditions. Following Wu et al. [28] and Ma et al. [29], the days with sunshine hours of ≥3 h and no precipitation were defined as sunny days, while those with sunshine hours of <3 h and no precipitation were defined as cloudy days, and the others are rainy and snowy days. It is noted that the observations of pavement conditions were not included in the transportation meteorological stations. Therefore, the pavement icing on rainy and snowy days needed to be recognized via the associated elements, e.g., the simultaneous conditions of low pavement temperature (≤0 °C) and precipitation occurrence at the neighboring national meteorological station [30], which have been examined and demonstrated to be effective in previous studies on transportation meteorological information at Lianyungang–Khorgos Expressway (G30) [31].

2.2.2. Nowcasting of Pavement Temperature

In this paper, the nowcasting of pavement temperature was constructed via the extraction of periodic features of local pavement temperature series and the subsequent regression procedures. Firstly, the power spectrum was employed to analyze the periodic oscillation characteristics of the local pavement temperature series. It is a frequency domain analysis method based on Fourier transform, which decomposes the total energy of the time series into components at different frequencies, and diagnoses the main cycle of the series according to the corresponding variance contributions, so as to determine the implied main frequency and cycles of the series [32,33]. The calculation formula of power spectrum intensity is as follows:

$$\hat{s_k} = \frac{1}{m}\left[r(0) + 2\sum_{j=1}^{m-1} r(j)cos\frac{k\pi j}{m} + r(m)cosk\pi\right],\ k = 0, 1, \ldots, m \quad (1)$$

$$r(j) = \frac{1}{n-j}\sum_{t=1}^{n-j}\left(\frac{x_t - \bar{x}}{s}\right)\left(\frac{x_{t+j} - \bar{x}}{s}\right) \quad (2)$$

where k is the wave number, $\bar{x}$ is the mean value of the sequence, s is the standard deviation of the sequence, and m is the maximum lag time length. The value of m is generally taken as $n/3$, where n is the sample.

On this basis, the red noise power spectrum test was utilized to examine the significance of the obtained periodic characteristics. The calculation formula of the red noise standard spectrum is as follows:

$$s_{0k} = \bar{s}\left[\frac{1 - r(1)^2}{1 + r(1)^2 + 2r(1)cos\frac{\pi k}{m}}\right],\ k = 0, 1, \ldots, m \quad (3)$$

$$\bar{s} = \frac{1}{2m}(s_0 + s_m) + \frac{1}{m}\sum_{k=1}^{m-1} s_k,\ k = 0, 1, \ldots, m \quad (4)$$

The power spectra ratio is defined as $\hat{s_k}/s_{0k}$ to detect the significant period. The significant periodic functions of the pavement temperature time series were then extracted using the mean-generation function model.

The mean-generation function was derived from the mean values at certain time intervals of the time series. The domain of the function definition was extended to the entire axis, which is called the periodic extension. The mean-generation function model could be well used in multi-step forecasts as well as extreme predictions and has been widely exploited in long-term weather forecasts and short-term climate predictions [34,35].

The sequence stationarity is one of the important prerequisites for analysis and modeling of the temporal series, while previous studies have revealed the significant non-stationary characteristics of the pavement temperature [25]. Therefore, the first-order difference method was used in this study to make the original sequence of pavement temperatures stabilized, in which the differential sequence of pavement temperature was taken as the dependent variable and all periodic function sequences passing the confidence level were used as the independent variables. The nowcasting model of pavement temperature can be constructed by the multiple linear regression method with the significant periodic function extended to the entire axis. The inversed differential calculation was, hence, carried out to obtain the forecast results of pavement temperatures for the next 6 h by every 10 min. Similar methods have achieved considerable effects in wind-nowcasting experiments [36].

For the two transportation meteorological observation stations V0001 and V0002, the pavement temperatures of every 10 min during the past 3 days before the current moment were used as the training dataset to predict the pavement temperatures at intervals of 10 min in the next 6 h. The forecast framework was updated with real-time observations with a rolling frequency of 10 min. Taking the first forecast experiment in this study (initialized at 23:00 3 December 2018) as an example, the pavement temperature series during 00:00 1 December to 23:00 3 December by every 10 min were firstly preprocessed by the first-order difference. Afterwards, the periodic functions passing the significance test were extracted through procedures of the power spectrum analysis, the red noise test, and the mean-generation function to construct the nowcasting model of the pavement temperature. The pavement temperature forecasts for the next 6 h at 10 min intervals since 00:00 4 December 2018 were then obtained based on the extended series of periodic functions and the inversed differential calculation.

Moreover, the same framework of nowcasting was also carried out and examined towards Station V0002 in Shanxi and the three transportation meteorological stations in Beijing, Hubei and Tibet using the 10-by-10 min observations in the winter of 2021/2022 to further investigate the applicability of the nowcasting method in multiple areas.

3. Results

3.1. Characteristics of Pavement Temperature Variations

3.1.1. Diurnal Variation Characteristics

In the two winters of 2018/2019 and 2019/2020 except the day with serious data deficiency (30 December 2018), there were 88 sunny days and 78 cloudy days. As for the rainy and snowy days, there were 2 (1) days with and 12 (13) days without icing at Station V0001 (V0002). Figure 2 shows the diurnal variations of pavement temperature on sunny days, cloudy days and rainy and snowy days (with unfrozen and frozen road surfaces). Under the multiple weather conditions, the pavement temperature is characterized by significant diurnal variations, showing the highest values in the early afternoon and lowest at around 07:00–08:00 in the morning. Furthermore, the sunny (rainy and snowy) days are featured by the highest (smallest) diurnal ranges of pavement temperature, and those in the cloudy days are located in between. The pavement temperature on sunny days is generally higher than cloudy and rainy and snowy days in the daytime. However, it decreases rapidly and becomes lower than that on cloudy and rainy and snowy days without icing in the nighttime and early morning but still higher than that on rainy and snowy days with icing. This is due to that the clouds on cloudy and non-icing rainy and snowy days tend to block the solar shortwave radiation reaching the surface, leading to relatively lower pavement temperatures in the daytime, while in the nighttime, they suppress the longwave radiation releasing out of the atmosphere from the surface, which induces the higher pavement temperatures than the sunny days with clear sky. On the other hand, the rainy and snowy days with icing show obviously lower pavement temperatures than the other conditions in both daytime and nighttime, which could be attributed to the joint impacts from the systematic cooling processes of the atmosphere and the reflection of solar radiation by snow or ice on the pavement surface. It may also be related to the definition of weather type in Section 2.2.1.

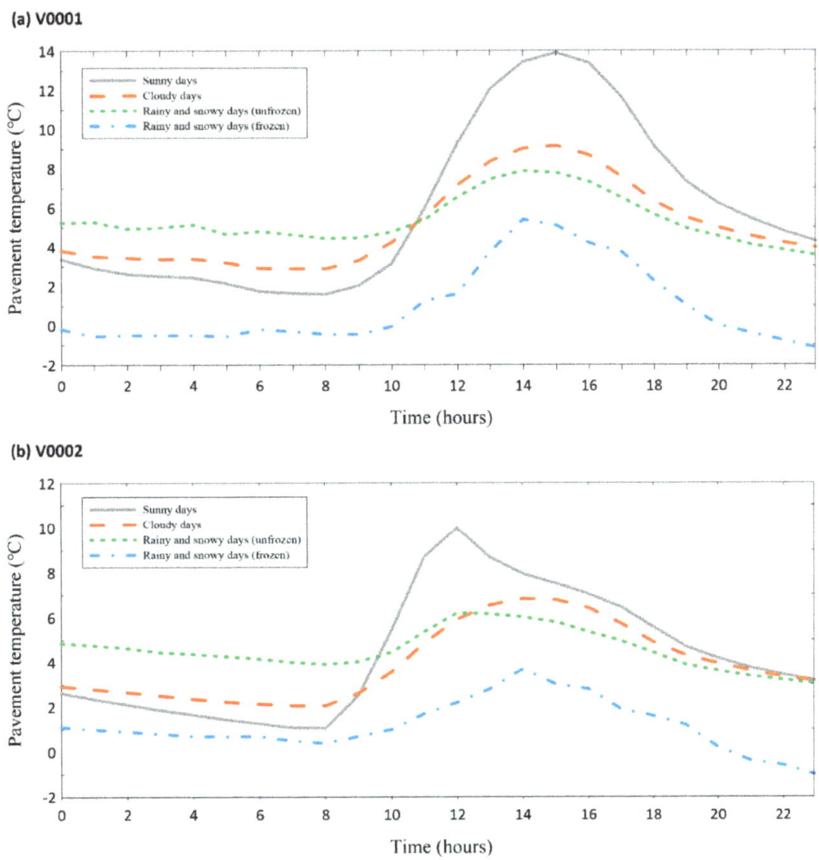

Figure 2. Diurnal variations of pavement temperatures at Stations V0001 (**a**) and V0002 (**b**) under different weather conditions.

Table 3 further displays the temporal averages and mean square deviations of pavement temperatures in the daytime and nighttime at Stations V0001 and V0002 under different weather conditions. According to the local characteristics of sunrise and sunset in winter, the daytime and nighttime are determined as 08:00–18:00 and 19:00–07:00 of the next day, respectively. It verifies the highest pavement temperatures in the daytime of sunny days among the several cases, which is meanwhile accompanied with the largest variability in temperature. In the nighttime, the pavement temperature is highest in non-icing rainy and snowy cases, and is the lowest (maintaining at around 0 °C) in icing rainy and snowy ones with the corresponding variability also being the smallest. Additionally, in the daytime, the pavement temperatures in the sunny days are significantly different from those under the other weather conditions passing the Student's *t*-test at the 95% confidence level, while the pavement temperatures in cloudy days and rainy and snowy days do not differ significantly. As for the nighttime, the differences in pavement temperatures among these several weather conditions are all non-significant according to the Student's *t*-test.

Table 3. Temporal averages (AVE; units: °C) and mean square deviations (MSD; units: °C) of pavement temperatures in the daytime and nighttime at Stations V0001 and V0002 under different weather conditions.

Weather Condition	V0001				V0002			
	Daytime		Nighttime		Daytime		Nighttime	
	AVE	MSD	AVE	MSD	AVE	MSD	AVE	MSD
Sunny	8.68	4.84	3.64	1.98	6.44	2.81	2.59	1.28
Cloudy	6.55	2.39	3.82	1.15	5.11	1.76	2.92	0.91
Rainy and snowy (unfrozen)	6.20	1.51	4.67	1.04	5.15	1.11	4.05	0.82
Rainy and snowy (frozen)	2.40	2.20	−0.36	0.61	1.98	1.04	0.45	0.69

3.1.2. Power Spectrum Features

In order to investigate the detailed periodic features of pavement temperature variations, the power spectrum analysis (Figure 3) was carried out towards the 12,960 data samples which consisted of the pavement temperature observations by every 10 min in the winter of 2018/2019 for Stations V0001 and V0002, respectively. The horizontal axis denotes the length of cycle period, and the vertical axis is the power spectral density of the corresponding cycle divided by the red noise standard spectrum at $\alpha = 0.05$ as described in Section 2.2.2. The spectral ratio of >1.0 represents the periodic significance at the 95% confidence level. The greater the ratio, the more significant the cycle.

As shown in Figure 3, the power spectra of pavement temperatures at the two stations display several significant peaks, locating at the cycle lengths of roughly 24 h, 12 h, 8 h, 6 h, 5 h and 4 h, which are all characterized by power spectrum ratios of >1.0. The first two leading cycles reflect the diurnal and semi-diurnal variations of pavement temperature, which are similar to the variations of surface air temperature and have been well revealed by previous studies. The obvious periodic pattern shows a single peak at noon and a single valley in the early morning [37]. However, the pavement temperatures are also characterized by significant intra-diurnal high-frequency periodic oscillations with periodic cycles of 8 h, 6 h, 5 h and 4 h, which have been seldom mentioned.

The change in the pavement temperature is affected by many factors, such as solar radiation, cloud cover, air temperature, wind speed, relative humidity, pressure, etc. Power spectrum analysis has been conducted for temperature, wind speed and humidity variables, and it was found that these variables also have significant short-period oscillation, especially for temperature and humidity. Studies on periodic variation of wind speed based on Morlet wavelet transformation indicates that on the daily time scale, the wind speed has a significant period of 8 h, 12 h and 24 h [38]. Smaller scale oscillations of the pavement temperature may be related with the short-period variation of other weather variables or the interaction of multiple variables. More comprehensive observation data are needed for a deeper analysis of this problem.

In order to study the intra-diurnal variations of pavement temperatures in more detail, taking the period from 1 December 2018 to 28 February 2019 as an example, power spectrum analysis is performed on the daily pavement temperature series by every 10 min (composing 144 samples in total). With reference to the intraseasonal oscillation analysis on the persistent heavy rainfall in Southern China by Wei et al. [35], the daily power spectral density is divided by the red noise standard spectrum at $\alpha = 0.05$, which are plotted as the distributions of daily power spectrum ratios of pavement temperatures (Figure 4).

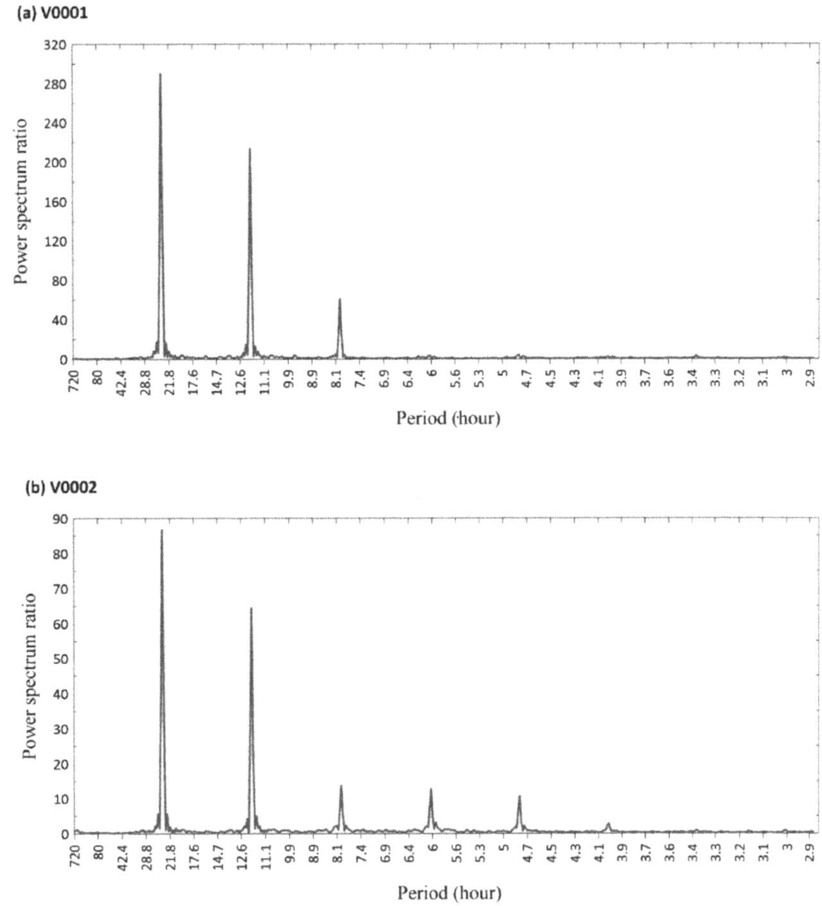

Figure 3. Power spectrum ratios of pavement temperatures by every 10 min in the winter of 2018/2019 for Stations V0001 (**a**) and V0002 (**b**).

It is indicated that there are both significant periodic oscillations with cycles of 6–12 h in the pavement temperature variations of the two transportation meteorological stations. The distribution of power spectrum ratios suggests that certain but limited differences exist in the periodic characteristics of pavement temperatures between the two stations and among different days. The significant periodic oscillations of 6–12 h are relatively stable throughout the winter. Among them, during the periods of 45th–46th days (14–15 January 2019), the 59th–60th days (28–29 January 2019), and the 70th day (8 February 2019), the periodic characteristics of pavement temperatures are generally weaker than those during other periods. The weather conditions are checked to be mainly cloudy during the above periods. In addition, the two pavement temperature series are also featured with periodic oscillations of 4–5 h, with more significant characteristics in Station V0002 than Station V0001 from the perspective of the power spectrum ratio. That is, although the two stations are located in the same climate region, obvious differences might occur between their periodic characteristics of pavement temperatures due to their differences in local environments. Therefore, it is of great importance to grasp the precise periodic features based on the observations of the station, which play a crucial role in the accurate predictions of local pavement temperature.

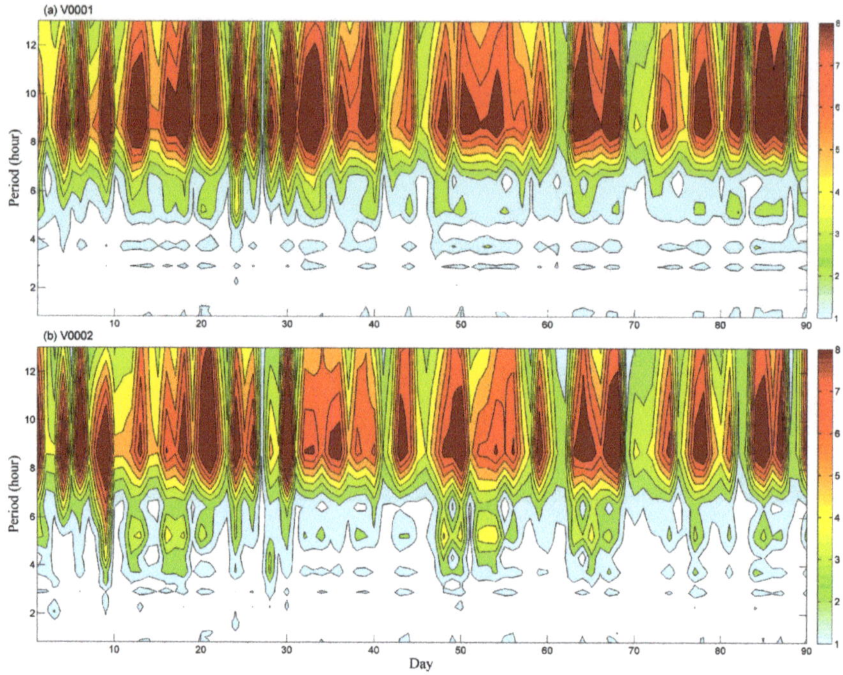

Figure 4. Distribution of daily power spectrum ratios of pavement temperatures in the winter of 2018/2019 for Stations V0001 (**a**) and V0002 (**b**). The shading denotes the significance at the 95% confidence level.

3.2. *Nowcasting of Pavement Temperatures and the Validation*

3.2.1. Impacts from Sample Sizes on Forecast Results

In this study, the nowcasting model is constructed based on the oscillation analysis of the pavement temperature series over certain periods in the past. Theoretically, it could describe more details of the pavement temperature variations with more observation samples used and provide more accurate predictions. However, the larger sample size would definitely result in greater computational complexity. Meanwhile, such a nowcasting model consisting of observations has high requirements for data quality and integrity. Due to the lack of maintenance with standardized processes, the transportation meteorological stations generally lack in quality compared with the normal meteorological observation stations, with the missing or abnormal data occurring more frequently. The data measures of quality control and numerical interpolation are necessary to improve the data application. Therefore, the workload of data preprocessing would also increase in correspondence with the larger sample size. In order to examine the impacts from sample sizes on the pavement temperature nowcasting, we used the 10-by-10 min observations of the first 3 days (432 samples), 4 days (576 samples), and 5 days (720 samples) for model constructions, respectively, with their forecast errors displayed and compared in Figure 5.

It can be seen that, for the two transportation meteorological stations, the increase in the sample sizes does not have obvious influences on the pavement temperature forecasts in the first 2 h. At Station V0001, the increasing sample sizes bring about only slight improvements on the forecasts in the first hour. After that, the forecast error even increases with more samples included, showing greater rising magnitudes at longer lead times. As for Station V0002, the forecast shows slightly higher skills in the first 4 h with the increase in sample numbers, while the effect of sample sizes becomes unstable at lead times longer than that, playing positive roles with 576 samples but negative roles with 720 samples. Overall, the optimal sample size is determined as 432 (i.e., 3 days before the forecast initialization moment) in the subsequent nowcasting experiments.

(a) V0001

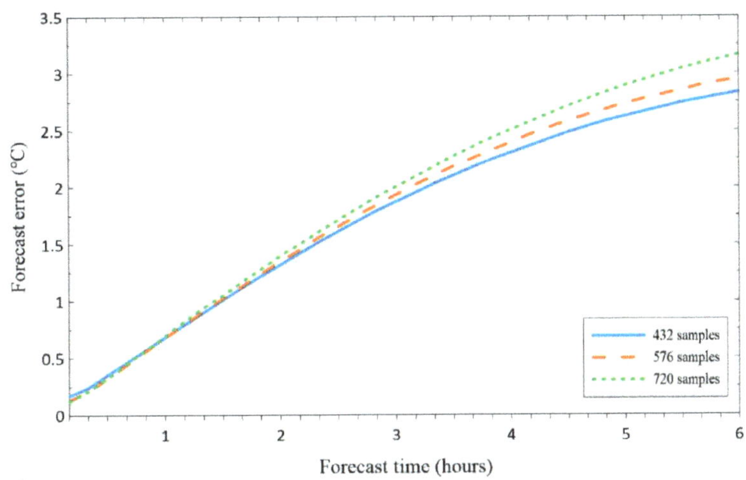

(b) V0002

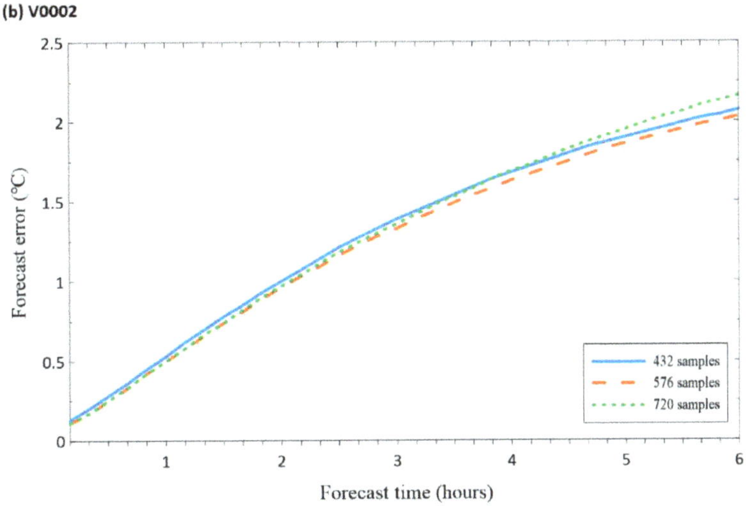

Figure 5. Impacts from sample sizes on forecast results of pavement temperatures for Stations V0001 (**a**) and V0002 (**b**).

3.2.2. General Evaluations of the Forecasts

Based on the methodology introduced in Section 2.2.2, rolling experiments of pavement temperature nowcasting were carried out for 23:00 on 3 December 2018 to 23:00 on 27 February 2019 using the pavement temperature observations in the previous 3 days. A total of 12,384 nowcasting experiment samples were obtained. Figure 6 displays the overall evaluations of pavement temperature nowcasting experiments for Stations V0001 and V0002 within lead times of 6 h by every 10 min. Forecasts for the two stations both show considerable skills within lead times of first few hours, while the forecast skills decrease significantly with the increasing lead time. The absolute errors of forecasts for

Stations V0001 and V0002 are no greater than 1 °C within lead times of 90 min and 120 min, respectively. Additionally, they show errors of <2 °C until lead times of 200 min and 330 min.

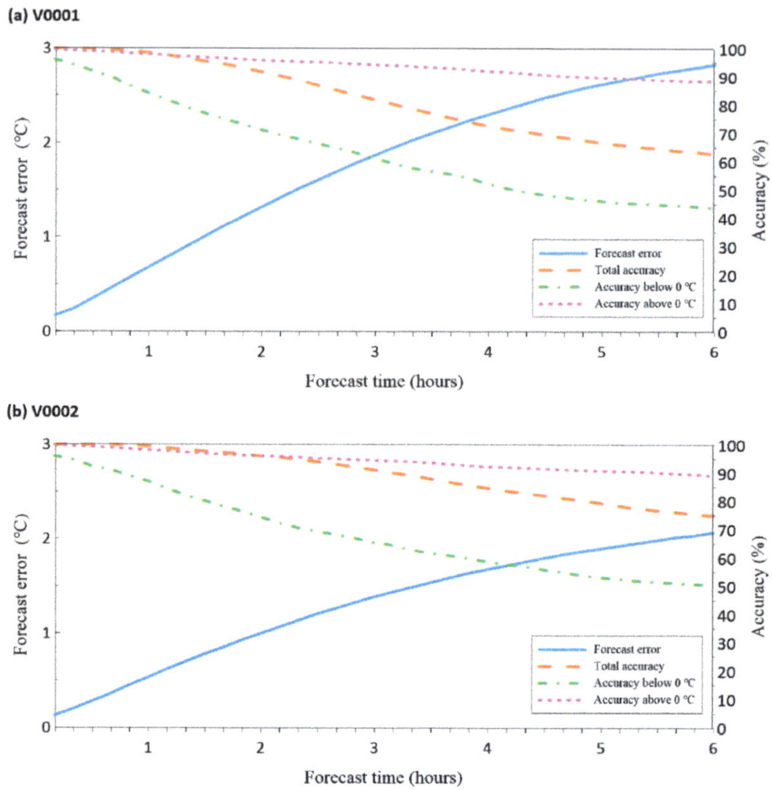

Figure 6. Evaluations of the pavement temperature nowcasting experiments by forecast errors (units: °C) and accuracies (units: %) for Stations V0001 (**a**) and V0002 (**b**), respectively.

Moreover, the statistics of ratios between the mean square deviations of the forecast errors at different lead times and those of the observed pavement temperature series are provided in Table 4. It is normally considered that the forecasts have practical application values with the ratio of <0.5. Forecasts for the two stations show similar skills with the ratios being no greater than 0.14, 0.26 and 0.36 for lead times of shorter than 1 h, 2 h and 3 h, respectively. It is implied that the proposed method could be well utilized in the nowcasting of pavement temperatures along the expressway.

Table 4. Ratios between the mean square deviations of the forecast errors at different lead times and those of the observed pavement temperature series.

Station	Lead Times (Units: 10 min)																	
	1	2	3	4	5	6	7	8	9	10	11	12	13	14	15	16	17	18
V0001	0.04	0.05	0.08	0.10	0.12	0.14	0.17	0.19	0.21	0.23	0.25	0.26	0.28	0.30	0.31	0.33	0.35	0.36
V0002	0.04	0.05	0.07	0.10	0.12	0.14	0.16	0.19	0.21	0.22	0.24	0.26	0.28	0.29	0.30	0.32	0.33	0.34

As for the forecast accuracy, when the pavement temperature observation and forecast are both positive (>0 °C) or both negative (<0 °C), it is considered a success, otherwise it is recorded as a failure. In general, the forecast accuracy rate for pavement temperatures above 0 °C reaches 90% at even lead times of 360 min, which is much higher than that below 0 °C. Although the two stations both show accuracy rates greater than 85% for negative pavement temperatures within lead times of the first hour, it decreases rapidly from 96% to 50% from lead times of 10 min to 360 min. It may be related to the small number of cases with temperatures below 0 °C. For the climate of the area where Xianyang Airport is located, the condition of temperature below 0 °C is very rare. This may affect the prediction accuracy of the statistical model based on the variation characteristics of observation data.

In terms of the previous studies on pavement temperature nowcasting, Wang et al. [26] used the random forest regression method to predict pavement temperatures in the next one hour at three transportation meteorological stations along the Ning-Su-Xu Expressway in Jiangsu province, with the obtained nowcasting results showing mean absolute errors of 0.92 °C, 0.61 °C, and 0.52 °C, respectively. Tang and Guo [25] examined the pavement temperature nowcasting experiments for the next 3 h at the Dongling Station on the Shenyang Third Ring Road (G1501) using the autoregressive summation moving average method with observations of the previous 23 h, exhibiting the mean absolute error of 0.26 °C. However, only 8 groups of experiments in 16 days are considered, and the forecast times mainly concentrate in the early morning, which might lead to some uncertainties of the results.

With respect to the current study with 12,384 sets of nowcasting experiments, the prediction results at Station V0001 (V0002) for lead times of every 10 min in 1 h show absolute errors of 0.17 °C (0.13 °C), 0.24 °C (0.20 °C), 0.35 °C (0.28 °C), 0.46 °C (0.36 °C), 0.57 °C (0.45 °C) and 0.68 °C (0.53 °C), respectively. Comparisons suggest that the pavement temperature nowcasting skills derived from the mean-generation function model in this study are generally equivalent to those of Wang et al. [26] using the random forest method. However, the experiment location of Xianyang Airport Expressway is characterized by a great complexity in the local environment, including the dense traffic flow and large amount of people as well as the heat island effect, making the corresponding nowcasting more difficult than the ordinary expressways. Moreover, the proposed nowcasting method provides the temporally refined results with the time interval being only 10 min in the present study.

Regarding the evaluations on forecast accuracy, the study follows the discrimination in Dong et al. [39] for intuitional comparisons of the forecast skills. They conducted a kind of qualitative categorized prediction on pavement temperatures using several statistical methods and revealed that the multivariate regression method is fairly effective in predicting the low pavement temperatures of <0 °C in northern Jiangsu with the forecast accuracies mostly being greater than 85%, while the support vector machine model is determined as the optimal for southern Jiangsu, and the corresponding accuracies reach almost 95%. By contrast, here in the current paper, the proposed nowcasting method for pavement temperatures using the mean-generation function model not only supplies effective and accurate nowcasting results for the decision makers, but also gives quantitative prediction results of pavement temperatures.

3.2.3. Nowcasting Skills Influenced by Weather Conditions and Initialization Time

To a great extent, the weather conditions have crucial impacts on not only the pavement temperature itself but also the corresponding nowcasting effect. Figure 7 displays the nowcasting evaluations of pavement temperatures under different weather conditions for the two stations. At the early stages, the forecast skills are generally similar with mean absolute errors being <0.2 °C for all the three conditions. When the lead time increases, the forecast errors increase significantly. The largest growing rates occur on sunny days, followed by the cloudy days, while the growing rates are slower on rainy and snowy days, which might be associated with the relatively small variability of pavement temperatures on rainy and snowy days. As the effect of the direct solar heating on the thermal sensors is not considered, the measurement error of pavement temperature on sunny days may increase. This might also be one reason for the lower accuracy of pavement temperature forecast on sunny days.

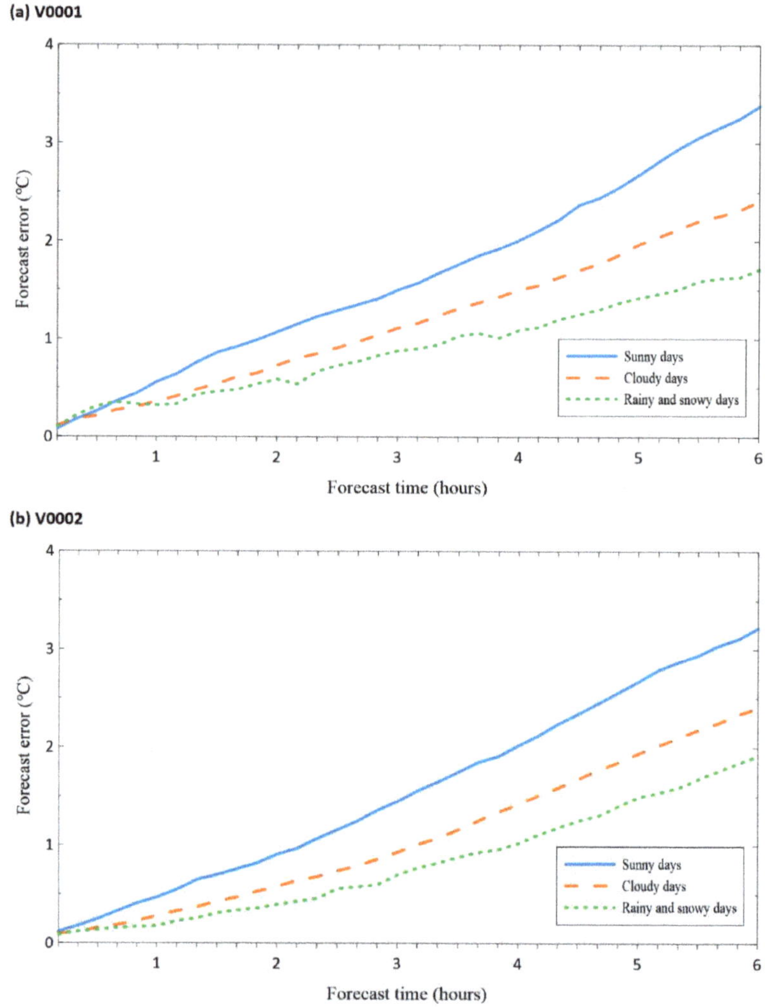

Figure 7. Errors of the pavement temperature nowcasting experiments under different weather conditions for Stations V0001 (**a**) and V0002 (**b**), respectively.

Meanwhile, prediction results in Wang et al. [26] based on the random forest regression method also indicate the highest forecast skills of pavement temperatures on rainy and snowy days, followed by cloudy days, and the worst on sunny days. When the pavement is covered by snow, or ice, etc., the heat exchanges between the ground and the atmosphere become weak or even disappears. The pavement temperature is rarely influenced by external factors and is relatively stable [2,31]. Such a phenomenon is also demonstrated by analyses on the transportation meteorological observations on roads and bridges [40,41].

Due to the significant diurnal variation characteristics of the pavement temperature, the nowcasting skills differ obviously among experiments initialized at different times (Figure 8). The mean absolute errors at the lead times of 1 h and 2 h for the two stations both show generally larger (smaller) values when initialized in the daytime (nighttime).

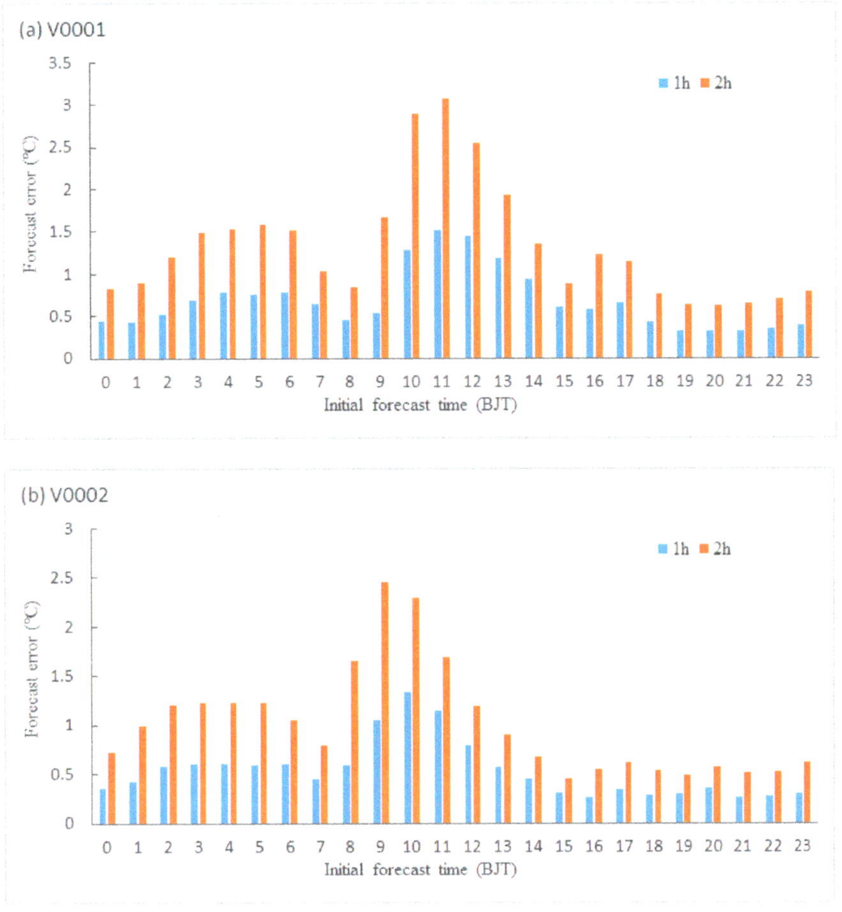

Figure 8. Errors of the pavement temperature nowcasting experiments initialized at different times for Stations V0001 (**a**) and V0002 (**b**), respectively.

For Station V0001, the absolute errors of the pavement temperature nowcasting initialized at the daytime (nighttime) at the lead time of 1 h range between 0.4 °C and 1.5 °C (0.3 °C and 0.7 °C), while they are distributed between 0.2 °C and 1.3 °C (0.2 °C and 0.6 °C) for Station V0002. The plot also suggests that the nowcasting errors of pavement temperature rise rapidly with the increase in lead times. At the lead time of 2 h, the nowcasting experiments initialized at daytime and nighttime show absolute errors of 0.8–3.0 °C and 0.6–1.5 °C, respectively, for Station V0001, while 0.4–2.4 °C and 0.4–1.2 °C for Station V0002.

3.2.4. Nowcasting Experiments at Different Regions

Aiming at exploring the applicability of the proposed method in different regions, similar nowcasting experiments on pavement temperature have been carried out towards the typical transportation meteorological stations in Beijing, Hubei and Tibet for the winter of 2021/2022. The nowcasting skills are compared with Station V0002, which are all displayed in Figure 9.

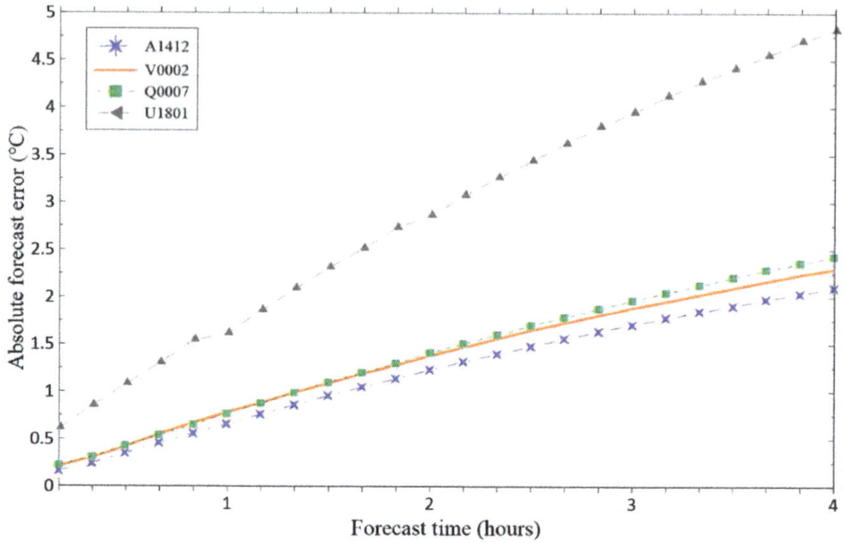

Figure 9. Errors of the pavement temperature nowcasting experiments at different stations in multiple areas.

The results show that the three stations in plain areas with lower altitudes (75 m, 41 m, and 708 m for Stations A1412, V0002, and Q0007, respectively) do not show much difference in the nowcasting errors within lead times of 4 h. Among them, Station A1412 in Beijing is characterized by the highest skills of pavement temperature nowcasting, with the absolute errors being 0.6 °C, 1.2 °C, 1.7 °C, and 2.1 °C for lead times of 1 h, 2 h, 3 h, and 4 h, respectively. However, the nowcasting errors of pavement temperature at Station U1801 in Tibet with the altitude of 3610 m display significantly larger values than the abovementioned three, which have already reached 1.6 °C at the lead time of 1 h. On the other hand, in terms of the ratios between the mean square deviations of the forecast errors and those of the observed pavement temperature series, the values are relatively small for Station U1801, i.e., 0.16, 0.25, and 0.34 at lead times of 1 h, 2 h and 3 h, respectively. It implies that although the pavement temperatures over the plateau area with complex terrain are generally more difficult to be predicted due to their larger variabilities, the proposed nowcasting method using the mean-generation function model in this study has certain reference values for the practical businesses of local early warnings and predictions of the low pavement temperature.

4. Conclusions

Using the pavement temperature observation data of the two transportation meteorological stations along the Xianyang Airport Expressway in Shanxi, China, as well as the datasets of precipitation and sunshine hours obtained from the nearby weather stations, the variation characteristics of local pavement temperatures were investigated for winter in this study. On this basis, a nowcasting method was proposed using a regression model via extracting the corresponding periodic features. Nowcasting experiments were then conducted and analyzed on the local pavement temperatures for the next 6 h with a rolling frequency of 10 min and a time interval of 10 min. The conclusions could be summarized as follows:

(1) Regardless of the weather condition, the observed temperature series are always characterized by significant diurnal variation characteristics, with the highest and lowest values occurring at around 14:00 and 07:00, respectively. Among the different weather conditions, during the daytime, the pavement temperature is the highest on sunny days,

followed by the cloudy days, with those on rainy and snowy days being the lowest. As for the nighttime, it shows highest pavement temperatures on non-icing rainy and snowy days, followed by the cloudy days, and the temperatures on sunny days are further lower, with those on icing rainy and snowy days being the lowest.

(2) The pavement temperatures at the Xianyang Airport Expressway are revealed with not only a significant 24 h periodic oscillation, but also high-frequency oscillation characteristics of about 12 h, 8 h, 6 h, 5 h and 4 h. The nowcasting based on the extraction of these periodic features using the mean-generation function model are demonstrated with considerable skills on predicting the local pavement temperatures. Examinations show the mean absolute errors of 0.2 °C, 0.6 °C, 1.2 °C and 1.5 °C for lead times of 20 min, 1 h, 2 h and 3 h, respectively.

(3) Comparisons among the pavement temperature nowcasting under different weather conditions indicate that the errors are smallest on rainy and snowy days, followed by cloudy days, and the skills are lowest on sunny days. This could be attributed to the more complex influence processes from factors such as clouds and radiation during the sunny days, as well as the larger variability of pavement temperatures. Furthermore, it also shows different nowcasting skills for different initialization times. The errors are generally higher (lower) for nowcasting experiments initialized in the daytime (nighttime), with the mean absolute errors of 0.7 °C and 1.4 °C (0.4 °C and 0.9 °C) for lead times of 1 h and 2 h, respectively.

(4) The expanded experiments for multiple transportation meteorological observation stations in Beijing, Hubei and Tibet demonstrate considerable universality and applicability of the method on pavement temperature nowcasting. It could achieve skillful nowcasting results in plain areas with lower altitudes but shows relatively insufficient performances in plateau areas with complex terrain. Nevertheless, considering the generally lower ratios between the mean square deviations of the forecast errors and those of the observed pavement temperature series, the nowcasting strategy can provide certain reference values for the practical businesses of local early warnings and predictions of the low pavement temperature.

5. Discussion

Changes in weather will affect the accuracy of the nowcast model. Two precipitation processes were picked to analyze the impact of weather change on the forecast skills. The results indicated that if the weather changes during the night, the impact on the forecast skill is smaller. However, if the weather changes during the day, the impact is larger, especially for the first several hours of weather change. With the rolling update of the nowcast model, the forecast results become closer to the observation, whereas the results need to be verified by more tests.

The statistical modeling method based on the extrapolation of high-frequency observations in pavement temperatures is demonstrated to be effective in the accurate nowcasting of pavement temperatures, which could provide an important basis for the identification of icing and snow covers on the pavement. In the road maintenance business, the advantages of an increase in accuracy of nowcasting products and a longer lead time of short-range forecast products are usually combined to arrange the road snow and ice removal. The prediction errors grow rapidly with the increase in lead times. Therefore, more detailed diagnoses and analyses on nowcasting errors are to be carried out in the future and would be favorable for more skillful nowcasting on the pavement temperatures [42,43]. Moreover, the statistical models could also be combined with the physical and dynamic models in the future [44,45].

Author Contributions: Methodology, L.F.; validation, L.F. and H.T.; data curation, M.L.; writing—original draft preparation, L.F. and L.M.; writing—review and editing, X.Y.; visualization, L.M.; project administration, H.T. All authors have read and agreed to the published version of the manuscript.

Funding: This research was funded by the National Key Research and Development program of China (Grant number 2020YFB1600103), the FengYun Application Pioneering Project (Grant number FY-APP-2021.0111) and the Innovation Fund Project of Public Meteorological Service Center of China Meteorological Administration (Grant number K2021006).

Institutional Review Board Statement: Not applicable.

Informed Consent Statement: Not applicable.

Data Availability Statement: The used data in this study are obtained from the data platform of China Meteorological Administration.

Conflicts of Interest: The authors declare no conflict of interest.

References

1. Brodsky, H.; Hakkert, A.S. Risk of a road accident in rainy weather. *Accid. Anal. Prev.* **1988**, *20*, 161–176. [CrossRef]
2. Sun, J.; Liang, F.; Chen, M.; Liao, X. An Analysis on Serious City Traffic Trouble Caused by Light Snow. *Chin. J. Atmos. Sci.* **2003**, *27*, 1057–1066.
3. Jiang, J.; Shi, L.; Ni, Y. A simulation of a high impact weather event. *J. Appl. Meteor. Sci.* **2005**, *16*, 231–237.
4. Zhu, S.; Yang, H.; Liu, D.; Wang, H.; Zhou, L.; Zhu, C.; Zu, F.; Wu, H.; Lyu, Y.; Xia, Y.; et al. Observations and Forecasts of Urban Transportation Meteorology in China: A Review. *Atmosphere* **2022**, *13*, 1823. [CrossRef]
5. Rayer, P.J. The Meteorological Office Road Surface Temperature Model. *Meteorol. Mag.* **1987**, *116*, 180–191.
6. Sass, B.H. A Numerical Model for Prediction of Road Temperature and Ice. *J. Appl. Meteorol.* **1992**, *31*, 1499–1506. [CrossRef]
7. Sass, B.H. A Numerical forecasting system for the prediction of slippery roads. *J. Appl. Meteorol.* **1997**, *36*, 801–817. [CrossRef]
8. Shao, J.; Lister, P. An Automated Nowcasting Model of Road Surface Temperature and State for Winter Road Maintenance. *J. Appl. Meteorol.* **1996**, *35*, 1352–1361. [CrossRef]
9. Crevier, L.P.; Delage, Y. METRo: A New Model for Road-Condition Forecasting in Canada. *J. Appl. Meteorol.* **2001**, *40*, 2026–2037. [CrossRef]
10. Shao, J.; Lister, P. The Prediction of Road Surface State and Simulation of the Shading Effect. *Bound.-Layer Meteorol.* **1995**, *73*, 411–419. [CrossRef]
11. Chapman, L.; Thornes, J. The influence of traffic on road surface temperatures: Implications for thermal mapping studies. *Meteor. Appl.* **2005**, *12*, 371–380. [CrossRef]
12. Bouilloud, L.; Martin, E.; Habets, F.; Boone, A.; Le Moigne, P.; Livet, J.; Marchetti, M.; Foidart, A.; Franchisteguy, L.; Morel, S.; et al. Road surface condition forecasting in France. *J. Appl. Meteorol. Climatol.* **2009**, *48*, 2513–2527. [CrossRef]
13. Kann, A.; Kršmanc, R.; Habrovský, R.; Šajn Slak, A.; Bujňák, R.; Schmid, F.; Tarjáni, V.; Wang, Y.; Wastl, C.; Bica, B.; et al. High-resolution nowcasting and its application in road maintenance: Experiences from the INCA central European area project. *IET Intell. Transp. Syst.* **2015**, *9*, 539–546. [CrossRef]
14. Feng, L.; Wang, X.; He, X.; Gao, J. Fine forecast of high road temperature along Jiangsu highways based on INCA system and METRo model. *J. Appl. Meteor. Sci.* **2017**, *28*, 109–118.
15. Liu, X.; Yu, Y.; Lei, G.; Liu, Z. Using radiant balance theory to calculate concrete road-surface temperature in summer. *J. Appl. Meteor. Sci.* **2004**, *15*, 623–628.
16. Zhu, C.; Xie, Z.; Yang, M.; Jing, Y. Study on numerical prediction model of extreme temperature on speedway-surface. *J. Meteorol. Sci.* **2009**, *29*, 645–650.
17. Tian, H.; Wu, H.; Zhao, L.; Chen, H.; Li, K.; Yang, X. Characteristics and statistical model of road surface temperature on Huning expressway. *J. Appl. Meteor. Sci.* **2009**, *20*, 737–744.
18. Niu, S.; Li, R.; Lyu, J.; Meng, L.; Ke, Y.; Yang, Z.; Xiong, S. Research on a numerical model for predicting three types of underlying surface temperature and ice. *Chin. J. Geophys.* **2011**, *54*, 909–917.
19. Meng, C.; Zhang, C. Development and verification of a numerical forecast model for road meteorological services. *J. Appl. Meteor. Sci.* **2012**, *23*, 451–458.
20. Feng, T.; Li, X.; Ding, D. Prediction of expressway surface temperature. *J. Highw. Transp. Res. Dev.* **2012**, *29*, 19–29.
21. Wu, H.; Ma, C.; Yang, R.; Zhang, J. Variation Characteristics of road surface temperature on highway of Hebei Province and its prediction model. *J. Arid Meteorol.* **2014**, *32*, 665–676.
22. Zhang, X.; Hao, Y.; Liang, J.; Pan, L.; Shen, J.; Lu, S. Characteristics of Road Surface Temperature of Shaanxi Expressway and Its Prediction Model. *J. Arid Meteorol.* **2019**, *37*, 1028–1034.
23. Wang, J.; Guo, C.; Gu, X. Study on Variation Characteristics and Prediction Model of Different Pavement Temperature in Baotou. *Desert Oasis Meteorol.* **2021**, *15*, 91–96.
24. Lin, Z.; Hu, J.; Zhu, C. A Method for Forecasting Hourly Expressway Surface Temperature during Night Time. *J. Highw. Transp. Res. Dev.* **2021**, *38*, 23–29.
25. Tang, J.; Guo, Z. Pavement Temperature Short-impending Prediction Based on ARIMA in Winter. *J. Tongji Univ. (Nat. Sci.)* **2017**, *45*, 1824–1829.

26. Wang, K.; Bao, Y.; Zhu, C.; Chen, C.; Yuan, C. Forecasts of road surface temperature in winter based on random forests regression. *Meteorol. Mon.* **2021**, *47*, 82–93.
27. Qu, X.; Qi, Y.; You, Q.; Wang, Y.; Wu, D.; Li, M. An approaching prediction method of road surface temperature of Winter Olympic Highway demonstration station based on METRo model. *J. Arid Meteorol.* **2020**, *38*, 497–503.
28. Wu, S.; Wu, D.; Deng, X.; Tan, H. Characteristics of road surface temperature on freeway over Nanling Hilly region. *Meteorol. Sci. Technol.* **2006**, *34*, 783–787.
29. Ma, J.; Liao, C.; Zhang, Y. Variation Characteristics of Road Surface Temperature on Highway of Changsha and Its Prediction Model. *Sci. Mosaic* **2017**, *2*, 22–25.
30. Bao, G.; Wang, W.; Zhang, J.; Zhang, J.; Ma, S.; Pei, S. Temporal-spatial distribution of road icing in Qinghai and its impact. *Meteorol. Sci. Technol.* **2016**, *44*, 104–110.
31. Zhang, H.; Lu, S.; Shen, J.; Zhang, X.; Dang, C. Temporal and Spatial Distribution Characteristics of Road Icing in Shaanxi and Its Risk Warning Model. *J. Arid Meteorol.* **2020**, *38*, 878–885.
32. Wei, F. *Modern Climate Statistical Diagnosis and Prediction Technology*; China Meteorological Press: Beijing, China, 2005; pp. 71–75.
33. Bai, Y.; Gao, X.; Zhang, W.; Xi, C.; Wang, L. Analysis of the array data power spectrum in Hailar. *Seismol. Res. Northeast China* **2020**, *36*, 49–55.
34. Feng, L.; Wei, F.; Zhu, Y. A predictive model for summer precipitation over China based on upper tropospheric temperature and North Atlantic Oscillation in the preceding spring. *Chin. J. Atmos. Sci.* **2011**, *35*, 963–976.
35. Wei, F.; Han, X.; Wang, Y.; Chen, G. *Physical Basis of Short-Term Climate Prediction in China and Short-Term Climate Prediction Methods*; China Meteorological Press: Beijing, China, 2015; pp. 211–222.
36. Chang, R.; Zhu, R.; Liu, Y.; He, X. Nowcasting model of wind speed based on mean generating function for wind farms. *Meteorol. Mon.* **2013**, *39*, 226–233.
37. Sun, Q.; Li, J.; Hu, T.; Han, W. Study on variation of road surface temperature. *J. China Foreign. Highw.* **2015**, *35*, 32–36.
38. Zhang, C.; Tan, Q.; Tang, S.; Gao, X. Classification and combination prediction model for wind power based on change period of wind speed. *Acta Energ. Sol. Sin.* **2017**, *4*, 999–1006.
39. Dong, T.; Bao, Y.; Yuan, C.; Zhou, L.; Jiao, S. Application of Three Statistical Forecast Models in Early Warning of Low Tem-perature on Road Surface in Jiangsu and Their Comparison. *Meteorol. Sci. Technol.* **2018**, *46*, 773–784.
40. Lyu, J.; Niu, S.; Zhou, Y.; Li, R.; Ke, Y.; Yang, Z. Characteristic of bridge and road surface temperature changes in winter and energy budget analysis. *Trans. Atmos. Sci.* **2013**, *36*, 546–553.
41. Wang, J.; Zhu, C.; Yuan, C.; Bao, Y. Study of low temperature differences between road surface and bridge surface on Yang-zhou-Liyang expressway in winter. *J. Trop. Meteorol.* **2018**, *34*, 279–288.
42. Lyu, Y.; Zhi, X.; Wu, H.; Zhou, H.; Kong, D.; Zhu, S.; Zhang, Y.; Hao, C. Analyses on the Multimodel Wind Forecasts and Error Decompositions over North China. *Atmosphere* **2022**, *13*, 1652. [CrossRef]
43. Fan, Y.; Zhu, S.; Wang, L.; Wang, X. Subseasonal dynamical prediction of South China Sea summer monsoon. *Atmos. Res.* **2022**, *278*, 106347. [CrossRef]
44. Zhu, S.; Zhi, X.; Ge, F.; Fan, Y.; Zhang, L.; Gao, J. Subseasonal forecast of surface air temperature using superensemble ap-proaches: Experiments over Northeast Asia for 2018. *Weather Forecast* **2021**, *36*, 39–51. [CrossRef]
45. Lyu, Y.; Zhi, X.; Zhu, S.; Fan, Y.; Pan, M. Statistical calibrations of surface air temperature forecasts over East Asia using pattern projection methods. *Weather Forecast* **2021**, *36*, 1661–1674. [CrossRef]

atmosphere

Article

Precipitation Nowcasting Based on Deep Learning over Guizhou, China

Dexuan Kong [1,2], Xiefei Zhi [1,*], Yan Ji [1], Chunyan Yang [2,*], Yuhong Wang [3], Yuntao Tian [1], Gang Li [4] and Xiaotuan Zeng [5]

1 Key Laboratory of Meteorology Disaster, Ministry of Education (KLME)/Joint International Research Laboratory of Climate and Environment Change (ILCEC)/Collaborative Innovation Center on Forecast and Evaluation of Meteorological Disasters (CIC-FEMD), Nanjing University of Information Science and Technology, Nanjing 210044, China
2 Meteorological Bureau of Qian Xinan Buyei and Miao Autonomous Prefecture, Xingyi 562400, China
3 Hebei Meteorological Observatory, Shijiazhuang 050021, China
4 Guizhou Meteorological Observatory, Guiyang 550002, China
5 Guangxi Meteorological Observatory, Nanning 530022, China
* Correspondence: zhi@nuist.edu.cn (X.Z.); 15519677649@163.com (C.Y.)

Abstract: Accurate precipitation nowcasting (lead time: 0–2 h), which requires high spatiotemporal resolution data, is of great relevance in many weather-dependent social and operational activities. In this study, we are aiming to construct highly accurate deep learning (DL) models to directly obtain precipitation nowcasting at 6-min intervals for the lead time of 0–2 h. The Convolutional Long Short-Term Memory (ConvLSTM) and Predictive Recurrent Neural Network (PredRNN) models were used as comparative DL models, and the Lucas–Kanade (LK) Optical Flow method was selected as a traditional extrapolation baseline. The models were trained with high-quality datasets (resolution: 1 min) created from precipitation observations recorded by automatic weather stations in Guizhou Province (China). A comprehensive evaluation of the precipitation nowcasting was performed, which included consideration of the root mean square error, equitable threat score (ETS), and probability of detection (POD). The evaluation indicated that the reduction of the number of missing values and data normalization boosted training efficiency and improved the forecasting skill of the DL models. Increasing the time series length of the training set and the number of training samples both improved the POD and ETS of the DL models and enhanced nowcasting stability with time. Training with the Hea-P dataset further improved the forecasting skill of the DL models and sharply increased the ETS for thresholds of 2.5, 8, and 15 mm, especially for the 1-h lead time. The PredRNN model trained with the Hea-P dataset (time series length: 8 years) outperformed the traditional LK Optical Flow method for all thresholds (0.1, 1, 2.5, 8, and 15 mm) and obtained the best performance of all the models considered in this study in terms of ETS. Moreover, the Method for Object-Based Diagnostic Evaluation on a rainstorm case revealed that the PredRNN model, trained well with high-quality observation data, could both capture complex nonlinear characteristics of precipitation more accurately than achievable using the LK Optical Flow method and establish a reasonable mapping network during drastic changes in precipitation. Thus, its results more closely matched the observations, and its forecasting skill for thresholds exceeding 8 mm was improved substantially.

Keywords: precipitation forecast; nowcasting; deep learning; ConvLSTM; PredRNN

Citation: Kong, D.; Zhi, X.; Ji, Y.; Yang, C.; Wang, Y.; Tian, Y.; Li, G.; Zeng, X. Precipitation Nowcasting Based on Deep Learning over Guizhou, China. *Atmosphere* **2023**, *14*, 807. https://doi.org/10.3390/atmos14050807

Academic Editor: Stefano Federico

Received: 19 February 2023
Revised: 22 April 2023
Accepted: 26 April 2023
Published: 28 April 2023

1. Introduction

Precipitation nowcasting means forecasting precipitation with a lead time of 0–2 h, focusing more on mesoscale–microscale weather systems at a high spatiotemporal resolution [1–3]. Highly accurate precipitation nowcasting is vital in support of various operational activities, e.g., disaster relief, energy management, and flood early warning systems. Consequently, the accuracy of precipitation nowcasting has a critical socioeconomic impact.

Currently, the primary means of improving the accuracy of precipitation nowcasting involve the integration of extrapolation and ensemble numerical weather prediction (NWP) [4–8]. However, literature shows that global coarse-resolution NWP models have challenges in generating accurate precipitation with a lead time of 0–2 h [9,10], affected by the spin-up issue and the difficulties in non-Gaussian data assimilation. Although convection-permitting models with high resolution could greatly improve performance, these models are computationally expensive. Traditional integration of extrapolation methods, such as cross-correlation tracking and optical flow, has been used widely in operational weather forecasting [11–19]. Still, they exhibit a marked reduction in forecasting skills with increasing lead time owing to their linear operating limits and their deficiencies in forecasting storm growth and decay [20].

In recent years, the application of deep learning (DL) [21] has become popular in the field of meteorology [22–24] owing to its nonlinear operation, and it has achieved substantial advances in terms of quantitative precipitation nowcasting using radar echo extrapolation [25–32]. In comparison with the traditional integration of extrapolation and NWP, DL models can realize relatively accurate precipitation nowcasting. For example, Shi et al. [27] involved the convolution operation in input-to-state and state-to-state transitions in the transformation of a two-dimensional image into a three-dimensional tensor and proposed the Convolutional Long Short-Term Memory (ConvLSTM) model. This approach, which was shown capable of effectively capturing spatial correlations and further realizing the extrapolation of spatiotemporal sequences, was applied to achieve quantitative precipitation nowcasting. Subsequently, they involved the convolution operation in the recurrent gated unit (GRU) and used a subnetwork to output a location-variant connection structure before performing state transitions. Then, they proposed the trajectory GRU that could handle the spatial correlations better and perform more accurately than previous methods [28]. Wang et al. [29] designed the Spatiotemporal LSTM (ST-LSTM) model that can transfer memories vertically and horizontally, and then they proposed the Predictive Recurrent Neural Network (PredRNN) model. To strengthen the power for modeling short-term dynamics and to alleviate the vanishing gradient problem, they improved the PredRNN model to PredRNN++ [30], which incorporates a cascaded dual-memory structure and a gradient highway unit. Ji et al. [31] exploited the advantages of different DL model architectures in combination with the ConvLSTM unit in the U-Net generator and proposed the CLGAN model, which can better capture the precipitation object and its characteristics. Chen et al. [32] compared and analyzed the extrapolation prediction of radar echoes using the ConvGRU method, cross-correlation method, optical flow method, and particle filter method and found that the output of the ConvGRU method more closely matched the location, intensity, and shape of actual radar echoes. Many other studies have modified existing techniques to obtain relatively better nowcasting performance in comparison with that achieved using the integration of extrapolation [33–47]. However, precipitation calculated from radar echoes is based on the Z–R relationship. We believe that the Z–R relationship cannot describe the nonlinear relationship between a radar echo and precipitation. Moreover, it is difficult to avoid calculation errors derived from the Z–R relationship. Factors such as radar model, detection range, and clutter interference will lead to poor universality of a DL model. Consequently, the objective of this study was to construct DL models for precipitation nowcasting based on precipitation observation data, which could directly obtain precipitation nowcasting.

The reason we use DL models is their nonlinear operation, and the ConvLSTM model is one of the classical models used in the precipitation nowcasting problem. It can extract the spatial characteristics while capturing the time characteristics efficiently. PredRNN improved memory information by transferring memories vertically and horizontally, and it can perform better for forecast time compared with ConvLSTM. The Lucas–Kanade Optical Flow (LK Optical Flow) is a good method that improved optical flow vector calculation for traditional optical flow [48], and it was selected as a traditional baseline. We hope to compare the two typical DL models with LK Optical Flow and try to construct highly accurate models based on precipitation observation data.

In this study, we formed two high-quality datasets: a pre-processed precipitation dataset (Pre-P dataset) and a heavy precipitation dataset (Hea-P dataset). Then, we constructed precipitation DL models that could directly obtain precipitation nowcasting based on precipitation observation data. Furthermore, we improved the forecasting skill of the DL models by increasing the length of the time series of the training set and by training with the Hea-P dataset. The study area selected was Guizhou Province in China (29.2–24.7° N, 103.6–109.6° E), which features highly complex terrain and frequently experiences geological disasters and floods [49,50]. Moreover, the region is also lacking in terms of high-accuracy precipitation nowcasting.

The remainder of this manuscript is organized as follows. Section 2 describes the data and methods used in the study. Section 3 comprehensively evaluates both the DL models trained with different datasets and the LK Optical Flow method. Finally, the conclusions and a discussion are presented in Section 4.

2. Data and Method

2.1. Guizhou Automatic Weather Station (AWS) Observations

High-quality real-time precipitation observation data (resolution: 1 min) recorded over a nine-year period from 20:00 (all times UTC+8) on 31 December 2012 to 20:00 on 31 December 2021 (observation day: from 20:00 on the previous day to 20:00 on the same day) by 1835 automatic weather stations (AWS) in Guizhou Province (as shown in Figure 1) were obtained from the Guizhou Meteorological Information Center.

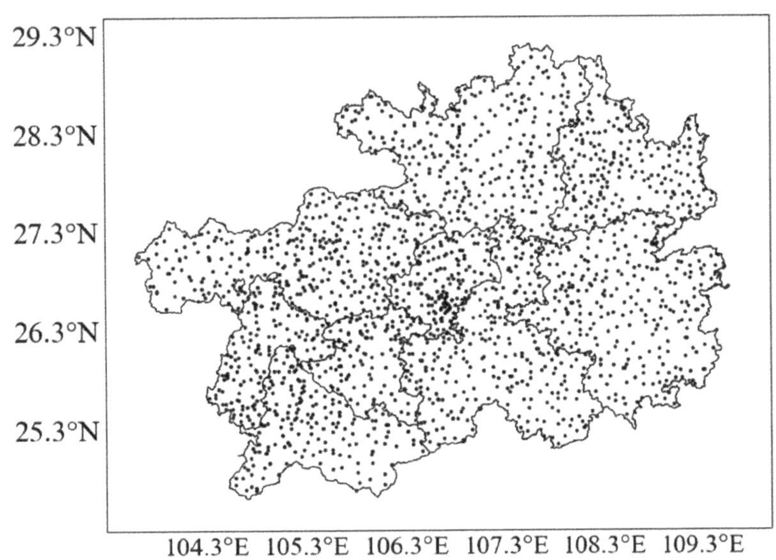

Figure 1. Distribution of the 1835 automatic weather stations in Guizhou Province, China.

2.2. Pre-Processed Precipitation Dataset (Pre-P Dataset)

The number of missing values in the observation data was reduced using the neighborhood method (where the time series before and after the missing values and the data of nearby stations were complete). Then, the bilinear interpolation method was used to interpolate station data within the region 24.5–29.5° N, 103.5–110° E to the gridded field (resolution: 0.125° × 0.125°). Data normalization resulted in a high-quality dataset with a range of variation of [0, 1]. Ultimately, the pre-processed(Pre-P) dataset was formed, comprising 788,400 samples.

2.3. Heavy Precipitation Dataset (Hea-P Dataset)

Data in the Pre-P dataset from May–September in each of the nine years from 2013–2021 were selected, and samples with little or no precipitation were removed. Consequently, the heavy precipitation (Hea-P) dataset was determined, comprising 332,640 samples.

2.4. Lucas–Kanade (LK) Optical Flow Method

The optical flow method is defined as a method for calculating the intensity of image pixel points over time to infer the speed and direction of object movement. It finds the correspondence between the previous and the current frame based on changes in the pixels of the image in the time series and the correlation between adjacent frames, from which it calculates the motion information of objects between adjacent frames. This study employed the LK Optical Flow method [48], the process of which can be divided into two steps:

In the first step, the optical flow vector field is calculated. The instantaneous velocity two-dimensional vector field of the changing trend of grayscale at each point on the image and the optical flow vector field are solved using the following three basic assumptions.

(1) The brightness is constant, i.e., when the same target moves between different frames, the brightness will not change in any way. According to this assumption, the basic equation of the optical flow method can be obtained:

$$I(x,y,t) = I(x+dx,y+dy,t+dt) \tag{1}$$

where x and y are the coordinate positions of a pixel in the image, t is the time series in which the image is located, and (dx, dy) is the distance moved to the next frame, using dt time, i.e., the light intensity of the pixel before and after motion has remained constant.

(2) The movement in continuous time is a 'small movement', which means that the change of time will not cause a drastic change in the target position. Expanding the Taylor section of Equation (1) yields:

$$(x,y,t) = I\left(x,y,t+\frac{\partial I}{\partial x}dx+\frac{\partial I}{\partial y}dy+\frac{\partial I}{\partial t}dt+\varepsilon\right), \tag{2}$$

where ε represents a second-order infinitesimal term that is negligible. Assuming the velocity vectors u and v of the optical flow in the x-axis and y-axis directions, the formulas will be as follows:

$$Iu = \frac{dx}{dt}, \quad v = \frac{dy}{dt}, \tag{3}$$

The partial derivatives of the gray scale of pixel points in the image on x, y, and t are as follows:

$$I_x = \frac{\partial I}{\partial x}, \quad I_y = \frac{\partial I}{\partial y}, \quad I_t = \frac{\partial I}{\partial t}, \tag{4}$$

Then, substituting Equation (2) into Equation (1) and dividing by dt gives the following:

$$I_x u + I_y v + I_t = 0, \tag{5}$$

where I_x, I_y, and I_z can be obtained from image data, and (u, v) is the optical flow vector.

(3) Spatial consistency (or spatial continuity) means that some points which around a specific point have the same optical flow field or velocity.

There are two unknowns in only one constrained equation. For analyzing the $n \times n$ region around a pixel, the pixel motion of the local region is assumed to be consistent. Then, the $n \times n$ equations can be established in the following matrix form:

$$\begin{bmatrix} I_{x1} & I_{y1} \\ I_{x2} & I_{y2} \\ \vdots & \vdots \end{bmatrix} \begin{bmatrix} u \\ v \end{bmatrix} = -\begin{matrix} I_{t1} \\ I_{t2}, \\ \vdots \end{matrix} \tag{6}$$

and written as vector:

$$A\vec{u} = \vec{b}, \tag{7}$$

where $\vec{u}$ is the velocity vector:

$$\vec{u} = (A^T A)^{-1} A^T \vec{b}, \tag{8}$$

Then, the equations are solved using the least squares method.

The second step is to use the real-time precipitation field at the initial time and the optical flow vector field calculated in the first step to extrapolate the precipitation field. In this study, the semi-Lagrangian advection scheme was used to extrapolate the precipitation field [48]. The semi-Lagrangian advection formula can be expressed as follows:

$$\overline{F}(t_0 + \tau, x) = F(t_0, x - \alpha), \tag{9}$$

where $\overline{F}$ is the extrapolated forecast precipitation field, F is the real-time precipitation observation at the initial time, t_0 is the lead time, τ is the forecast time, and α is the displacement vector within the lead time.

The semi-Lagrangian advection scheme divides the lead time τ into N steps for extrapolation, in this study, the lead time Δt was 6 min. The displacement amount α can be obtained from the following iteration:

$$\alpha^{(n+1)} \Delta t \vec{V}(t_0, x - \frac{\alpha^{(n)}}{2}), \tag{10}$$

where $\vec{V}\left(t_0, x - \frac{\alpha^{(n)}}{2}\right)$ is the velocity vector of the precipitation at point $x - \frac{\alpha}{2}$, and $\vec{V}$ is the optical flow vector field, assuming the initial value of α is 0. The total displacement in the lead time is the sum of the displacements in N steps, and n is the number of iterations.

The specific extrapolation scheme uses the remapped particle-mesh Semi-Lagrangian scheme improved by Reich [51].

2.5. *Convolutional Long Short-Term Memory (ConvLSTM) Model*

The ConvLSTM model can extract the two-dimensional spatial characteristics of precipitation while capturing the time characteristics. The convolution operation and pooling operation are integrated into the LSTM, which adds the "input gate," "output gate," and "forget gate" based on the RNN model and can record historical data over a longer time and improve forecast accuracy. Then, realizing the spatiotemporal sequence prediction of the precipitation, the main calculation equations in the model are as follows [27]:

$$i_t = \sigma(W_{xi} * X_t + W_{hi} * H_{t-1} + W_{ci} \odot C_{t-1} + b_i), \tag{11}$$

$$f_t = \sigma\left(W_{xf} * X_t + W_{hf} * H_{t-1} + W_{cf} \odot C_{t-1} + b_f\right), \tag{12}$$

$$C_t = f_t \odot C_{t-1} + i_t \odot tanh(W_{xc} * X_t + W_{hc} * H_{t-1} + b_c), \tag{13}$$

$$O_t = \sigma(W_{xo} * X_t + W_{ho} * H_{t-1} + W_{co} \odot C_t + b_o), \tag{14}$$

$$H_t = O_t \odot tanh(C_t), \tag{15}$$

where X_t and H_t represent input data and output data, respectively; i_t, f_t, and O_t represent the input gate, forget gate, and output gate, respectively; C_t is the cell state; "$*$" denotes the convolution operator; and "$\odot$" denotes the Hadamard product.

2.6. *Predictive Recurrent Neural Network (PredRNN)*

For more accurate transmission of time information, the new ST-LSTM was proposed, which can transfer memories vertically and horizontally. Then, the PredRNN model was obtained based on the ConvLSTM model, and the complete formula of the ST-LSTM model can be expressed as follows [29]:

$$i_t = \sigma\left(W_{xi} * X_t + W_{hi} * H_{t-1}^l + b_i\right), \tag{16}$$

$$f_t = \sigma\left(W_{xf} * X_t + W_{hf} * H_{t-1}^l + b_f\right), \tag{17}$$

$$C_t^l = f_t \odot C_{t-1}^l + i_t \odot tanh\left(W_{xg} * X_t + W_{hg} * H_{t-1}^l + b_g\right), \tag{18}$$

$$i_t' = \sigma\left(W_{xi}' * X_t + W_{mi} * M_t^{l-1} + b_i'\right), \tag{19}$$

$$f_t' = \sigma\left(W_{xf}' * X_t + W_{mf} * M_t^{l-1} + b_f'\right), \tag{20}$$

$$M_t^l = f_i' \odot M_t^{l-1} + i_t' \odot tanh\left(W_{xg}' * X_t + W_{mg} * M_t^{l-1} + b_g'\right), \tag{21}$$

$$O_t = \sigma\left(W_{xo} * X_t + W_{ho} * H_{t-1}^l + W_{co} * C_t^l + W_{mo} * M_t^l + b_o\right), \tag{22}$$

$$H_t^l = O_t \odot tanh\left(W_{1\times1} * \left[C_t^l, M_t^l\right]\right), \tag{23}$$

W_{xi}, W_{hi}, W_{mi}, W_{xf}, W_{hf}, W_{mf}, W_{xg}, W_{hg}, W_{ho}, W_{mo}, $W_{xi'}$, $W_{xf'}$, $W_{xg'}$ are all weight parameters, and b_i, b_f, b_g, b_o, b_i', $b_{f'}$, $b_{g'}$ are learnable offset parameters. With C_t^l representing time memory, M_t^l representing spatial memory, H_t^l representing the value of the hidden layer, subscript t representing the time step, and superscript representing the kth hidden layer existing in the ST-LSTM network.

For the parameters of DL models in this paper, we set the initial learning rate and the epochs as 0.001 and 20, respectively. The root mean square error (RMSE) was selected as the loss function.

2.7. Verification Metrics

2.7.1. Root Mean Square Error (RMSE)

The expression for the RMSE is as follows:

$$RMSE = \frac{1}{n}\sum_{i=1}^{n}[(f_i - O_i)^2]^{\frac{1}{2}}, \tag{24}$$

where n represents the number of grid samples of the space field, f_i represents the forecast value of the i–th sample, and O_i represents the observation of the i–th sample. The smaller the *RMSE* value, the smaller the difference between the forecast value and the observed value, i.e., the smaller the forecast error.

2.7.2. Probability of Detection (POD), False Alarm Ratio (FAR), Probability of False Detection (POFD), and Equitable Threat Score (ETS)

Expressions for the probability of detection (POD), false alarm ratio (FAR), probability of false detection (POFD), and equitable threat score (ETS) are as follows [52–54]:

$$\text{POD} = \frac{N_A}{N_A + N_C}, \tag{25}$$

$$\text{FAR} = \frac{N_B}{N_A + N_B}, \tag{26}$$

$$\text{POFD} = \frac{N_C}{N_B + N_E}, \tag{27}$$

$$E = \frac{(N_A + N_B)(N_A + N_C)}{N_A + N_B + N_C - E}, \tag{28}$$

$$\text{ETS} = \frac{N_A - E}{N_A + N_B + N_C - E}, \tag{29}$$

where N_A represents the number of forecast events that correspond to observed events (forecast has, observation has); N_B represents the number of events that are null (forecast has, observation has not); N_C represents the number of events that are missed (forecast has not, observation has); N_E represents the number of events that both forecast, and observation do not occur. POD refers to the proportion of the predicted actual precipitation area in the total actual precipitation area; the larger the value, the higher the forecast accuracy. FAR refers to the proportion of the area with no actual precipitation in the forecast precipitation area in the total forecast precipitation area; the smaller the value, the smaller the forecast null rate. POFD refers to the proportion of the area that is missed in the actual precipitation area in the area where all actual precipitation does not occur; the smaller the value, the lower the FAR of the forecast. Additionally, the higher the ETS score, the better the forecast performance.

In this study, we applied five thresholds (i.e., 0.1, 1, 2.5, 8, and 15 mm) to calculate the ETS of the hourly graded precipitation. The forecasting skill of 6-min precipitation was calculated based on the threshold of 0.01 mm.

2.7.3. Method for Object-Based Diagnostic Evaluation (MODE)

The Method for Object-Based Diagnostic Evaluation (MODE) is a spatial evaluation method based on object attribute characteristics, which mainly evaluates a prediction by comparing and analyzing the attributes and similarities of the main observed and forecast areas of precipitation [55,56].

Through the operation of spatial convolution and then under different precipitation thresholds (i.e., 0.01, 0.1 and 2.5 mm), the important areas of precipitation to be studied are calculated and identified as follows [57]:

$$C(x,y) = \sum_{u,v} \varphi(u,v) f(x-u, y-v), \tag{30}$$

$$\varphi(u,v) = \begin{cases} \frac{1}{\pi R^2}, & u^2+v^2 \leq R^2 \\ 0, & u^2+v^2 > R^2 \end{cases}, \tag{31}$$

where *f* represents the original data field, C represents the convolutional field, φ represents the filter function, and (*x,y*) and (*u,v*) represent grid coordinates. The mask field *M* is obtained by threshold control of the convolutional field, i.e., the precipitation area in the convolutional field where precipitation intensity is greater than or equal to threshold *T* is calculated [55,56]:

$$M(x,y) = \begin{cases} 1, & C(x,y) \leq T \\ 0, & C(x,y) > T \end{cases}, \tag{32}$$

By assigning grid points in the continuous region of $M = 1$ to the value of the corresponding grid points in the original precipitation field, the reconstruction field *F* can be obtained, which not only retains most of the original precipitation information of each object (precipitation without convolution processing), but also identifies the main area of falling precipitation when the precipitation threshold is reached. The formula for calculating *F* is as follows [55,56]:

$$F(x,y) = M(x,y) f(x,y), \tag{33}$$

Then, according to Equation (33), certain important properties of the observation field and the precipitation field in these aeras of falling precipitation are calculated. In this study, area, angle, aspect ratio, centroid position of longitude, and centroid position of latitude were selected as the attributes for assessment and analysis of the area of falling precipitation.

According to Davis et al. [55,56], the matching rule proposed by uses the calculated attributes to match objects, i.e., all matches in the process come from the important areas of falling precipitation in the forecast field and in the observation field identified as needing study, and the calculation formula used in this process is as follows [55–58]:

$$D < \frac{Area_O^{\frac{1}{2}} + Area_f^{\frac{1}{2}}}{2}, \tag{34}$$

where $Area_o$ and $Area_f$ are the areas of the main areas of falling precipitation identified in the observation and forecast fields, and *D* is the centroid distance between them.

Finally, according to the weight and the confidence factor of the attribute, the total similarity between the observation field and the important area of falling precipitation in the precipitation field is calculated using the fuzzy logic method [57]:

$$I = \frac{\sum_{i=1}^{n} \omega_i c_i G_i}{\sum_{i=1}^{n} \omega_i c_i}, \tag{35}$$

where c_i and ω_i represent the confidence factor and the weight for property *i*, respectively, which depend only on the specific properties of the subject of tax reduction, and where the confidence factor varies with the area size and distance of the area of falling precipitation; and G_i is the similarity factor of the *i*-th attribute that is a monotonous recursive function with a value range of between 1 and 0.

3. Results

3.1. Data Quality Control (DQC) Evaluation

We selected May–September 2019 as the verification period because several extreme precipitation events occurred during this period. For example, intense rainstorms that occurred during 5–11 June 15–19 June, 20–25 June, and 27 June to 1 July caused notable geological and flood disasters [59].

Reducing the number of missing values in the original AWS data produced a complete time series and improved the spatial resolution of the dataset, thereby making the training samples more representative of the observed precipitation. Data normalization narrowed the range of extreme precipitation and improved the learning efficiency of the DL models with respect to the features of precipitation observations during the training process. Both processes helped DL models that exhibited improved performance in terms of ETS. The Hea-P dataset before and after data quality control (DQC) was divided into a training set and a validation set with a ratio of 8:1, and ETS evaluations were performed for the verification period.

The ETSs of the DL models for 6-min precipitation nowcasting before and after DQC are shown in Figure 2. The average ETS of the ConvLSTM model and of the PredRNN model increased by 0.1458 and 0.0660, respectively. The ConvLSTM model showed greater improvement in comparison with the PredRNN model; however, the PredRNN model also exhibited reasonable improvement beyond the 12-min lead time.

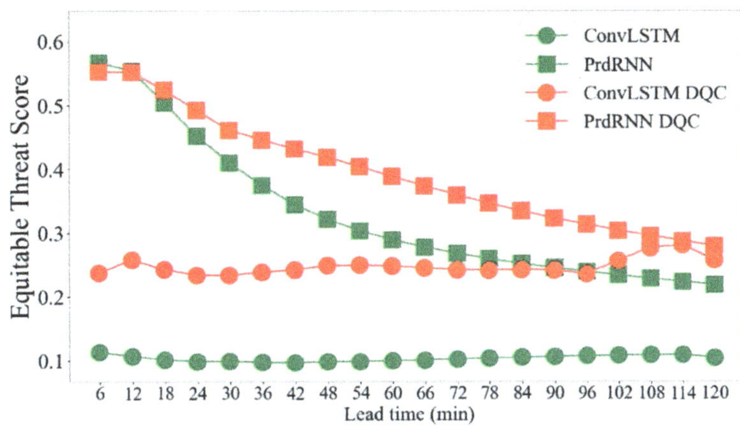

Figure 2. Comparison of the ETSs of the DL models before and after data quality control (DQC).

Currently, there is no unified evaluation metric for minute-level precipitation. Consequently, we calculated the ETS of hourly graded precipitation accumulated from 6-min precipitation for different thresholds (i.e., 0.1, 1, 2.5, 8, and 15 mm), as listed in Table 1. The forecasting skill of the ConvLSTM model with thresholds of 0.1 and 1 mm was well improved, and the ETS for the threshold of 0.1 mm for the 1-h lead time increased by 0.2286. Overall, the PredRNN model exhibited greater improvement in comparison with that of the ConvLSTM model, i.e., the average ETS for thresholds exceeding 2.5 mm for the 1-h lead time and for thresholds below 2.5 mm for the 2-h lead time increased by more than 0.09 and by 0.1305, respectively.

Table 1. ETSs of hourly graded precipitation for DL models before and after DQC for different thresholds (i.e., 0.1, 1, 2.5, 8, and 15 mm).

Model	ConvLSTM		ConvLSTM DQC		PredRNN		PredRNN DQC	
Lead time	1 h	2 h	1 h	2 h	1 h	2 h	1 h	2 h
≥0.1 mm	0.1009	0.1658	0.3295	0.2523	0.4868	0.2121	0.5097	0.3425
≥1 mm	0.2223	0.1406	0.2637	0.2346	0.3574	0.1302	0.3694	0.2363
≥2.5 mm	0.1033	0.0922	0.0687	0.0390	0.2341	0.1046	0.3319	0.1700
≥8 mm	0.0039	0.0037	0.0032	0	0.0858	0.0276	0.1783	0.0586
≥15 mm	0	0	0.0004	0	0.1289	0.0283	0.2329	0.0594

3.2. DL Models Trained Using Datasets with Different Time Series Lengths

Pre-P datasets with DQC and different time series lengths (i.e., 1, 3, 5, and 8 years) were divided into training sets, testing sets, and validation sets with ratios of 2:1:1, 6:1:1, 10:2:1, and 16:3:1, respectively, and the verification period for all evaluations was May–September 2019. Then, DL models were constructed using the training sets with different time series lengths (i.e., 1, 3, 5, and 8 years). Verifications including RMSE, ETS, POD, FAR, and POFD were applied to assess the 6-min precipitation nowcasting for the threshold of 0.01 mm generated using the DL models, as shown in Figure 3. With increasing length of the time series of the training set, the DL models showed a marked increase in terms of POD and ETS, and the stability of the nowcasts with time was enhanced, especially for the ConvLSTM model trained with the training set with the 8-year time series length, i.e., the average POD increased by 0.1066. Additionally, the PredRNN model trained with the training set with the 8-year time series length performed better than the ConvLSTM model, i.e., the RMSE decreased sharply, the FAR was low, and the ETS increased by 0.0763.

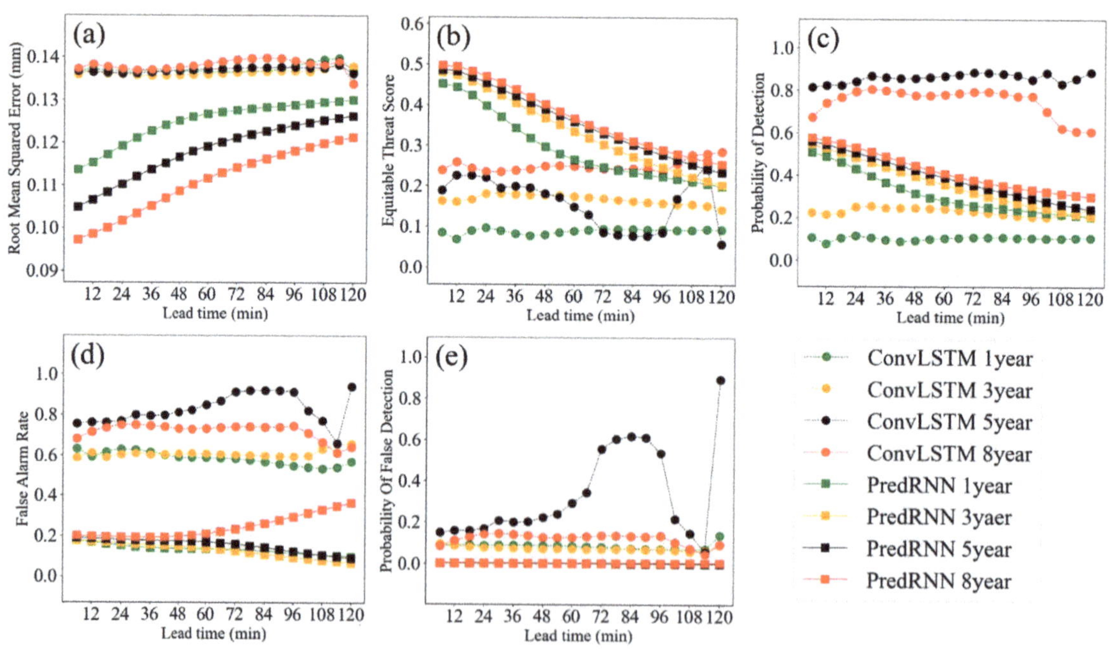

Figure 3. Forecasting skill of DL models trained with different time series lengths: (**a**) RMSE, (**b**) ETS, (**c**) POD, (**d**) FAR, and (**e**) POFD.

Taking the same verification period as above, we calculated the ETS of hourly graded precipitation (Table 2). The ETSs of the DL models were improved with the increasing length of the time series of the training set. For the training set with the 8-year time series length, the ETS of the ConvLSTM model for the threshold of 0.1 mm increased sharply by 0.2484 for the 2-h lead time, while that for thresholds below 2.5 mm for the 1-h lead time increased by more than 0.05. However, the PredRNN model showed better improvement in terms of ETS. For example, the average ETS for the 0–2-h lead time for different thresholds (i.e., 0.1, 1, 2.5, 8, and 15 mm) increased by 0.0852, while that for the 1-h lead time increased by 0.1067, and that for the threshold of 2.5 mm increased by 0.1464.

Table 2. Hourly graded precipitation ETSs for the two DL models trained with Pre-P datasets with different time series lengths for different thresholds (i.e., 0.1, 1, 2.5, 8, and 15 mm).

Model	**ConvLSTM 1 Year**		**ConvLSTM 3 Years**		**ConvLSTM 5 Years**		**ConvLSTM 8 Years**	
Lead time	1 h	2 h	1 h	2 h	1 h	2 h	1 h	2 h
≥0.1 mm	0.2553	0.0039	0.3015	0.0139	0.3232	0.3256	0.3295	0.2523
≥1 mm	0.2030	0.2113	0.2352	0.1300	0.2531	0.2534	0.2637	0.2346
≥2.5 mm	0.0135	0.0078	0.0313	0.0001	0.0170	0.0223	0.0687	0.0390
≥8 mm	0.0004	0	0.0067	0	0.0007	0.0013	0.0032	0
≥15 mm	0	0	0.0024	0	0	0.0006	0.0004	0
Model	PredRNN 1 Year		PredRNN 3 Years		PredRNN 5 Years		PredRNN 8 Years	
Lead time	1 h	2 h	1 h	2 h	1 h	2 h	1 h	2 h
≥0.1 mm	0.4302	0.1257	0.4352	0.2455	0.4680	0.2560	0.4868	0.2121
≥1 mm	0.2991	0.0352	0.3112	0.1535	0.3468	0.1517	0.4183	0.1287
≥2.5 mm	0.0877	0.0211	0.1485	0.0209	0.1757	0.0389	0.2341	0.1046
≥8 mm	0.0011	0.0002	0.0062	0.0005	0.0122	0.0001	0.0858	0.0276
≥15 mm	0.0021	0.0003	0.0076	0.0028	0.0064	0.0025	0.1289	0.0283

To summarize, increasing the length of the training set time series and the number of the training samples both improved the forecasting skill of the DL models. Overall, the PredRNN model exhibited greater improvement in comparison with the ConvLSTM model, especially for the 1-h lead time, and the PredRNN model trained with the Pre-P dataset with the 8-year time series length obtained the best performance in terms of ETS.

3.3. Deep Learning Model Training Using the Heavy Precipitation (Hea-P) Dataset

Although increasing the length of the time series of the training set could improve the forecasting skill of the DL models, the model with the best performance still exhibited no improvement in terms of the ETS for hourly graded precipitation when compared with the results obtained using the LK Optical Flow method (Table 3). Even though the ETS of the PredRNN model with the best performance was slightly higher than that of the traditional LK Optical Flow method for the thresholds of 0.1, 2.5, and 8.0 mm for the 1-h lead time, the ETS for the threshold of 15.0 mm for the 1-h lead time was lower than that derived using the LK Optical Flow method. Moreover, the ETSs of the PredRNN model with different thresholds (i.e., 0.1, 2.5, 8, and 15 mm) decreased rapidly with time, resulting in a lower overall ETS in comparison with that derived using the LK Optical Flow method.

Table 3. ETSs of hourly graded precipitation for the LK Optical Flow method and the DL models trained using the two types of datasets for different thresholds (i.e., 0.1, 1, 2.5, 8, and 15 mm).

Model	**ConvLSTM 8 Years**			**PredRNN 8 Years**			**LK Optical Flow**		
Lead time	1 h	2 h	0–2 h	1 h	2 h	0–2 h	1 h	2 h	0–2 h
≥0.1 mm	0.3295	0.2523	0.2909	0.4868	0.2121	0.3495	0.4502	0.2914	0.3708
≥1 mm	0.2637	0.2346	0.2492	0.4183	0.1287	0.2735	0.3322	0.2011	0.2666
≥2.5 mm	0.0687	0.0390	0.0539	0.2341	0.1046	0.1694	0.2178	0.1061	0.1620
≥8 mm	0.0032	0.0000	0.0016	0.0858	0.0276	0.0567	0.1081	0.0422	0.0751
≥15 mm	0.0004	0.0000	0.0002	0.1289	0.0283	0.0786	0.1681	0.0407	0.1044

To further improve the forecasting skill of the DL models, we removed samples with little or no precipitation and left the extreme precipitation to form the Hea-P dataset, which we hoped would represent a more effective training sample. Taking the same verification period as described in Section 3.2, Hea-P datasets with different time series lengths (i.e., 1, 3, 5, and 8 years) were divided into training sets, and validation sets with a ratio of 1:1, 3:1, 5:1, and 8:1, respectively. We constructed DL models using the training sets with time series of different lengths. Then, evaluation of the 6-min precipitation nowcasting was performed based on the RMSE, ETS, POD, FAR, and POFD, which allowed comparative assessment of the LK Optical Flow method and the DL models trained with the two types of datasets (i.e., Pre-P dataset and Hea-P dataset), as shown in Figure 4.

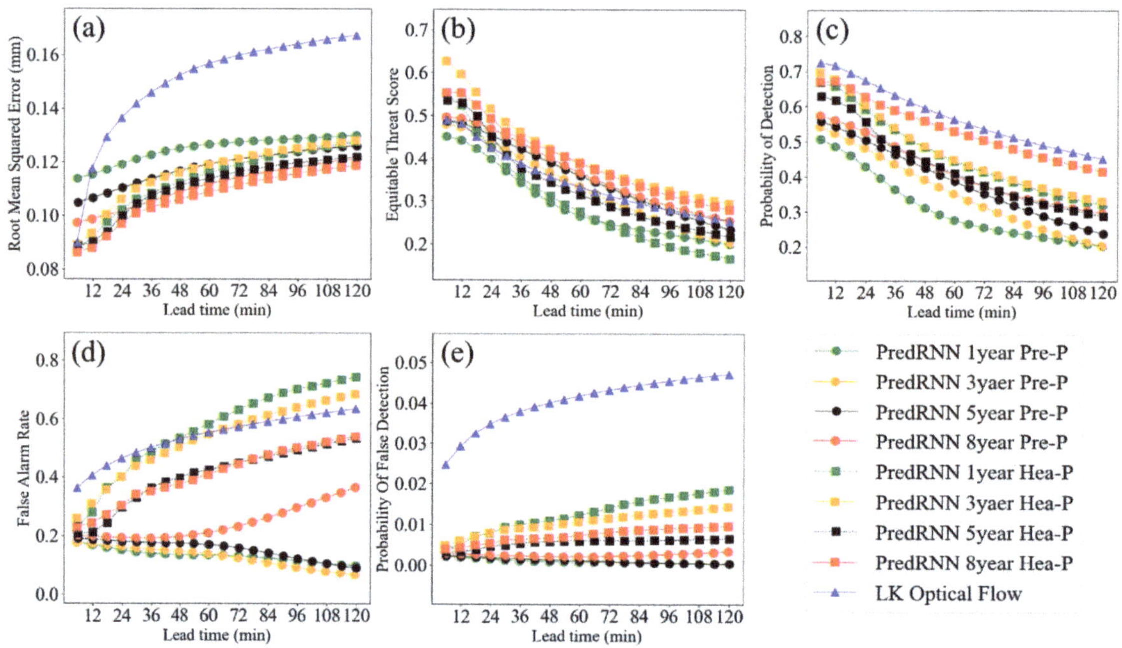

Figure 4. Forecasting skill of the LK Optical Flow model and the PredRNN model trained with two types of datasets: (**a**) RMSE, (**b**) ETS, (**c**) POD, (**d**) FAR, and (**e**) POFD.

With a sharp increase in POD and reduction in RMSE for the 8-year time series length, the DL models trained with the Hea-P dataset performed better than those trained with the Pre-P dataset in terms of 6-min precipitation nowcasting; specifically, the PredRNN model outperformed the LK Optical Flow method in terms of forecasting skill. For the 1- and 3-year time series lengths, the DL models trained with the Hea-P dataset exhibited reduced ETSs and sharply increased FAR scores. Meanwhile, the DL models trained with the Hea-P datasets with the 5- and 8-year time series lengths also sharply increased in FAR scores. Overall, use of the Hea-P datasets improved the POD and resulted in sharply increased FAR values for the DL models because the DL models tended to generate more high-value precipitation nowcasting owing to training samples with high-value precipitation in the Hea-P dataset.

The ETSs of hourly graded precipitation for different thresholds are listed in Table 4. The DL models trained with the Hea-P dataset improved markedly. The PredRNN model trained with the training set with the 8-year time series length obtained notable improvement for thresholds exceeding 2.5 mm, i.e., the ETS of the PredRNN model for thresholds of 2.5, 8, and 15 mm for a 1-h lead time increased sharply by 0.1141, 0.0702, and 0.0648, respectively. The average ETS for thresholds below 2.5 mm for the 2-h lead time also increased by more than 0.05. Importantly, the ETS for the threshold of 8.0 mm increased by an average of 0.0676 for a 0–2-h lead time and by 0.0925 for a 1-h lead time. Additionally, the ETS of the ConvLSTM model for the threshold of 2.5 mm increased by 0.0801, while that for below the threshold of 1.0 mm decreased (note: the ETSs of the ConvLSTM model are not listed in the table).

Table 4. ETSs of hourly graded precipitation for the PredRNN model trained with two types of datasets for different thresholds (i.e., 0.1, 1, 2.5, 8, and 15 mm).

Model		PredRNN 1 Year		PredRNN 3 Years		PredRNN 5 Years		PredRNN 8 Years	
	Lead time	1 h	2 h	1 h	2 h	1 h	2 h	1 h	2 h
	≥0.1 mm	0.4302	0.1257	0.4352	0.2455	0.4680	0.2560	0.4868	0.2121
Pre-P	≥1 mm	0.2991	0.0352	0.3112	0.1535	0.3468	0.1517	0.4183	0.1287
dataset	≥2.5 mm	0.0877	0.0211	0.1485	0.0209	0.1757	0.0389	0.2341	0.1046
	≥8 mm	0.0011	0.0002	0.0062	0.0005	0.0122	0.0001	0.0858	0.0276
	≥15 mm	0.0021	0.0003	0.0076	0.0028	0.0064	0.0025	0.1289	0.0283
	Lead time	1 h	2 h	1 h	2 h	1 h	2 h	1 h	2 h
	≥0.1 mm	0.4612	0.2744	0.4492	0.2862	0.4743	0.3351	0.5097	0.3425
Hea-P	≥1 mm	0.2571	0.1279	0.3168	0.1891	0.3513	0.2322	0.3694	0.2363
dataset	≥2.5 mm	0.2498	0.1035	0.2506	0.0978	0.2664	0.1074	0.3319	0.1700
	≥8 mm	0.1320	0.0442	0.1334	0.0482	0.1384	0.0411	0.1783	0.0586
	≥15 mm	0.1177	0.0312	0.1774	0.0658	0.1828	0.0459	0.2329	0.0594

Comprehensive evaluation revealed the DL models with the performance in terms of ETS. The PredRNN model trained with the Hea-P dataset with the 8-year time series length obtained the best performance of all models in terms of ETS for hourly graded precipitation (Table 5). The PredRNN model outperformed the LK Optical Flow method for all thresholds (i.e., 0.1, 1, 2.5, 8, and 15 mm), especially thresholds exceeding 2.5 mm. Although the ConvLSTM model was improved by training with the Hea-P dataset, it still performed a lower ETS than that of the other two models for all thresholds.

Table 5. ETSs of hourly graded precipitation for the LK Optical Flow method and the DL models trained with the Hea-P dataset for different thresholds (i.e., 0.1, 1, 2.5, 8, and 15 mm).

Model	ConvLSTM 8 Years		PredRNN 8 Years		LK Optical Flow	
Lead time	1 h	2 h	1 h	2 h	1 h	2 h
≥0.1 mm	0.2526	0.2016	0.5097	0.3425	0.4502	0.2914
≥1 mm	0.1452	0.1108	0.3694	0.2363	0.3322	0.2011
≥2.5 mm	0.1337	0.1342	0.3319	0.1700	0.2178	0.1061
≥8 mm	0.0294	0.0283	0.1783	0.0586	0.1081	0.0422
≥15 mm	0.0012	0.0042	0.2329	0.0594	0.1681	0.0407

The spatial distributions of ETSs for hourly graded precipitation derived using the LK Optical Flow method and the DL models trained with the Hea-P dataset with the 8-year time series length for the 1-h lead time are shown in Figure 5. For thresholds of 0.1, 1, 2.5, and 8 mm, the higher the threshold, the lower the ETS of hourly graded precipitation; however, the ETS for the threshold of 8 mm exhibited a sharp decrease. Areas with high ETSs for thresholds of 0.1 and 1 mm mainly occurred in eastern, southern, and southwestern parts of Guizhou, whereas high ETSs for thresholds of 2.5, 8, and 15 mm mainly occurred in north-central and southern parts of Guizhou. For thresholds of 8 and 15 mm, the ETS of the ConvLSTM model was close to zero owing to many false predictions in areas of high-value precipitation and high values of POFD and FAR. This can result from cumulative error magnified by iterative calculation. For the threshold of 15 mm, the ETSs of the LK Optical Flow method and the PredRNN model showed improvement mainly in the southern and north-central parts of Guizhou. Additionally, the ETSs of the PredRNN model with different thresholds (i.e., 0.1, 1, 2.5, 8 and 15 mm) indicate that the PredRNN model outperformed both the ConvLSTM model and the LK Optical Flow method.

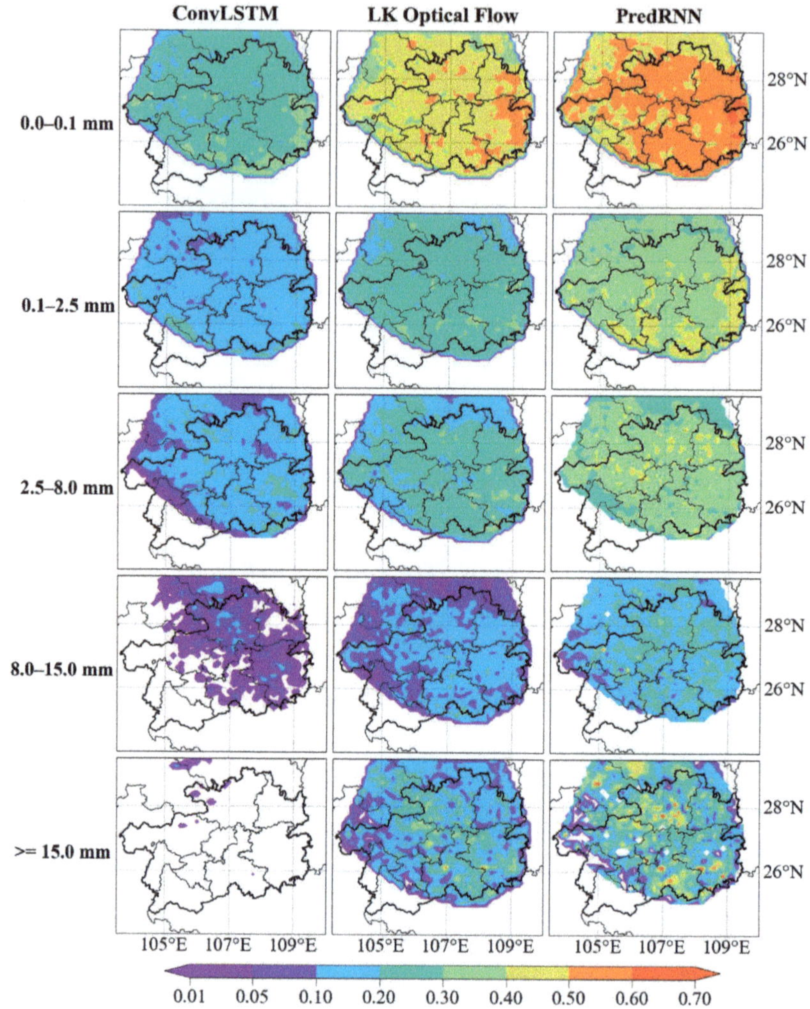

Figure 5. Spatial distributions of ETSs for hourly graded precipitation for the 1-h lead time, derived from the LK Optical Flow method and the DL models trained with the Hea-P dataset with the 8-year time series length.

Figure 6 presents the spatial distributions of ETSs for the 2-h lead time. The ETSs of both the LK Optical Flow method and the PredRNN model for different thresholds (i.e., 0.1, 1, 2.5, 8, and 15 mm) decreased sharply for the 2-h lead time with the same pattern of spatial distribution as that shown for the 1-h lead time. The ETS of the PredRNN model for thresholds of 0.1 and 15 mm maintained values of 0.3–0.5 and 0.2–0.3, respectively, which were superior to the values of the other two models.

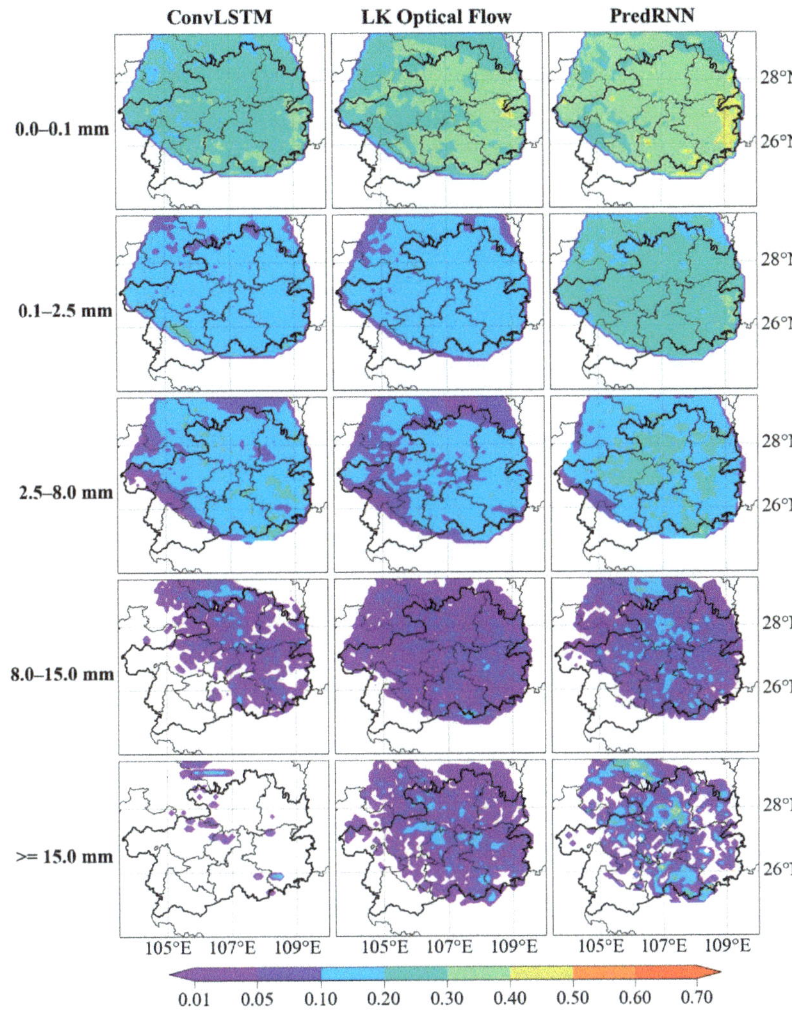

Figure 6. Spatial distributions of ETSs for hourly graded precipitation for the 2-h lead time, derived from the LK Optical Flow method and the DL models trained with the Hea-P dataset with the 8-year time series length.

In summary, the DL models trained with the Hea-P dataset improved notably in terms of ETS for 6-min and hourly graded precipitation, especially for thresholds of 2.5, 8, and 15 mm. We determined that the PredRNN model trained with the Hea-P dataset with an 8-year time series length performed best, outperforming the LK Optical Flow method for all thresholds. Furthermore, it performed well for thresholds of 2.5, 8, and 15 mm in the north-central and southern parts of Guizhou.

3.4. Structure Evaluation on a Rainstorm Case

As a case for evaluation, we selected a severe rainstorm event that occurred on 6 June 2019, which produced a two-hour period (01:30–03:24) of intense precipitation [59]. We used the MODE approach to further assess the performance of both the LK Optical Flow method and the PredRNN model trained with the Hea-P dataset with the 8-year time series length in relation to this rainstorm case. The MODE consisted of an evaluation of the

6-min precipitation for the threshold of 0.01 mm and an evaluation of the hourly graded precipitation for thresholds of 0.1, 1, and 2.5 mm.

The attribute values of the 6-min precipitation objects for the observations and nowcasts, calculated for a threshold of 0.01 mm, included the area, axis angle, aspect ratio, zonal centroid, and meridional centroid, from which we obtained the total similarity of the precipitation objects. All attribute values and the total similarity of the 6-min precipitation objects for both the LK Optical Flow method and the PredRNN model are shown in Figure 7. The attribute values of the observations have a wide range and show rapid fluctuation during the rainstorm, resulting in both the shape and the position of the area of precipitation varying wildly and rapidly with time. Meanwhile, the zonal centroid and the meridional centroid indicate that the precipitation system first evolved toward the southwest and then moved rapidly toward the northeast. With the characteristic of linear variation, the attribute values of the LK Optical Flow method change smoothly with time, whereas the attribute values of the PredRNN model fluctuate widely and are much closer to those of the observations.

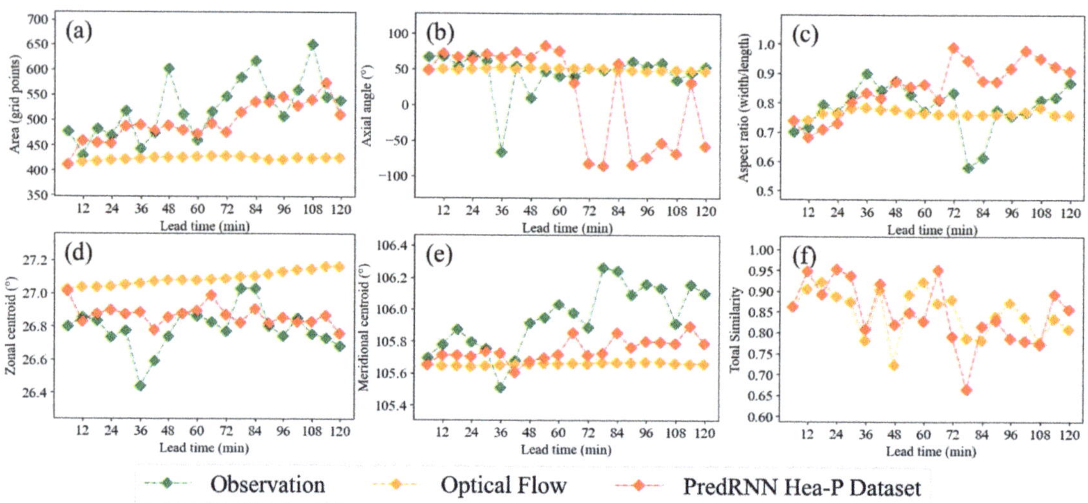

Figure 7. Attribute values of 6-min precipitation objects for the LK Optical Flow method and the PredRNN model trained with the Hea-P dataset with the 8-year time series length: (**a**) area, (**b**) axis angle, (**c**) aspect ratio, (**d**) zonal centroid, (**e**) meridional centroid, and (**f**) total similarity.

It is evident from Figure 8 that the precipitation nowcasting of the LK Optical Flow method for 6–12-min lead times is highly similar to the observations and shows only minimal linear movement in position and shape. Nevertheless, with such characteristics, the deviation between the observations and the nowcasting of the LK Optical Flow method increases from 18 to 30 min. Conversely, the nowcasting generated using the PredRNN model for 18 to 30 min is closer to the observations because of the ability of the model to capture the nonlinear changes in the observations, demonstrating especially good performance for thresholds of 8 and 15 mm.

We calculated attribute values of hourly graded precipitation for thresholds of 0.1, 1, and 2.5 mm (Table 6). The deviations between the observations and the nowcasting generated using the LK Optical Flow method are more pronounced for a 2-h lead time owing to the axial angle, aspect ratio, zonal centroid, and meridional centroid. Meanwhile, with an increase in the precipitation threshold, the PredRNN model trained with the Hea-P dataset with the 8-year time series length gradually increased in total similarity, substantially outperforming the LK Optical Flow method. For the threshold of 2.5 mm, the

PredRNN model maintains a value of total similarity of >0.9, whereas the value for the LK Optical Flow method rapidly drops below 0.75.

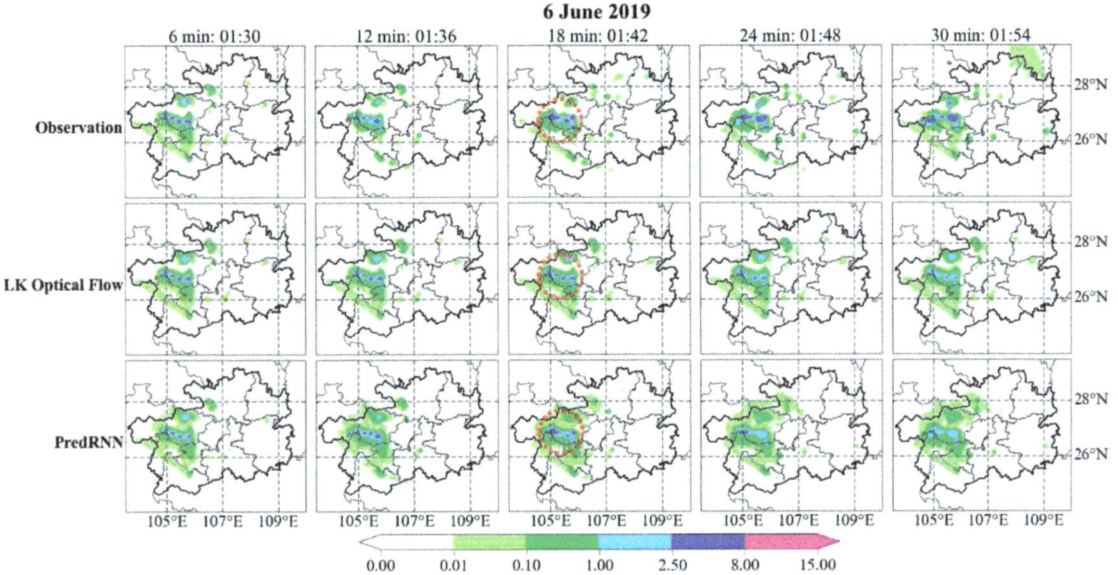

Figure 8. Distributions of observed and predicted precipitation for 6–30-min lead times for the evaluated rainstorm case.

Table 6. Hourly graded precipitation MODE for different thresholds (i.e., 0.1, 1, and 2.5 mm) for the LK Optical Flow method and the PredRNN model trained with the Hea-P dataset with the 8-year time series length.

Precipitation Threshold	Observation or Forecast	Area (Grid Points)		Axial Angle (°)		Aspect Ratio (Width/Length)		Zonal Centroid (°)		Meridional Centroid (°)		Total Similarity	
	Lead time	1 h	2 h	1 h	2 h	1 h	2 h	1 h	2 h	1 h	2 h	1 h	2 h
≥0.1 mm	Observation	474	499	47.98	50.59	0.73	0.81	26.79	26.78	105.86	106.04	—	—
	Optical Flow	372	377	65.17	64.77	0.72	0.74	26.94	27.03	105.67	105.68	0.85	0.81
	PredRNN	418	487	72.74	74.84	0.77	0.89	26.86	26.87	105.7	105.83	0.85	0.88
≥1 mm	Observation	178	265	30.47	−60.85	0.84	0.78	26.75	26.53	105.56	105.9	—	—
	Optical Flow	199	198	58.19	63.99	0.9	0.92	26.93	26.99	105.45	105.44	0.83	0.7
	PredRNN	175	205	41.36	−42.91	0.95	0.7	26.86	26.76	105.54	105.75	0.91	0.81
≥2.5 mm	Observation	105	150	21.5	−48.17	0.7	0.59	26.78	26.63	105.57	105.9	—	—
	Optical Flow	135	135	49.95	55.55	0.88	0.89	26.99	27.08	105.41	105.41	0.72	0.68
	PredRNN	92	123	20.11	−44.88	0.68	0.59	26.82	26.73	105.49	105.81	0.94	0.92

The results of the above quantitative evaluation are well reflected in the spatial distributions of hourly precipitation shown in Figure 9. It is evident that the PredRNN model performs better than the LK Optical Flow method for thresholds exceeding 2.5 mm, i.e., it can well capture the nonlinear movement and evolution of the precipitation for a 2-h lead time. The LK Optical Flow method shows poor performance owing to the minimal changes in the shape and position of the area of precipitation with time, resulting in false predictions exceeding the threshold of 8.0 mm in northern Guizhou.

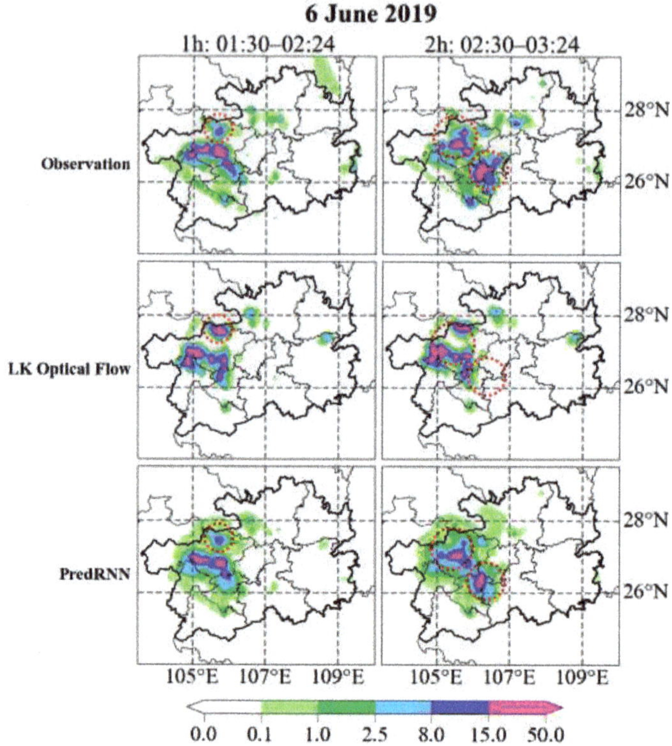

Figure 9. Distributions of observed and predicted hourly precipitation for 6–30 min lead times for the evaluated rainstorm case.

The characteristics of tiny movement and linear evolution mean that the precipitation nowcasting of the LK Optical Flow method is unable to capture nonlinear evolution that does not follow the rule of the latest optical flow vector in the observations. The errors also increased over time owing to the inability of the LK Optical Flow method to adapt to the rapid and extreme evolution of precipitation. However, the PredRNN model trained with the Hea-P dataset can overcome the deficiencies of the traditional LK Optical Flow method, capturing complex nonlinear evolution and establishing a reasonable mapping network during the drastic evolution of precipitation. Thus, the nowcasting of the PredRNN model is closer to the observations, especially for thresholds of 2.5, 8, and 15 mm.

4. Conclusions and Discussion

In this study, we formed two high-quality datasets (the Pre-P dataset and the Hea-P dataset) based on AWS precipitation observation data. Precipitation nowcasting at 6-min intervals for the lead times of 0–2 h was generated for Guizhou using the traditional LK Optical Flow method and the ConvLSTM and PredRNN DL models trained with the Pre-P and Hea-P datasets. Evaluations based on the RMSE, ETS, POD, and FAR were used to assess the performance of the different models in generating precipitation nowcasting. A rainstorm case was evaluated using the MODE approach to further examine the performance of the LK Optical Flow method and the PredRNN model trained with the Hea-P dataset with the 8-year time series length. The results obtained and the conclusions derived were as follows.

Reducing the number of missing values in the AWS observation data improved the quality of the DL training sample data. Data normalization also improved the training efficiency of the DL models. Both processes helped improve the forecasting skill of the DL models. The greater the length of the time series of the training dataset, the better the forecasting skill of the DL model. Increasing the time series length and the number of samples in the training data improved the POD and ETS of the DL models and enhanced the stability of the nowcasting over time. The PredRNN model was most improved for hourly graded precipitation, especially for the 1-h lead time.

Training with the Hea-P dataset further improved the forecasting skill of the DL models and sharply increased the ETSs for thresholds of 2.5, 8, and 15 mm, especially for the 1-h lead time. The improvement in the PredRNN model was greater than that in the ConvLSTM model. The PredRNN model trained with the Hea-P dataset with the 8-year time series length outperformed the traditional LK Optical Flow method for all thresholds (i.e., 0.1, 1, 2.5, 8, and 15 mm) and obtained the best performance in terms of ETS in comparison with the other DL models examined in this study. It also performed high ETSs for thresholds exceeding 2.5 mm in the north-central and southern parts of Guizhou. Additionally, the DL models tended to predict high-value precipitation owing to the high-value precipitation training samples in the Hea-P dataset, which is why most DL models showed a sharp increase in FAR values. The ETS of the ConvLSTM model for the threshold of 2.5 mm increased, but the ETS for the thresholds of 0.1 and 1 mm decreased owing to a sharp increase in the FAR. Owing to the magnified cumulative error through iterative calculation, the ConLSTM model performed many false predictions in high-value precipitation, and the ETS of the ConvLSTM model for thresholds of 8 and 15 mm was close to zero. Unlike radar data, observation instruments and measurement standards for precipitation observation data are unified in the industry. Generally, the spatial resolution of AWS data is higher than that of weather radar. Consequently, the DL models considered in this study could have reasonable transferability to other regions. Thus, the same approach could be used to construct highly accurate precipitation nowcasting DL models based on high-quality observation data.

The rainstorm case considered for evaluation revealed the characteristics of minimal movement and linear evolution in the traditional LK Optical Flow method. The generated precipitation nowcasting exhibited increasing error over time because the latest optical flow vector could not capture subsequent nonlinear evolution. Conversely, the PredRNN model trained with the Hea-P dataset could overcome the deficiencies of the traditional LK Optical Flow method and could capture the complex nonlinear evolution. Thus, the generated precipitation nowcasting was much closer to the observations. Specifically, the PredRNN model outperformed the traditional LK Optical Flow method for the threshold of 2.5 mm in the evaluated rainstorm case.

Currently, DL models using multisource observation data can overcome certain physical constraints and generate radar echoes that are physically more reasonable and of reference significance. For example, Li et al. [26] used a DL model and multisource data to produce radar echoes with physical characteristics that were improved in comparison with those derived using single-source observation data. In future work, DL models with multisource observations that include parameters such as temperature, pressure, and wind speed should be investigated. The inclusion of such dynamic and thermodynamic meteorological information will further improve the forecasting skill of DL models. Additionally, a comparative discussion with radar echo and more reasonable evaluation metrics will be added, such as radially averaged power spectral density [26,60,61].

Author Contributions: D.K. and X.Z. (Xiefei Zhi). contributed to the conception and design of the study. C.Y. and Y.T. contributed to the analysis. X.Z. (Xiaotuan Zeng), G.L., Y.J., D.K. and Y.W. organized the database. All authors contributed to writing and revising the manuscript. All authors have read and agreed to the published version of the manuscript.

Funding: The study was jointly supported by the Natural Science Foundation of Hebei Province, China (Grant No.: D2021304003) and The Meteorological Science, Technology Open Research Fund of Guizhou Meteorological Bureau (Grant No.: KF201704), the Provincial and Municipal Joint Fund

Project of Guizhou Province Meteorological Bureau "Research on precipitation forecasting of short-term in Southwest Guizhou", the Provincial and Municipal Joint Fund Project of Guizhou Province Meteorological Bureau "Research on AI-based intelligent weather monitoring and dispatching technology".

Institutional Review Board Statement: Not applicable.

Informed Consent Statement: Not applicable.

Data Availability Statement: The observation data used in this study were provided by the Guizhou Meteorological Information Center and are not public in nature. The dataset was newly established in this study.

Acknowledgments: The authors thank the Guizhou Meteorological Information Center for its cooperation and data support.

Conflicts of Interest: The authors declare that the research was conducted in the absence of any commercial or financial relationships that could be construed as potential conflict of interest.

References

1. Sun, J. Convective-scale assimilation of radar data: Progress and challenges. *Q. J. R. Meteorol. Soc.* **2005**, *131*, 3439–3463. [CrossRef]
2. Bachmann, K.; Keil, C.; Craig, G.C.; Weissmann, M.; Welzbacher, C.A. Predictability of deep convection in idealized and operational forecasts: Effects of radar data assimilation, orography, and synoptic weather regime. *Mon. Weather Rev.* **2020**, *148*, 63–81. [CrossRef]
3. Wang, T.; Liu, Y.; Dong, C.; Li, J. Review of short-term precipitation forecasting methods and their applications. *Electron. World* **2019**, *10*, 11–13.
4. Bowler, N.E.; Pierce, C.E.; Seed, A.W. STEPS: A probabilistic precipitation forecasting scheme which merges an extrapolation nowcast with downscaled NWP. *Q. J. R. Meteorol. Soc.* **2006**, *132*, 2127–2155. [CrossRef]
5. Pulkkinen, S. PySTEPS: An open-source Python library for probabilistic precipitation nowcasting (v1. 0). *Geosci. Model Dev.* **2019**, *12*, 4185–4219. [CrossRef]
6. Atencia, R.T.; Sairouni, A. Improving QPF by blending techniques at the meteorological service of catalonia. *Nat. Hazards Earth Syst. Sci.* **2010**, *7*, 1443–1455. [CrossRef]
7. Haiden, T.; Kann, A.; Wittmann, C.; Pistotnik, G.; Bica, B. The integrated nowcasting through comprehensive analysis (INCA) system and its validation over the eastern alpine region. *Weather Forecast.* **2011**, *2*, 166–183. [CrossRef]
8. Liang, Q.; Feng, Y.; Deng, W.; Hu, S.; Huang, Y.; Zeng, Q.; Chen, Z. A composite approach of radar echo extrapolation based on TREC vectors in combination with model predicted winds. *Adv. Atmos. Sci.* **2010**, *5*, 1119–1130. [CrossRef]
9. Sun, J.; Xue, M.; Wilson, J.W.; Zawadzki, I.; Ballard, S.P.; Onvlee-Hooimeyer, J.; Joe, P.; Barker, D.M.; Li, P.-W.; Golding, B.; et al. Use of NWP for nowcasting convective precipitation: Recent progress and challenges. *Bull. Am. Meteorol. Soc.* **2014**, *95*, 409–426. [CrossRef]
10. Ravuri, S.; Lenc, K.; Willson, M.; Kangin, D.; Lam, R.; Mirowski, P.; Fitzsimons, M.; Athanassiadou, M.; Kashem, S.; Madge, S.; et al. Skillful precipitation nowcasting using deep generative models of radar. *Nature* **2021**, *597*, 672–677. [CrossRef]
11. Crane, R.K. Automatic cell detection and tracking. *IEEE Trans. Geosci. Electron.* **1979**, *17*, 250–262. [CrossRef]
12. Rasmussen, R.; Dixon, M.; Hage, F.; Cole, J.; Rehak, N. Weather support to deicing decision making (WSDDM): A winter weather nowcasting system. *Bull. Amer. Meteorol. Soc.* **2001**, *82*, 579–596. [CrossRef]
13. Mueller, C.; Saxen, T.; Roberts, R.; Wilson, J.; Betancourt, T.; Dettling, S.; Oien, N.; Yee, J. NCAR auto-nowcast system. *Weather Forecast.* **2003**, *18*, 545–561. [CrossRef]
14. Cheung, P.; Yeung, H.Y. Application of optical-flow technique to significant convection nowcast for terminal areas in Hong Kong. In Proceedings of the 3rd WMO International Symposium on Nowcasting and Very Short-Range Forecasting (WSN12), Rio de Janeiro, Brazil, 6 August 2012; pp. 6–10.
15. Cao, C.Y.; Chen, Y.Z.; Liu, D.H. The optical flow method and its application to nowcasting. *Acta Meteor. Sinica* **2015**, *73*, 471–480.
16. Han, L.; Wang, H.Q.; Lin, Y.J. Application of optical flow method to nowcasting convective weather. *Acta Sci. Nat. Univ. Pekin.* **2008**, *44*, 751–755.
17. Feng, Y.; Kitzmiller, D. A short-range quantitative precipitation forecast algorithm using back-propagation neural network approach. *Adv. Atmos. Sci.* **2006**, *23*, 405–414. [CrossRef]
18. Dixon, M.; Wiener, G. TITAN: Thunderstorm identification, tracking, and nowcasting—A radar-based methodology. *J. Atmos. Ocean. Technol.* **1993**, *10*, 785–797. [CrossRef]
19. Johnson, J.T.; MacKeen, P.L.; Witt, A.; Mitchell, E.D.; Stumpf, G.J.; Eilts, M.D.; Thomas, K.W. The storm cell identification and tracking algorithm: An enhanced WSR-88D algorithm. *Weather Forecast.* **1998**, *13*, 263–276. [CrossRef]
20. Wilson, J.W.; Feng, Y.; Chen, M.; Roberts, R.D. Nowcasting challenges during the Beijing Olympics: Successes, failures, and implications for future nowcasting systems. *Weather Forecast.* **2010**, *25*, 1691–1714. [CrossRef]
21. LeCun, Y.; Bengio, Y.; Hinton, G. Deep learning. *Nature* **2015**, *521*, 436–444. [CrossRef]

22. Peng, T.; Zhi, X.; Ji, Y.; Ji, L.; Tian, Y. Prediction Skill of Extended Range 2-m Maximum Air Temperature Probabilistic Forecasts Using Machine Learning Post-Processing Methods. *Atmosphere* **2020**, *11*, 823. [CrossRef]
23. Zhu, Y.; Zhi, X.; Lyu, Y.; Zhu, S.; Tong, H.; Mamtimin, A.; Zhang, H.; Huo, W. Forecast calibrations of surface air temperature over Xinjiang based on U-net neural network. *Front. Environ. Sci.* **2022**, *10*, 2296–2665. [CrossRef]
24. Zhi, X.; Wang, J. Editorial: AI-based prediction of high-impact weather and climate extremes under global warming: A perspective from the large-scale circulations and teleconnections. *Front. Earth Sci.* **2023**, *11*, 2296–6463. [CrossRef]
25. Pan, X.; Lu, Y.; Zhao, K.; Huang, H.; Wang, M.; Chen, H. Improving Nowcasting of Convective Development by Incorporating Polarimetric Radar Variables Into a Deep-Learning Model. *Geophys. Res. Lett.* **2021**, *48*, e2021GL095302. [CrossRef]
26. Li, D.; Liu, Y.; Chen, C. MSDM v1.0: A machine learning model for precipitation nowcasting over eastern China using multisource data. *Geosci. Model Dev.* **2021**, *14*, 4019–4034. [CrossRef]
27. Shi, X.; Chen, Z.; Wang, H.; Yeung, D.Y.; Wong, W.K.; Woo, W.C. Convolutional LSTM network: A machine learning approach for precipitation nowcasting. *Adv. Neural Inf. Process. Syst.* **2015**, *28*, 1–12.
28. Shi, X.; Gao, Z.; Lausen, L.; Wang, H.; Yeung, D.Y.; Wong, W.K.; Woo, W.C. Deep learning for precipitation nowcasting: A benchmark and a new model. *Adv. Neural Inf. Process. Syst.* **2017**, *30*, 1–17.
29. Wang, Y.; Long, M.; Wang, J.; Gao, Z.; Yu, P.S. Predrnn: Recurrent neural networks for predictive learning using spatiotemporal lstms. *Adv. Neural Inf. Process. Syst.* **2017**, *30*, 1–10.
30. Wang, Y.; Gao, Z.; Long, M.; Wang, J.; Philip, S.Y. PredRNN++: Towards a resolution of the deep-in-time dilemma in spatiotemporal predictive learning. In Proceedings of the International Conference on Machine Learning, Stockholm, Sweden, 10–15 July 2018; pp. 5123–5132.
31. Ji, Y.; Gong, B.; Langguth, M.; Mozaffari, A.; Zhi, X. CLGAN: A GAN-based video prediction model for precipitation nowcasting. *EGUsphere* **2022**, 1–23. [CrossRef]
32. Chen, X.; Liu, J.; Zheng, Q.; Li, X.; Liu, J. A Study on Radar Echo Nowcasting Based on Convolutional Gated Recurrent Unit Neural Network. *Plateau Meteorol.* **2021**, *40*, 1–13.
33. Foresti, L.; Sideris, I.; Nerini, D.; Beusch, L.; Germann, U. Using a 10-year radar archive for nowcasting precipitation growth and decay: A probabilistic machine learning approach. *Weather Forecast.* **2017**, *34*, 1547–1569. [CrossRef]
34. Agrawal, S.; Barrington, L.; Bromberg, C.; Burge, J.; Gazen, C.; Hickey, J. Machine learning for precipitation nowcasting from radar images. *arXiv* **2019**, arXiv:1912.12132.
35. Mao, Y.; Sorteberg, A. Improving Radar-Based Precipitation Nowcasts with Machine Learning Using an Approach Based on Random Forest. *Weather Forecast* **2020**, *35*, 2461–2478. [CrossRef]
36. Germann, U.; Zawadzki, I. Scale dependence of the predictability of precipitation from continental radar images. Part II: Probability forecasts. *J. Appl. Meteorol.* **2004**, *43*, 74–89. [CrossRef]
37. Sun, N.; Zhou, Z.; Li, Q.; Zhou, X. Spatiotemporal Prediction of Monthly Sea Subsurface Temperature Fields Using a 3D U-Net-Based Model. *Remote Sens.* **2022**, *14*, 4890. [CrossRef]
38. Trebing, K.; Stańczyk, T.; Mehrkanoon, S. SmaAt-UNet: Precipitation nowcasting using a small attention-UNet architecture. *Pattern Recognit. Lett.* **2021**, *145*, 178–186. [CrossRef]
39. Gong, B.; Langguth, M.; Ji, Y.; Mozaffari, A.; Stadtler, S.; Mache, K.; Schultz, M.G. Temperature forecasting by deep learning methods. *Geosci. Model Dev. Discuss.* **2022**, *15*, 8931–8956. [CrossRef]
40. Chkeir, S.; Anesiadou, A.; Mascitelli, A.; Biondi, R. Nowcasting extreme rain and extreme wind speed with machine learning techniques applied to different input datasets. *Atmos. Res.* **2022**, *282*, 106548. [CrossRef]
41. Huang, H.; Lin, L.; Tong, R.; Hu, H.; Zhang, Q.; Iwamoto, Y.; Han, X.; Chen, Y.W.; Wu, J. Unet 3+: A full-scale connected unet for medical image segmentation. In Proceedings of the ICASSP 2020–2020 IEEE International Conference on Acoustics, Speech and Signal Processing (ICASSP), Barcelona, Spain, 4–8 May 2020; pp. 1055–1059.
42. Akbari Asanjan, A.; Yang, T.; Hsu, K.; Sorooshian, S.; Lin, J.; Peng, Q. Short-term precipitation forecast based on the persian system and lstm recurrent neural networks. *J. Geophys. Res. Atmos.* **2018**, *123*, 12543–12563. [CrossRef]
43. Ayzel, G.; Heistermann, M.; Sorokinc, A.; Nikitinc, O.; Lukyanovacet, O. All convolutional neural networks for radar-based precipitation nowcasting. *Procedia Comput. Sci.* **2019**, *150*, 186–192. [CrossRef]
44. Su, A.; Li, H.; Cui, L.; Chen, Y. A convection nowcasting method based on machine learning. *Adv. Meteorol.* **2020**, *2020*, 5124274. [CrossRef]
45. Tian, L.; Li, X.; Ye, Y.; Xie, P.; Li, Y. A generative adversarial gated recurrent unit model for precipitation nowcasting. *IEEE Geosci. Remote Sens. Lett.* **2020**, *17*, 601–605. [CrossRef]
46. Han, L.; Sun, J.; Zhang, W. Convolutional neural network for convective storm nowcasting using 3-d doppler weather radar data. *IEEE Trans. Geosci. Remote Sens.* **2019**, *58*, 1487–1495. [CrossRef]
47. Suzanna, M.; Alexandre, E. Precipitation Nowcasting with Weather Radar Images and Deep Learning in São Paulo, Brasil. *Atmosphere* **2020**, *11*, 1157.
48. Yu, L.; Du, X.; Zhao, L.; Yang, H. A new methodology for pixel-quantitative precipitation nowcasting using a pyramid Lucas Kanade optical flow approach. *J. Hydrol.* **2015**, *529*, 354–364.
49. Gan, W.; Li, G.; Wan, X. Variation Characteristics of Extreme Precipitation During May-September in Guizhou Province in Recent 57 Years. *J. Arid. Meteorol.* **2018**, *36*, 617–623.

50. Bai, H.; Gao, H. Influences of the Somaliacross-equatorial flow on the beginning date of rainy season in Southwest China. *Chin. J. Atmos. Sci.* **2017**, *41*, 702–712. (In Chinese)
51. Reich, S. An explicit and conservative remapping strategy for semi-Lagrangian advection. *Atmos. Sci. Lett.* **2017**, *8*, 58–63. [CrossRef]
52. Zhu, S.; Zhang, L.; Jiang, H.; Lyu, Y.; Fan, Y.; Guo, Z.; Zhi, X. Pattern projection calibrations on subseasonal forecasts of surface air temperature over East Asia. *Weather. Forecast.* **2023**. [CrossRef]
53. Zhu, S.; Zhi, X.; Ge, F.; Fan, Y.; Zhang, L.; Gao, J. Subseasonal forecast of surface air temperature using superensemble approaches: Experiments over Northeast Asia for 2018. *Weather. Forecast.* **2021**, *36*, 39–51. [CrossRef]
54. Lyu, Y.; Zhi, X.; Zhu, S.; Fan, Y.; Pan, M. Statistical calibrations of surface air temperature forecasts over East Asia using pattern projection methods. *Weather. Forecast.* **2021**, *36*, 1661–1674. [CrossRef]
55. Davis, C.; Brown, B.; Bullock, R. Object-based verification of precipitation forecasts. Part I: Methodology and application to mesoscale rain areas. *Mon. Weather Rev.* **2006**, *134*, 1772–1784. [CrossRef]
56. Davis, C.; Brown, B.; Bullock, R. Object-based verification of precipitation forecasts. Part II: Application to convective rain systems. *Mon. Weather Rev.* **2006**, *134*, 1785–1795. [CrossRef]
57. Johnson, A.; Wang, X. Object-Based Evaluation of a Storm-Scale Ensemble during the 2009 NOAA Hazardous Weather Testbed Spring Experiment. *Mon. Weather Rev.* **2013**, *141*, 1079–1098. [CrossRef]
58. Ji, L.; Zhi, X.; Simmer, C.; Zhu, S.; Ji, Y. Multimodel ensemble forecasts of precipitation based on an object-based diagnostic evaluation. Mon. *Weather Rev.* **2020**, *148*, 2591–2606. [CrossRef]
59. Min, A.; Liao, Y.; Deng, W.; Zhang, C. A brief description of the main rainstorm weather process in China from April to October 2019. *Torrential Rain Disasters* **2020**, *39*, 539–548.
60. Ayzel, G.; Scheffer, T.; Heistermann, M. RainNet v1.0: A convolutional neural network for radar-based precipitation nowcasting. *Geosci. Model Dev.* **2020**, *13*, 2631–2644. [CrossRef]
61. Huang, Q.; Chen, S.; Tan, J. TSRC: A Deep Learning Model for Precipitation Short-Term Forecasting over China Using Radar Echo Data. *Remote Sens.* **2022**, *15*, 142. [CrossRef]

atmosphere

Article

Variation Characteristics of Temperature and Rainfall and Their Relationship with Geographical Factors in the Qinling Mountains

Yangna Lei [1,2], Rongwei Liao [3,*], Yumeng Su [4], Xia Zhang [1,2], Duanyang Liu [5,*] and Lei Zhang [6]

1 Shaanxi Climate Center, Xi'an 710014, China
2 Key Laboratory of Eco-Environment and Meteorology for the Qinling Mountains and Loess Plateau, Shaanxi Meteorological Bureau, Xi'an 710014, China
3 State Key Laboratory of Severe Weather, Chinese Academy of Meteorological Sciences, Beijing 100081, China
4 Liaoning Meteorological Bureau, Shenyang 110001, China
5 Key Laboratory of Transportation Meteorology of China Meteorological Administration, Nanjing Joint Institute for Atmospheric Sciences, Nanjing 210041, China
6 Tianjin Meteorological Information Centre, Tianjin 300074, China
* Correspondence: liaorw@cma.gov.cn (R.L.); liuduanyang@cma.cn (D.L.)

Abstract: The Qinling Mountains (QMs) are considered to be the division in geology, geochemistry, and physical geography between northern China and southern China. They have crucial effects on regional climate, especially on rainfall and temperature, and have shown great scientific relevance to climate change research in China. Using the observational daily and monthly rainfall and temperature data derived from meteorological and regional automatic stations—as well as the methods of correlation analysis, climate trend analysis, and Mann–Kendal and *t* tests—we revealed the spatiotemporal change characteristics of temperature and rainfall and their correlation with elevation, longitude, and latitude. The results show that the annual mean temperature (AMT) underwent a significant increasing trend in the QMs. The maximum AMT increase occurred in spring, and the minimum occurred in summer. Positive anomalies of annual mean rainfall amount (AMRA) occurred in the 1960s, 1980s, and 2010s, and negative anomalies occurred in the 1970s, 1990s, and 2000s. In the QMs, the amount of moderate rainfall (MR) occupied the maximum proportion and accounted for 27.9% of the AMRA, whereas the torrential rainfall (TR) occupied the minimum proportion and accounted for 12.8%. The AMRA amount significantly decreased by 130.1 mm from the 1980s to the 1990s and accounted for 13.5% of the measure in the 1980s. The AMT and AMRA showed consistent change trends with increases in elevation and latitude and showed the opposite trend as the longitude increased. The results offer a further understanding of the meteorological background of the QMs, helping us in further investigating the potential physical mechanisms that influence the spatiotemporal distribution characteristics of temperature and rainfall in the QMs. This study will provide a scientific basis for rainfall and temperature forecasts, with relevance to local ecosystems, agriculture, soil erosion, and the prevention and mitigation of floods in the future.

Keywords: Qinling mountains; temperature; rainfall; change characteristics; geographical factors

Citation: Lei, Y.; Liao, R.; Su, Y.; Zhang, X.; Liu, D.; Zhang, L. Variation Characteristics of Temperature and Rainfall and Their Relationship with Geographical Factors in the Qinling Mountains. *Atmosphere* **2023**, *14*, 696. https://doi.org/10.3390/atmos14040696

Academic Editor: Tomeu Rigo

Received: 31 January 2023
Revised: 17 March 2023
Accepted: 31 March 2023
Published: 7 April 2023

1. Introduction

Global warming has increased the scope of climate system changes, leading to frequent extreme rainfall events, high temperature events, and drought events, which seriously threaten the safety of global ecological and environmental systems [1–4]. According to the *Sixth Assessment Report* of the Intergovernmental Panel on Climate Change (IPCC), human activities have affected the climate to various degrees in different ways. Compared to pre-industrialization levels (1850–1900), the increase in global temperature from 2010 to 2019 was about 0.8–1.3 °C, which has had a profound impact on social development [5].

The Qinling Mountains (QMs), located predominantly in the south of Shanxi province in central China, are considered to be a division in geology, geochemistry, and physical geography between northern China and southern China [6]. Furthermore, as an important geographical boundary between central and eastern China, the QMs are situated at the edge of the Asian monsoon region and are sensitive to climate change [7,8]. Thus, they represent a climate transition belt where the typical subtropical zone gradually shifts into a warm temperate zone from south to north and the humidity shifts towards a semi-humid zone from east to west [9].

The effects of the QMs on regional climate, especially rainfall and temperature, have been shown to have great scientific relevance for climate change research in China [10–12]. As a key region in the study of regional climate change in China, the changes experienced in the climate elements in the QMs in the past decades have been the research focus of many scholars. In previous studies, both increasing and decreasing trends in annual rainfall have been reported in northwest China [13–15]. Meng et al. [16] indicated a declining trend in annual rainfall in the QMs. Seasonally, a decreasing trend was also observed in spring and autumn, while increasing trends were observed in summer and winter. Spring and autumn rainfall significantly contributed to this observed decline in annual rainfall. Li et al. [17] concluded that the intensity of extreme rainfall increased in the Qinling–Daba Mountains. Shao et al. [18] pointed out that most extreme rainfall indices decreased in spring, autumn, and winter and increased in summer in the Qinling–Daba Mountains. Li et al. [6] revealed that the Qinling–Daba Mountains have an obvious effect on both the spatial–temporal distribution and diurnal cycle of regional rainfall. Zhang et al. [4] used CMIP 6 data to estimate future rainfall changes in the QMs and revealed the basic characteristics of the atmospheric water cycle in mountainous areas under the action of monsoons as well as the temporal and spatial variation mechanisms of water resources in the "central water tower".

Mo et al. [19] simulated the temperature field in the QMs by constructing a digital elevation model (DEM) map. Liu et al. [20] analyzed and compared the temperature of the QMs over the past 200 years using the tree ring statistical method. Bai et al. [21] concluded that the isotherm found in January in the QMs has gradually moved northward in the past 50 years. Bai et al. [22] found that the trend of climate change and the time points of abrupt climate change were consistent over the northern and southern slopes in the QMs. Li et al. [23] indicated that the total rainfall amount has decreased, whereas the amount of extreme rainfall has increased, based on eight extreme rainfall indexes in the northern and southern QMs. According to the study of Zhang et al. [24], the rainfall in the QMs has shown a downward trend in the last 50 years, and the 800 mm rainfall contour line has clearly moved. Zhang et al. [25] found that the rainfall belt in the QMs has moved over the last 40 years compared with the standard period. The study of Gao et al. [26] showed that the climate in the QMs has undergone a warming and humidifying trend.

In addition, these studies on climate change in the QMs are mainly based on the data obtained from more than 30 meteorological stations in recent years, and different results show that the temperature has undergone an upward shift [25,27,28]. Furthermore, there is a huge elevation difference of more than 3000 m in the QMs, and the meteorological data for the high-elevation regions are based on the conventional vertical lapse rate of temperature and the data derived from low-altitude stations. However, these extrapolation methods cannot fully reflect the complex variability in temperature and rainfall in the QMs, and it is necessary to obtain data from higher-elevation stations for supplementary correction [29].

Therefore, studying the rules of trends, as well as their attributions at different altitudes and different time scales in the QMs, is essential to exploring climate change in China. In this study, climate trend analysis, mutation tests, spatial interpolation, etc., were applied to determine the rainfall and temperature trends as well as their hidden values and to analyze the correlations with geographic factors, such as altitude, longitude, and latitude in the QMs so as to determine the influence of geographical factors on mountain climate. The aim of the work is to understand the response of regional climate change to global warming; the results of this study will enhance the scientific basis of guidelines regarding how to deal with future climate change and promoting sustainable development and ecological protection.

This paper is organized as follows: The details of the datasets and methodology are given in the Materials and Methods section. The rainfall and temperature trends on annual and seasonal scales, and analyses of their correlations with geographic factors, are given in the Results section. The Discussion and Conclusions sections are given at the end.

2. Data and Methods

2.1. Data

In this study, the QM region refers to the mountains between the Weihe River and the Hanjiang River in the south of Shaanxi Province, bounded by the Bahe River and the Danjiang River Valley in the east and ending at the Jialing River in the west. The range of the QMs is 32.42° N–35.27° N, 103.8° E–113.07° E. We used the daily and monthly mean temperature and rainfall gauge data from 32 national surface weather stations during the period 1961–2021 and annual mean temperature and rainfall gauge data from 406 regional automated stations during the period 2020–2021. All the data were used to analyze climate change in the QMs and were subjected to quality control (QC). The QC procedures for the current gauge data include the station information check, the missing value and eigenvalue check, the time consistency check, the climate extreme value behavior check, the spatial consistency check, and the interior consistency check. The spatial distribution of meteorological stations and the study area are shown in Figure 1. DEM data at a 30 m resolution were downloaded from the National Science Data Mirroring Website of the Computer Network Information Center, Chinese Academy of Science (http://www.gscloud.cn, accessed on 1 March 2022) [30]. Moreover, the periods of March–May, June–August, September–November, and December–February represent spring, summer, autumn, and winter, respectively.

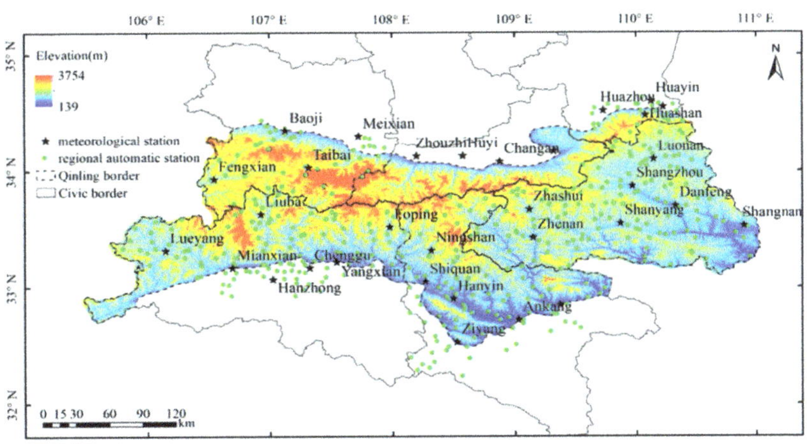

Figure 1. Spatial distribution of meteorological stations in the Qinling Mountains.

2.2. Methods

2.2.1. Climate Trend Analysis

The function of climate trend analysis is primarily to calculate and analyze the tendencies and rates of meteorological elements via the following formula [31]:

$$Y_i = a_0 + a_1 t_1 \quad (1)$$

In this study, Y_i is the value of the meteorological element, t_1 is the time (1961–2021), a_1 is the linear trend (that is, the annual climate tendency rate), and a_0 is the constant term.

2.2.2. Mann–Kendall (M-K) Test

The Mann–Kendall (M-K) method is a detection method on the basis of non-parametric statistics proposed by Mann et al. [32] and Kendall et al. [33]. It can infer overall distribution through the analysis of sample data, introducing the inverse sequence calculation, which can be applied to the detection of the mutation and can reflect the exact location of the mutations [34]. It is widely employed to detect monotonic trends in the time series of hydrometeorological variables, including temperature [35], streamflow [36], and rainfall [37]. This method does not require that the detected data adhere to a specific distribution, and as such, this method requires no assumptions about the data that need to be tested [34].

In this study, the M-K method is used to test the abrupt change in temperature and rainfall series. Its principle is to construct a rank sequence S_k in chronological order for element sequence x:

$$S_k = \sum_{i=1}^{k} r_i, \qquad k = 2, 3, \cdots, n \tag{2}$$

where

$$r_i = \begin{cases} 1, & for\ x_i > x_j, \\ 0, & for\ x_i \leq x_j, \end{cases} \qquad j = 1, 2, \cdots, i.$$

The statistics UF_k are defined under the assumption of random independence of time series:

$$UF_k = \frac{|S_k - E(S_k)|}{\sqrt{var(S_k)}} \tag{3}$$

where $UF_1 = 0$, $E(S_k)$, is the mean of the cumulative S_k; $var(S_k)$ is the variance of the cumulative S_k.

$$\begin{cases} E(S_k) = \dfrac{k(k-1)}{4}, \\ var(S_k) = \dfrac{k(k-1)(2k+5)}{72}. \end{cases} \tag{4}$$

Variable UF_k obeys normal distribution, and different significance levels are set to determine whether the trend of variable UF_k is significant in the confidence interval. The element sequence x is arranged in reverse chronological order, and the above process is repeated with $UB_k = -UF_k(k = n, n-1, \cdots, 1), UB_1 = 0$. By analyzing the trend of UF_k and UB_k, the trend of element sequence x can be obtained and the time of its mutation can be determined. If $UF_k > 0$, it indicates that the sequence tends to rise; otherwise, it declines. If the values of UF_k and UB_k are greater than the critical value of a significance level, the sequence shows a significant trend. If there is an intersection point between UF_k and UB_k, the position of the intersection point is the place at which the mutation occurs [38]. However, this method has some drawbacks. In the case of multiple mutation points or multiple scale mutations in the sequence, this method is not suitable to be applied [34].

2.2.3. Running *t*-Test

Considering some disadvantages of the Mann–Kendall test, we also used the running *t*-test method to test the abrupt change of temperature and rainfall series at the same time. The basic idea of the running *t*-test is based on the significance test; to determine if two samples will occur as mutations, one must analyze whether the difference in the two samples' mean values is obvious or not [34]. If the difference is greater than the given significance level, the two samples exist with obvious qualitative changes.

The principle of this operation is as follows: x is the time series, and n is the number of samples; a certain time point is artificially set as the reference point, and n_1 and n_2 are the numbers of samples before (x_1) and after (x_2) the reference point; t meets the distribution of $t(n_1 + n_2 - 2)$. $\overline{x_1}$ and $\overline{x_2}$ are the mean values of x_1 and x_2, respectively; S_1^2 and S_2^2 are the variances of x_1 and x_2, respectively. The test statistics (t) are calculated as follows [34,38]:

$$t = \frac{\overline{x_1} - \overline{x_2}}{S\sqrt{\frac{1}{n_1} + \frac{1}{n_2}}} \sim t(n_1 + n_2 - 2) \quad (5)$$

$$S = \sqrt{\frac{(n_1 - 1)S_1^2 + (n_2 - 1)S_2^2}{n_1 + n_2 - 2}} \quad (6)$$

For a significance level of α, we can calculate the test statistic t_α, and if $|t| > t_\alpha$, a mutation exists. However, according to the descriptive analysis of test methods and the previous research experience, the running *t*-test is relatively appropriate for the recognition of mean value type mutations [34,38].

2.2.4. Kriging Interpolation

Kriging interpolation is the method of interpolation deriving from regionalized variable theory, it is a geostatistical interpolation method and an optimal method for estimating regional spatial differences based on the spatial variation of the property in terms of the variogram [39–41]. Kriging interpolation obtains the estimated values of unknown points using known point data by considering the spatial relationship between sample points, using the variogram calculation and structural information. There is no boundary effect in the region and the output surface is smooth. The formula of calculation is as follows [39,41–43]:

$$Z(x_0) = \sum_{i=1}^{k} \lambda_i Z(x_i) \quad (7)$$

where $Z(x_0)$ is the estimated value of the meteorological point; λ_i is the weight coefficient of the measured sample point *i*; $Z(x_i)$ is the value of the known meteorological station. Kriging interpolation is widely used in the study of variables with spatial correlation [44–46].

In this study, monthly and annual $0.1° \times 0.1°$ grid data are generated using Kriging interpolation, and a correlation analysis between temperature, precipitation data, and the geographical factors is subsequently applied. Statistical significance has been assessed using Student's *t*-test. All the significance values are at the 95% confidence level unless otherwise stated.

3. Results

3.1. Temporal Characteristics of Temperature and Rainfall

3.1.1. Interannual and Interdecadal

The time series of regionally averaged annual mean temperature (AMT) (Figure 2) in the QMs during the period from 1961–2021 shows that the AMT in the QMs has increased significantly over the past 61 years, and the temperature tendency rate (TTR) is 0.22 °C/10a ($p = 0.05$). The maximum annual temperature was 14.4 °C, which occurred in 2013. In that year, an anomalous long-term high temperature occurred in the QMs, the subtropical high was stronger to the north (Figures not shown), and the atmospheric circulation anomaly may be the reason for the temperature increase in the QMs [47]. On the other hand, the minimum value was 12.2 °C, which occurred in 1984. In that year, the La Niña event triggered a strong East Asian winter monsoon, the Siberian high pressure was strengthened, and the cold air was active in East Asia (Figures not shown), so the minimum temperature occurred in the QMs [48,49]. According to the five-year moving mean temperature, the upward trend of temperature was relatively slow before the 1980s, and the TTR value was 0.09 °C/10a. Then, the AMT gradually increased from the 1980s, and the TTR was 0.54 °C/10a during the period 1981–1998. After that, the increasing trend slowed down, and the TTR was 0.13 °C/10a during the period 1999–2021.

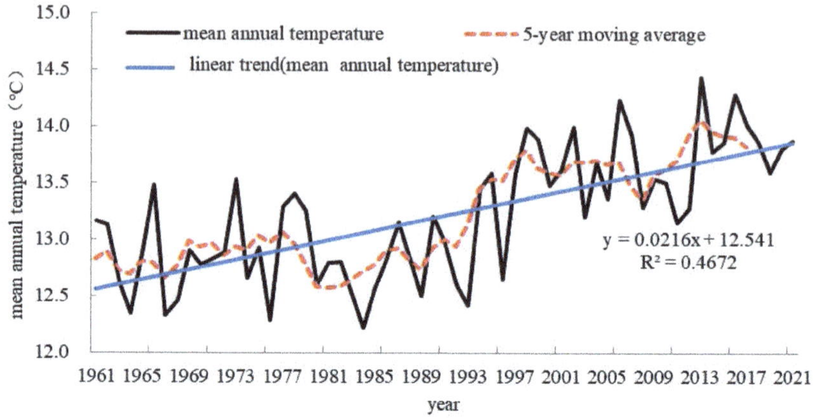

Figure 2. Time series of regionally averaged annual mean temperature (AMT) (unit: °C) from 32 stations in the QMs during the period 1961–2021. The black line denotes the temperature curve, the red dashed line denotes the 5-year moving average curve, and the blue line denotes the linear trend curve.

Figure 3 shows the time series of regionally averaged annual mean rainfall amount (AMRA) in the QMs during the period 1961–2021; it is noted that the trend of AMRA in the QMs over the last 61 years was not significant. The maximum AMRA was 1184.2 mm, which occurred in 2021. In that year, the anomalous plateau upper trough and the anomalous subtropical high affected the rainfall in the QMs [50]. On the other hand, the minimum value was 492.1 mm, which occurred in 1997. In that year, the occurrence of El Niño phenomenon caused the Western Pacific subtropical high to move northward and retreat southward rapidly, which was the reason for the negative rainfall anomaly in the QMs [51]. The variability in types of rainfall is consistent with the variability in AMRA, with large interannual fluctuation, but this trend is not significant (Figures not shown).

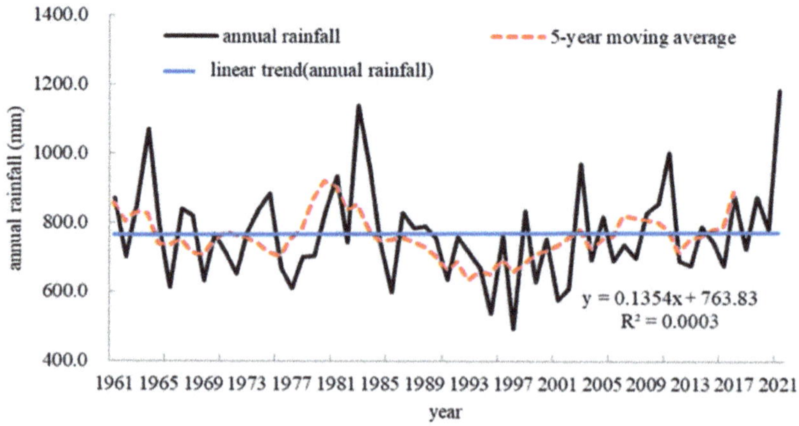

Figure 3. Time series of regionally averaged annual mean rainfall amount (AMRA) (unit: mm) from 32 stations in the QMs during the period 1961–2021. The black line denotes the rainfall curve, the red dashed line denotes the 5-year moving average curve, and the blue line denotes the linear trend curve.

Table 1 shows the interdecadal statistics of regionally averaged AMT in the QMs, indicating that negative temperature anomalies occurred in the 1960s, 1970s, and 1980s. However, a positive temperature anomaly occurred after the 1990s. The differences seen in the temperature anomalies in the 1960s, 1970s, 1980s, and 2010s were 1.0 °C ($p = 0.05$), 0.9 °C ($p = 0.05$), and 1.1 °C ($p = 0.05$), respectively. This indicates that the temperature has increased significantly across the QMs in the past 51 years.

Table 1. The regionally averaged AMT in QMs for different periods.

Variable	1960s	1970s	1980s	1990s	2000s	2010s
AMT (°C)	12.8	13.0	12.7	13.3	13.6	13.8
AMT anomaly (°C)	−0.4	−0.2	−0.5	0.1	0.4	0.6

To further understand the features of rainfall, we have examined the different classifications of rainfall in the QM region. Here, we have divided the daily rainfall into five types: light rainfall (LR; 0.1–9.9 mm), moderate rainfall (MR; 10.0–24.9 mm), heavy rainfall (HR; 25.0–49.9 mm), torrential rainfall (TR; 50.0–99.9 mm), and downpour rainfall (DR; ≥100.0 mm); after this, we calculated the monthly mean values of LR, MR, HR, TR, and DR during the period from 1961–2021.

As shown in Table 2, regarding the regionally averaged differences in types of rainfall in the QMs during the period from 1961–2021, the MR type represents the maximum proportion, accounting for 27.9% of the AMRA, while the TR type represents the minimum proportion, accounting for 12.8% of the AMRA. Moreover, LR, HR, and DR accounted for 24.0%, 21.5%, and 13.8%, respectively.

Table 2. The regionally averaged light rainfall (LR), moderate rainfall (MR), heavy rainfall (HR), torrential rainfall (TR), and downpour rainfall (DR) in the QMs during the period 1961–2021.

Period (Years)	LR (mm)	MR (mm)	HR (mm)	TR (mm)	DR (mm)
1961–2021	218.9	254.3	196.2	116.9	125.4

As shown in Table 3, positive AMRA anomalies occurred in the 1960s, 1980s, and 2010s, and the value of the difference exceeded 50 mm in the 1980s. Negative AMRA anomalies occurred in the 1970s, 1990s, and 2000s, and the value of the difference was −76.4 mm in the 1990s. The AMRA significantly decreased by 130.1 mm from the 1980s to 1990s, accounting for 13.5% in the 1980s; this indicates a declining trend in the AMRA in the QMs from the 1980s to the 1990s, which is also supported by previous studies, which reported an obvious change in the QMs after the 1980s, with a declining rainfall trend over the QMs [16]. In addition, a trend of increase in the AMRA was seen from the 1990s to 2010s, and it accounted for 13.0% in the 1990s. Furthermore, a positive LR anomaly occurred during the period 1960–1980, after which a negative LR anomaly occurred. Negative MR and TR anomalies occurred in the 1990s and 2000s, and positive anomalies occurred in the 2010s. Negative HR and TR anomalies occurred in the 1990s, and positive anomalies occurred after the 2000s. The above analysis shows that the increase in the AMRA across the QMs in the 2010s was mainly caused by increases in MR, HR, TR, and DR; in addition, the LR had been lower than the mean rainfall since the 1990s.

According to the above analysis, the results suggest that the annual rainfall trend was declining in the QMs before the 2000s; this result is consistent with those of Meng et al. [52] and Wang et al. [14]. In addition, we found an increasing trend in the AMRA in the QMs from the 1990s to the 2010s and a declining trend from the 1980s to the 1990s. Global warming [53], Pacific decadal oscillation (PDO) [53], Atlantic multidecadal oscillation (AMO) [53], and Asian–Pacific oscillation (APO) [54] might be the reason for AMRA change in the QMs.

Table 3. The regionally averaged annual mean rainfall amount (AMRA), light rainfall (LR), moderate rainfall (MR), heavy rainfall (HR), torrential rainfall (TR), and downpour rainfall (DR) in the QMs in different periods.

Variable	1960s	1970s	1980s	1990s	2000s	2010s
AMRA (mm)	941.9	885.8	965.5	835.4	898.4	943.9
AMRA anomaly (mm)	30.1	−26.0	53.7	−76.4	−13.5	32.1
LR (mm)	236.0	219.3	227.7	203.2	214.6	212.8
LR anomaly (mm)	17.1	0.4	8.8	−15.7	−4.4	−6.2
MR (mm)	279.7	246.4	266.3	233.4	240.7	259.5
MR anomaly (mm)	25.5	−7.9	12.1	−20.8	−14.2	5.3
HR (mm)	195.3	180.1	219.8	167.5	199.1	215.5
HR anomaly (mm)	−0.9	−16.1	23.6	−28.7	2.9	19.2
TR (mm)	111.6	112.4	131.1	109.3	112.2	125.0
TR anomaly (mm)	−5.3	−4.5	14.1	−7.7	−4.7	8.1
DR (mm)	119.3	127.6	120.6	122.0	131.8	131.1
DR anomaly (mm)	−6.1	2.2	−4.8	−3.4	6.4	5.7

3.1.2. Seasonal

To further determine the seasonal features of temperature and rainfall, we have examined the different seasonal features of AMT and AMRA in the QMs (Table 4). The results show that the AMT increased from spring to winter, and the tendency rates were 0.33 °C/10a (p = 0.01), 0.07 °C/10a, 0.20 °C/10a (p = 0.05) and 0.28 °C/10a (p = 0.05), respectively. The maximum increase in AMT occurred in spring, and the minimum occurred in summer. In addition, the polar vorticity index and Atlantic SST index were the most relevant circulation index and climate index, respectively, to seasonal AMT. These two indices revealed the trend of climate warming in the past 50 years and might be related to the significant warming in the QMs [55]. Furthermore, the significant warming might also be associated with the warm phase of Atlantic multidecadal oscillation (AMO) [56].

Table 4. The regionally averaged AMT and AMRA in the QMs during different seasons in the period 1961–2021.

Variable	Spring	Summer	Autumn	Winter
AMT (°C)	13.8	23.9	13.2	2.0
Tendency rate (°C/10a)	0.33 ***	0.07	0.20 **	0.28 **
AMRA (mm)	240.0	512.9	401.0	34.8
Tendency rate (mm/10a)	−6.35	8.14	−1.10	0.66
LR (mm)	61.2	68.9	69.0	20.3
LR tendency rate (mm/10a)	−1.77	−1.53	−1.46	0.06
MR (mm)	66.1	107.0	80.3	14.5
MR tendency rate (mm/10a)	−3.36	2.06	−2.78	0.10
HR (mm)	48.6	113.5	70.6	–
HR tendency rate (mm/10a)	−0.83	3.03	0.33	–
TR (mm)	64.1	98.8	72.9	–
TR tendency rate (mm/10a)	−0.36	2.06	0.41	–
DR (mm)	–	124.7	108.1	–
DR tendency rate (mm/10a)	–	1.16	–	–

(Note: the superscript "*" indicates the following—*** p = 0.01, ** p = 0.05; the 10a stands for 10 years).

Furthermore, the trend of increase in the AMRA manifested in winter and summer in the QMs, with the maximum increase in summer and the minimum increase in winter. In spring and autumn, decreases in AMRA occurred, with the maximum decreasing trend seen in spring and the minimum decreasing trend seen in autumn. Zuo et al. [57] and Zhao et al. [58] noted that the spring AMRA decreased in the QMs might be caused by a significant decrease in snow cover in Eurasia and a strengthening northerly wind from East Asia in spring. Gu et al. [59] indicated that the decrease in AMRA in autumn in the QMs might be closely related to SST anomalies in the tropical Pacific. In addition, the increased AMRA in winter and summer were related to global warming, Pacific decadal oscillation (PDO), and Atlantic multidecadal oscillation (AMO) [53]. Regarding the seasonal distributions of the different types of rainfall, MR, HR, TR, and DR showed increasing trends in summer, but LR did not, so we can infer that the increase in AMRA in summer was mainly caused by the increases in MR, HR, TR, and DR. In winter, LR and MR showed increasing trends. In spring, MR, HR, TR, and DR showed decreasing trends, which is consistent with the decreasing trend of AMRA. In autumn, LR and MR showed decreasing trends, while HR and TR showed increasing trends; as such, we can infer that the decrease in AMRA in autumn was mainly caused by the decreases in LR and MR. Moreover, the change in temperature and specific humidity in the lower atmosphere were strongly associated with the decrease in the frequency of LR occurred in summer. The increasing temperature was considered to reduce the frequency of LR occurred [60].

In addition, previous studies suggested that the annual and seasonal rainfall in the QMs was influenced by climate anomalies and geographical factors [61], which could be due to atmospheric circulation anomalies, the Asian monsoon anomaly, or a combination of factors (i.e., vegetation cover percentage, direction of slope, degree of slope, and so on) [62,63].

3.2. Spatial Characteristics of Temperature and Rainfall

3.2.1. Spatial Distribution of Temperature

Figure 4A shows a "dipole-type" spatial pattern in the AMT, which was distributed from southeast to northwest in the QMs. The highest AMT was measured in Ankang and its neighboring regions, with maximum values of above 15 °C. Mean temperatures of 14.0–15.0 °C were measured in Mianxian, Yangxian, Hanzhong, Shangnan, Shangluo, and Danfeng. Taibai in the northwestern region of the QMs and Luonan in the northeast yielded the minimum AMT, with values of about 8.0 °C; the difference between the maximum and minimum AMT was about 7.0 °C. In the QM region, the AMT showed an increasing trend (Figure 4B). The greatest increases in AMT occurred in Zhen'an, Zhashui, Fuping, and Baoji, with a rate of over 0.3 °C/10a, and the smallest increases were measured in Shiquan, Ankang, Hanyin, and Ziyang, primarily distributed in the southern region of the QMs, with increase rates below 0.1 °C/10a. The above results show obvious regional differences in the increases in AMT; the increase rate was higher in the northwestern and central regions of the QMs and could be associated with Figure 1 for elevation. That is, the AMT increase rate appeared to be greater at higher elevations, which was basically consistent with the analysis results of Dong et al. [64].

Over the past 50 years, the changing temperature trends over the northern and southern regions of the QMs have been obvious and synchronous, with the warming process manifesting a "non-smooth, nonlinear, and ladder-shaped" pattern. The spatial variation in temperature is characterized by "synchronous warming and differential north–south change" [63]. The QM region displays differences in temperature in response to global warming over the north and south. The northern boundary of the north subtropical zone extends upwards along the southern QMs, whereas the warming zone extends in the form of an enclave into the northern QMs due to rapid urbanization and mountain blocking [63,65].

3.2.2. Spatial Distribution of Rainfall

As can be seen in Figure 5A, the greatest AMRA was measured in the southwest, while lower levels occurred in the northeast of the QMs. The maximum AMRA values were measured in Foping, Ningshan, Shiquan, Hanyin, and Ziyang, which were above 1000 mm. On the contrary, the minimum AMRA values were measured in Baoji, Xi'an, and Weinan, located in the north of the QMs, and were below 600 mm. The difference between the maximum and minimum AMRA was over 400 mm. The maximum trends of increase in AMRA were measured in Shangnan and Danfeng, with rates of 10.1–18.8 mm/10a (Figure 5B).

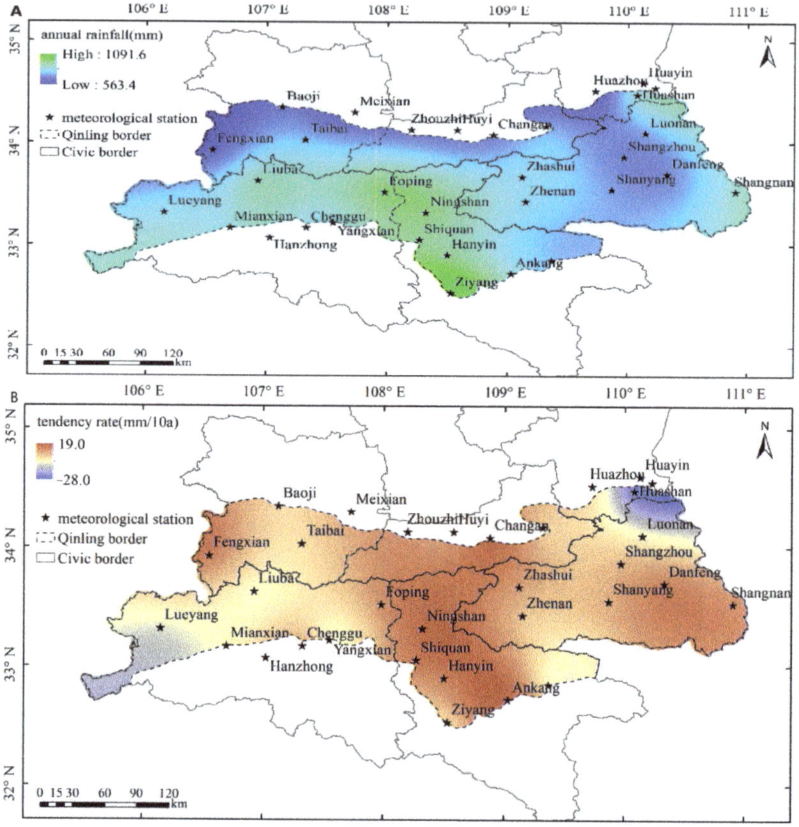

Figure 4. (**A**) The spatial distribution of AMT (unit: °C) in the QMs averaged from 1961 to 2021. (**B**) The spatial distribution of linear variation in AMT trends (unit: °C/10a) in the QMs.

As shown in Figure 6, the greatest distribution of LR occurred in the southwest of the QMs, and the least occurred in the northeast. The total amount of LR type in the QM region was above 170.8 mm, with the highest measurements in Liuba, Ziyang, Foping, and Ningshan and the smallest in Huayin, Huazhou, and Tongguan. More MR type was measured in the central QM region, and less occurred in the eastern and western regions; its total value was above 195.6 mm, with the highest values measured in Ankang and Foping in the east of Hanzhong and Zhashui and Zhen'an in the west of Shangluo. The smallest values measured were in the west of Baoji and the east of Shangluo. More HR and TR were measured in the south of the QMs, and less were measured in the remaining regions; the highest values occurred in Ziyang, Foping, Ningshan, and Hanyin, and the smallest values occurred in the north and east of the QMs. In addition, a decreasing LR trend was seen across most of the QM region, while increasing trends were seen in Zhashui and Chenggu. The increases in MR, HR, and TR occurred in the central QM region, with decreases in the remaining regions.

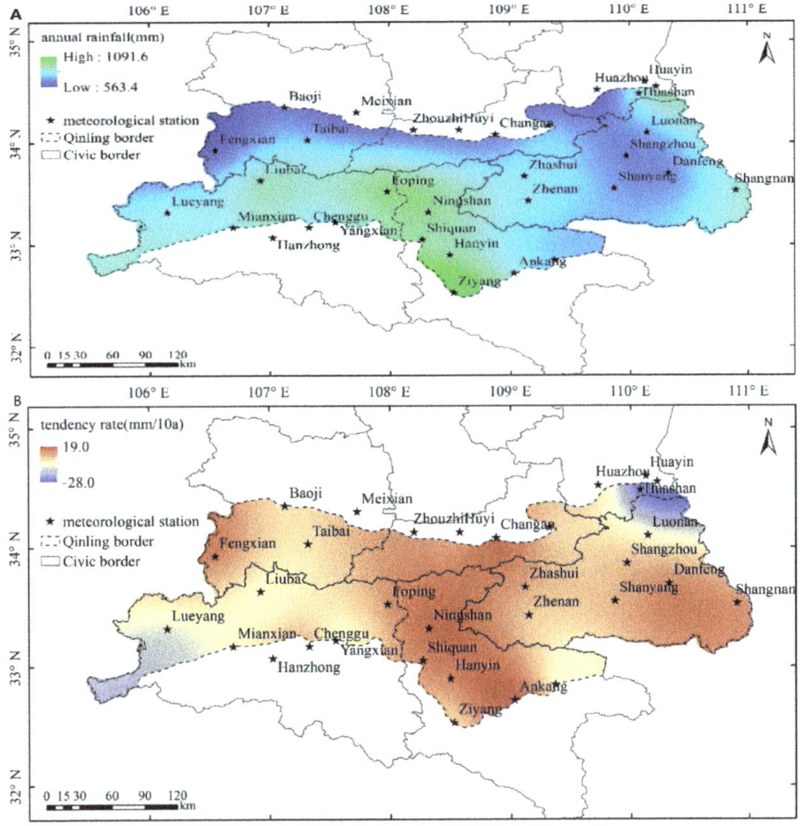

Figure 5. As in Figure 4, but for AMRA (**A**) (unit: mm) and AMRA trends (**B**) (unit: mm/10a).

3.3. The Abrupt Changes in Temperature and Rainfall

In this study, we used the M-K test and running *t*-test to examine dramatic changes in the spatiotemporal characteristics of temperature. As can be seen from the results of the M-K test (Figure 7A) and running *t*-test (Figure 7B) applied to the AMT time series across the QMs, the most abrupt change occurred in 1997—that is, a significant shift in the AMT occurred in 1997. The AMT showed an increasing trend after 1997. In fact, the most significant increase trend was seen after 2002 ($p = 0.05$), signifying that the AMT in the QMs has increased significantly since this time. In 1997, the occurrence of El Niño phenomenon led to a global atmospheric circulation anomaly, subtropical high anomaly, and hot and dry weather [51]. After that, solar activity, volcanic activity, human factors, greenhouse gas emissions, and land use changes played a major role in climate warming, and the increase in AMT in the QMs continued [66]. It can be seen from Figure 8, regarding the M-K test (Figure 8A) and running *t*-test (Figure 8B) applied to the annual DR time series across the QMs, that the year of most abrupt change was 1984; that is, a significant shift in the annual DR occurred in 1984. The rainfall anomaly in the QMs might be associated with sea surface temperature (SST) anomaly over the central-eastern equatorial Pacific and the North Atlantic [67]. In 1984, an SST anomaly occurred in the east-central equatorial Pacific, accompanied by the Walker circulation and meridional circulation anomalies. Additionally, the DR increased in the QMs [67]. In addition, the time series of AMRA, LR, MR, and HR showed no obvious year of abrupt change (Figures not shown), and they generally remained stable. Other previous studies have indicated that various external forcing factors, such as solar radiation, greenhouse gases, and land use, could cause the abrupt changes in the QMs during the period from 1961–2021 [68].

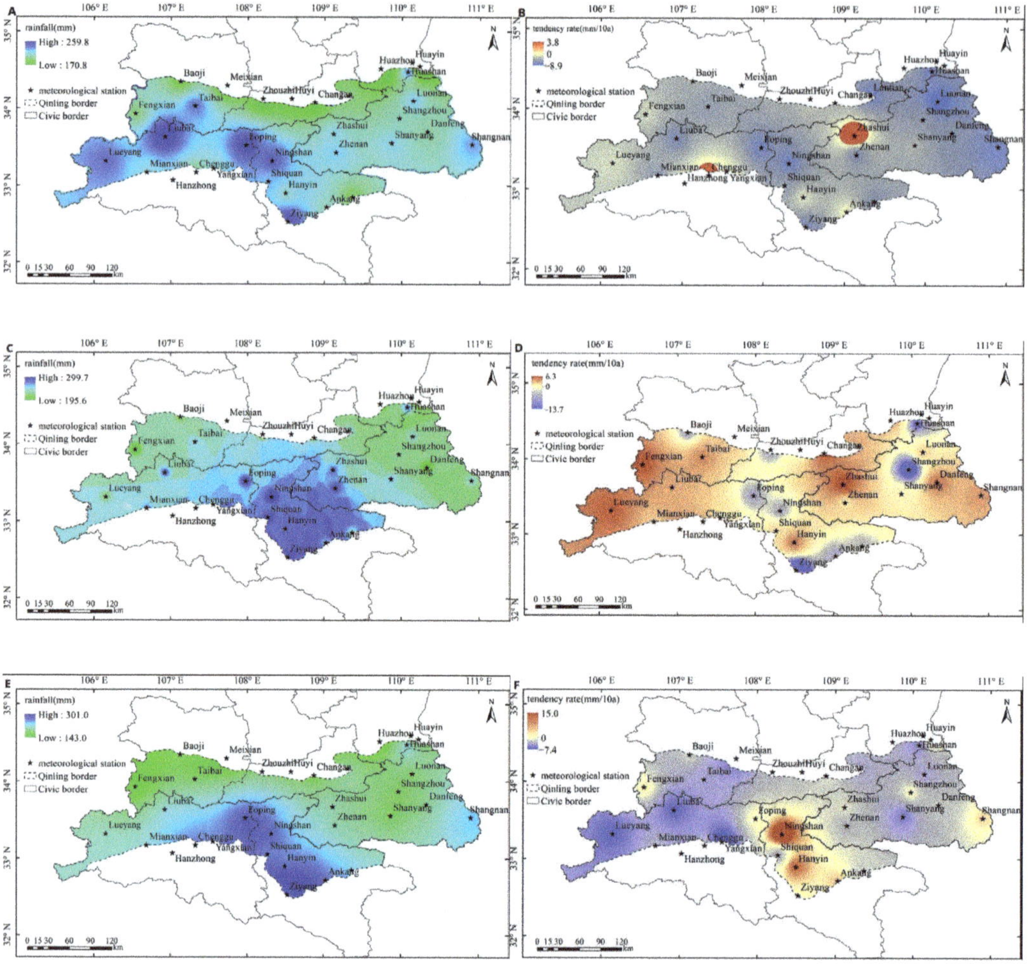

Figure 6. As in Figure 5A, but for LR (**A**), MR (**C**), HR (**E**), and TR (**G**) (unit: mm). As in Figure 5B, but for LR (**B**), MR (**D**), HR (**F**), and TR (**H**) trends (unit: mm/10a).

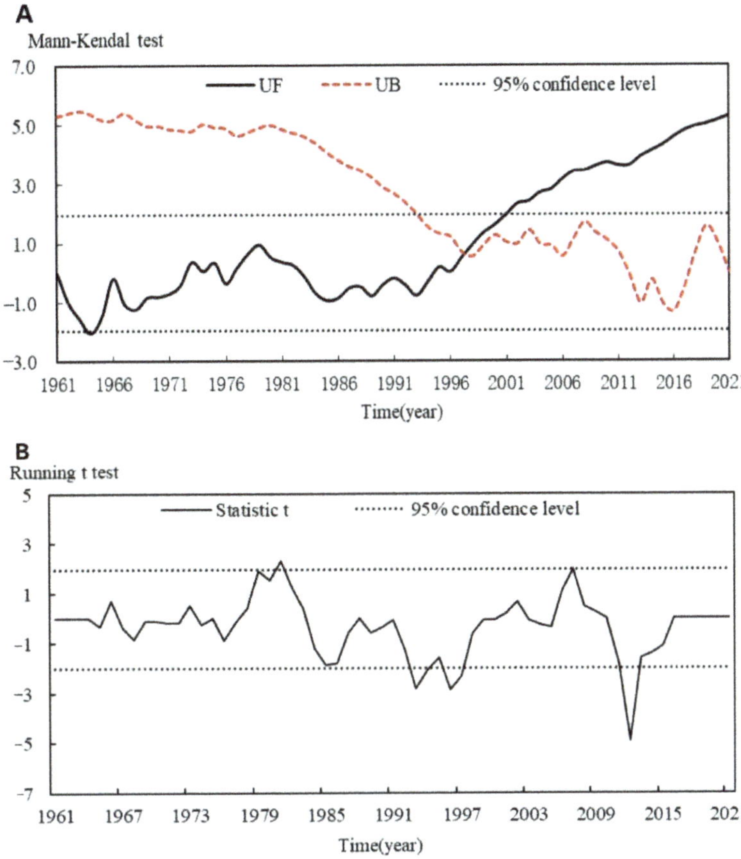

Figure 7. Mann–Kendall (**A**) and running *t*-tests (**B**) for the time series of the AMT over the QMs during the period from 1961–2021. The two black dashed lines indicate the 95% confidence level of the two tests. In (**A**), the black line denotes the sequential statistical curve, UF, and the red dashed line denotes the reverse statistical curve, UB. The black lines in (**B**) denote the sequential statistical curve t.

3.4. Correlation Analysis between Temperature, Rainfall, and Geographical Factors

3.4.1. Correlation Analysis of Temperature, Rainfall, and Elevation

In order to better understand the distribution and variation of temperature and rainfall in the QMs, we used the research methods of Huang et al. [69] and Bi [70] to analyze the correlation between AMT and AMRA with geographic factors in this paper. The correlation coefficients of AMT and AMRA with altitude, longitude, and latitude and the variation characteristics of AMT and AMRA with geographical factors were studied. Since the highest elevation of the QMs is over 3700 m and the highest national meteorological station is 2064.9 m, we combined the data from regional automatic stations with data from the national station to complete the spatial interpolation analysis. This better reflects the actual distribution of meteorological elements in the high-elevation region, helping us to understand the relationship between temperature, rainfall, and geographical elements.

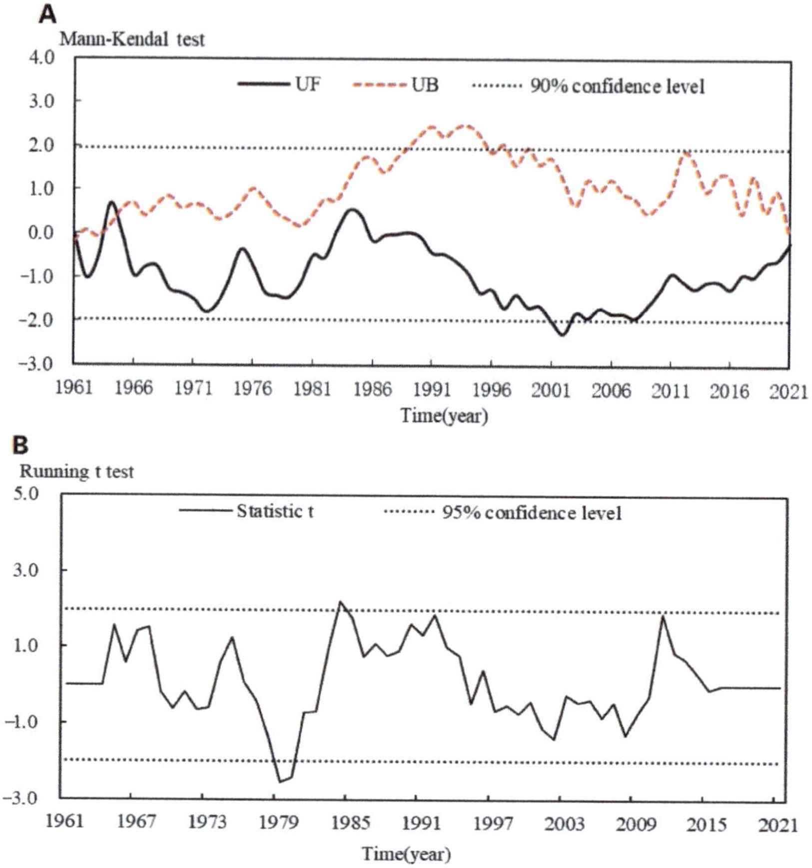

Figure 8. As in Figure 7, but for DR.

As can be seen from Table 5, there was a significant negative correlation between AMT and elevation, with a correlation coefficient of −0.700 ($p = 0.01$). On the other hand, the correlation coefficient between AMRA and elevation was significantly positive, with a value of 0.142 ($p = 0.01$). Furthermore, there was a significant positive correlation between AMT and longitude, with a correlation coefficient of 0.147 ($p = 0.01$). The correlation coefficient between AMRA and longitude was significantly negative, with a value of −0.233 ($p = 0.01$). In addition, there was a significant negative correlation of AMT and AMRA with latitude, with correlation coefficients of −0.617 ($p = 0.01$) and −0.868 ($p = 0.01$).

Table 5. Correlation coefficient between the AMT and AMRA and various geographical factors in the QMs during the period from 1961–2021.

	Elevation (m)	Longitude (°E)	Latitude (°N)
AMT (°C)	−0.700 ***	0.147 ***	−0.617 ***
AMRA (mm)	0.142 ***	−0.233 ***	−0.868 ***

(Note: the superscript "*" indicates the following—*** $p = 0.01$).

In order to explore the changes in air temperature and rainfall with elevation in greater detail, the mean temperature and rainfall within the corresponding elevation ranges were calculated at intervals of 100 m [69,70]. As Figure 9 shows, the AMT exhibited a decreasing trend with elevation increase, at a rate of 0.45 °C/100 m ($p = 0.01$). In the elevation range of 0 m (16.4 °C) to 3730 m (−1.6 °C), the mean temperature decreased by 18 °C. In the elevation range of 0 to 2400 m, the temperature decreased slowly as the elevation increased, but above 2400 m, the temperature decreased rapidly. In particular, the maximum temperature decreased as the elevation increased from 3300 to 3700 m.

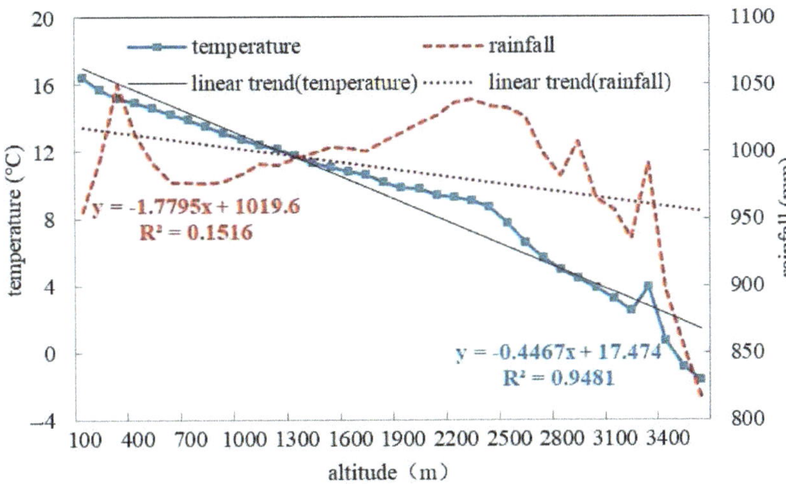

Figure 9. The AMT (unit: °C) and AMRA (unit: mm) change with elevation in the QMs. The brown dashed line denotes the AMRA curve, and the blue dashed line denotes the AMT curve. The black line denotes the linear trend of the temperature curve, and the black dashed line denotes the linear trend of the rainfall curve.

The AMRA showed an increasing trend as the elevation increased. In the elevation range of 0 to 3730 m, the changes in AMRA with increased elevation were more complicated, with an alternating trend of "increase–decrease–increase–decrease". In the elevation range of 200 to 300 m, the maximum AMRA reached was 1050.0 mm. Above 900 m, the AMRA began to increase gradually, and another maximum value was measured at the elevation of 2200 m to 2300 m—1038.8 mm. Then, above 2400 m, as the elevation increased, the AMRA decreased in a fluctuating pattern; in the elevation range of 3600 to 3730 m, the minimum AMRA was 817.6 mm. In addition, the AMRA rapidly dropped above 3400 m. These results signify consistent change trends in the AMT and AMRA with elevation increases, while the AMT and AMRA decreased from low to high elevation in the QM region.

3.4.2. Correlation Analysis of Temperature, Rainfall, and Longitude

In order to explore the changes in air temperature and rainfall at different longitudes in more detail, the mean temperature and rainfall values within corresponding longitudinal ranges were calculated at intervals of 0.2° [69,70]. As can be seen from Figure 10, with changes in longitude, the AMT showed an increasing trend at a rate of 0.04 °C/0.2° (no significant). The minimum temperature was 9.8 °C, which was measured in the longitude range of 107.4°–107.6° E, and the maximum value was 14.4 °C in the longitude range of 110.6°–110.8° E. Furthermore, the AMRA increased in the western QM region and decreased in the east with longitudinal increase. In the longitude range of 107.6°–107.8° E, the maximum AMRA was measured with a value of 1109.2 mm. To the east of 107.8° E, the AMRA showed a decreasing trend, and the minimum value of 875.8 mm appeared in the longitude range of 110.6°–110.8° E. These results further prove that the AMT and AMRA showed different change trends with longitudinal increase; the AMT increased and the AMRA decreased from the west to the east in the QM region.

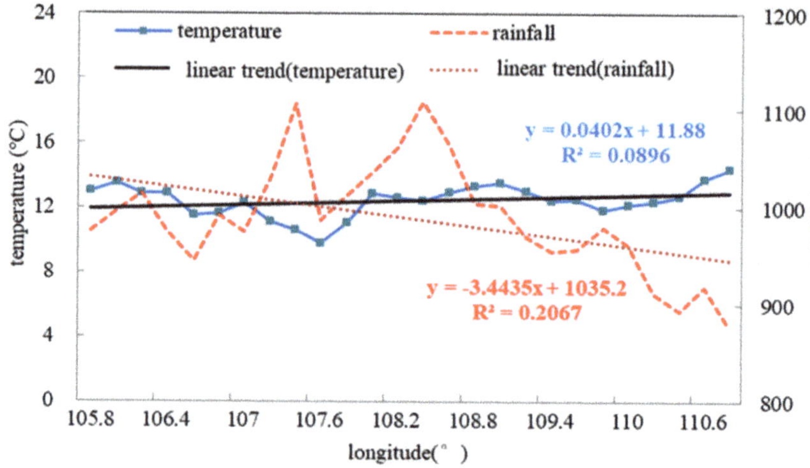

Figure 10. As in Figure 9, but for longitude in the QMs.

3.4.3. Correlation Analysis of Temperature, Rainfall, and Latitude

In order to explore the changes in air temperature and rainfall that occurred with latitude in more detail, the mean values of temperature and rainfall within the corresponding latitude range were calculated at intervals of 0.1° [69,70]. Figure 11 shows that with increases in latitude, the AMT decreased with a linear trend of 0.23 °C/0.1° ($p = 0.01$). The minimum AMT was 10.0 °C, measured at 34° N, and the maximum AMT was 15.9 °C at 32.8° N. Furthermore, with increases in latitude, the AMRA showed a decreasing trend of up to −12.6 mm/0.1° ($p = 0.01$). The minimum AMRA was 842.4 mm, measured at 34.5° N, and the maximum AMRA was 1165.0 mm at 32.5° N. This indicates significant differences in AMRA between the north and south of the QM region. The AMT and AMRA showed consistent change trends with increases in latitude, with both decreasing gradually from the south to the north of the QM region.

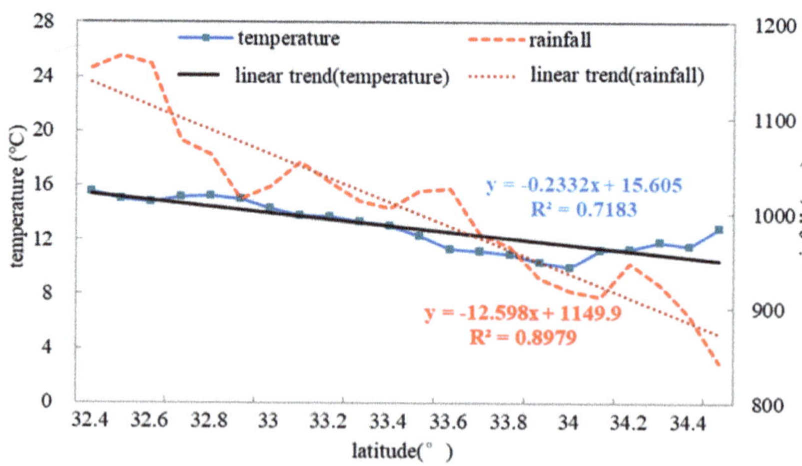

Figure 11. As in Figure 9, but for latitude in the QMs.

4. Discussion and Conclusions

In this study, we investigated the spatiotemporal characteristics of temperature and rainfall in the QMs during the period 1961–2021, and the correlations between temperature and rainfall and geographical factors were studied in detail. The results show that the AMT has significantly increased in the QMs. This increase has been significant in all seasons, with the maximum increasing tendency in spring and the minimum increasing tendency in summer. Positive AMRA anomalies occurred in the 1960s, 1980s, and 2010s, and negative anomalies occurred in the 1970s, 1990s, and 2000s. In the last ten years, MR, HR, TR, and DR showed increasing trends, but LR did not. The AMT increase rate was greater in the northwest and central regions of the QMs, whereas this rate was smaller in the southwestern and eastern regions of the QMs. Considering the elevation distribution of the QMs, the AMT increase rate appeared to be greater at higher elevations, whereas this rate was lower at lower elevations. This result indicates that higher-elevation regions have a more respond positive to climate change than lower-elevation regions [71]. The AMRA showed a decreasing trend in the southwestern and northeastern regions of the QMs, as there were fewer regions with an increasing AMRA trend compared to those where a decreasing AMRA trend occurred.

In the QMs, MR represented the maximum proportion and accounted for 27.9% of the AMRA, whereas TR represented the minimum proportion and accounted for 12.8%. The AMRA significantly decreased by 130.1 mm from the 1980s to the 1990s and accounted for 13.5% of the total in the 1980s. The increase in AMRA in the QMs in the 2010s was mainly caused by the increases in MR, HR, TR, and DR. The years with the most abrupt changes in AMT and DR were 1997 and 1984, respectively, in the QMs. The AMT and AMRA showed consistent change trends with elevation and latitude increases—the AMT and AMRA decreased from a low elevation to high elevation and from the south to the north of the QM region, respectively. The AMT and AMRA showed different change trends with longitude increases—the AMT increased and the AMRA decreased from the west to the east of the QM region.

In this paper, different types of rainfall have been analyzed. Most of the literature has mainly focused on the TR type, indicating that the change trend of TR over the last 50 years has not been obvious, but we suggest an increasing trend in the TR in this century, which is basically consistent with the research conclusions of Kang [45] and Huang et al. [72].

It should be noted that although this study emphasizes the spatiotemporal characteristics of temperature and rainfall in the QMs, several issues remain unclear. For example, the primary limitations of this study are the limited number of surface weather stations in the QMs, the limited duration of the data collection period and the complex topography of the area, which require further exploration the correlation between meteorological factors and geographical factors in the QMs. Furthermore, it was difficult to analyze all the human and natural factors that could affect rainfall and temperature in this area because of the complex topography [16]. Additionally, solar radiation, aerosol, ENSO, and snow depth may also affect rainfall and temperature in the QMs [73]. In the future, studies should focus on the potential physical mechanisms that influence the spatiotemporal distribution characteristics of temperature and rainfall in order to obtain more accurate trends and perform attribution analyses. With increases in observational data, more in-depth research should be carried out on the spatiotemporal distribution characteristics of temperature and rainfall and their correlations with geographical factors. Our results are expected to enhance the understanding of the meteorological background of the QMs.

Author Contributions: Conceptualization, formal analysis, methodology, writing—original draft, and writing—review and editing, Y.L. and R.L.; project administration, Y.S., X.Z., D.L. and L.Z.; supervision, D.L., R.L. and Y.L. All authors have read and agreed to the published version of the manuscript.

Funding: This work was sponsored by the National Key Research and Development Program of Shaanxi (2021SF-493) and the Transverse Item-research Project of the Chinese Academy of Meteorological Sciences (Grant No. IN_JS_2022031).

Institutional Review Board Statement: This study was approved and guided by the Ethics Committee of the Chinese Academy of Meteorological Sciences, China Meteorological Administration.

Informed Consent Statement: Informed consent was obtained from all subjects involved in the study.

Data Availability Statement: DEM data with 30 m resolution are downloaded from the National Science Data Mirroring Website of the Computer Network Information Center, Chinese Academy of Science (http://www.gscloud.cn).

Acknowledgments: The authors acknowledge, with deep gratitude, the contribution of the anonymous reviewers whose time and effort in providing comments have improved this paper substantially.

Conflicts of Interest: The authors of this article declare that there is no conflict of interests regarding the publication of this article. The authors of the manuscript do not have a direct financial relation that might lead to a conflict of interest for any of the authors.

References

1. Nijssen, B.; O'Donnell, G.M.; Hamlet, A.F.; Lettenmaier, D.P. Hydrologic sensitivity of global rivers to climate change. *Clim. Chang.* **2001**, *50*, 143–175. [CrossRef]
2. Trenberth, K.E. Atmospheric moisture residence times and cycling: Implications for precipitation rates and climate change. *Clim. Chang.* **1998**, *39*, 667–694. [CrossRef]
3. Babaousmail, H.; Hou, R.; Ayugi, B.; Sian, K.T.C.L.K.; Ojara, M.; Mumo, R.; Chehbouni, A.; Ongoma, V. Future changes in mean and extreme precipitation over the Mediterranean and Sahararegions using bias-corrected CMIP6 models. *Int. J. Climatol.* **2022**, *42*, 7280–7297. [CrossRef]
4. Zhang, Z.; Duan, K.; Liu, H.; Meng, Y.; Chen, R. Spatio-temporal variation of precipitation in the Qinling mountains from 1970 to 2100 based on CMIP6 data. *Sustainability* **2022**, *14*, 8654. [CrossRef]
5. Fan, X.; Qin, Y.Y.; Gao, X. Interpretation of the main conclusions and suggestions of IPCC AR6 working group I report. *Environ. Prot.* **2021**, *49*, 44–48. (In Chinese) [CrossRef]
6. Li, X.; Wang, N.; Wu, Z. Terrain effects on regional precipitation in a warm season over Qinling-Daba mountains in central China. *Atmosphere* **2021**, *12*, 1685. [CrossRef]
7. Yang, F.M.; Wang, N.; Shi, F.; Ljungqvist, F.C.; Wang, S.; Fan, Z. Multi-proxy temperature reconstruction from the West Qinling Mountains, China, for the past 500 years. *PLoS ONE* **2013**, *8*, e57638. [CrossRef] [PubMed]
8. Yang, F.M.; Wang, N.; Shi, F.; Fredrik, C.L.; Zhao, S.; Liu, T. The spatial distribution of precipitation over the West Qinling region, China, AD 1470–2000. *Palaeogeogr. Palaeoclimatol. Palaeoecol.* **2016**, *443*, 278–285. [CrossRef]
9. Deng, C.H.; Bai, H.Y.; Ma, X.P.; Zhao, T.; Gao, S.; Huang, X.Y. Spatiotemporal differences in the climatic growing season in the Qinling Mountains of China under the influence of global warming from 1964 to 2015. *Theor. Appl. Climatol.* **2019**, *138*, 1899–1911. [CrossRef]
10. Liu, K.; Ma, N.X.; Xu, Y.L.; Sun, G.N. Protection and construction of eco-environment in Qinling Mountainous area. *Chin. J. Eco.* **2004**, *23*, 157–160. [CrossRef]
11. Sun, H.; Bai, H.Y.; Zhang, Q.Y.; Ge, X.P.; Zhang, S.H. Responses of vegetation coverage to regional environmental change in the northern and southern Qinling Mountains. *Acta Sci. Circumstantiae* **2009**, *29*, 2635–2641. [CrossRef]
12. Zhou, Q.; Bian, J.J.; Zheng, J.Y. Variation of air temperature and thermal resources in the northern and southern regions of the Qinling Mountains from 1951 to 2009. *Acta Geogr. Sin.* **2011**, *66*, 1211–1218. [CrossRef]
13. Han, X.; Xue, H.; Zhao, C.; Lu, D. The roles of convective and stratiform precipitation in the observed precipitation trends in Northwest China during 1961–2000. *Atmos Res.* **2016**, *169*, 139–146. [CrossRef]
14. Wang, X.; Wang, B.; Xu, X. Effects of large-scale climate anomalieson trends in seasonal precipitation over the Loess Plateau of China from 1961 to 2016. *Ecol Indic.* **2019**, *107*, 105643. [CrossRef]
15. Li, B.; Chen, Y.; Shi, X.; Chen, Z.; Li, W. Temperature and precipitation changes in different environments in the arid region of northwest China. *Theor. Appl. Climatol.* **2013**, *112*, 589–596. [CrossRef]
16. Meng, Q.; Bai, H.Y.; Sarukkalige, R.J.; Fu, G.B.; Jia, W.H.; Zhang, C.H. Annual and seasonal precipitation trends and their attributions in the Qinling Mountains, a climate transitional zone in China. *Theor. Appl. Climatol.* **2021**, *144*, 401–413. [CrossRef]
17. Li, S.; Yang, S.; Liu, X. Spatiotemporal variability of extreme precipitation in north and south of the Qinling-Huaihe region and influencing factors during 1960–2013. *Prog. Geogr.* **2015**, *3*, 354–363. [CrossRef]
18. Shao, Y.; Mu, X.; He, Y.; Sun, W.; Zhao, G.; Gao, P. Spatiotemporal variations of extreme precipitation events at multi-time scales in the Qinling-Daba mountains region, China. *Quat. Int.* **2019**, *525*, 89–102. [CrossRef]

19. Mo, S.G.; Zhang, B.P. Simulation of temperature fields based on DEM in Qinling Mts. *Mountain Res.* **2007**, *25*, 406–411. [CrossRef]
20. Liu, Y.; Liu, N.; Song, H.M.; Cai, Q.F.; Bao, G.; Wang, W.P. Reconstructed mean air temperature from January to July at the divide sampling site in the mid-Qinling Mountains with tree-ring widths. *Adv. Clim. Chang. Res.* **2009**, *5*, 260–265. [CrossRef]
21. Bai, H.Y.; Ma, X.P.; Gao, X.; Hou, Q.L. Variations in January temperature and 0 °C isothermal curve in Qinling Mountains based on DEM. *Acta Geogr. Sin.* **2012**, *11*, 1443–1450. [CrossRef]
22. Bai, J.; Yan, J.P.; Su, K.H. Differential analysis of abrupt climate change between southern and northern Qinling Mountains in the past 50 years. *J. Shaanxi Norm. Univ.* **2010**, *38*, 98–105. [CrossRef]
23. Li, F. *Spatial-Temporal Changes and the Causes of Extreme Precipitation Events to the North and South of Qinling Mountain during 1962~2011*; Northwest Normal University: Lanzhou, China, 2014. [CrossRef]
24. Zhang, S.H. Spatial and temporal variation of precipitation in Qinling Mountains in the last 50 years. *J. Shangluo Univ.* **2014**, *28*, 62–66. [CrossRef]
25. Zhang, L.W.; Yan, J.P.; Geng, H.J.; Cao, Y.J. The shifts of annual average temperature and precipitation belts in the south and north region of Qinling Mountains, Shaanxi province. *J Shaanxi Norm. Univ.* **2011**, *39*, 81–84. [CrossRef]
26. Gao, X. *Study on Climate Change Trend and Spatial Distribution Characteristics of Qinling-Daba Mountains in Recent 50 Years*; Northwest University: Xi'an, China, 2011; Available online: https://kns.cnki.net/kcms2/article/abstract?v=3uoqIhG8C475KOm_zrgu4lQARvep2SAkhskYGsHyiXksOA0FurgRFfHae_iXgbL9XHTnulT3WpybW6AjUBv9dOj-Y2ZAJvPY&uniplatform=NZKPT (accessed on 1 September 2021).
27. Song, T.X.; Yan, J.P.; Ma, L. Study on Climatic Differentiation in the South and North Qinling Mountains in Recent 50 Years. *Arid. Zone Res.* **2011**, *28*, 492–498. [CrossRef]
28. Zhang, Y.; Bai, H.Y.; Huang, X.Y.; Su, K. Variation of extreme temperature and its impact on regional warming in Qinling Mountains during recent 55 a. *Mt. Res.* **2018**, *36*, 23–33. [CrossRef]
29. Zhai, D.P.; Bai, H.Y.; Qin, J.; Deng, C.H.; Liu, R.J.; He, H. Temporal and spatial variability of air temperature lapse rates in Mt. Taibai, Central Qinling Mountains. *Acta Geogr. Sin.* **2016**, *71*, 1587–1595. [CrossRef]
30. Wang, T.; Yang, M.H.; Yan, S.J.; Geng, G.P.; Li, Q.H.; Wang, F. Effects of temperature and precipitation on spatiotemporal variations of net primary productivity in the Qinling Mountains, China. *Pol. J. Environ. Stud.* **2021**, *30*, 409–422. [CrossRef]
31. Wang, T. Analysis of the influence factors of topography on regional precipitation supported by GIS- A case study of Tianshui City, Gansu Province. *Gansu Sci. Technol.* **2012**, *28*, 58–61. [CrossRef]
32. Mann, H.B. Nonparametric tests against trend. *Econometrica* **1945**, *13*, 245–259. [CrossRef]
33. Kendall, M.G. Rank correlation methods. *Br. J. Psychol.* **1990**, *25*, 86–91. [CrossRef]
34. Du, R.S.; Shang, F.H.; Ma, N. Automatic mutation feature identification from well logging curves based on sliding *t* test algorithm. *Clust Comput.* **2019**, *22* (Suppl. 6), 14193–14200. [CrossRef]
35. Dogan, M.; Ulke, A.; Cigizoglu, H.K. Trend direction changes of Turkish temperature series in the first half of 1990s. *Theor. Appl. Climatol.* **2015**, *121*, 23–39. [CrossRef]
36. Zhang, Q.; Liu, C.L.; Xu, C.Y. Observed trends of annual maximum water level and streamflow during past 130 years in the Yangtze River basin, China. *J. Hydrol.* **2006**, *324*, 255–265. [CrossRef]
37. Song, X.Y.; Song, S.B.; Sun, W.Y. Recent changes in extreme precipitation and drought over the Songhua River Basin, China, during 1960–2013. *Atmos. Res.* **2015**, *157*, 137–152. [CrossRef]
38. Wei, F.F. *Modern Climate Statistical Diagnosis and Prediction Techniques*; China Meteorological Press: Beijing, China, 2007; ISBN 9787502942991.
39. Matheron, G. Principles of geostatistics. *Econ. Geol.* **1963**, *58*, 1246–1266. [CrossRef]
40. Zhang, J.; Yu, X.L. Analysis of land use change and its influence on runoff in the Puhe River Basin. *Environ. Sci. Pollut. Res.* **2021**, *28*, 40116–40125. [CrossRef]
41. Tang, G.A.; Yang, X. *Experimental Course of Spatial Analysis of Geographic Information System*; Science Press: Beijing, China, 2006; ISBN 9787030338969.
42. Olivwe, M.A.; Webster, R. Kriging: A method of interpolation for geographical information systems. *Int. J. Geogr. Inf. Sci.* **1990**, *4*, 313–332. [CrossRef]
43. Shi, Z.; Jin, H.M.; Li, Y.; Li, H.Y. Development and Application of Soft Package for Geostatistics. *J. Soil Water Conserv.* **2005**, *19*, 170–173. [CrossRef]
44. Li, X.J.; Li, C.K.; Yin, Z.H. ArcGIS based kriging interpolation and its application. *Bull. Surv. Mapp.* **2013**, *9*, 87–90. Available online: http://tb.chinasmp.com/CN/Y2013/V0/I9/87 (accessed on 1 September 2021).
45. Kang, L.W. *Spatial and Temporal Distribution of Rainstorm and Risk Assessment of Rainstorm Disasters in North and South Qinling Mountains of Shaanxi Province*; Shaanxi Normal University: Xi'an, China, 2014; Available online: http://cdmd.cnki.com.cn/article/cdmd-10718-1014403027.htm (accessed on 1 September 2021).
46. Pan, G.Y.; Cao, X.Y.; Zhang, X.; Tao, S.Y. Spatial and temporal characteristics of extreme precipitation in Ganjiang River Basin in the past 50 years. *Torrential Rain Disasters* **2020**, *39*, 102–108. [CrossRef]
47. Tang, T.; Jin, R.H.; Peng, X.Y.; Niu, R.Y. The Analysis of Causes about Extremely High Temperature in the summer of 2013 Year in the Southern Region. *J. Chengdu Univ. Inf. Technol.* **2014**, *29*, 652–659. [CrossRef]
48. Yan, J.P.; Huang, C.C. Statistical analysis on influence of ENSO event on the climate in Shaanxi. *J. Catastrophology* **1998**, *4*, 39–42.

49. Liu, M.X. *The Characteristics and Atmospheric Circulation Causes of Two Types of Extreme Cold Events in China;* Lanzhou University: Lanzhou, China, 2021. [CrossRef]
50. Li, D.; Gu, W. Analysis of characteristics and causes of precipitation anomalies over northern China in autumn 2021. *Meteor Mon.* **2022**, *48*, 494–503. (In Chinese) [CrossRef]
51. Hou, M.Q.; Fang, J.G.; Bai, A.J. Analysis of the weather and circulation characteristics of high temperature and drought in summer of 1997. *J. Shaanxi Meteorol.* **1999**, *1*, 4–7. Available online: http://www.cnki.com.cn/Article/CJFDTotal-SXQI901.001.htm (accessed on 1 September 2021).
52. Meng, Q.; Gao, X.; Bai, H.Y.; Zhang, Y.; Wang, H.Y. Temporal and spatial varirtions and trends of extreme precipitation in Qinling Mountains during the period 1960-2015. *Res. Soc. Water Conserv.* **2019**, *26*, 171–183. [CrossRef]
53. Du, J.Y.; Tao, L.; Xu, C.Y. Interdecadal variation of land precipitation in China and relative contributions of global warming, IPO and AMO. *Acta Meteorol. Sin.* **2022**, *80*, 685–700. [CrossRef]
54. Zhao, P.; Zhu, Y.N.; Zhang, R.H. An Asian-Pacific teleconnection in summer tropospheric temperature and associated Asian climate variability. *Clim Dynam.* **2007**, *29*, 293–303. [CrossRef]
55. Qi, Y.D. *Analysis of Temperature Change Trend and Its Urbanization and Sea-Atmosphere Impact Factors;* Wuhan University: Wuhan, China, 2020. [CrossRef]
56. Li, S.; Wang, Y.M.; Gao, Y.Q. A review of researches on the Atlantic Multidecadal Oscillation (AMO). *Trans. Atmos. Sci.* **2009**, *32*, 458–465. [CrossRef]
57. Zuo, Z.Y.; Zhang, R.H.; Wu, B.Y. Interdecadal variation of spring precipitation over southern China and its association with snow cover over Eurasia from 1979 to 2004. *Sci. Sin.* **2011**, *41*, 134–142. [CrossRef]
58. Zhao, H.Y.; Yang, Y.F.; Guo, J.Q.; Wang, X.; Zhao, Q.Y. Effect of Abnormal East Asia North Wind on Spring Precipitation in the Eastern Region of Northwest China. *Arid. Zone Res.* **2013**, *30*, 457–461. [CrossRef]
59. Gu, W.; Li Wj Chen, L.J.; Jia, X.L. Interannual variations of autumn precipitation in China and their relations to the distribution of tropical Pacific sea surface temperature. *Clim. Environ. Res.* **2011**, *17*, 467–480. [CrossRef]
60. Yang, J.H.; Zhang, Q.; Liu, X.Y.; Yue, P.; Shang, J.L.; Ling, H.; Li, W.J. Spatial-temporal characteristics and causes of summer precipitation anomalies in the transitional zone of typical summer monsoon, China. *Chin. J. Geophys.* **2019**, *62*, 4120–4128. [CrossRef]
61. Mu, X.M.; Chen, G.L.; Xu, X.X. Analysis on the relationship between precipitation and geographical factors in Loess Plateau. *Res. Soil Water Conserv.* **1992**, *16*, 80–86. Available online: http://stbcyj.paperonce.org/oa/DArticle.aspx?type=view&id=19920210 (accessed on 1 June 2022).
62. Li, C.; Zhang, H.; Singh, V.P.; Fan, J.; Wei, X.; Yang, J.T.; Wei, X.C. Investigating variations of precipitation concentration in the transitional zone between Qinling Mountains and Loess Plateau in China: Implications for regional impacts of AO and WPSH. *PLoS ONE* **2020**, *15*, e0238709. [CrossRef] [PubMed]
63. Li, S.S.; Lu, J.Y.; Yan, J.P.; Liu, X.F.; Kong, F.; Wang, J. Spatiotemporal variability of temperature in northern and southern Qinling Mountains and its influence on climatic boundary. *Acta Geogr. Sin.* **2018**, *73*, 13–24. [CrossRef]
64. Dong, D.H.; Huang, G. Relationship between Altitude and Variation Characteristics of the Maximum Temperature, Minimum Temperature, and Diurnal Temperature Range in China. *Chin. J. Atmos. Sci.* **2015**, *39*, 1011–1024. [CrossRef]
65. Schulze, H.; Langenberg, H. Climate science: Urban heat. *Nat. Geosci.* **2014**, *7*, 553. [CrossRef]
66. Dong, Q.Q.; Wang, H.X. Variation and reason analysis of temperature and precipitation of Weihe River in Guanzhong area in the last 60 years. *South–North Water Transf. Water Sci. Technol.* **2016**, *14*, 33–38. [CrossRef]
67. Li, M.J. *Analysis of Multi-Scale Spatial-Temporal Variability of Precipitation and Causes of Regional Drought and Flood in China;* University of Chinese Academy of Sciences: Beijing, China, 2021; Available online: https://d.wanfangdata.com.cn/thesis/Y3864553 (accessed on 1 June 2022).
68. Qiu, Y.H. *A Dissertation Submitted in Partial Fulfillment of The Requirements for the Degree of Master of Science;* Nanjing Normal University: Nanjing, China, 2019; Available online: https://mall.cnki.net/magazine/Article/CMFD/1019250006.htm (accessed on 1 June 2022).
69. Huang, M.D.; Zhang, P. Influences of geographic and topographic factors on spatial distribution of precipitation in Urumchi. *Meteorol. Sci. Technol.* **2009**, *37*, 25–28. [CrossRef]
70. Bi, X. *Analysis on Climate Change Characteristics of Temperature and Precipitation and the Correlation Analysis with Geographical Factors in Hubei Province;* Central China Normal University: Wuhan, China, 2013. [CrossRef]
71. Wang, P.; Tang, G.; Cao, L.; Liu, Q.; Ren, Y. Surface air temperature variability and its relationship with altitude & latitude over the Tibetan Plateau in 1981–2010. *Adv. Clim. Change Res.* **2012**, *8*, 313–319. [CrossRef]
72. Huang, X.C.; Wang, X.F.; Kang, L.W.; Wang, S. Temporal and spatial distribution characters of torrential rain in Qinling Area of Shaanxi Province. *Acta Agric. Jiangxi.* **2016**, *28*, 103–108. [CrossRef]
73. Su, J.Z.; Wen, M.; Ding, Y.H.; Gao, Y.Q.; Song, Y.F. Hiatus of global warming: A review. *Chin. J. Atmos. Sci.* **2016**, *40*, 1143–1153. [CrossRef]

atmosphere

Communication

Study of Relative Humidity Vertical Distribution Characteristics before Precipitation by Microwave Radiometer Data over Southeast China

Yongjiang Yu [1,2,*], Yan Zou [3] and Weihua Pan [1,2]

1 Fujian Key Laboratory of Severe Weather, Fuzhou 350007, China
2 Fujian Institute of Meteorological Science, Fuzhou 350007, China
3 Fujian Climate Centre, Fuzhou 350007, China
* Correspondence: fjqxyyj@163.com

Abstract: We investigated the relative humidity (RH) vertical distribution characteristics before precipitation using microwave radiometer measurements over southeast China in 2021. The superposed epoch method is used to analyze the profile and vertical statistical characteristics and evolution of RH during precipitation events. There is a shallow, high-humidity area on the ground, with a thickness of about 0.1–0.2 Km, from 12 to 8 h before precipitation. An obvious dry layer appears in the lower layer near the ground 8–0 h before precipitation, with a thickness of about 1 km and humidity of less than 80%, which continues until the appearance of precipitation. The water vapor content in the air begins to accumulate and the humidity increases before the occurrence of LRs, MRs, and HRs, classified by total rainfall. The SDPs, MDPs, and LDPs, which are classified by precipitation duration, showed more obvious and significant characteristics of humidity increase. The statistical analysis of the 44 precipitation cases shows that the relative humidity on the ground and in the air increases significantly before precipitation, and the vertical distribution of the relative humidity and the increase in the water vapor content in the air have a more direct and obvious impact on the precipitation duration. The deep and high-humidity area of 2–4 km is conducive to maintaining the precipitation process for a long time.

Keywords: relative humidity; microwave radiometer data; total rainfall; precipitation duration; vertical distribution

Citation: Yu, Y.; Zou, Y.; Pan, W. Study of Relative Humidity Vertical Distribution Characteristics before Precipitation by Microwave Radiometer Data over Southeast China. *Atmosphere* **2023**, *14*, 513. https://doi.org/10.3390/atmos14030513

Academic Editors: Duanyang Liu, Hongbin Wang and Shoupeng Zhu

Received: 5 February 2023
Revised: 26 February 2023
Accepted: 1 March 2023
Published: 7 March 2023

1. Introduction

Moisture plays an important role in many weather processes, especially in continuous precipitation process. The root cause of abnormally heavy precipitation is closely related to the water vapor supply and transportation under the background of large-scale circulation [1,2]. Atmospheric water vapor content and vertical distribution are crucial meteorological parameters for understanding atmospheric thermodynamic processes [3]. On the one hand, the water vapor content directly influences the global hydrological cycle, which is closely related to atmospheric energy transmission, radiation budget balance, cloud and rain formation, and climate change [4–6]. On the other hand, the microphysical features of precipitation, which can be reflected by its vertical structure, are helpful in understanding the thermodynamic and dynamic properties of precipitation systems.

The influence of water vapor at different heights on surface precipitation is different [7,8]. The sensitivity of the structure and strength of a squall line in the initial and early development stages to initial low-level humidity and environment vertical wind shear are investigated in a two-dimensional idealized squall line simulation using the WRF model. The results of the sensitivity test of low-level humidity indicate that increasing low-level humidity is favorable for convective triggering and stronger convective systems [9]. The radar reflectivity and precipitation rate from the Precipitation Radar onboard

Tropical Rainfall Measuring Mission satellite are obtained over the global tropical regions (35° S–35° N) during the period from 2007 to 2012, combined with coincident vertical velocity at 400 hPa, relative humidity at 850 hPa, and lower tropospheric stability from European Centre for Medium-Range Weather Forecasts reanalysis. The results imply that coincident vertical velocity at 400 hPa and relative humidity at 850 hPa are most likely to play a dominant role in dictating the vertical development of convection [10]. The analysis of cyclone heavy rain processes with different low-level humidity conditions in Qingdao indicated that the relative humidity at 850 hPa before the heavy rain was higher and the duration of rainfall was longer [11].

The traditional method of detecting atmospheric water vapor and temperature profiles is based on the radiosonde (RS) carried by a sounding balloon. It has the advantages of high accuracy and reliability, low power consumption, light weight, and small size. However, it is influenced by limits of equipment and manpower costs, the launch frequency of RS is relatively low [3]. Additionally, the motion path of the sounding balloon can vary, largely due to the horizontal advection and variable ascent rate [12], which will cause measured drift in the atmospheric profile. The microwave radiometer (MWR) is a passive remote sensing instrument that provides vertical profiles of atmospheric temperature and water vapor content by measuring the thermal radiation emitted by the atmosphere. Compared with the RS, it has applicability due to the advantages of high temporal resolution, reasonable vertical resolution, and ability to automatically measure under almost all-weather conditions [13–15]. In addition, the MWR can also provide other atmospheric parameters, such as cloud base height, integrated water vapor (IWV), and liquid water path (LWP), which are important for evaluating cloud water resources, climate change, and precipitation [16–18]. Performance of MWR was also estimated by comparing with the RS data. The temperature profile measured by the microwave radiometer was better under cloudy conditions, while the RH profile had higher accuracy under cloud-free conditions [19–21]. The comparison of temperature and vapor density obtained from MWR and RS observations during the Integrative Monsoon Frontal Rainfall Experiment Show that: for all sky conditions, the temperature of MWR has a good agreement with that of RS. The vapor density from two measurements also shows reasonable agreement [22–24].

In the past, research on microwave radiometer retrieval data mainly focused on the evolution characteristics of the total accumulated water vapor and cloud liquid water in the whole layer before the beginning of precipitation. There are few studies on the distribution and evolution characteristics of the vertical profile of water vapor and relative humidity, and the analysis of the contribution of moisture to precipitation duration is also less involved.

Jian'ou city is located to the south of Wuyi Mountains. Subtropical monsoon climate prevailing at this city brings significant seasonal differences in temperature and rainfall. Jian'ou has abundant rainfall, with an average annual precipitation of 1753.7 mm, and the disasters caused by precipitation in this city are very serious. In this study, Jian'ou is adopted as a representative city for the southeast China to investigate the vertical structure characteristics of humidity in different rainfall regimes using continuous MWR observations in 2021.

The remainder of this paper is organized as follows. Section 2 introduces the dataset and method used in this study. Our analyses of RH variation and its vertical distribution before precipitation are presented and discussed in Section 3, and conclusions are given in Section 4.

2. Materials and Methods

The observation station selected in this study is located at Jian'ou National Reference Climate Station (118°19′23″ E, 27°03′10″ N, UTC +8, 150 m above sea level) in southeast China from Jan to Dec 2021 to explore the humidity properties of the atmosphere before different rainfall regimes (Figure 1).

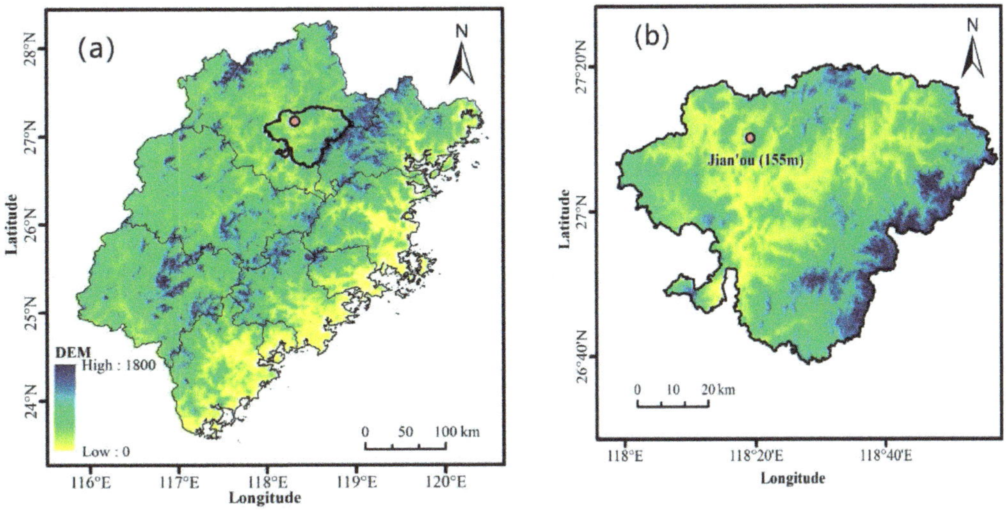

Figure 1. Geographical map of study area ((**a**) is Fujian Province, (**b**) is Jian'ou Station).

2.1. Instruments

The MWR is a QFW6000 unit independently manufactured by China Research Institute of Radiowave Propagation. It observes brightness temperature in 48 channels, including 24 K bands (22–32 GHz) and 24 V bands (51–59 GHz). Combined with ambient temperature, pressure, and relative humidity from the meteorological sensors attached to MWR, atmospheric profiles (temperature, vapor density, and relative humidity) are retrieved from these brightness temperatures. Atmospheric profiles derived from MWR have a temporal resolution of $\sim$3 min and vertical resolutions of 50 m from the surface to 0.5 km, 100 m from 0.5 to 2.0 km, and 250 m from 2.0 to 10.0 km.

RS is usually used as a standard criterion to evaluate other measurements because it detects atmospheric parameters directly with various correction methods for improving its accuracy. Atmospheric temperature, water vapor density, and relative humidity profiles retrieved from the MWR were verified using radiosonde soundings.

Comparison of temperature and relative humidity obtained from MWR and RS were operated at Fu'zhou station during December 2016, which have the same climate with Jian'ou in the north of Fujian Province. For all sky conditions, the temperature of MWR has good agreement with that of RS and a correlation coefficient (R) of 0.995, with a root mean square error of 0.3–0.22 °C. The relative humidity from two measurements also shows reasonable agreement, with an R of 0.743 and a root mean square error of 2.5–12.55%. The relative humidity data obtained from MWR at Jian'ou station was filtered strictly by removing missing data and singular value number during 2021. So, the atmospheric profiles data used in this paper were successive, stable, and credible.

2.2. Data and Methods

The MWR dataset has the temporal resolution of 3 min. Since the time step of the data is not an integer, in order to facilitate the analysis and maintain the data accuracy as much as possible, before the individual statistics, this paper takes the average value of the physical quantity in a total of 5 min before and after a certain time as the physical quantity value at that time, and it is reasonable to set the temporal resolution as 5 min.

Therefore, select and classify rainfall samples according to the following criteria:

1. Investigate the hourly precipitation data of Jian'ou Station from 1 January to 31 December 2021, and continuous precipitation as a sample.

2. Eliminate the case in which precipitation occurred in the 12 h period before the start of precipitation, and eliminate the impact of early precipitation on the inversion data of ground-based microwave radiometer.
3. Eliminate the weak rainfall samples with 1 mm rainfall in the first hour and less than 6 h duration.
4. The rainfall samples are divided into light rain (LR, cumulative rainfall is 0–5 mm), moderate rain (MR, cumulative rainfall is 5–10 mm), and heavy rain (HR, cumulative rainfall > 10 mm) according to the cumulative rainfall.
5. The rainfall samples are divided into short-duration precipitation (SDP, precipitation duration is 1–3 h), medium-duration precipitation (MDP, precipitation duration is 4–8 h), and long-duration precipitation (LDP, precipitation duration is >9 h) according to the duration of rainfall.

The occurrence time of precipitation is set as the 0-hour time, and the 1 h before precipitation is set as −1 h while the 1 h after precipitation is set as +1 h. If no precipitation occurs during the 12 h before 0-hour time, and the accumulated precipitation within 1 h after 0 time is more than 1 mm or the precipitation duration is greater than 6 h, then the case is picked. Using this principle, 44 samples were selected to investigate the temporal variation in relative humidity (see Table 1). According to the total rainfall, there are 12 LRs, 13 MRs, and 19 HRs. According to the precipitation duration, there are 23 SDPs, 13 MDPs, and 8 LDPs.

Table 1. Cases information of precipitation at Jian'ou station during 2021.

No.	Start Time of Precipitation	Precipitation Duration/h	Cumulative Rainfall/mm	Classification by Duration	Classification by Total Rainfall
1	2021/2/16 18:00	6	4.8	MDP	LR
2	2021/3/1 23:00	5	7.3	MDP	MR
3	2021/3/4 23:00	2	2.9	SDP	LR
4	2021/3/11 5:00	6	8.3	MDP	MR
5	2021/3/30 18:00	1	10.0	SDP	HR
6	2021/4/22 13:00	1	4.5	SDP	LR
7	2021/4/24 20:00	13	19.1	LDP	HR
8	2021/4/27 8:00	4	2.5	MDP	LR
9	2021/5/4 16:00	3	7.9	SDP	MR
10	2021/5/7 17:00	11	27.0	LDP	HR
11	2021/5/13 18:00	1	8.1	SDP	MR

Table 1. *Cont.*

No.	Start Time of Precipitation	Precipitation Duration/h	Cumulative Rainfall/mm	Classification by Duration	Classification by Total Rainfall
12	2021/5/16 17:00	1	5.3	SDP	MR
13	2021/5/23 19:00	4	8.8	MDP	MR
14	2021/5/27 5:00	8	15.3	MDP	HR
15	2021/5/31 16:00	8	8.2	MDP	MR
16	2021/6/4 0:00	5	10.6	MDP	HR
17	2021/6/19 14:00	2	2.9	SDP	LR
18	2021/6/21 16:00	7	23.8	MDP	HR
19	2021/6/28 2:00	16	56.9	LDP	HR
20	2021/6/29 16:00	2	21.8	SDP	HR
21	2021/7/20 15:00	1	6.1	SDP	MR
22	2021/7/23 15:00	2	7.0	SDP	MR
23	2021/7/24 14:00	2	3.4	SDP	LR
24	2021/7/27 20:00	2	26.5	SDP	HR
25	2021/7/28 16:00	3	36.0	SDP	HR
26	2021/7/31 0:00	1	1.0	SDP	LR
27	2021/7/31 20:00	3	55.9	SDP	HR
28	2021/8/1 17:00	9	35.6	LDP	HR
29	2021/8/2 16:00	1	9.4	SDP	MR
30	2021/8/3 15:00	3	16.0	SDP	HR
31	2021/8/11 18:00	1	5.5	SDP	MR
32	2021/8/14 16:00	2	3.1	SDP	LR
33	2021/8/15 17:00	5	27.8	MDP	HR

Table 1. *Cont.*

No.	Start Time of Precipitation	Precipitation Duration/h	Cumulative Rainfall/mm	Classification by Duration	Classification by Total Rainfall
34	2021/8/20 16:00	3	12.0	SDP	HR
35	2021/8/28 18:00	1	1.2	SDP	LR
36	2021/9/8 20:00	3	10.5	SDP	HR
37	2021/9/12 16:00	1	3.6	SDP	LR
38	2021/10/19 18:00	16	12.1	LDP	HR
39	2021/11/2 4:00	6	6.4	MDP	MR
40	2021/11/13 1:00	6	1.9	MDP	LR
41	2021/11/21 19:00	14	17.9	LDP	HR
42	2021/12/16 5:00	8	5.8	MDP	MR
43	2021/12/20 20:00	21	15.7	LDP	HR
44	2021/12/25 12:00	13	3.2	LDP	LR

In order to study the change characteristics of relative humidity and its vertical structure before the beginning of different types of precipitation, temporal variations in the relative humidity were analyzed with superposed epoch method, which highlights the effect of the factor in the key time periods while weakening the effects of other factors [25].

The superposed epoch is a row–column array in which the "response" index values filling any row are data pertaining to a single key event. Thus, the number of rows is the sample size of such events. The columns line up the index values in fixed time relation to the key times; column averages comprise the "superposed epoch analysis." By this averaging method, any fluctuations in the response index that are locked in time relative to the key time column are preserved in the average, whereas fluctuations shifting in time from row to row are averaged out [26].

3. Results and Discussion

Because the RH can provide information about hydrometeors in the atmosphere, it is a key parameter used in numerical models and precipitation forecast. We analyzed the variation in the RH before and after precipitation using the superposed epoch method. This section first presents statistical characteristics of RH from MWR before precipitation. Moreover, retrieval applications of MWR in rainy environments are further studied to support nowcasting and precipitation duration estimation.

3.1. Variation in RH before and after Precipitation Classify by Total Rainfall

Figure 2 shows statistical characteristics of relative humidity from 12 precipitation cases of LR. The relative humidity on the ground and in the air began to increase 12–6 h before the start of precipitation, and the relative humidity increased in the near ground layer (below 0.2 km) and the middle and low layer (1–3 km), but the humidity of these two layers did not reach saturation (exceeded 80%). There is an obvious dry layer (humidity is lower than 70%) between the two layers of accumulated water vapor growth, and the thickness of the dry layer is about 0.8 km; 6–0 h before the start of precipitation, the wet layer near the ground began to rise, and a dry layer with a thickness of 1 km appeared near the ground until the occurrence of precipitation. The humidity continues to increase between 1–3 km during −6 to 0 h, forming a high-humidity area.

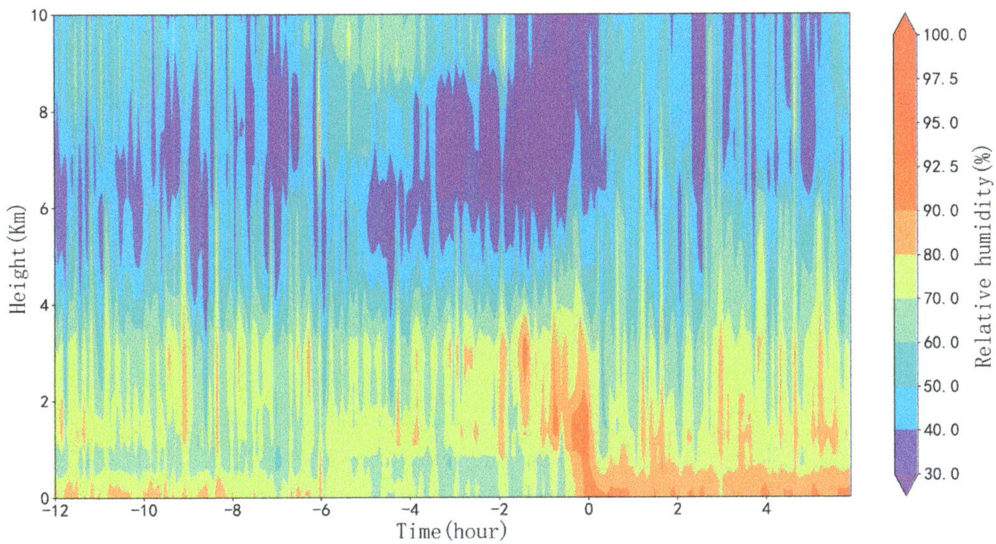

Figure 2. Vertical distributions of relative humidity during 12 h before and 6 h after the start of LRs in 2021.

The statistical analysis results of 13 precipitation cases of MR are shown in Figure 3. It can be seen that the ground and air humidity increased 12–8 h before the precipitation, especially the water vapor content near the ground increased significantly, the thickness of the high-humidity area on the ground reached 0.2 km, and the relative humidity exceeded 90%. After −8 h, the humidity in the lower layer decreased, an obvious dry layer appeared below 2 km, and the water vapor began to accumulate in the high humidity area around 3 km in the middle layer, and the wet layer thickened and became nearly saturated, until the appearance of precipitation.

Figure 4 illustrates the vertical distributions of relative humidity during 12 h before and 6 h after the start of 19 HR cases. From 12 to 8 h before the beginning of precipitation, there is a process of significant humidity increase on the ground, with humidity exceeding 80% and thickness only about 0.1 km, and then the humidity near the ground decreases, and a dry layer with humidity less than 70% appears below 1 km. From 12 h before the precipitation, the process of external water vapor transport and humidity increase in the middle and lower layers (1–4 km) began to occur, and the wet layer was thick.

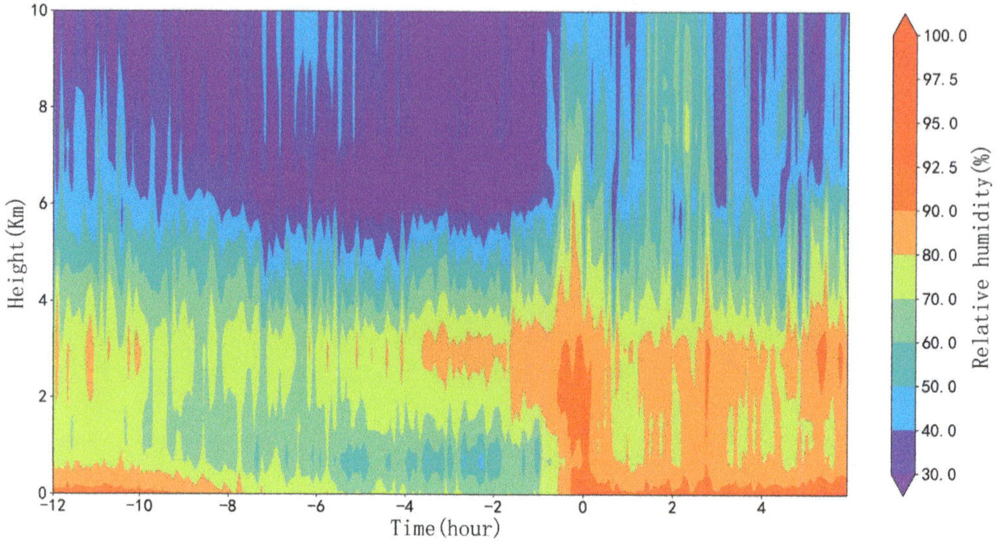

Figure 3. Vertical distributions of relative humidity during 12 h before and 6 h after the start of MRs in 2021.

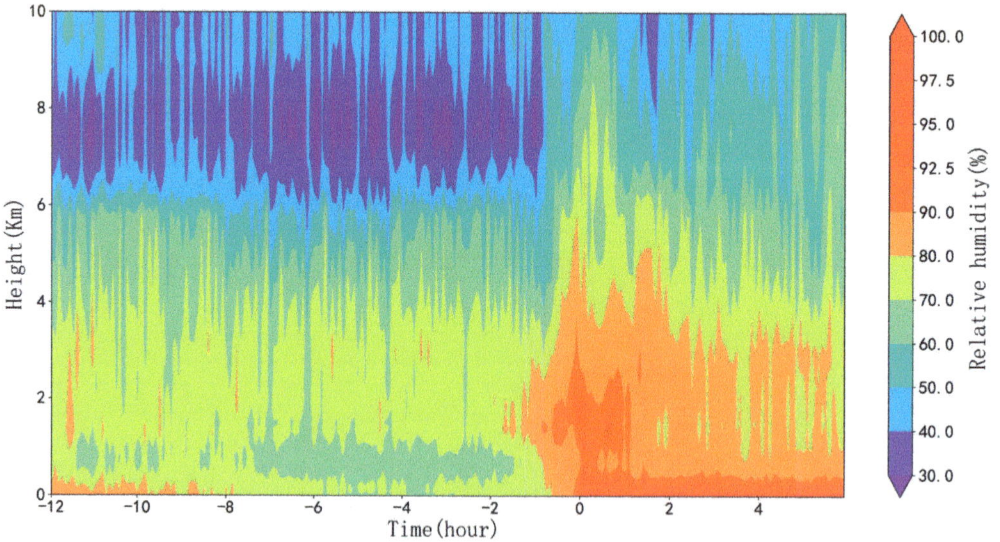

Figure 4. Vertical distributions of relative humidity during 12 h before and 6 h after the start of HRs in 2021.

3.2. Variation in RH before and after Precipitation Classify by Duration

Precipitation is when water vapor in the atmosphere condenses and becomes heavy enough to fall to the ground. Clouds are made of water vapor under different conditions, including variations in relative humidity and air pressure, the vapor particles can begin to combine and form much larger droplets. The increase in relative humidity, which in turn leads to the accumulation of water vapor, is a necessary condition for the formation of precipitation. It not only affects the total rainfall in the rainfall regimes, but also helps to maintain the rainfall regimes for a long time.

Figure 5 presents the mean RH profile from 0 to 10 km during 12 h before and 6 h after the occurrence of 23 SDP cases at Jian'ou station. It is shown that the ground relative humidity began to increase significantly 12–8 h before the start of precipitation, and average relative humidity exceeds 80%, but the high humidity area near the ground is shallow, about 0.2 km thick. From −8 to 0 h before the beginning of precipitation, with the increase in water vapor, the relative humidity on the ground decreases, and a dry layer appears near the ground. From −4 to 0 h before the precipitation, the water vapor in the air began to increase gradually, and the humidity layer at the height of 1–4 km began to thicken. Within 2 h before the precipitation, the humidity increased rapidly and approached saturation, with the average humidity layer thickness reaching 3 km.

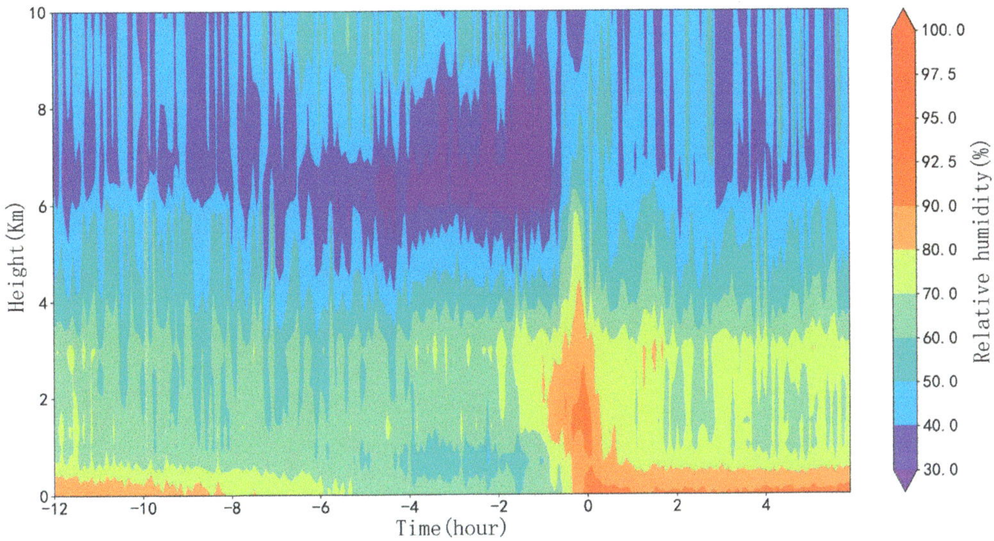

Figure 5. Vertical distributions of relative humidity during 12 h before and 6 h after the start of SDPs in 2021.

The temporally averaged RH profiles during 12 h before and 6 h after the occurrence time of the 13 MDP events are presented in Figure 6. It is shown that the ground relative humidity began to increase significantly 12 h before the beginning of precipitation, and a shallow and discontinuous high humidity layer appeared, with the average relative humidity reaching more than 80%. Continuous and deep wet areas appear, with average humidity of more than 80% and average thickness of 2.2 Km, between 1.8–4 Km in the air. There is an obvious dry layer between the high humidity layer on the ground and in the air, with a thickness of about 1.5 km. With the further transportation of water vapor within 2 h before the precipitation, the dry layer between the ground and the air disappears, the thickness of the humidity layer reaches 5–6 Km, and the average humidity exceeds 90%, which is close to saturation.

Figure 7 presents the statistical characteristics of the relative humidity profile during the 8 LDP cases at Jian'ou station before precipitation. It can be seen that 12–8 h before the start of precipitation, along with the water vapor transport, a deep moisture layer begins to appear in the middle and low layers, with a thickness of 5–6 Km and an average humidity of more than 80%; the layer between 2 and 4 km is the water vapor concentration zone, which air close to saturation and humidity is more than 90%. From −8 to −1 h before precipitation, the bottom atmospheric humidity began to decrease. The atmospheric moisture content is low within 1 km above the ground, and the humidity is less than 80%. The moisture is mainly concentrated in the middle and upper layers of 1.5–5.5 km, the relative humidity is more than 80%, and the local humidity exceeds 90%.

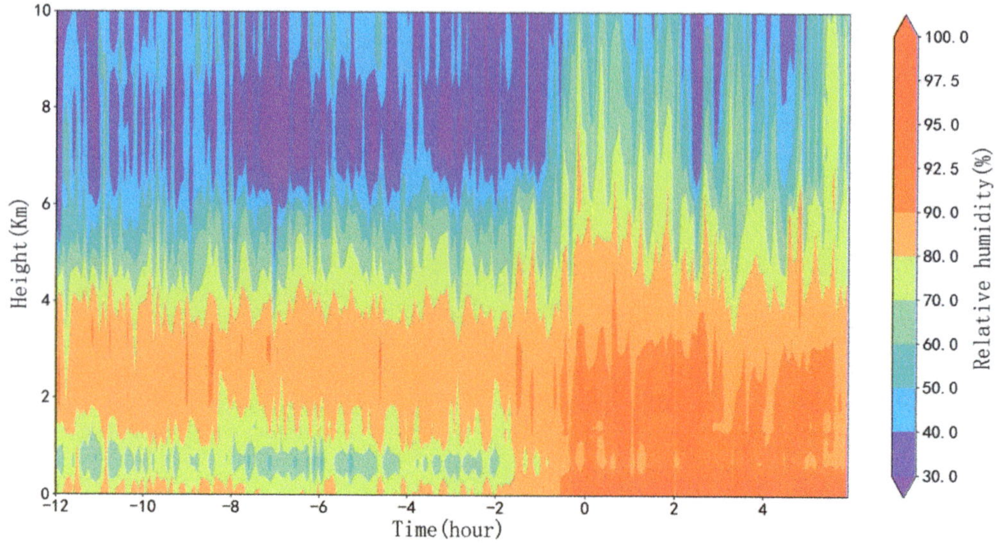

Figure 6. Vertical distributions of relative humidity during 12 h before and 6 h after the start of MDPs in 2021.

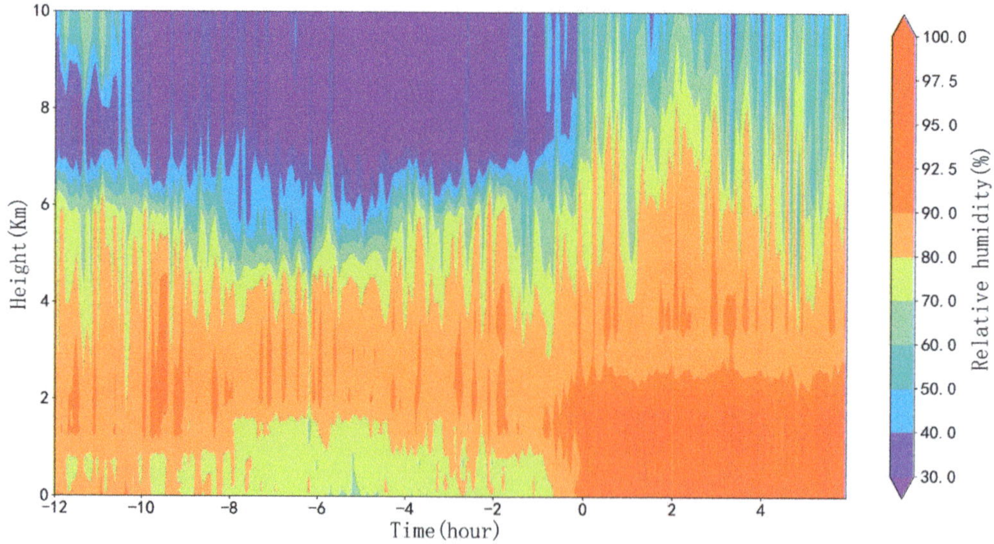

Figure 7. Vertical distributions of relative humidity during 12 h before and 6 h after the start of LDPs in 2021.

4. Summary and Conclusions

The main objective of this study was to investigate the variation and vertical distributions of RH before precipitation. We used the superposed epoch method to analyze the profile of RH during precipitation events.

There is a shallow high humidity area on the ground, with a thickness of about 0.1–0.2 km, from 12 to 8 h before precipitation. An obvious dry layer appears in the lower layer near the ground 8–0 h before precipitation, with a thickness of about 1 km and humidity of less than 80%, which continues until the appearance of precipitation.

The water vapor content in the air began to accumulate and the humidity increased before the occurrence of LRs, MRs, and HRs, which is classified by total rainfall. The increase in air humidity continues from 12 h before precipitation to the beginning of precipitation. The wet layer thickness of LRs is the thinnest, and the height of wet layer is also low. Before precipitation of MRs and HRs, the characteristics of water vapor accumulation in the atmosphere are similar. High humidity areas with relative humidity greater than 80% are continuous and deep. Before the start of HRs, the thickness of the humidity layer (RH > 80%) can reach 4 km and the height can reach 6 km.

The SDPs, MDPs, and LDPs, which are classified by precipitation duration, also showed statistical characteristics of significant increase in humidity. There is a high humidity area in the middle and low layers until 2 h before SDPs. The 12–0 h before the occurrence of MDPs, a continuous and deep humidity layer appeared between 1.8 and 4 Km on the ground, with an average thickness of 2.2 Km. The humidity layer became thicker as the precipitation approached. The 12–6 h before LDPs, water vapor content below 6 km is very rich, and the thickness of high humidity area reaches 5–6 Km. After that, the humidity layer rises, and the high humidity layer between 1.5 and 5.5 km is maintained until the appearance of precipitation.

The statistical analysis of the filtered 44 precipitation cases shows that the relative humidity on the ground and in the air increases significantly before precipitation, and the vertical distribution of the relative humidity and the increase of the water vapor content in the air have a more direct and obvious impact on the precipitation duration. The deep and high humidity area of 2–4 km is conducive to maintaining the precipitation process for a long time.

A large number of numerical model simulations and reanalysis data analysis show that the physical quantities that affect the atmospheric thermal and dynamic environment during precipitation mainly include rising speed, relative humidity, temperature, and vertical wind shear. The transportation of warm and humid air leads to the increase in the convective available potential energy over the precipitation area. The increase in low-level humidity is favorable for convective triggering and stronger convective systems. The increased convective intensity and upward movement make it easier for new convective cells to form at the leading edge of the cold poll. The increased low-level humidity also leads to more precipitation, which raises the intensity of cold poor. Coincident vertical velocity at 400 hPa and relative humidity at 850 hPa most likely play a dominant role in dictating the vertical development of convection. The important causes of longer duration and the larger accumulated of precipitation was the higher relative humidity conditions and long-lasting upward motion, as well as continuous water vapor transport. Therefore, it is of great significance to strengthen the analysis of the of low-level jet and relative humidity vertical distribution by microwave radiometer data for the prediction of precipitation.

Author Contributions: Conceptualization, Y.Y.; Y.Z.; W.P.; methodology, Y.Y.; Y.Z.; software, Y.Y.; validation, Y.Z.; formal analysis, Y.Y.; Y.Z.; investigation, Y.Y.; Y.Z.; resources, Y.Y.; data curation, Y.Y.; writing—original draft preparation, Y.Y.; writing—review and editing, Y.Y.; visualization, W.P.; supervision, Y.Y.; project administration, Y.Y.; funding acquisition, Y.Y.; Y.Z. All authors have read and agreed to the published version of the manuscript.

Funding: This research was funded by the Open Research Fund of Fujian Provincial Meteorological Bureau (2021KX02), the National Key Research and Development Program of China (2019YFC1510303).

Institutional Review Board Statement: Not applicable.

Informed Consent Statement: Not applicable.

Data Availability Statement: The data were observed by Fujian Provincial Meteorological Bureau. If you want to obtain it, please apply through email (fjqxyyj@163.com).

Acknowledgments: The authors acknowledge the free use of the microwave radiometer data provided by Fujian Provincial Meteorological Bureau. We also appreciate the anonymous reviewers for their constructive comments and thoughtful suggestions.

Conflicts of Interest: The authors declare that they have no conflict of interest.

References

1. Zhang, L.; Du, L.; Chen, L. Analysis of Water Vapor Source and Transport Path of Abnormal Heavy Rain in Wuhan. *Meteorol. Environ. Sci.* **2014**, *37*, 69–74.
2. Wang, X.; Xu, G.; Yuan, K. Different characteristic analysis of inversion parameters for heavy rainfall and weak rainfall by microwave radiometer data. *Torrential Rain Disasters* **2016**, *35*, 227–233.
3. Jin, S.; Ma, Y.; Gong, W.; Liu, B.; Lei, L.; Fan, R. Characteristics of vertical atmosphere based on five-year microwave remote sensing data over Wuhan region. *Atmos. Res.* **2021**, *260*, 105710. [CrossRef]
4. Ramanathan, V.; Crutzen, P.J.; Kiehl, J.T.; Rosenfeld, D. Atmosphere—Aerosols, climate, and the hydrological cycle. *Science* **2001**, *294*, 2119–2124. [CrossRef] [PubMed]
5. Rosenfeld, D.; Lohmann, U.; Raga, G.B.; O'Dowd, C.D.; Kulmala, M.; Fuzzi, S.; Reissell, A.; Andreae, M.O. Flood or drought: How do aerosols affect precipitation? *Science* **2008**, *321*, 1309–1313. [CrossRef]
6. Obregon, M.A.; Costa, M.J.; Silva, A.M.; Serrano, A. Impact of aerosol and water vapour on SW radiation at the surface: Sensitivity study and applications. *Atmos. Res.* **2018**, *213*, 252–263. [CrossRef]
7. Qi, Y.; Fan, S.; Mao, J.; Li, B.; Guo, C.; Zhang, S. Impact of Assimilating Ground-Based Microwave Radiometer Data on the Precipitation Bifurcation Forecast: A Case Study in Beijing. *Atmosphere* **2021**, *12*, 551. [CrossRef]
8. Qi, Y.; Fan, S.; Li, B.; Mao, J.; Lin, D. Assimilation of Ground-Based Microwave Radiometer on Heavy Rainfall Forecast in Beijing. *Atmosphere* **2022**, *13*, 74. [CrossRef]
9. Gao, Y.; Sun, L.; Ma, X.; Meng, Z.; Cheng, K. The sensitivity of structure and strength of squall line to low-level humidity and environment vertical wind shear. *Trans. Atmos. Sci.* **2022**, *45*, 938–947. (In Chinese)
10. Liu, H.; Guo, J.; Chen, T.; Zhai, P. On the seasonal variation of various types of precipitation over global tropical ocean region: A perspective from TRMM measurements. *Chin. Sci. Bull.* **2017**, *62*, 90–104. (In Chinese) [CrossRef]
11. Liu, X.; Ma, Y.; Ling, Y.; Jiang, D.; Wan, F. Analysis of Two cyclone heavy rain processes with different low-level humidity conditions in Qingdao. *Meteorol. Sci. Technol.* **2019**, *47*, 818–829.
12. Bedoya-Velásquez, A.E.; Navas-Guzmán, F.; Moreira, G.D.A.; Román, R.; Cazorla, A.; Ortiz-Amezcua, P.; Benavent-Oltra, J.A.; Alados-Arboledas, L.; Olmo-Reyes, F.J.; Foyo-Moreno, I.; et al. Seasonal analysis of the atmosphere during five years by using microwave radiometry over a mid-latitude site. *Atmospheric Res.* **2019**, *218*, 78–89. [CrossRef]
13. Li, Q.; Wei, M.; Wang, Z.; Chu, Y. Evaluation and Improvement of the Quality of Ground-Based Microwave Radiometer Clear-Sky Data. *Atmosphere* **2021**, *12*, 435. [CrossRef]
14. Liu, M.; Liu, Y.-A.; Shu, J. Characteristics Analysis of the Multi-Channel Ground-Based Microwave Radiometer Observations during Various Weather Conditions. *Atmosphere* **2022**, *13*, 1556. [CrossRef]
15. Ma, R.; Li, X. Sounding Data from Ground-Based Microwave Radiometers for a Hailstorm Case: Analyzing Spatiotemporal Differences and Initializing an Idealized Model for Prediction. *Atmosphere* **2022**, *13*, 1535. [CrossRef]
16. Cui, C.; Wan, R.; Wang, B. The Mesoscale Heavy Rainfall Observing System (MHROS) over the middle region of the Yangtze River in China. *J. Geophys. Res. Atmos.* **2015**, *120*, 10399–10417. [CrossRef]
17. Huang, J.; Minnis, P.; Lin, B.; Yi, H.; Sun-Mack, S.; Fan, T.-F.; Ayers, J. Determination of ice water path in ice-over-water cloud systems using combined MODIS and AMSR-E measurements. *Geophys. Res. Lett.* **2006**, *33*. [CrossRef]
18. Lei, L.; Wang, Z.; Ma, Y.; Zhu, L.; Qin, J.; Chen, R.; Lu, J. Measurement of Solar Absolute Brightness Temperature Using a Ground-Based Multichannel Microwave Radiometer. *Remote Sens.* **2021**, *13*, 2968. [CrossRef]
19. Shi, Y.; Wei, J.; Qiao, Z.; Zhao, J.; Wang, G. Atmospheric Exploration of the Qinghai–Tibet Plateau during the East Asian Winter Monsoon (EAWM) from a Ground-Based Microwave Radiometer. *Atmosphere* **2022**, *13*, 549. [CrossRef]
20. Wei, J.; Shi, Y.; Ren, Y.; Li, Q.; Qiao, Z.; Cao, J.; Ayantobo, O.O.; Yin, J.; Wang, G. Application of Ground-Based Microwave Radiometer in Retrieving Meteorological Characteristics of Tibet Plateau. *Remote Sens.* **2021**, *13*, 2527. [CrossRef]
21. Xu, G.; Xi, B.; Zhang, W.; Cui, C.; Dong, X.; Liu, Y.; Yan, G. Comparison of atmospheric profiles between microwave radiometer retrievals and radiosonde soundings. *J. Geophys. Res. Atmos.* **2015**, *120*, 10313–10323. [CrossRef]
22. Xu, G.; Xi, B.; Zhang, W.; Cui, C.; Dong, X.; Liu, Y.; Yan, G. Cloud occurrence frequency and cloud liquid water path for non-precipitating clouds using ground-based measurements over central China. *J. Atmos. Sol.-Terr. Phys.* **2021**, *215*, 105575. [CrossRef]
23. Zhang, W.; Xu, G.; Liu, Y.; Yan, G.; Li, D.; Wang, S. Uncertainties of ground-based microwave radiometer retrievals in zenith and off-zenith observations under snow conditions. *Atmos. Meas. Tech.* **2017**, *10*, 155–165. [CrossRef]

24. Zhang, W.; Xu, G.; Xi, B.; Ren, J.; Wan, X.; Zhou, L. Comparative study of cloud liquid water and rain liquid water obtained from microwave radiometer and micro rain radar observations over central China during the monsoon. *J. Geophys. Res. Atmos.* **2020**, *125*, e2020JD032456. [CrossRef]
25. Zheng, Z.; Xu, G.; Li, Q.; Chen, C.; Li, J. Effect of precipitation on reducing atmospheric pollutant over Beijing. *Atmos. Pollut. Res.* **2019**, *10*, 1443–1453. [CrossRef]
26. Haurwitz, M.; Brier, G. A Critique of the Superposed Epoch Analysis Method: Its Application to Solar–Weather Relations. *Mon. Weather. Rev.* **1981**, *109*, 2074–2079. [CrossRef]

atmosphere

Communication

Association between the Rail Breakage Frequency in Beijing–Tianjin–Hebei High-Speed Railway and the Eurasian Atmospheric Circulation Anomaly

Liwei Huo [1], Linman Xiao [1], Ji Wang [2,*], Dachao Jin [1], Yinglong Shi [3] and Qian Zhang [4]

1 Key Laboratory of Meteorological Disaster, Ministry of Education (KLME)/Joint International Research Laboratory of Climate and Environment Change (ILCEC)/Collaborative Innovation Center on Forecast and Evaluation of Meteorological Disasters (CIC-FEMD), Nanjing University of Information Science & Technology, Nanjing 210044, China
2 Beijing Regional Climate Center, Beijing 100089, China
3 College of Meteorology and Oceanography, National University of Defense Technology, Changsha 410003, China
4 College of Atmosphere and Remote Sensing, Wuxi University, Wuxi 214105, China
* Correspondence: wangji_zl@163.com

Abstract: The spatiotemporal variations in the frequency of rail breakage (FRB) in the high-speed railway of the Beijing–Tianjin–Hebei (BTH) region and its relationship with atmospheric circulation anomalies and surface temperature are analyzed in this study, based on the monthly FRB data of BTH region and the ERA5 reanalysis data from 2010 to 2020. The frequency of rail breaking in the BTH region varies significantly depending on the season, with winter having the highest incidence. In fact, more than 60% of the total FRB in the BTH region occur during the winter season. Both the annual total and winter FRB in BTH region are very unevenly distributed in time and space, and both are relatively similar in spatial distribution patterns. The FRB in Beijing railway section is the most frequent, followed by Tianjin, and the lowest frequency is observed in Chengde. It is found that the increasing winter FRB in BTH region and the intensified Siberian high are related. When the Siberian high is strong, the East Asian winter monsoon and the East Asian Trough in the middle troposphere could be enhanced through atmospheric teleconnection, which is conducive to the cold air advection from northern high latitudes to the BTH region, resulting in an abnormally cold winter in BTH region, thus providing low temperatures for broken rails on high-speed railways, and vice versa. The research results might provide a scientific basis for monitoring and predicting the broken rails in BTH high-speed railway during winter, thereby providing a guarantee for the safe operation of the high-speed railway.

Keywords: Beijing–Tianjin–Hebei region; rail breakage; frequency; high-speed railway; Siberian high; teleconnection; temperature

Citation: Huo, L.; Xiao, L.; Wang, J.; Jin, D.; Shi, Y.; Zhang, Q. Association between the Rail Breakage Frequency in Beijing–Tianjin–Hebei High-Speed Railway and the Eurasian Atmospheric Circulation Anomaly. *Atmosphere* **2023**, *14*, 561. https://doi.org/10.3390/atmos14030561

Academic Editor: Pak-Wai Chan

Received: 14 December 2022
Revised: 28 February 2023
Accepted: 11 March 2023
Published: 15 March 2023

1. Introduction

The Beijing–Tianjin–Hebei (BTH) region is China's capital economic circle, which includes the capital city of China, Beijing, and the municipality of Tianjin, and is densely populated. High-speed railway plays an important role in the coordinated development of the Beijing–Tianjin–Hebei region. For example, the Beijing-Zhangjiakou high-speed railway provided a strong foundation for the transportation service guarantee of the 2022 Beijing Winter Olympics.

Railway transportation has become the main mode of modern transportation due to its advantages of low climate impact, strong transportation capacity, and energy saving [1]. In recent years, China's economy has been booming, and with the implementation of "the Belt and Road initiative", the construction of public infrastructure has played a significant role in rapid economic development, and railway development is one of the top priorities [2].

Railways with a speed of 250 km/h or more are called high-speed railways [3]. At present, high-speed railway has become one of the main ways to travel in China, so the safety of high-speed railway is related to the safety of passengers' lives and property. The rails are exposed to the open air for a long time and are prone to break under the repeatedly exerted axle loads [4]. Broken rail accidents have always been a serious hazard to a train's operation safety, even causing train derailing and overturning, bringing a large number of casualties and property damage.

Generally speaking, the fracture of rail is directly related to the internal defects. Some of these defects are formed in the process of rail production, such as micro-cracks formed in the steel rolling [5]; some are produced during the installation and connection of rails, such as cracks generated during welding and cooling [6]. During the use of steel rails, the initial crack expands under the dynamic load of the train until fracture occurs, especially at low temperatures when the possibility of fracture increases greatly [6]. The main reason is the large seasonal temperature difference. The locked rail temperature is high in track laying construction, when the stress desperation is higher than the locked temperature due to high temperature, whereas winter cooling increases the tensile stress of track temperature. Moreover, there are stress peaks during cooling process, and thus rails cannot withstand the temperature pull generated by cooling, leading to their breakage.

For example, in the early morning of 19 December 2004, the rails at the Fata Temple crossing in Beijing were frozen and cracked, causing 10 trains to be delayed. The main reason for the rail freezing and cracking this time is that the temperature difference between day and night on the 18th is too large. Coupled with the sleet during the day, the rain on the rails freezes after a sudden drop in temperature at night, causing the rails to freeze and crack.

There are many factors influencing winter temperature anomalies in the BTH region. The East Asian winter monsoon, for instance, has a significant impact on temperature anomalies in BTH region [7,8]. The Arctic Oscillation (AO) is one of the major teleconnection patterns in the northern hemisphere [9,10]. AO is characterized by the phenomenon of a zonal band-like seesaw structure in the sea level pressure field at northern mid- and high latitudes. AO can affect winter temperature anomalies in North China by regulating atmospheric circulation [11]. The Eurasian telecorrelation [12] is an important telecorrelation pattern in the Northern Hemisphere winter, and the Eurasian telecorrelation pattern is negatively correlated with winter temperature anomalies in the BTH region [13]. Ural blocking activity in the Urals can also influence winter temperature anomalies in the BTH by modulating atmospheric circulation anomalies in Eurasia [14,15]. There is also a link between the circumglobal teleconnection [16] and winter temperature anomalies in eastern China [17].

The Siberian high also plays an important role in winter temperature anomalies in Eurasia [18–20]. Could the Siberian high influence the winter temperature anomalies in the BTH region, thereby affecting the frequency of rail breakage (FRB) in high-speed railway? To clarify this problem might provide scientific basis for monitoring and predicting the frequency of broken rails and guarantee the safe operation of high-speed railway in the BTH region.

2. Data and Methods

The monthly FRB in high-speed railway of BTH region from 2010 to 2020 is used in present study. In order to reveal the relationship between the FRB and atmospheric circulation anomalies affecting the BTH region in winter, the monthly ERA5 reanalysis data from the European Centre for Medium-Range Forecasting (ECMWF) with a horizontal resolution of 1.25° × 1.25° for 2010–2020 are also selected, including surface pressure, 2 m air temperature, and isobaric wind and geopotential height fields with a vertical resolution of 25 hPa at 1000–100 hPa [21]. CN05.1 average temperature data [22] are also utilized. The winter season referred to in this article is from November to February of the following year.

Considering the limited the length of monthly FRB data in BTH region, referring to the method applied by Huo et al. [23], November, December, January, and February are all regarded as different samples instead of adopting seasonal average. In this way, there are 40 months in a total of 10 winters from 2010 to 2020, and a relatively long time series can be constructed. Pearson correlation analysis, linear regression methods, and Student's t test are adopted in the statistical analysis of this work.

3. Spatio-Temporal Variability of FRB in BTH Region during Winter

The spatial distribution of FRB in BTH high-speed railway from 2010 to 2020 (Figure 1a) shows that the FRB is very unevenly distributed. The maximum of rail breaking occurs 51 times in Beijing, and the second maximum occurs 20 times in Tianjin. The minimum of rail breaking is only once in Chengde, and the sub-minimum is five times in Baoding. In addition to the uneven spatial distribution, there are significant seasonal differences in the FRB of the BTH region. From the time series of monthly FRB in BTH region (Figure 1b), it can be found that the rail breakage of high-speed railway mostly occurs in winter, and the FRB in each month of winter exceeds 20. The FRB in spring (March, April, and May) and autumn (September and October) is second. Except for the FRB in September of 4, the FRB in other months is between 5 and 10. The lowest FRB occurs in summer (June, July, and August). During the 11-year period from 2010 to 2020, the FRB in summer is only 5, including 2 in June, 1 in July, and 2 in August.

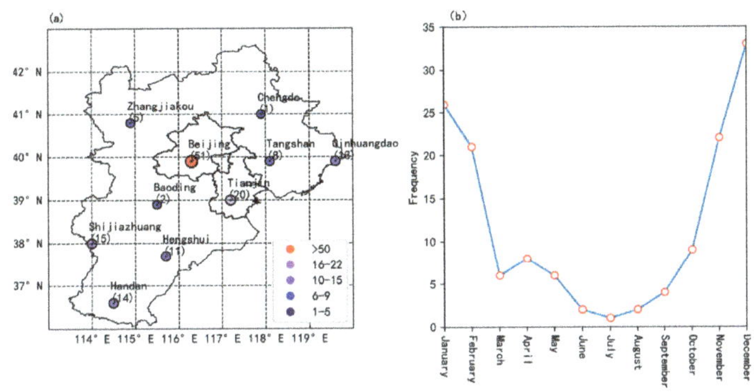

Figure 1. (**a**) Spatial distribution of annual total frequency of rail breakage (FRB) and (**b**) time series of monthly FRB of high-speed railway in Beijing–Tianjin–Hebei (BTH) region during 2010–2020.

Since winter is the season with the highest FRB in BTH high-speed railway. The average FRB during winter is obtained to further analyze its spatial and temporal distribution pattern. The spatial distribution of winter FRB in BTH region is also very uneven (Figure 2a), with the maximum and sub-maximum values still occurring in Beijing and Tianjin with 32 and 16 breaks, respectively, while the minimum and sub-minimum values are still located in Chengde and Baoding with one and two breakages, respectively.

In the BTH region, the percentage of FRB in winter accounts for more than 60% of the total annual FRB (Figure 2b). In Chengde and Baoding, the percentages of winter FRB accounts for 100% of the total FRB, i.e., the rail breaks in these two areas all occur in winter. The winter FRB in Qinhuangdao and Hengshui both exceeds 90% of the total FRB. The percentages of the winter FRB in Beijing and Tianjin are 63% and 80% of the total frequency, respectively. The average temperature in the BTH region during 2010–2020 is less than 0 °C in December, January, and February, except for 3.1 °C in November, which might be the reason for the higher FRB in winter.

The interannual variation in the winter FRB of BTH region is also great (Figure 2c). The winter of 2011/2012 has the highest FRB, with 15, accounting for 17% of the total. The FRB in winter of 2015/2016 and 2016/2017 is the least, with two rail breaks in both years, each accounting for 2% of the total rail breakage, respectively. The FRB also varies considerably in the remaining years.

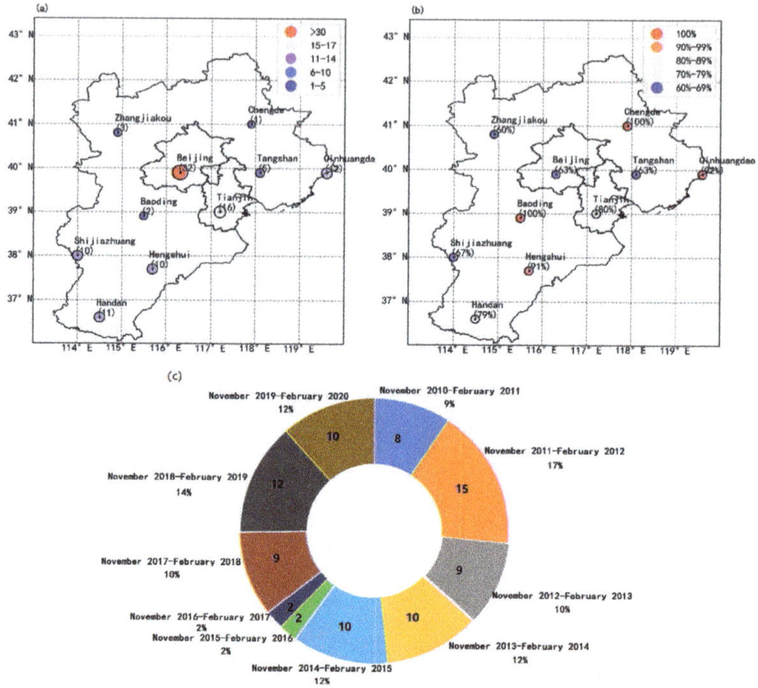

Figure 2. (**a**) Spatial distribution of the winter FRB, and (**b**) percentage of the FRB in winter against the FRB of whole year in BTH region during 2010–2020. (**c**) The doughnut chart of winter FRB in BTH region of each year. The integer on doughnut chart represents the FRB of each winter, and the percentage around doughnut chart represents the percentage of the winter FRB against the annual total FRB in each year.

4. Relationship between the FRB in BTH Region and the Siberian High Anomaly

In order to reveal the influencing factors of the winter FRB in BTH region, the spatial distribution of correlation coefficient between the FRB and the surface air pressure was calculated (Figure 3a). It can be found that there is a significant positive correlation between the winter FRB in BTH region and the surface pressure in the region (44–53° N, 60–95° E) adjacent to Balkhash Lake, where the significantly correlated region is located in the southern part of the West Siberian Plain, indicating that the winter FRB in BTH region is closely related to the Siberian high anomaly. Indeed, there are some differences between the above-mentioned significantly correlated region and the region selected for the Siberian high intensity index defined by previous studies [24]. Therefore, the average surface pressure in the region [44–53° N, 60–95° E] is defined as the index of Siberian high (I_{SH}) in present study. The correlation coefficient between the I_{SH} and the FRB is calculated to be 0.46, which can pass the 95% significance test. In other words, when the Siberian high strengthens, the winter FRB in BTH high-speed rail increases, and vice versa. It is also noted that there is a certain positive correlation between the winter FRB in BTH region and the surface air pressure in Mongolia Plateau, Hetao region, and the middle and lower reaches of the Yangtze River.

The winter spatial distribution of correlation coefficients between I_{SH} and 2 m temperature (as shown in Figure 3b) reveals that 2 m temperatures in the Siberian Plain, Mongolian Plateau, and most regions of China, except the Tibetan Plateau, are significantly and negatively correlated with the I_{SH}. In other words, when the Siberian high is strong (weak), temperatures in the aforementioned regions become abnormally low (high). Consequently, a strengthened Siberian high causes winter temperatures in the BTH region to drop abnormally, resulting in external temperature conditions that are favorable for the occurrence of rail breakage. The spatial distribution of correlation coefficients between CN05.1 surface temperature observations and I_{SH} in the Chinese region is consistent with the reanalysis data (Figure is omitted).

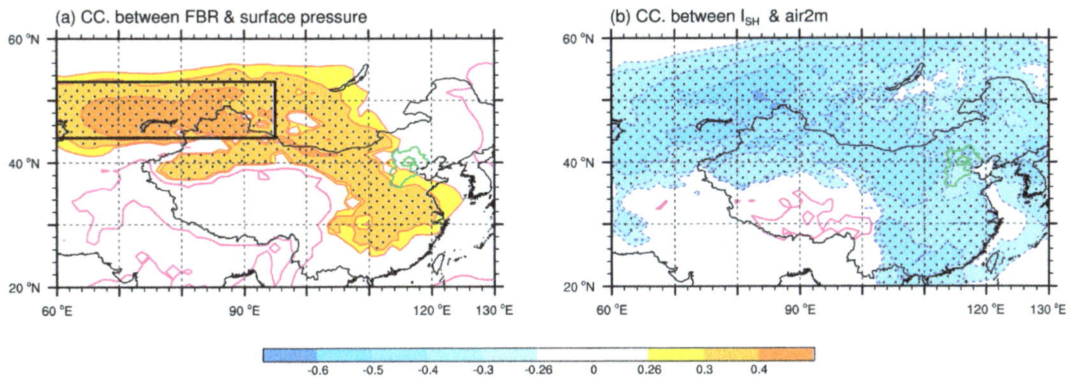

Figure 3. (**a**) Spatial distribution of correlation coefficients between FRB in BTH region and surface pressure in winter, where the rectangular box is the area selected to define Siberian high index (I_{SH}). (**b**) Correlation coefficient between I_{SH} and 2 m temperature field. The solid magenta line denotes the zero line. The shaded area/dot shading is the area that passes the 90/95% significance test. Green curve indicates the BTH region.

To elucidate how the Siberian high affects the FRB in BTH high-speed railway, the potential height and wind fields are regressed using the I_{SH} (Figure 4). It can be seen that that when the Siberian high anomaly is positive, significant positive potential height anomalies and anomalous anticyclonic circulation are observed over the Siberian plain in the lower (Figure 4a), middle (Figure 4b) and upper troposphere (Figure 4c). On the contrary, negative potential height anomalies and anomalous cyclonic circulation are found over the northeastern Asian region. The above anomalous circulation shows a "seesaw oscillation" in the geopotential anomaly field over the Siberian plain and northeastern Asia. Note that the northwest-southeast tilting dipole in 500 hPa potential height anomaly is similar to that of the Eurasian-Pacific (EUP) teleconnection [12] over Asian continental and Pacific regions. During the positive phase of EUP teleconnection, the equivalent barotropic Rossby wave can be transmitted from Europe along the great circle route into Asia-Pacific region [25], and enhance the Siberian high and the East Asian winter monsoon [26,27] (Takaya and Nakamura, 2005; Maeda et al., 2021). Our results also indicate that when the Siberian high is anomalously strong, on the one hand, anomalous northwesterly winds exist at the middle and high latitudes and anomalous northeasterly winds at low latitudes over East Asia in the lower troposphere (Figure 4a), which could enhance the East Asian winter monsoon [7,8]. On the other hand, the negative anomalies of potential height and anomalous cyclonic circulation in the middle troposphere of Northeast Asia (Figure 4b) strengthen the East Asian trough, thus favoring the cold air advection into BTH region from northern high latitudes. The regressed whole-layer integrated temperature advection field by I_{SH} (Figure 5) shows that when the Siberian high is strong, cold air advection exists in the Siberian plain and the Northeast Asian region including the BTH region. The above reasons together lead to the abnormally low temperature in the BTH region, which provides favorable conditions for the occurrence of broken rails.

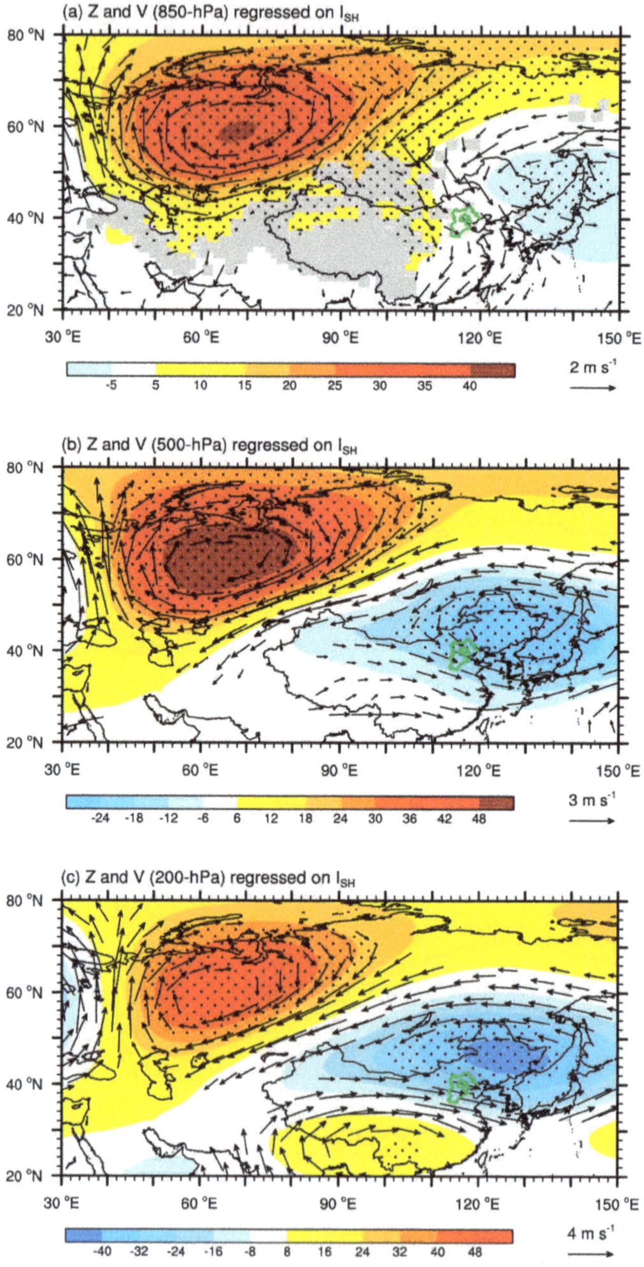

Figure 4. Regressions of I_{SH} with the (**a**) 850 hPa, (**b**) 500 hPa and (**c**) 200 hPa geopotential height onto (shading, units: dgpm) and wind field (arrows, units: ms^{-1}). Regression coefficients exceeding the 95% confidence level are stippled. Only the vectors at the 95% confidence level or higher are shown. Green curve indicates the BTH region.

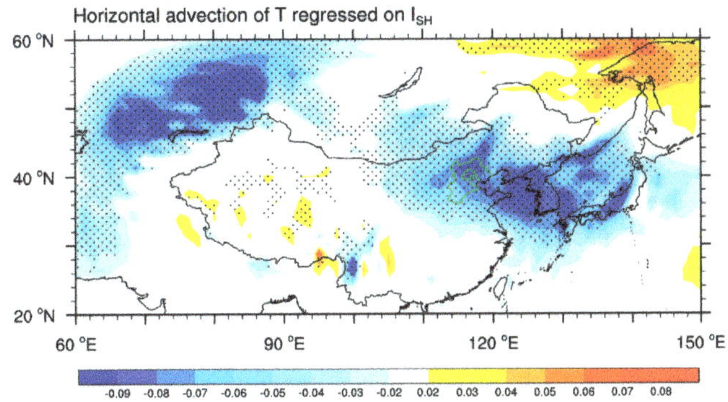

Figure 5. Regressed the vertically integrated temperature advection from the surface to 100 hPa by I_{SH}. Regression coefficients exceeding the 95% confidence level are stippled. Green curve indicates the BTH region.

5. Discussion and Conclusions

Based on monthly TBF data of high-speed railway in BTH region from 2010 to 2020 and ERA5 atmospheric reanalysis data, the spatiotemporal variations of the winter FRB in BTH region and its relationship with Siberian high anomaly are analyzed. The main conclusions are as follows:

1. The spatial and temporal distribution of the total FRB in BTH is very uneven. The maximum FRB in Beijing is 51, and the minimum FRB in Chengde is 1. There are obvious seasonal differences in FRB, with the most frequent rail breaks in winter, followed by spring and autumn, and the least in summer.
2. The spatial and temporal variability in winter FRB of BTH region is also obvious, and its spatial distribution is similar to that of the annual total FRB. The highest and lowest winter FRB are located in Beijing and Chengde respectively, with 32 and 1 broken rails respectively. The percentage of winter FRB in all regions exceeds 60% of the total annual FRB, and some regions, such as Baoding and Chengde, have 100% of the total FRB in winter. There is also a very obvious interannual variation in the winter FRB.
3. A significant positive correlation between the winter FRB in BTH region and the Siberian high is found. When the Siberian high is strong, it is accompanied by positive potential height anomalies and anomalous anticyclonic circulation over the Siberian region, as well as negative potential height anomalies and anomalous cyclonic circulation over the northeastern Asian region. These circulation anomalies show an equivalent barotropic feature in the vertical direction and produce the anomalous northwesterly prevailing over BTH region in the lower troposphere, enhance the East Asia winter monsoon, and deepen East Asian trough in the middle troposphere. All the above circulation anomalies are conducive to the cold air advection from high latitudes to BTH region, causing winter cooling in the BTH region, thus providing low temperatures for rail breakage.

It should be pointed out that this study is limited to the length of FRB data, with the period of 2010–2020. Therefore, it is necessary to analyze the relationship between Siberian high and winter temperature anomaly in BTH region using longer data records, and to clarify the dynamic and thermodynamic mechanisms. Using the meteorological element fields under different future shared socio-economic paths [28] proposed by phase 6 of the Coupled Model Intercomparison Project (CMIP6), the future winter temperatures in BTH region in the context of high and medium forcing scenarios could be predicted, which can provide a background field of meteorological elements for the prediction of future FRB in BTH high-speed railway. In the future, how will the intensity of Siberian high change under different shared socio-economic paths? Will the change of Siberian high intensity cause the winter temperature variations in BTH region? These are also questions worth studying. The FRB is influenced by a number of factors beyond meteorological conditions, such as the sum of track length in a given subregion, age and hardness of the rails, maximum vehicle axle load, total load on the rail line [29], the occurrence of subgrade vibration isolation, type of superstructure [30], objects, such as bridges and tunnels, the value of the neutral temperature assumed in the continuous weld rail (CWR) design, and that actually prevailing during rail jointing [31]. It is also important to investigate the impact of the above factors on FRB in the BTH region. The length of the railway railroads in individual cities of BTH region is uneven. It would be an interesting question to study the temporal variation of FRB in individual cities of the BTH region and their linkage with meteorological factors if more detailed data on high-speed railroads data could be obtained.

In addition, monthly averaged rail breakage data are used in this study. However, it is also important to use the day-by-day rail breakage data to analyze the features of temperature and daily difference of temperature on the day when the rail breakage occurred and to explore the related atmospheric circulation anomaly, so as to provide a scientific basis for predicting the meteorological background of FRB in BTH high-speed railway. These issues will be further studied in the future.

Author Contributions: L.H., J.W. and D.J. conceived and designed the study. L.X. prepared Figures 1 and 2, and D.J. prepared Figures 3–5. L.H. and L.X. wrote the early draft. L.H., J.W., D.J., Y.S. and Q.Z. contributed to writing the manuscript. All authors have read and agreed to the published version of the manuscript.

Funding: This work is jointly supported by the National Natural Science Foundation of China (Grants 42030605, 42005016), and the Natural Science Fundamental Research Project of Jiangsu Colleges and Universities (19KJB170024).

Institutional Review Board Statement: Not applicable.

Informed Consent Statement: Not applicable.

Data Availability Statement: The monthly ERA5 reanalysis data is provided by the European Centre for Medium-Range Forecasting (ECMWF) from their website at https://cds.climate.copernicus.eu/#!/search?text=ERA5&type=dataset on 23 March 2021. The monthly FRB in high-speed railway of BTH region could be obtained through the corresponding author for reasonable request.

Conflicts of Interest: The authors declare no competing interest.

References

1. Wang, Y.; Li, K.; Xu, X.; Zhang, Y. Transport energy consumption and saving in China. *Renew. Sustain. Energy Rev.* **2014**, *29*, 641–655. [CrossRef]
2. Li, X.; Yan, J. *Research on the Development of Railway Industry in China and the Correlation between Railway and National Economy Based on Input-Output Analysis*; Beijing Jiaotong University: Beijing, China, 2016; pp. 1–3.
3. Zhao, H.; Liang, J.; Liu, C. High-Speed EMUs: Characteristics of technological development and trends. *Engineering* **2020**, *6*, 234–244. [CrossRef]
4. Xiao, Q.; Zheng, J.; Liu, J.; Fang, J. Analysis of the wheel/rail rolling contact fatigue of a high-speed train under the transient mechanism. *J. Mech. Sci. Technol.* **2017**, *31*, 2235–2242. [CrossRef]
5. Magel, E.; Mutton, P.; Ekberg, A.; Kapoor, A. Rolling contact fatigue, wear and broken rail derailments. *Wear* **2016**, *366–367*, 249–257. [CrossRef]
6. Shur, E.; Borts, A.; Zakharov, S. Rails for Low Operating Temperature and High Speed. *Transp. Soil Eng. Cold Reg.* **2020**, *1*, 221–232.
7. Chen, W.; Graf, H.F.; Huang, R. The Interannual Variability of East Asian Winter Monsoon and Its Relation to the Summer Monsoon. *Adv. Atmos. Sci.* **2000**, *17*, 48–60.
8. Ding, Y.H. *Monsoons over China*; Springer Science & Business Media: Berlin/Heidelberg, Germany, 1994; p. 432.
9. Thompson, D.W.J.; Wallace, J.M. The Arctic Oscillation signature in the wintertime geopotential height and temperature fields. *Geophys. Res. Lett.* **1998**, *25*, 1297–1300. [CrossRef]
10. Thompson, D.W.J.; Wallace, J.M. Annular modes in the extratropical circulation. Part I: Month-to-month variability. *J. Clim.* **2000**, *13*, 1000–1016. [CrossRef]
11. He, C.; He, J. Relation between Arctic Oscillation and North China Air Temperature in Winter. *J. Nanjing Inst. Meteorol.* **2003**, *26*, 1–7. (In Chinese)
12. Wallace, J.M.; Gutzler, D.S. Teleconnections in the geopotential height field during the Northern Hemisphere winter. *Mon. Weather Rev.* **1981**, *109*, 784–812. [CrossRef]

13. Wang, N.; Zhang, Y. Evolution of Eurasian teleconnection pattern and its relationship to climate anomalies in China. *Clim. Dyn.* **2015**, *44*, 1017–1028. [CrossRef]
14. Luo, B.H.; Luo, D.H.; Dai, A.G.; Simmonds, I.; Wu, L. A connection of winter Eurasian cold anomaly to the modulation of Ural blocking by ENSO. *Geophys. Res. Lett.* **2021**, *48*, e2021GL094304. [CrossRef]
15. Yao, Y.; Zhang, W.; Luo, D.; Zhong, L.; Pei, L. Seasonal Cumulative Effect of Ural Blocking Episodes on the Frequent Cold events in China during the Early Winter of 2020/21. *Adv. Atmos. Sci.* **2022**, *39*, 609–624. [CrossRef]
16. Branstator, G. Circumglobal Teleconnections, the Jet Stream Waveguide, and the North Atlantic Oscillation. *J. Clim.* **2002**, *15*, 1893–1910. [CrossRef]
17. Huang, W.; Yang, Z.; He, X.; Lin, D.; Wang, B.; Wright, J.S.; Chen, R.; Ma, W.; Li, F. A possible mechanism for the occurrence of wintertime extreme precipitation events over South China. *Clim. Dyn.* **2019**, *52*, 2367–2384. [CrossRef]
18. Cohen, J.; Saito, K.; Entekhabi, D. The role of the Siberian high in Northern Hemisphere climate variability. *Geophys. Res. Lett.* **2001**, *28*, 299–302. [CrossRef]
19. Gong, D.; Ho, C.H. The Siberian High and climate change over middle to high latitude Asia. *Theor. Appl. Climatol.* **2002**, *72*, 1–9. [CrossRef]
20. Lu, C.; Zhang, Y.; Guan, Z. Winter anticyclone activities in Siberia and their relationship to the regional temperature anomaly. *Int. J. Climatol.* **2022**, *42*, 6429–6440. [CrossRef]
21. Hersbach, H.; Bell, B.; Berrisford, P.; Hirahara, S.; Horány, A.; Muñoz-Sabater, J.; Nicolas, J.; Peubey, C.; Radu, R.; Scheper, D.; et al. The ERA5 global reanalysis. *Q. J. R. Meteorol. Soc.* **2020**, *146*, 1999–2049. [CrossRef]
22. Wu, J.; Gao, X. A gridded daily observation dataset over China region and comparison with the other datasets. *Chin. J. Geophys.* **2013**, *56*, 1102–1111. (In Chinese)
23. Huo, L.; Wang, J.; Jin, D.; Luo, J.; Shen, H.; Zhang, X.; Min, J.; Xiao, Y. Increased summer electric power demand in Beijing driven by preceding spring tropical North Atlantic warming. *Atmos. Ocean. Sci. Lett.* **2022**, *15*, 100146. [CrossRef]
24. Hasanean, H.M.; Almazroui, M.; Jones, P.D.; Alamoudi, A.A. Siberian high variability and its teleconnections with tropical circulations and surface air temperature over Saudi Arabia. *Clim. Dyn.* **2013**, *41*, 2003–2018. [CrossRef]
25. Hoskins, B.J.; Karoly, D.J. The steady linear response of a spherical atmosphere to thermal and orographic forcing. *J. Atmos. Sci.* **1981**, *38*, 1179–1196. [CrossRef]
26. Takaya, K.; Nakamura, H. Mechanisms of intraseasonal amplification of the cold Siberian high. *J. Atmos. Sci.* **2005**, *62*, 4423–4440. [CrossRef]
27. Maeda, S.; Takemura, K.; Kobayashi, C. Planetary Wave Modulations Associated with the Eurasian Teleconnection Pattern. *J. Meteorol. Soc. Jpn. Ser. II* **2021**, *99*, 449–458. [CrossRef]
28. O'Neill, B.C.; Tebaldi1, C.; Van Vuuren, D.P.; Eyring, V.; Friedlingstein, P.; Hurtt, G.; Knutti, R.; Kriegler, E.; Lamarque, J.F.; Lowe, J.; et al. The Scenario Model Intercomparison Project (ScenarioMIP) for CMIP6. *Geosci. Model Dev.* **2016**, *9*, 3461–3482. [CrossRef]
29. Reddy, V. Modelling and Analysis of Rail Grinding & Lubrication Strategies for Controlling Rolling Contact Fatigue (RCF) and Rail Wear. Ph.D. Thesis, Queensland University of Technology, Brisbane City, QLD, Australia, 2004; p. 43.
30. Ciotlaus, M.; Kollo, G.; Marusceac, V.; Orban, Z. Rail-wheel interaction and its influence on rail and wheels wear. *Procedia Manuf.* **2019**, *32*, 895–900. [CrossRef]
31. Gao, Y.; Wang, P.; Wang, K.; Xu, J.; Dong, Z. Damage tolerance of fractured rails on continuous welded rail track for high-speed railways. *Rail. Eng. Sci.* **2021**, *29*, 59–73. [CrossRef]

atmosphere

MDPI

Article

Time-Series Prediction of Intense Wind Shear Using Machine Learning Algorithms: A Case Study of Hong Kong International Airport

Afaq Khattak [1,*], Pak-Wai Chan [2], Feng Chen [1,*] and Haorong Peng [3]

1 The Key Laboratory of Infrastructure Durability and Operation Safety in Airfield of CAAC, Tongji University, 4800 Cao'an Road, Jiading, Shanghai 201804, China
2 Hong Kong Observatory, 134A Nathan Road, Kowloon, Hong Kong, China
3 Shanghai Research Center for Smart Mobility and Road Safety, Shanghai 200092, China
* Correspondence: khattak@tongji.edu.cn (A.K.); fengchen@tongji.edu.cn (F.C.)

Abstract: Machine learning algorithms are applied to predict intense wind shear from the Doppler LiDAR data located at the Hong Kong International Airport. Forecasting intense wind shear in the vicinity of airport runways is vital in order to make intelligent management and timely flight operation decisions. To predict the time series of intense wind shear, Bayesian optimized machine learning models such as adaptive boosting, light gradient boosting machine, categorical boosting, extreme gradient boosting, random forest, and natural gradient boosting are developed in this study. The time-series prediction describes a model that predicts future values based on past values. Based on the testing set, the Bayesian optimized-Extreme Gradient Boosting (XGBoost) model outperformed the other models in terms of mean absolute error (1.764), mean squared error (5.611), root mean squared error (2.368), and R-Square (0.859). Afterwards, the XGBoost model is interpreted using the SHapley Additive exPlanations (SHAP) method. The XGBoost-based importance and SHAP method reveal that the month of the year and the encounter location of the most intense wind shear were the most influential features. August is more likely to have a high number of intense wind-shear events. The majority of the intense wind-shear events occurred on the runway and within one nautical mile of the departure end of the runway.

Keywords: wind shear; time-series modeling; machine learning; Bayesian optimization

Citation: Khattak, A.; Chan, P.-W.; Chen, F.; Peng, H. Time-Series Prediction of Intense Wind Shear Using Machine Learning Algorithms: A Case Study of Hong Kong International Airport. *Atmosphere* **2023**, *14*, 268. https://doi.org/10.3390/atmos14020268

Academic Editors: Duanyang Liu, Hongbin Wang and Shoupeng Zhu

Received: 27 December 2022
Revised: 25 January 2023
Accepted: 27 January 2023
Published: 28 January 2023

1. Introduction

Wind shear is a potentially hazardous meteorological occurrence characterized by sudden changes in wind speed and/or direction. If this event occurs below 500 m (1600 feet) above the ground, it is classified as low-level wind shear; if its magnitude exceeds 30 knots, it is known as intense wind shear [1]. It is one of the most worrisome phenomena for an aircraft because it creates violent turbulence and eddies as well as dramatic shifts in the aircraft's horizontal and vertical progression, which can ultimately result in a frequent missed approach, touching down short of the runway (loss of lift), or deviation from the true flight path during landing descent, as depicted in Figure 1. The intense wind shear has two potentially dangerous effects on landing aircraft: aberration of the flight path and deviation from the set approach speed [2]. Due to unanticipated changes in wind speed or direction, the pilot may perceive immense pressure during the landing phase when the engine power is low and the airspeed is close to stall speed.

Numerous airports around the world have reaped substantial benefits from the availability of precise, high-resolution, remote sensing technologies such as the Terminal Doppler Weather Radar (TDWR) [3] and the Doppler Light Detection and Range (LiDAR) [4,5]. By a significant margin, the most prevalent methods for detecting wind shear are TDWR, ground-based anemometer networks, and wind profilers. Since the mid-1990s, this method has proved effective for alerting airports to wind shear, particularly during the passage of tropical cyclones and thunderstorms. Clear weather prevents the TDWR system from providing accurate wind data. However, certain wind-shear events are associated with airflow reaching the airport from rugged terrain. To address these circumstances, a new method of detection independent of humidity must be developed. For this purpose, the LiDAR system has been added to the TDWR as a booster in order to detect and warn of wind shear in clear skies. Doppler LiDAR can detect return signals from aerosols and provide precise Doppler wind measurements when the air is clear. Although these tracking or observation-based technological advances are effective at detecting wind shear in the vicinity of an airport, they are unable to predict when the next wind-shear event will occur, or which risk factors contribute to its occurrence [6]. Forecasting intense wind shear in the vicinity of the airport runway and the factors that contribute to the occurrence of intense wind shear are of the utmost importance, as their occurrence can cause significant challenges for departing and approaching flights.

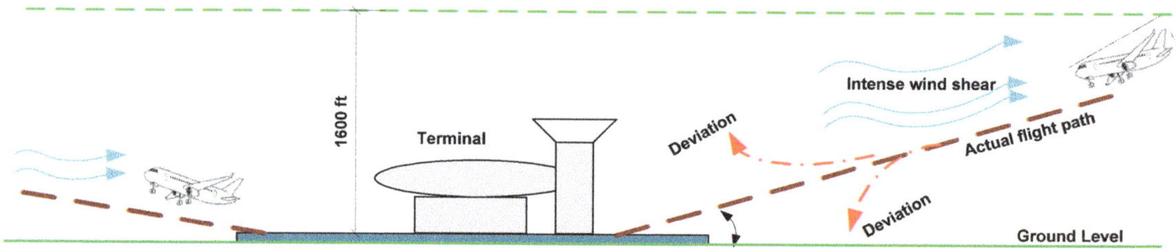

Figure 1. Intense wind shear effect on landing aircraft.

The development of a framework for the prediction of intense wind shear requires a substantial amount of historical data on wind-shear events. Despite the fact that numerous researchers in the power and energy domain have attempted to forecast wind speed due to the demand for wind energy electricity generation and advancements in wind energy competitiveness [7–9], few researchers have attempted to forecast wind-shear events in the vicinity of airport runways [10,11]. For time-series modeling, several statistical and mathematical techniques have been employed in the past, such as autoregressive integrated moving average (ARIMA) [12–14], Kolmogorov–Zurbenko filters [15,16], exponential smoothing [17,18], and others. These often result in good forecasting accuracy. However, machine learning algorithms have recently been applied in various domains due to their high forecasting precision and improved operational efficiency [19–24]. Therefore, in this study, we propose the development of time-series prediction models of intense wind shear using machine learning algorithms. The study employed Doppler LiDAR data from 2017 to 2010 and machine learning algorithms including the Adaptive Boosting (AdaBoost) [25], Light Gradient Boosting Machine (LightGBM) [26], Categorical Boosting (CatBoost) [27], Gradient Boosting (XGBoost) [28], Random Forest [29], and Natural Gradient Boosting (NGBoost) [30] methods, optimized via a Bayesian optimization approach [31], as shown in Figure 2.

In addition to evaluating the performance of models in order to select the optimal model, crucial factors that contribute to the occurrence of intense wind shear are also revealed. Researchers in the field of civil aviation safety should seize this opportunity as understanding the complex interactions between multiple risk factors that determine the occurrence of intense wind shear is essential for aviation and meteorological applications.

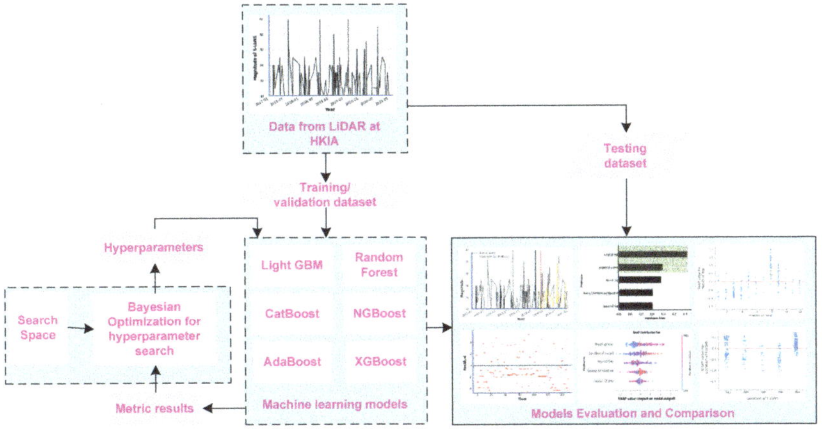

Figure 2. Framework for the time-series prediction of intense wind-shear event.

2. Data and Methods

2.1. Study Location

Hong Kong International Airport (HKIA) is among the most susceptible airports in the world to the occurrence of wind-shear events, and from 1998 to 2015 a significant number of intense wind-shear events were documented. Wind-shear events occur once every 400 to 500 flights, according to HKIA-based pilot flight reports [32]. The airport is situated on Lantau Island, surrounded on three sides by open sea water and by mountains to the south that reach heights of more than 900 m above sea level. As is illustrated in Figure 3, the mountainous terrain to the south of the HKIA exacerbates wind shear by disrupting the flow of air and producing turbulence along the HKIA flight paths. Previously, HKIA had two runways: the north and south runways. However, a newly constructed runway (third runway) implies that the former north runway is now designated as the central runway. These are oriented at 070 degrees and 250 degrees. There are a total of eight possible configurations because each runway can be utilized for takeoffs and landings in either direction. For instance, runway '07LA' indicates landing ('A' refers to arrival), with a heading angle of 070° (abbreviated to '07') utilizing the left runway (hence 'L'). This depiction demonstrates aircraft landing on the North Runway from the western side of the HKIA. Similarly, an aircraft taking off from the South Runway in the west would use runway 25LD.

Figure 3. HKIA and surrounding terrain.

2.2. Data Processing from Doppler LiDAR

The Doppler LiDAR at the HKIA detects the magnitude and reports the location of occurrence of wind-shear events. Figure 4 depicts an illustration of a radial velocity plot obtained from a Plan Position Indicator (PPI) scan of the HKIA's south runway LIDAR at an elevation angle of 3° from the horizon. To the west and south of the location, three nautical miles (5.6 km) west-southwest of the western end of the south runway, there was a huge area of winds in the opposite direction (colored green in Figure 4) to the dominant east–southeast airflow.

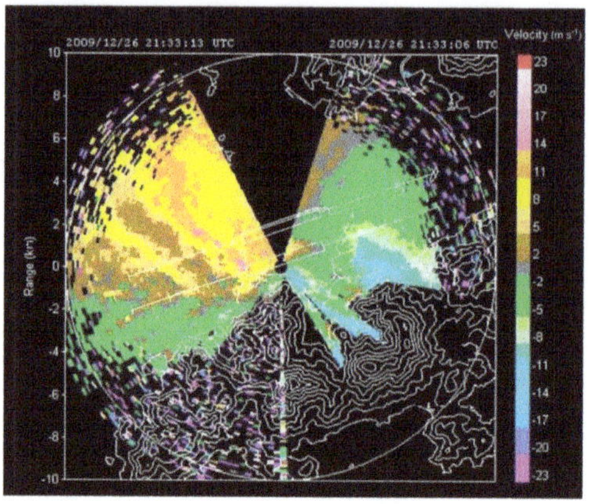

Figure 4. Wind shear detection by LIDAR.

The development of our time-series prediction models required a substantial amount of intense wind shear data for our research. Therefore, we first extracted the 2017 to 2020 wind shear data from LiDAR and filtered it to obtain only intense wind-shear events, i.e., wind shear with a magnitude greater than or equal to 30 knots. The filtration produced 3781 intense wind shear data points, which are presented in Table 1. Previous research [11] on the wind shear prediction utilized hourly data from pilot reports and weather reports, which resulted in lower accuracy due to the transient and sporadic nature of wind shear. In several instances at the HKIA, the Doppler LiDAR reported intense wind shear intervals as short as 1 min; consequently, we have considered these instances. As an example, from Table 1, we can observe that on 29 March 2019 intense wind-shear events of 37 knots and 39 knots were detected at 10:12 PM and 10:14 PM (at a 2 min interval) on runways 07CA and 07RA, respectively. The encounter locations are designated as either RWY, MD, or MF, as is shown in Figure 5. The rectangle in gray denotes the runway (RWY). On the right side of the runway, the rectangles indicate the distance in miles to the final approach (1-MF is equal to 1 nautical mile to the final approach). Likewise, the rectangles on the left indicate the distance from the runway's departure end. For instance, 2-MD indicates two nautical miles from the runway's edge at the departure end.

Table 1. Sample of extracted data from HKIA-based LiDAR.

Date	Time	Runway	Intense Wind Shear Magnitude	Encounter Location
16 May 2017	5:17 PM	07RA	35 knots	RWY
19 June 2017	5:19 PM	25LA	32 knots	1-MD
—	—	—	—	—
—	—	—	—	—
29 March 2019	10:12 PM	07CA	37 knots	RWY
29 March 2019	10:14 PM	07RA	39 knots	RWY
—	—	—	—	—
—	—	—	—	—
21 September 2020	3:58 AM	07RA	30 knots	2-MF

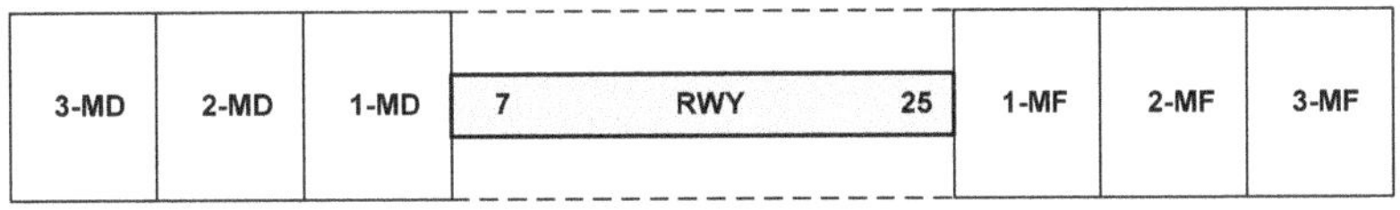

Figure 5. Schematic diagram for the representation of intense wind shear encounter locations.

2.3. *Machine Learning Regression Algorithms*

In this study, six machine learning regression algorithms were employed for the time-series prediction of intense wind-shear events, including LightGBM, XGBoost, NGBoost, AdaBoost, CatBoost, and RF. The fundamentals of the regression algorithm are described as follows:

2.3.1. Light Gradient Boosting Machine (LightGBM) Regression

LightGBM is a gradient learning framework that is based on decision trees and the concept of boosting. It is a variant of gradient learning. Its primary distinction from the XGBoost model is that it employs histogram-based schemes to expedite the training phase while lowering memory usage and implementing a leaf-wise expansion strategy with depth constraints. The fundamental concept of the histogram-based scheme is to partition continuous, floating-point eigenvalues into '*k*' bins and build a histogram with a width of *k*. It does not require the additional storage of presorted outcomes and can also save the value after the partitioning of features, which is usually adequate to store with 8-bit integers, thereby lowering memory consumption to 1/8 of the original. This imprecise partitioning has no effect on the model's precision. It is irrelevant whether the segmentation point is accurate or not because the decision tree is a weak study model. The regularization effect of the coarser segmentation points can also successfully prevent over-fitting.

Several hyperparameters must be adjusted for the LightGBM regression model to prevent overfitting, reduce model complexity, and achieve generalized performance. These hyperparameters are *n_estimators*, which is the number of boosted trees to fit, *num_leaves*, which is the maximum number of tree leaves for the base learners, *learning_rate*, which controls the estimation changes, *reg_alpha*, which is the L1 regularization term on weights, and *reg_lambda*, which is the L2 regularization term on model weights.

2.3.2. Extreme Gradient Boosting (XGBoost) Regression

XGBoost is a tree-based boosting technique variant. Fundamentally, XGBoost reveals the functional relationship, Γ, between the input factors x and the response y via an iterative procedure wherein individual, independent trees are trained in a sequential manner on the residuals from the preceding tree. The mathematical expression for the tree-based estimates is given by Equation (1).

$$\hat{Y} = \Gamma(X) = \frac{1}{n}\sum_{k=1}^{n} \Gamma_k(X) \quad (1)$$

where $\hat{Y}$ represents the predictions and n illustrates the total number of trees. The regularized objective function, $\Psi(\Omega)$, is minimized to learn the set of functions Γ_k, which are employed in the model, as shown by Equations (2) and (3).

$$\Psi(\Omega) = \sum_i \lambda(\hat{y}_i, y_i) + \sum_k \Pi(\Gamma_k) \quad (2)$$

$$\Pi(\Gamma_k) = \phi T + \frac{1}{2} l \|\omega\|^2 \quad (3)$$

where λ represents the differentiable convex loss function that estimates the difference between the prediction and actual response. The term Π is an additional regularization expression that panelizes the growth of further trees in the model to reduce intricacies and over-fitting. The term ϕ represents the leaf's complexity, and T is the total number of leaves in a tree. Likewise, for the XGBoost regression model, hyperparameters including the *n_estimators, num_leaves, learning_rate, reg_alpha,* and *reg_lambda* must be optimized to prevent overfitting and reduce model complexity.

2.3.3. Natural Gradient Boosting (NGBoost) Regression

NGBoost is a supervised learning technique with basic probabilistic prediction capabilities. A probabilistic prediction generates a complete probability distribution over a whole outcome space, allowing users to evaluate the uncertainty in the model's predictions. In conventional point prediction configurations, the object of concern is an estimate of the scalar function, $\Phi(y|x)$, in which x represents a vector of different factors and y is the response, but uncertainty estimates are not considered. In a probabilistic prediction context, on the other hand, a stochastic forecast with a probability distribution, $\Theta_\theta(y|x)$, is generated by predicting the parameters θ. Provided that NGBoost is intended to be scalable and modular with respect to the base estimator (for instance the decision trees), probability distribution parameter (for instance, normal, Laplace, etc.), and scoring rule, NGBoost can perform probabilistic forecasts with flexible, tree-based models (for instance, the Maximum Likelihood Estimation). As is depicted in Figure 6, the input vector of the different factors x in the hybrid NGBoost model is forwarded to the base estimator (decision trees) to generate a probability distribution, $\Theta_\theta(y|x)$, over the a whole outcome space, y. The models are then improved using a scoring rule, $S(\Theta_\theta, y)$, that produces calibrated uncertainty and point predictions using a maximum likelihood estimation function. Prior to evaluation, the NGBoost regression model parameters *n_estimators* and the *learning_rate* must be optimized.

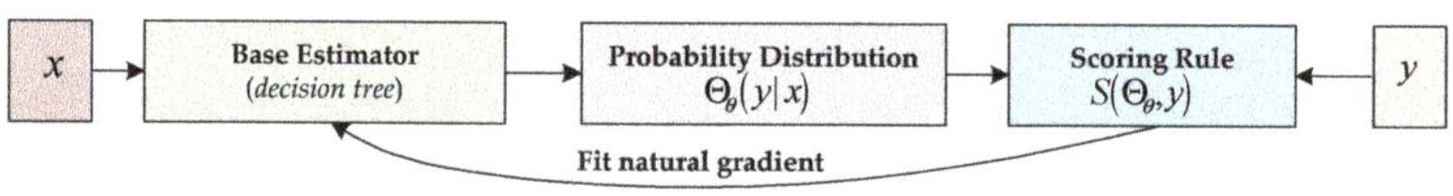

Figure 6. Mechanism of NGBoost regression algorithm.

2.3.4. Categorical Boosting (CatBoost) Regression

CatBoost is an innovative, gradient-boosting decision tree technique. It is capable of handling categorical factors and employ them in the training phase rather than in preprocessing phase. CatBoost's advantage is that it utilizes a new pattern to determine the leaf values while choosing the tree structure, which aids in reducing over-fitting and enables the utilization of the entire training data set, i.e., it organizes the data of each instance randomly and quantifies the mean value of the instances. For the regression problem, the average of the acquired data must be utilized for a priori estimations. The parameters for the CatBoost regression model that must be optimized prior to evaluation are *n_estimators, max_depth,* and the *learning_rate.*

2.3.5. Adaptive Boosting (AdaBoost) Regression

Adaptive Boosting Regression is a straightforward ensemble learning model which creates a powerful regressor by integrating several weak learners, resulting in a high-accuracy model. The core concept is to establish the weights of weak regressors and train the dataset at each iteration such that reliable projections of unusual observations may be made. The working principle of AdaBoost is provided below:

- The weight distribution (π) is initialized as $\pi = 1/m$;
- At iteration t, the weak learning is trained, i.e., h_t: $x \rightarrow \Re$, using the weight distribution;
- The weight distribution is updated in accordance with previous instances of the training dataset as $\pi_k = \frac{\pi_{k-1} exp^{-\psi_k h(x_k)}}{\Omega}$;
- The final output over all the iterations $t = 1, 2, \ldots, T$ is returned as $f(X) = \sum_{t=1}^{T} \pi_t h_t(X)$ and $H(X) = sign(f(X))$.

The AdaBoost model uses a decision stump as a weak learner. The critical hyperparameters that need to be tuned during the learning process are the *n_estimators* and *learning_rate*. The *n_estimators* are the number of decision stump to train iteratively and the *learning_rate* controls the contribution of each learner. There is required to be a trade-off between both the *n_estimators* and *learning_rate*.

2.3.6. Random Forest (RF) Regression

The RF is an ensemble of tree-based predictors in which each tree is trained with values of an independently sampled random vector that has the same distribution for all other trees in the forest. The kth tree is conceptually trained using an independent random vector, ζ_k, with the same distribution as previous random vectors, ζ_{k-1}, resulting in a tree, $\psi(X, \zeta_k)$, in which X is the input vector of different factors. When a large number of trees are grown in a forest, their mean predictions are obtained, which improves the accuracy of predictions and prevents over-fitting. Mathematically, it can be illustrated as Equation (4).

$$\hat{Y} = \frac{1}{l}\sum_{k=1}^{l} \psi_k(X) \tag{4}$$

where $\hat{Y}$ represent the response and l is the total amount of generated trees $(1 \leq k \leq l)$. The mean squared generalization error of any tree $\psi(X)$ is illustrated as $E_{X,Y}(Y - \psi(X))^2$ for the input vector of difference (X) and the response vector (Y). As the number of trees in the forest approaches the infinity, the mean squared generalization almost certainly becomes:

$$E_{X,Y}(Y - \Lambda_k \psi(X, \zeta_k))^2 \rightarrow E_{X,Y}(Y - E_\zeta \psi(X, \zeta))^2 \tag{5}$$

A few crucial hyperparameters must be tuned during the learning phase in order to achieve an optimized prediction score for the RF model. These hyperparameters are the *n_estimators*, which is the number of trees in the forest, and the *max_depth*, which is the maximum number of levels, or branches between the root node and the deepest leaf node.

2.4. *Principle of Bayesian Optimization*

The structure parameters of a machine learning model are its hyperparameters. Adapting a machine learning model to multiple situations requires adjusting the hyperparameters [33,34]. In this study, a Bayesian hyperparameter optimization method is implemented. The goal is to establish the mapping, $f(x, \theta)$, in which y is the response, x is the input vector, and the θ vector determines the size of the mapping. The core principle of Bayesian optimization is adjusting the hyperparameter of a given model in order to formulate a model of the loss function. It utilizes a loss function to efficiently search for and select the optimal set of hyperparameters. Employing the hyperparameter θ in a tree-based machine learning model as one of the points in the multidimensional search space for the optimization, the hyperparameter that minimizes the loss function value, $f(\theta)$, can be found in the set $A \in X^d$, as shown by Equation (6).

$$\theta^* = \underset{\theta \in A}{\operatorname{argmin}} f(\theta) \tag{6}$$

Usually, there is no prior information about the model's structure; therefore, it is assumed that the noise in the observation is shown by Equation (7).

$$y(\theta) = f(\theta) + \varepsilon, \text{ and } \varepsilon \sim N\left(0, \sigma_{noise}^2\right) \tag{7}$$

The Bayesian framework offers two fundamental options. First, a hypothesis function. $p(f|\mathcal{D})$ (also known as a prior function). must be chosen to represent the hypothesis of the function to be optimized. Second, the posterior model determines the acquisition function for determining the subsequent test point. Using the prior function,$p(f|\mathcal{D})$, the Bayesian framework constructs a loss function model based on an observed data sample, $\mathcal{D}$. The prior function model, $p(f|\mathcal{D})$, chooses between optimization and development based on its characteristics.

2.5. Performance Assessment

The generalization capacity of various machine learning regression models could be synthetically quantified using four different metrics: the mean absolute error (MAE), mean squared error (MSE), root mean squared error (RMSE), and the R-square (R^2, coefficient of determination). According to Equation (8), the MAE is the average of the individual prediction errors' absolute values across all instances. The average squared difference between observed and predicted values, as shown in Equation (9) is how the MSE computes regression model error. According to Equation (10), the RMSE is the square root of the difference between the observed and predicted values. A regression model's ability to accurately predict values is indicated by R^2, which ranges from 0 to 1. R^2 is provided by Equation (11).

$$\mathrm{MAE} = \sum_{\chi=1}^{\Phi} \frac{|y_\chi - \hat{y}_\chi|}{\chi} \tag{8}$$

$$\mathrm{MSE} = \frac{1}{\chi}\sum_{\chi=1}^{\Phi} (y_\chi - \hat{y}_\chi)^2 \tag{9}$$

$$\mathrm{RMSE} = \sqrt{\sum_{\chi=1}^{\Phi} \frac{(y_\chi - \hat{y}_\chi)^2}{\chi}} \tag{10}$$

$$R^2 = 1 - \frac{\sum_{\chi=1}^{\Phi} (y_\chi - \hat{y}_\chi)^2}{\sum_{\chi=1}^{\Phi} (y_\chi - y_{avg})^2} \tag{11}$$

where χ is the total number of observations, y represents the actual observation value, and $\hat{y}$ represents the predicted value.

3. Results and Discussion

The LiDAR data of 2017 to 2020 from the Hong Kong Observatory and the aviation weather forecast department at HKIA were used to train and test six different machine learning regression models with the goal of determining how well these models can predict the occurrence of intense wind-shear events. Figure 7a depicts the total LiDAR-obtained intense wind-shear data from 1 January 2017 to 31 December 2020. The data from 1 January 2017 to 31 December 2019 are the training set, which is depicted by the black line in Figure 7b, while the data from 1 January 2020 to 31 December 2020 are the test set, which is depicted by the green line. The vertical red line with dashes divides the training data from the test data.

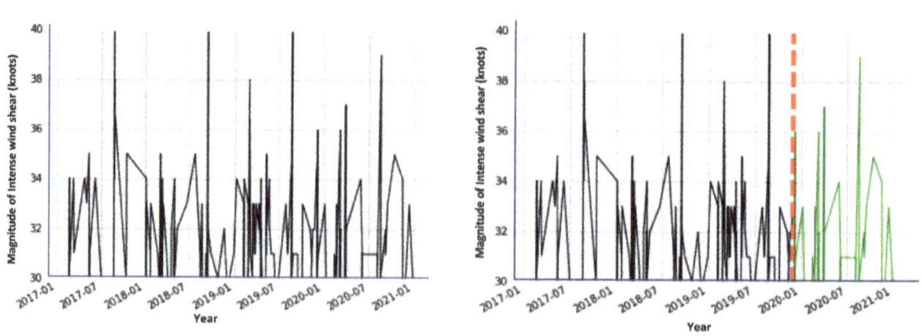

(a) (b)

Figure 7. LiDAR data: (a) 2017–2020 intense wind shear data; (b) splitting data into train and test sets.

The statistical information of the intense wind shear dataset is shown in Table 2. The machine learning models, coupled with Bayesian optimization and a 5-fold cross validation, provide the predicted results based on the optimal hyperparameters. The Hyperopt python package was used for the implementation of Bayesian optimization. The optimal hyperparameters with search space are shown in Table 3. Table 4 shows the comparison of the prediction performance of the machine learning regression algorithms. The predicted intense wind shear values, based the on machine learning regression algorithms, are plotted in Figure 8, and the residual errors by the machine learning models are shown by the scatter plots (Figure 9). In addition, feature importance and contribution are illustrated by Figure 10, and the effect of important factors is shown by Figure 11.

Table 2. Statistical information of intense wind shear from HKIA-based LIDAR.

Dataset	Max	Median	Min	Mean	St. Dev
Entire dataset	40	33	30	33.881	2.596
Train dataset	40	33	30	33.743	2.455
Test dataset	40	34	30	33.921	2.366

Table 3. Optimal hyperparameters of machine learning regression algorithms.

Algorithm	Hyperparameters	Range	Optimal Values
LightGBM	{(*n_estimators*), (*num_leaves*), (*learning rate*), (*reg_lambda*), (*reg_alpha*)}	{(100–1500), (30–100), (0.001–0.2), (1.1–1.5), (1.1–1.5)}	{1180, 28, 0.10, 1.19, 1.01}
CatBoost	{(*n_estimators*), (*max_depth*), (*learning rate*)}	{(200–1500), (2–15), (0.001–0.2)}	{1060, 8, 0.08}
AdaBoost	{(*n_estimators*), (*learning rate*)}	{(100–1500), (0.001–0.2)}	{790, 0.04}
RF	{(*n_estimators*), (*max_depth*)}	{(50–1000), (2–15)}	{955, 5}
XGBoost	{(*n_estimators*), (*num_leaves*), (*learning rate*), (*reg_lambda*), (*reg_alpha*)}	{(100–1500), (30–100), (0.001–0.2), (1.1–1.5), (1.1–1.5)}	{880, 65, 0.05, 1.18, 1.40}
NGBoost	{(*n_estimators*), (*learning rate*)}	{(100–1500), (0.001–0.2)}	{1130, 0.03}

Table 4 demonstrates that the Bayesian optimized-XGBoost model outperforms other machine learning models with a minimum MAE value of 1.764, an MSE value of 5.611, an RMSE value of 2.368, and a maximum R-square value of 0.859. The AdaBoost model, with an MAE of 1.863, MSE of 6.815, RMSE of 2.610, and an R-square value of 0.549, performs the worst. In addition, an analysis of Figure 8 reveals that XGBoost appears to provide a better fit of the actual test intense wind shear time-series and a smaller residual error, represented by red dots closer to horizontal line, when compared to other forecasting results (Figure 9).

Table 4. Performance assessment of Bayesian optimized machine learning models.

Models	Performance Metrics			
	MAE	MSE	RMSE	R-Square
LightGBM	1.813	5.840	2.416	0.711
NGBoost	1.858	6.298	2.509	0.619
Random Forest	1.851	6.194	2.488	0.647
CatBoost	1.795	5.783	2.404	0.753
XGBoost	1.764	5.611	2.368	0.859
AdaBoost	1.863	6.815	2.610	0.549

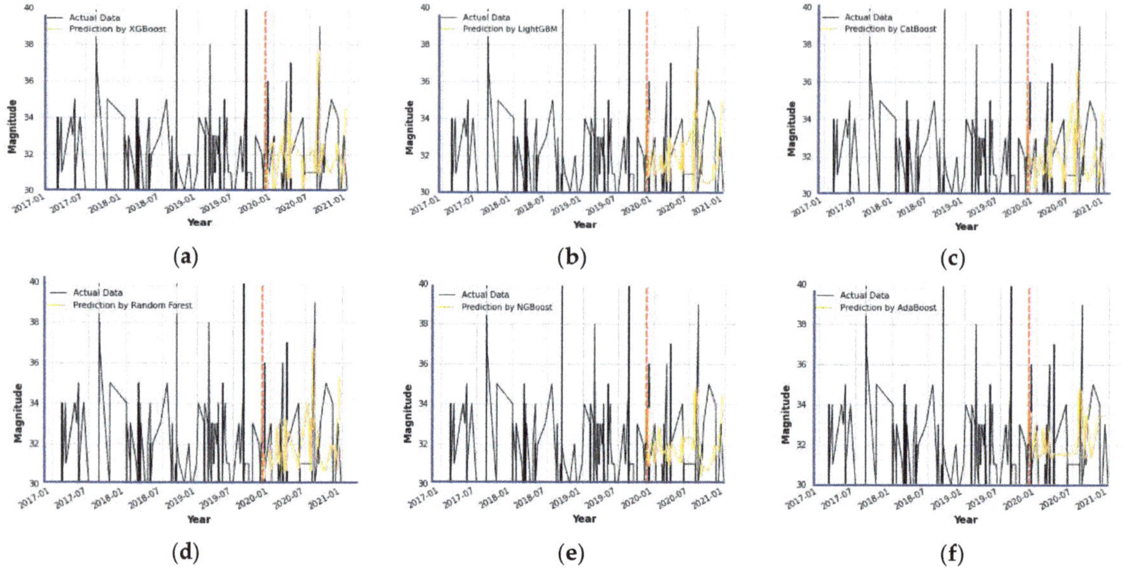

Figure 8. Predictions using machine learning models: (**a**) prediction of intense wind shear by XGBoost; (**b**) prediction of intense wind shear by LightGBM; (**c**) prediction of intense wind shear by CatBoost; (**d**) prediction of intense wind shear by Random Forest; (**e**) prediction of intense wind shear by NGBoost; and (**f**) prediction of intense wind shear by AdaBoost.

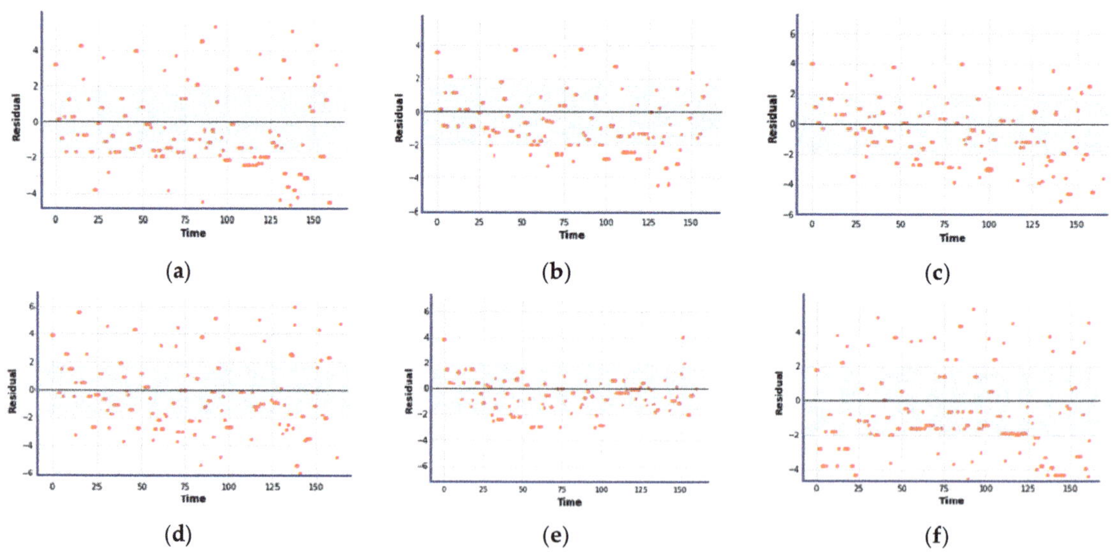

Figure 9. Residual analysis by machine learning regression models; (**a**) NGBoost; (**b**) LightGBM; (**c**) CatBoost; (**d**) Random Forest; (**e**) XGBoost; and (**f**) AdaBoost.

The importance and contribution of the factors are depicted in Figure 10 and are based on the importance score that was determined by the Bayesian optimized-XGBoost model and the XGBoost-based SHAP contribution plot, respectively. In both cases, it was observed that the month of year was the most significant feature, with an importance score of 0.33, followed by the location of intense wind shear (0.19), the hour of the day (0.18), and runway orientation (0.16). Figure 10b revealed that months of the year coded by lower values are less likely to cause intense wind shear, in contrast to those with medium values. Similarly, the location of an encounter with intense wind shear, represented by higher values, is more likely to cause intense wind shear. In the following section, each important feature that plays a role in the occurrence of intense wind shear is discussed in more detail.

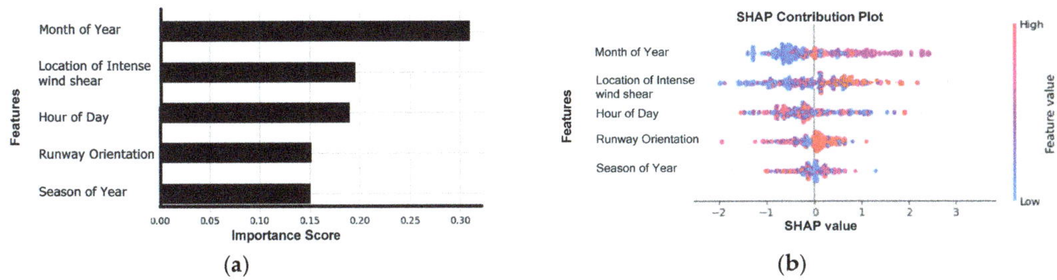

Figure 10. Importance and contribution plots: (**a**) XGBoost-based feature importance plot and (**b**) XGBoost-based SHAP contribution plot.

Figure 11a,b depict the scatter plot of two significant factors. Figure 11a illustrates that the highest number of intense wind-shear events were recorded in August. The intense wind shear in August might be due to cross-mountain airflow, which occurs over the HKIA in August and September, during the south-west monsoon, or during passages of tropical cyclones. These terrain-disrupted airflows cause a number of intense wind-shear events, which negatively impact HKIA's flight safety and operations. This is also consistent with the previous study [11,35].

On the RWY and 1-MD from the edge of the RWY, a large number of intense wind-shear events are observed, as shown in Figure 11b. A small number of intense wind-shear events were observed as the distance increases from the RWY. To the best of our knowledge, none of the previous studies have pinpointed the location where intense wind shear is most prevalent. Nevertheless, our research indicates that RWY and 1-MD from edge of RWY are crucial to the occurrence of intense wind shear. Pilots must maintain vigilance at 1-MD during takeoff.

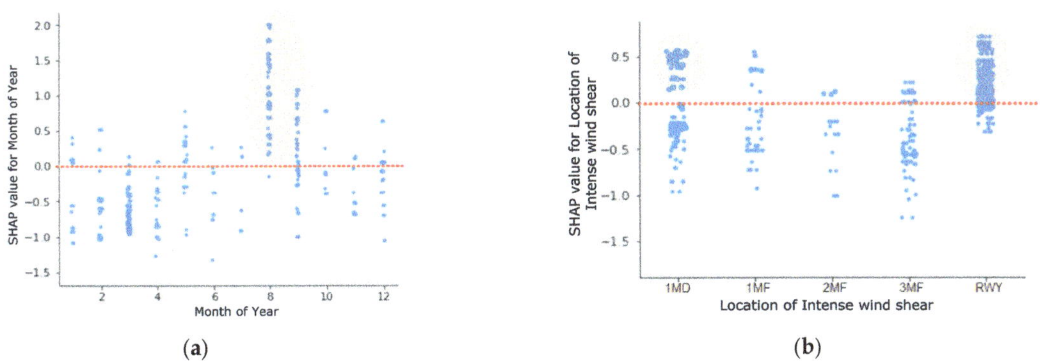

Figure 11. Effect of factors on the Intense wind shear: (**a**) month of year and (**b**) encounter location of intense wind shear.

4. Conclusions and Recommendations

This study is a first attempt at developing a time-series prediction model of intense wind-shear events based on HKIA-based LiDAR data. Six state-of-the-art machine learning regression algorithms, optimized via the Bayesian optimization approach, were employed in this regard. The HKIA-based LiDAR data from 2017 to 2020 was used as the input. From this study, the following conclusions can be drawn:

- On the testing dataset (intense wind-shear data of HKIA-based LiDAR from 1 January 2020 to 31 December 2020), the Bayesian optimized-XGBoost model had the best overall performance of all the optimized machine learning regression models, with an MAE (1.764), MSE (5.611), RMSE (2.368), and R-square (0.859), which was followed by Bayesian optimized-CatBoost model, which had an MAE (1.795), MSE (5.783), RMSE (2.404), and R-square (0.753);
- The AdaBoost regression model demonstrated the lowest performance in terms of MAE (1.863), MSE (6.815), RMSE (2.610), and R-square (0.549);
- The Bayesian optimized-XGBoost model demonstrated that the month of year was the most influential factor, followed by distance of occurrence of intense wind shear from the RWY;
- August is more likely to have intense wind-shear events. Similarly, most of the intense wind-shear events are expected to occur at RWY and 1-MD from the runway departure end. The pilots are required to be cautious during takeoff.

For aviation authorities and researchers interested in aviation safety, the methodology put forth in this study can be used to conduct an extensive investigation of intense wind shear. The study covered in this paper was the time-series prediction of intense wind shear using six machine learning models coupled with a Bayesian optimization approach. Future research might use an amalgamation of a stacking ensemble and various other machine learning ensemble algorithms with a number of additional risk factors, such as the impact of atmospheric pressure and temperature. In addition, the causes of the occurrence of wind shear (weather- or terrain-induced) could be used in future research.

Author Contributions: Conceptualization, A.K. and P.-W.C.; data curation, P.-W.C.; formal analysis, A.K.; funding acquisition, F.C.; investigation, P.-W.C.; methodology, A.K.; project administration, F.C.; resources, H.P.; software, F.C.; supervision, P.-W.C.; validation, F.C.; visualization, H.P.; writing—original draft, A.K.; writing—review & editing, H.P. All authors have read and agreed to the published version of the manuscript.

Funding: This research was supported by National Natural Science Foundation of China (U1733113), the Shanghai Municipal Science and Technology Major Project (2021SHZDZX0100), the Research Fund for International Young Scientists (RFIS) of the National Natural Science Foundation of China (NSFC) (Grant No. 52250410351) and the National Foreign Expert Project (Grant No. QN2022133001L).

Institutional Review Board Statement: Not applicable.

Informed Consent Statement: Not applicable.

Data Availability Statement: The data are not publicly available due to restrictions.

Acknowledgments: We are grateful to the Hong Kong International Airport Observatory for providing us with LiDAR data.

Conflicts of Interest: The authors declare no conflict of interest.

References

1. Chan, P.W. Severe wind shear at Hong Kong International airport: Climatology and case studies. *Meteorol. Appl.* **2017**, 3, 397–403. [CrossRef]
2. Bretschneider, L.; Hankers, R.; Schönhals, S.; Heimann, J.M.; Lampert, A. Wind Shear of Low-Level Jets and Their Influence on Manned and Unmanned Fixed-Wing Aircraft during Landing Approach. *Atmosphere* **2021**, *13*, 35. [CrossRef]
3. Michelson, M.; Shrader, W.; Wieler, J. Terminal Doppler weather radar. *Microw. J.* **1990**, *33*, 139.

4. Shun, C.; Chan, P. Applications of an infrared Doppler lidar in detection of wind shear. *J. Atmos. Ocean. Technol.* **2008**, *25*, 637–655. [CrossRef]
5. Li, L.; Shao, A.; Zhang, K.; Ding, N.; Chan, P.-W. Low-level wind shear characteristics and LiDAR-based alerting at Lanzhou Zhongchuan International Airport, China . *J. Meteorol. Res.* **2020**, *34*, 633–645. [CrossRef]
6. Hon, K.-K. Predicting low-level wind shear using 200-m-resolution NWP at the Hong Kong International Airport. *J. Appl. Meteorol. Climatol.* **2020**, *59*, 193–206. [CrossRef]
7. Zhang, Y.; Pan, G.; Chen, B.; Han, J.; Zhao, Y.; Zhang, C. Short-term wind speed prediction model based on GA-ANN improved by VMD. *Renew. Energy* **2020**, *156*, 1373–1388. [CrossRef]
8. Zhang, Z.; Ye, L.; Qin, H.; Liu, Y.; Wang, C.; Yu, X.; Yin, X.; Li, J. Wind speed prediction method using shared weight long short-term memory network and Gaussian process regression. *Appl. Energy* **2019**, *247*, 270–284.
9. Cai, H.; Jia, X.; Feng, J.; Li, W.; Hsu, Y.M.; Lee, J. Gaussian Process Regression for numerical wind speed prediction enhancement. *Renew. Energy* **2020**, *146*, 2112–2123. [CrossRef]
10. Khattak, A.; Chan, P.W.; Chen, F.; Peng, H. Prediction and Interpretation of Low-Level Wind Shear Criticality Based on Its Altitude above Runway Level: Application of Bayesian Optimization–Ensemble Learning Classifiers and SHapley Additive explanations. *Atmosphere* **2022**, *12*, 2102. [CrossRef]
11. Chen, F.; Peng, H.; Chan, P.W.; Ma, X.; Zeng, X. Assessing the risk of windshear occurrence at HKIA using rare-event logistic regression. *Meteorol. Appl.* **2020**, *27*, e1962. [CrossRef]
12. Singh, R.K.; Rani, M.; Bhagavathula, A.S.; Sah, R.; Rodriguez-Morales, A.J.; Kalita, H.; Nanda, C.; Sharma, S.; Sharma, Y.D.; Rabaan, A.A.; et al. Prediction of the COVID-19 pandemic for the top 15 affected countries: Advanced autoregressive integrated moving average (ARIMA) model. *JMIR Public Health Surveill.* **2020**, *6*, e19115. [CrossRef] [PubMed]
13. Singh, S.; Parmar, K.S.; Kumar, J.; Makkhan, S.J. Development of new hybrid model of discrete wavelet decomposition and autoregressive integrated moving average (ARIMA) models in application to one month forecast the casualties cases of COVID-19. *Chaos Solitons Fractals* **2020**, *135*, 109866. [CrossRef] [PubMed]
14. Dansana, D.; Kumar, R.; Das Adhikari, J.; Mohapatra, M.; Sharma, R.; Priyadarshini, I.; Le, D.N. Global forecasting confirmed and fatal cases of COVID-19 outbreak using autoregressive integrated moving average model. *Front. Public Health* **2020**, *8*, 580327. [CrossRef] [PubMed]
15. Plitnick, T.A.; Marsellos, A.E.; Tsakiri, K.G. Time series regression for forecasting flood events in Schenectady, New York. *Geosciences* **2018**, *8*, 317. [CrossRef]
16. Sadeghi, B.; Ghahremanloo, M.; Mousavinezhad, S.; Lops, Y.; Pouyaei, A.; Choi, Y. Contributions of meteorology to ozone variations: Application of deep learning and the Kolmogorov-Zurbenko filter. *Environ. Pollut.* **2022**, *310*, 119863. [CrossRef]
17. Smyl, S. A hybrid method of exponential smoothing and recurrent neural networks for time series forecasting. *Int. J. Forecast.* **2020**, *36*, 75–85. [CrossRef]
18. Sinaga, H.; Irawati, N. A medical disposable supply demand forecasting by moving average and exponential smoothing method. In Proceedings of the 2nd Workshop on Multidisciplinary and Applications (WMA) 2018, Padang, Indonesia, 24–25 January 2018.
19. Zhang, S.; Khattak, A.; Matara, C.M.; Hussain, A.; Farooq, A. Hybrid feature selection-based machine learning Classification system for the prediction of injury severity in single and multiple-vehicle accidents. *PLoS ONE* **2022**, *17*, e0262941. [CrossRef]
20. Dong, S.; Khattak, A.; Ullah, I.; Zhou, J.; Hussain, A. Predicting and analyzing road traffic injury severity using boosting-based ensemble learning models with SHAPley Additive exPlanations. *Int. J. Environ. Res. Public Health* **2022**, *19*, 2925. [CrossRef]
21. Khattak, A.; Almujibah, H.; Elamary, A.; Matara, C.M. Interpretable Dynamic Ensemble Selection Approach for the Prediction of Road Traffic Injury Severity: A Case Study of Pakistan's National Highway N-5. *Sustainability* **2022**, *14*, 12340. [CrossRef]
22. Goodman, S.N.; Goel, S.; Cullen, M.R. Machine learning, health disparities, and causal reasoning. *Ann. Intern. Med.* **2018**, *169*, 883–884. [CrossRef]
23. Guo, R.; Fu, D.; Sollazzo, G. An ensemble learning model for asphalt pavement performance prediction based on gradient boosting decision tree. *Int. J. Pavement Eng.* **2022**, *23*, 3633–3644. [CrossRef]
24. Zhao, Y.; Deng, W. Prediction in Traffic Accident Duration Based on Heterogeneous Ensemble Learning. *Appl. Artif. Intell.* **2022**, *36*, 2018643. [CrossRef]
25. Freund, Y.; Schapire, R.; Abe, N. A short introduction to boosting. *J. Jpn. Soc. Artif. Intell.* **1999**, *14*, 1612.
26. Ke, G.; Meng, Q.; Finley, T.; Wang, T.; Chen, W.; Ma, W.; Ye, Q.; Liu, T.Y. Lightgbm: A highly efficient gradient boosting decision tree. *Adv. Neural Inf. Process. Syst.* **2017**, *30*, 52.
27. Prokhorenkova, L.; Gusev, G.; Vorobev, A.; Dorogush, A.V.; Gulin, A. CatBoost: Unbiased boosting with categorical features. *Adv. Neural Inf. Process. Syst.* **2018**, *31*.
28. Chen, T.; Guestrin, C. Xgboost: A scalable tree boosting system. In Proceedings of the 22nd Acm Sigkdd International Conference on Knowledge Discovery and Data Mining, San Francisco, CA, USA, 13–17 August 2016; pp. 785–794.
29. Breiman, L. Random forests. *Mach. Learn.* **2001**, *45*, 5–32. [CrossRef]
30. Duan, T.; Anand, A.; Ding, D.Y.; Thai, K.K.; Basu, S.; Ng, A.; Schuler, A. Ngboost: Natural gradient boosting for probabilistic prediction. In Proceedings of the 37th International Conference on Machine Learning, Virtual Event, 13–18 July 2020; pp. 2690–2700.

31. Snoek, J.; Larochelle, H.; Adams, R.P. Practical Bayesian optimization of machine learning algorithms. In *Advances in Neural Information Processing Systems 25 (NIPS 2012)*; 2012; Volume 25. Available online: https://proceedings.neurips.cc/paper/2012/hash/05311655a15b75fab86956663e1819cd-Abstract.html (accessed on 4 August 2022).
32. Chan, P.; Hon, K. Observation and numerical simulation of terrain-induced windshear at the Hong Kong International Airport in a planetary boundary layer without temperature inversions. *Adv. Meteorol.* **2016**, *2016*, 1454513. [CrossRef]
33. Wu, J.; Chen, X.Y.; Zhang, H.; Xiong, L.D.; Lei, H.; Deng, S.H. Hyperparameter optimization for machine learning models based on Bayesian optimization. *J. Electron. Sci. Technol.* **2019**, *17*, 26–40.
34. Victoria, A.H.; Maragatham, G. Automatic tuning of hyperparameters using Bayesian optimization. *Evol. Syst.* **2021**, *12*, 217–223. [CrossRef]
35. Chan, P.W.; Hon, K.K. Observations and numerical simulations of sea breezes at Hong Kong International Airport. *Weather* **2022**, *78*, 55–63. [CrossRef]

atmosphere

MDPI

Article

Analyses on the Multimodel Wind Forecasts and Error Decompositions over North China

Yang Lyu [1], Xiefei Zhi [1,*], Hong Wu [2,*], Hongmei Zhou [3], Dexuan Kong [4], Shoupeng Zhu [2], Yingxin Zhang [5] and Cui Hao [5]

1 Key Laboratory of Meteorology Disaster, Ministry of Education (KLME)/Joint International Research Laboratory of Climate and Environment Change (ILCEC)/Collaborative Innovation Center on Forecast and Evaluation of Meteorological Disasters (CIC-FEMD), Nanjing University of Information Science and Technology, Nanjing 210044, China
2 Key Laboratory of Transportation Meteorology of China Meteorological Administration, Nanjing Joint Institute for Atmospheric Sciences, Nanjing 210000, China
3 Dongtai Meteorological Bureau, Yancheng 224200, China
4 Meteorological Bureau of Qian Xinan Buyei and Miao Autonomous Prefecture, Xingyi 562400, China
5 Beijing Meteorological Observatory, Beijing 100016, China
* Correspondence: zhi@nuist.edu.cn (X.Z.); wuhong@cma.gov.cn (H.W.)

Abstract: In this study, wind forecasts derived from the European Centre for Medium-Range Weather Forecasts (ECMWF), the National Centers for Environmental Prediction (NCEP), the Japan Meteorological Agency (JMA) and the United Kingdom Meteorological Office (UKMO) are evaluated for lead times of 1–7 days at the 10 m and multiple isobaric surfaces (500 hPa, 700 hPa, 850 hPa and 925 hPa) over North China for 2020. The straightforward multimodel ensemble mean (MME) method is utilized to improve forecasting abilities. In addition, the forecast errors are decomposed to further diagnose the error sources of wind forecasts. Results indicated that there is little difference in the performances of the four models in terms of wind direction forecasts (DIR), but obvious differences occur in the meridional wind (U), zonal wind (V) and wind speed (WS) forecasts. Among them, the ECMWF and NCEP showed the highest and lowest abilities, respectively. The MME effectively improved wind forecast abilities, and showed more evident superiorities at higher levels for longer lead times. Meanwhile, all of the models and the MME manifested consistent trends of increasing (decreasing) errors for U, V and WS (DIR) with rising height. On the other hand, the main source of errors for wind forecasts at both 10 m and isobaric surfaces was the sequence component (SEQU), which rose rapidly with increasing lead times. The deficiency of the less proficient NCEP model at the 10 m and isobaric surfaces could mainly be attributed to the bias component (BIAS) and SEQU, respectively. Furthermore, the MME tended to produce lower SEQU than the models at all layers, which was more obvious at longer lead times. However, the MME showed a slight deficiency in reducing BIAS and the distribution component of forecast errors. The results not only recognized the model forecast performances in detail, but also provided important references for the use of wind forecasts in business departments and associated scientific researches.

Keywords: wind forecast; error decomposition; bias; distribution; sequence

Citation: Lyu, Y.; Zhi, X.; Wu, H.; Zhou, H.; Kong, D.; Zhu, S.; Zhang, Y.; Hao, C. Analyses on the Multimodel Wind Forecasts and Error Decompositions over North China. *Atmosphere* **2022**, *13*, 1652. https://doi.org/10.3390/atmos13101652

Academic Editors: Jimy Dudhia and Leonardo Primavera

Received: 10 August 2022
Accepted: 7 October 2022
Published: 10 October 2022

1. Introduction

Wind, the movement of air, is one of the most important meteorological elements, and plays a significant role in determining and controlling climate and weather [1]. It has various impacts on human life and economic society, in both positive and negative ways. Appropriate wind conditions can help many industries, such as wind power production, whereas high winds can cause downed trees and power lines, flying debris and buildings to collapse, which may lead to power outages, transportation disruptions, damage to buildings and vehicles, and injury or death [2]. With respect to transportation fields at the near surface, windy conditions can create dangerous driving situations on highways [3]. As for at higher levels, abnormal winds can increase risks in terms of unstable aircraft, posing profound threats to aviation safety [4]. Thus, accurate and reliable forecasts of winds play an important role in both reducing traffic accidents and improving the efficiency of traffic operations [5,6].

So far, due to improved understanding of atmospheric physical processes and the rapid development of computer technology, numerical weather prediction (NWP) has been greatly developed and used in various predictions of weather and climate [7–9]. Taking wind as an example, subjective forecasts are always limited in ability because of the lack of enough observations, while the NWP could enrich wind forecasts with multiple lead times and multiple levels, as required [10]. In addition, it has been demonstrated that the NWP models are generally capable of reasonably forecasting atmospheric conditions. However, obvious differences in forecasting abilities always feature different NWP models in different regions. Comprehensive assessments are necessary for the rational application of NWP products and for further enhancing forecast ability [11–13].

On the other hand, considering the chaotic characteristics of atmosphere dynamics, even the best NWP model has inevitable systematic biases. Therefore, it is important to further post-process NWP model outputs to effectively improve forecasting abilities [14–16]. Correspondingly, many statistical post-processing methods, which enhance forecast abilities by learning a function derived from the historical performances of models, have been developed and widely utilized in recent years. Such as the frequency matching method [17,18], the mean bias removal [19], the pattern projection methods [20,21] and the decaying average method [22,23]. Moreover, due to the inherent limitation and uncertainty of an individual NWP model, the multimodel ensemble methods, including the straightforward ensemble mean, the bias-removed ensemble mean and other advanced superensemble algorithms, have been proposed to calibrate forecast errors of temperature, precipitation, wind and other variables, making full use of valid information from various NWP models [24–27].

Over the past few decades, the multimodel ensemble forecasts based on various algorithms have been demonstrated as capable of effectively improving single NWP results, which is always featured with lower root mean square errors, higher correlation coefficients and many other metrics with higher abilities [28–30]. However, most of these assessments could only provide composite scores, which lack certain physical interpretabilities and give little insight into which aspects of the forecasts are good or bad. In this regard, decomposing performance measures into multiple interpretable elements has been considered an intelligent option to obtain more realistic and insightful assessments, and comparisons between different forecast systems [31–33]. At present, error decomposition has been widely utilized to analyze the sources of errors and to indicate future directions for improvement [34,35]. Taking the metric of mean square error (MSE) as an example [32], Murphy et al. [36] decomposed the MSE into correlation, conditional bias, unconditional bias and possible other contributions. Afterwards, Geman et al. [37] decomposed the MSE into bias and variance. More recently, Hodson et al. [38] have further decomposed it into components of bias, distribution and sequence.

In this study, the wind forecasts derived from the European Centre for Medium-Range Weather Forecasts (ECMWF), the National Center for Environmental Prediction (NCEP), the United Kingdom Meteorological Office (UKMO) and the Japan Meteorological Agency (JMA), accompanied by a multimodel ensemble mean (MME), are evaluated and compared for multiple layers including ground (10 m) and isobaric surfaces (500 hPa, 700 hPa, 850 hPa and 925 hPa). The study area selected is North China (46° N–36° N, 111° E–119° E; NC), which features the most populous region and a major agricultural and industrial sector [39,40]. Meanwhile, forecast errors are decomposed to diagnose the error sources of wind forecasts in NWP models, and analyzed to determine which aspects of the forecasts are improved by the MME. The manuscript is organized as follows. The datasets and methods are briefly described in Section 2. Section 3 displays the comprehensive evaluation of the wind forecast abilities of ECMWF, NCEP, UKMO, JMA and MME. Finally, a summary and discussion are presented in Section 4.

2. Data and Method

2.1. Data

The used forecast datasets of meridional wind (u) and zonal wind (v) at ground (10 m) and isobaric surfaces (500 hPa, 700 hPa, 850 hPa, 925 hPa) with lead times of 1–7 days were derived from ECMWF, NCEP, UKMO and JMA in the the Observing System Research and Predictability Experiment (THORPEX) Interactive Grand Global Ensemble (TIGGE).

In addition, ERA5 reanalysis is selected for verification. ERA5 is a product of the Integrated Forecast System (IFS) release 41r2, which was operational at ECMWF during the period March 2016 to November 2016. ERA5 therefore benefits from a decade of developments in model physics, core dynamics and data assimilation [41]. Various considerations have to be made when choosing the verification dataset to evaluate the performance of NWP models. Station observation has the advantage of being independent of all models, but wind observational datasets over isobaric surfaces are difficult to obtain. Meanwhile, reanalysis provides consistent "maps without gaps" of essential climate variables by optimally combining observations and models [42]. Moreover, ERA5 data has been demonstrated to be capable of effectively reflecting and describing the local atmospheric conditions in observations, and has been widely used in associated studies including forecast error evaluation, analyzing the thermodynamic characteristics of warm sector heavy rainfall, etc. [43–46]. On the other hand, a previous study has proved that whether the verification data consist of reanalysis or observations, it has little impact on final assessment results [47]; therefore, we chose ERA5 for verification in this study.

Correspondingly, the study area is unified as North China (46° N–36° N, 111° E–119° E; NC), with a horizontal resolution of 0.5° × 0.5°, and the entire year of 2020 is selected for evaluation. Both forecast and verification datasets are obtained from the ECMWF archive at https://apps.ecmwf.int/datasets/, accessed on 1 August 2022. The topography of North China and its surrounding area is described in Figure 1.

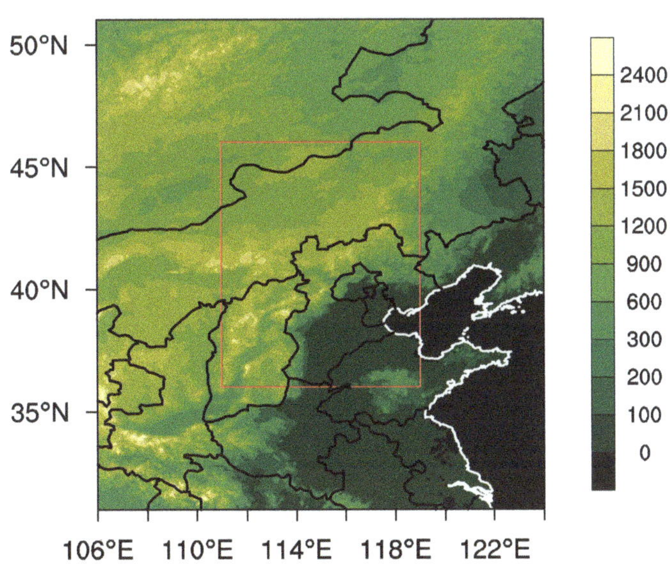

Figure 1. Topography (m) of North China (marked region) and its surrounding area.

2.2. Verification Metrics

Aimed at quantitative assessments of forecast results of different NWP models and the MME method over North China for assessed period, several metrics are employed; including the root mean square error ($RMSE$) and temporal correlation coefficient (TCC):

$$RMSE = \sqrt{\frac{1}{n}(f_i - o_i)^2} \tag{1}$$

$$TCC = \frac{\sum_{i=1}^{n}\left(f_i - \overline{f}\right)(o_i - \overline{o})}{\sqrt{\sum_{i=1}^{n}\left(f_i - \overline{f}\right)^2}\sqrt{\sum_{i=1}^{n}(o_i - \overline{o})^2}} \tag{2}$$

where n indicates the total number of samples. The term f_i and o_i represent the forecast and observation of sample i, respectively. The terms $\overline{f}$ and $\overline{o}$ refer to the average forecast and observation, respectively.

In addition, the error decomposition proposed by Hodsonal et al. [38] is utilized to diagnose the sources of error for both NWP models and the MME method. Firstly, the MSE at each grid can be calculated by Equation (3):

$$MSE = \frac{1}{n}\sum_{i=1}^{n}(f_i - o_i)^2 \tag{3}$$

where f_i and o_i represent the forecast and observation of sample i, respectively. According to the decomposing method proposed by Geman et al. [37], the MSE can be decomposed into bias and variance:

$$\begin{aligned} MSE(e) &= (E(e^2) - E(e)^2) + E(e)^2 \\ &= Var(e) + Bias(e)^2 \end{aligned} \tag{4}$$

where e represents the forecast error of the model as the difference between the forecast and observation, while $E(e)$ represents the mean of the forecast error which is equal to $Bias(e)$ and $Var(e)$ represents the variance of the forecast error. The variance component quantifies the extent to which the model reproduces the observed variability, while the bias component quantifies the ability of the model to reproduce the average characteristics of the observations. Meanwhile, the variance component can be further decomposed to obtain a deeper understanding of model performance [38]. The derivation begins by monotonically sorting the model predictions and observations, then decomposing the MSE of the result:

$$w = sort(f) - sort(o) \tag{5}$$

$$MSE(w) = Bias(w)^2 + Var(w) \tag{6}$$

where $sort(f)$ and $sort(o)$ represent the sorted observations and forecasts, respectively, and w represents the forecast error after sorting. Considering that changing the sequence of the data does not change the mean error of the data, bias before and after the sorting is equal. Meanwhile, the sorted observations and forecasts share the same time series, and the variance at this point, $Var(w)$, describes the error caused by the data distribution ($Dist(e)$); thus, the Equations (7) and (8) can be obtained:

$$Var(w) = Dist(e) \tag{7}$$

$$MSE(w) = Bias(e)^2 + Dist(e) \tag{8}$$

Furthermore, the difference between $MSE(e)$ and $MSE(w)$ can be attributed to the time series variation, Sequence(e); thus, the following equation can be obtained:

$$\begin{aligned} MSE(e) - MSE(w) &= Var(e) - Var(w) \\ &= Sequence(e) \end{aligned} \tag{9}$$

In conclusion, the MSE can be decomposed into the bias element, the distribution element and the sequence element as follows:

$$\begin{aligned} MSE(e) &= Bias(e)^2 + Var(e) \\ &= Bias(e)^2 + (Var(e) - Var(w)) + Var(w) \\ &= Bias(e)^2 + Sequence(e) + Distribution(e) \end{aligned} \quad (10)$$

where $Bias(e)^2$ is the bias component, which characterizes the ability of the forecast to reproduce the average characteristics of the observations, $Sequence(e)$ is the sequence error component, which characterizes the error due to the forecast being ahead of (or lagging behind) the observations. $Distribution(e)$ is the distribution error component, which characterizes the error due to the difference in data distribution between the forecasts and the observations. In order to transfer the units of associated error components from $\left(\frac{m}{s}\right)^2$ into m/s, we divide both sides of the equation by RMSE at the same time and obtain the error decomposition of the RMSE.

3. Result

3.1. Evaluation of Multiple NWP Models and the MME

Figure 2 describes the regional averaged RMSE and TCC of ECMWF, NCEP, UKMO, JMA and MME for wind forecasts at the 10 m level over North China (NC) during a validation period of 1–7 lead days, including the meridional wind (U10), zonal wind (V10), wind speed (WS10) and wind direction (DIR10). Generally, multiple forecasts are characterized by consistent trends of increasing RMSE and decreasing TCC with growing lead times. The ECMWF shows the best performance, but with limited superiorities to UKMO and JMA, while the NCEP shows the lowest ability among the four NWP models. Specifically, the ECMWF features the lowest RMSEs and the highest TCCs at most lead times for all the elements. On the other hand, NCEP tends to show the highest RMSEs and the lowest TCCs, but it does not show much difference in comparison to other models in terms of WS10 forecasts. Furthermore, the MME is significantly superior to the individual NWP models, which is more evident for longer lead times. The RMSEs of the MME are lower than ECMWF by 0.3–0.5 m/s (12°–35°) for U10, V10 and WS10 (DIR10) for all lead times, and the MME shows TCCs of 0.1–0.15 higher than ECMWF for wind forecasts.

For assessments of the spatial distribution of forecast abilities for the NWP models and MME, with the lead time of 1 day taken as an example, Figure 3 describes the spatial distributions of RMSE for U10, V10, WS10 and DIR10 derived from ECMWF, NCEP and MME, which denote the best NWP model, the worst NWP model and the multimodel ensemble mean, respectively. In terms of U10 and V10, the lower RMSEs are continuously seen around central NC, whereas the highest RMSEs occur around northwestern NC. Meanwhile, the RMSEs of NCEP are higher than ECMWF over the whole area, and the advantages of MME to ECMWF are mainly reflected over the southwest NC. As for DIR10, the RMSE spatial distribution of ECMWF, NCEP and MME are generally consistent, with the largest RMSEs reaching up to 120° occurring at central NC, while the lowest RMSEs of lower than 40° are seen at northwestern NC. It is worth noting that the RMSEs are obviously lower over all regions in the MME than ECMWF.

In order to assess the wind forecasts at multiple isobaric surfaces, Figure 4 describes the regional averaged RMSE of U, V, WS and DIR at 500 hPa, 850 hPa, 700 hPa and 925 hPa, derived from ECMWF, NCEP, UKMO, JMA and MME over NC, with lead times of 1, 4 and 7 days taken as examples. Generally, the multiple forecasts are characterized by consistent trends of increasing RMSE (decreasing RMSE) for U, V and WS (DIR) with the rising height. Among them, the RMSE of U, V and WS show the highest growth rates between 925 hPa and 850 hPa, and the highest growth rate of DIR is seen between 700 hPa and 500 hPa. Furthermore, the ECMWF shows lower RMSE than the other NWP models at all isobaric surfaces, which is more evident at higher levels. The advantages of ECMWF diminish with increased lead times. Furthermore, the MME tends to show lower RMSE for U, V and WS

(DIR) than ECMWF at all levels for all lead times, which is more obvious at higher (lower) levels for longer lead times.

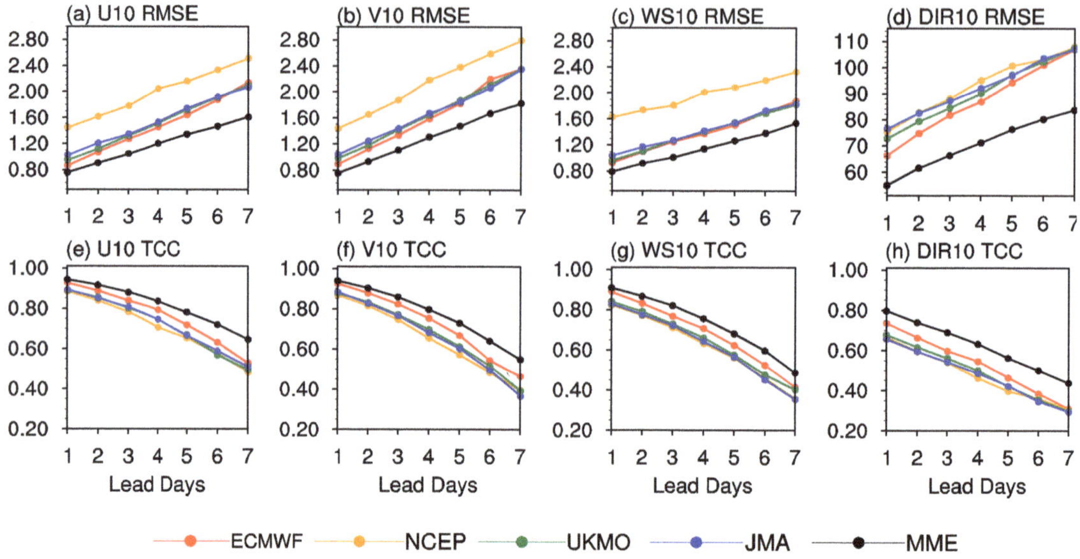

Figure 2. Variations in RMSE and TCC of U10, V10, WS10 and DIR10, at lead times of 1–7 days derived from ECMWF, NCEP, UKMO, JMA and MMA averaged over North China.

To reveal the spatial distribution of wind forecast abilities at the isobaric surfaces for NWP models and the MME, Figure 5 describes the RMSE spatial distribution for U500, V500, WS500 and DIR500 derived from ECMWF, NCEP and MME, with the lead time of 1 day taken as an example. Generally, multiple forecasts show similar error distribution characteristics for U500, V500 and WS500. Specifically, the lower RMSEs are seen at central and northeastern NC, while the largest RMSEs occur at northwestern NC. Furthermore, NCEP shows limited forecast ability, with RMSEs reaching up to 2.2 m/s at most areas for U500, V500 and WS500, while the RMSEs of MME are mostly lower than 2 m/s. In terms of DIR500, the lowest RMSEs are seen at central NC for ECMWF, NCEP and MME, while the largest occurs at the northwestern and southern NC. Furthermore, the MME shows clear superiority to the two NWP models, with its RMSEs of lower than 60° for most areas.

To summarize, there is little difference in the performances of the four NWP models in terms of wind direction forecasts, but clear differences occur in the meridional wind, zonal wind and wind speed forecasts. The ECMWF shows general advantages over the other three at both 10 m and isobaric surfaces, which are more pronounced at isobaric surfaces. Furthermore, the forecast abilities of MME are superior to ECMWF for U, V, WS and DIR, which are more distinct at higher levels for longer lead times. It is worth noting that multiple forecasts manifest with the consistent trends of increasing (decreasing) RMSE for U, V and WS (DIR) with rising height. In addition, all the NWP models and MME tend to show higher forecast abilities at central NC, while they manifest with lower ability at northwestern NC for both ground and isobaric surfaces.

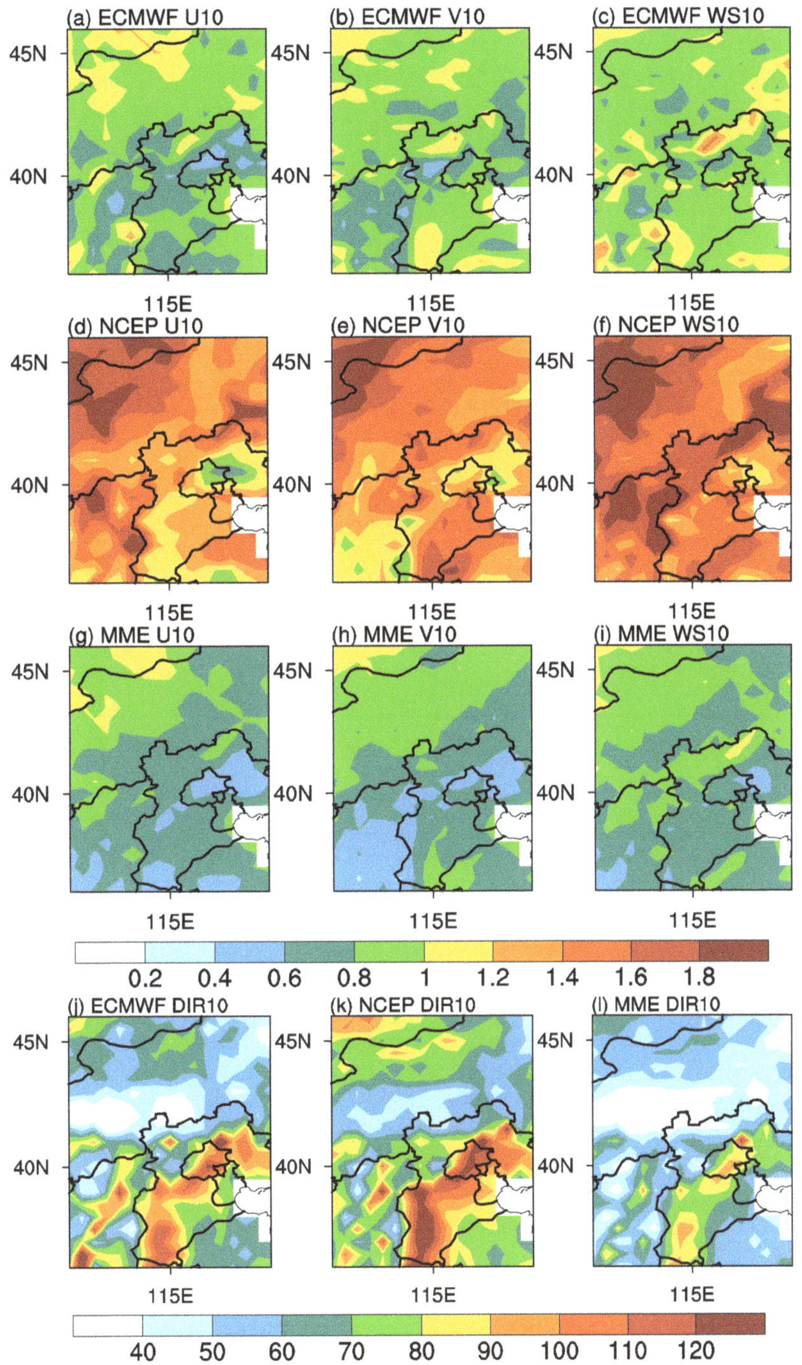

Figure 3. Spatial distributions of RMSEs for U10, V10, WS10 and DIR10 with a lead time of 1 day derived from ECMWF, NCEP and MME.

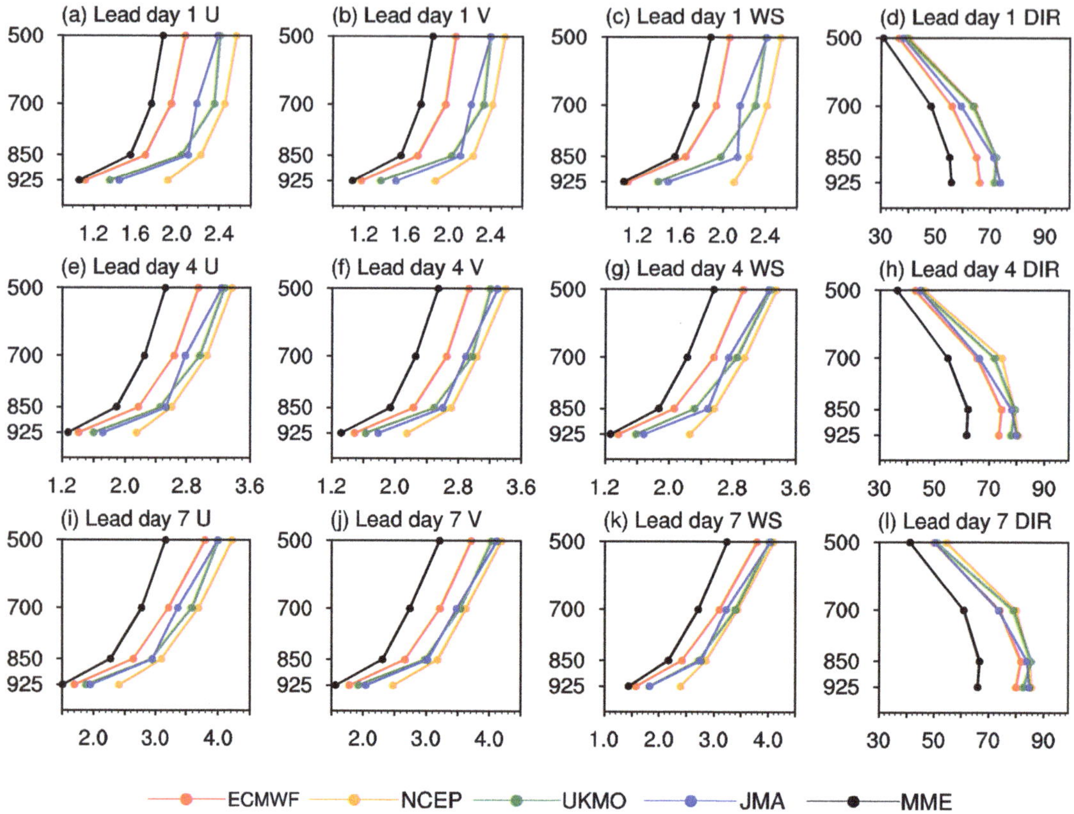

Figure 4. Variations in RMSE for U, V, WS and DIR at isobaric surfaces (500 hPa, 700 hPa, 850 hPa, 925 hPa) for lead times of 1–7 days, derived from ECMWF, NCEP, UKMO, JMA and MMA, averaged over North China.

3.2. Error Decompositions of the Wind Forecasts

Although the forecast abilities of NWP models and MME have been assessed in Section 3.1 via metrics, including RMSE and TCC, they tend to provide overall ability scores and give little insight into which aspects of the models are good or bad. Thus, the error decomposition method is utilized in this section to diagnose the error sources of wind forecasts in NWP models, and to analyze which aspects of the forecasts are improved by the MME method.

Figure 6 describes the regional-averaged RMSE, the decomposed bias component (BIAS), the distribution error component (DIST) and the sequence error component (SEQU) of the 10 m wind speed (WS10) and direction (DIR10) over NC derived from ECMWF, NCEP, UKMO, JMA and MME for lead times of 1–7 days. Generally, SEQU is the main source of error for both WS10 and DIR10, and rises rapidly with increasing lead times. While BIAS and DIST account for a relatively small proportion of the total error and do not increase with growing lead times. It implies that the 10 m wind forecast errors are mainly attributed to the forecasts being ahead of (lagging behind) the observations. However, the deficiency of NCEP for WS10, compared with other NWP models, could mainly be attributed to the BIAS and DIST. Furthermore, the MME tends to generate lower SEQU than four NWP models for both WS10 and DIR10, which is more evident at longer lead times, while the BIAS and DIST of the MME could not show obvious superiority over the best NWP model.

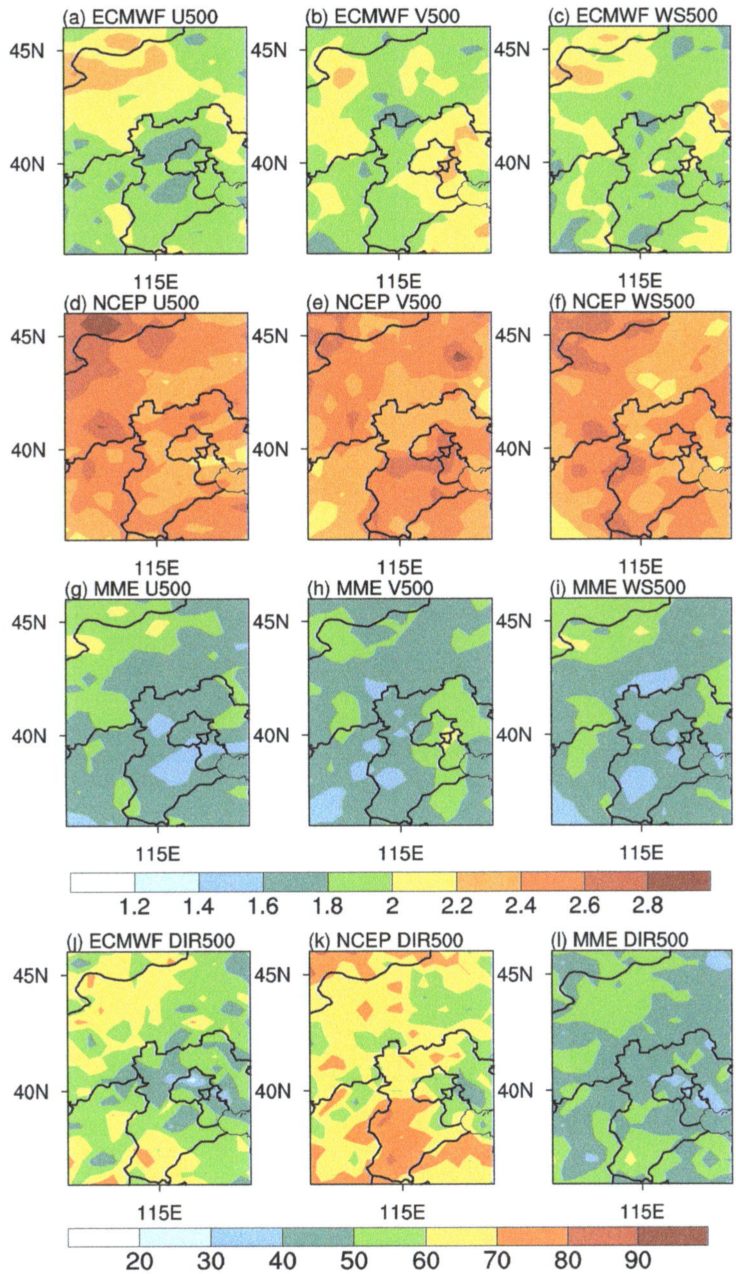

Figure 5. Spatial distributions of RMSEs for U500, V500, WS500 and DIR500 with a lead time of 1 day derived from ECMWF, NCEP and MME.

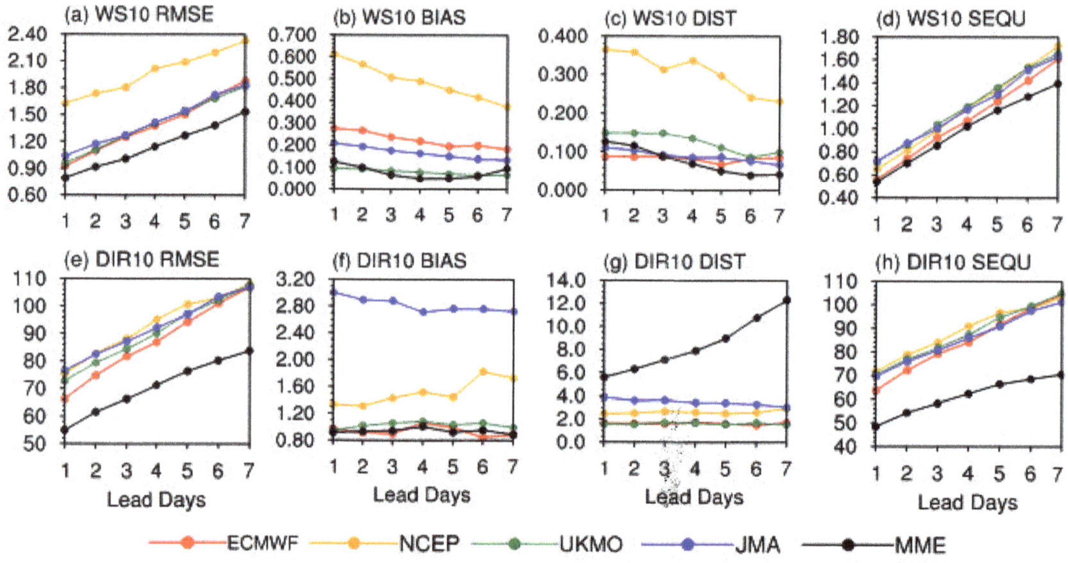

Figure 6. Variations in RMSE, decomposed BIAS, DIST and SEQU for WS10 and DIR10 at lead times of 1–7 days derived from ECMWF, NCEP, UKMO, JMA and MMA, averaged over North China.

To assess the spatial distributions of each error component, Figures 7 and 8 describe the BIAS, DIST and SEQU spatial distributions derived from ECMWF, NCEP and MME over NC for WS10 and DIR10, respectively, with the lead time of 1 day taken as an example. Generally, multiple forecasts perform with consistent spatial distribution for both WS10 and DIR10. In terms of WS10, the largest BIASs and DISTs occur at central NC, while also characterized by the lowest SEQUs. In addition, the largest SEQUs of up to 1 m/s can be seen at northwestern and southeastern NC. Although MME is generally superior to ECMWF, its DISTs at northwestern NC are obviously higher than the ECMWF results. For DIR10, the largest BIASs, DISTs and SEQUs mainly occur at central NC, and the lowest DISTs and SEQUs can be seen at northwestern NC. Moreover, the MME shows lower SEQUs than the two NWP models over most areas, but the DISTs of the MME are generally higher than the two NWP models, which is more distinct at southeastern NC. It is worth noting that the higher BIASs and DISTs tend to occur in the regions characterized with high altitudes, while SEQUs are less affected. This implies that the BIASs and DISTs might be associated with the deficiency of NWP models in simulating real terrain.

Aiming at diagnoses of the wind forecast errors at the isobaric surface, Figure 9 shows the regional averaged RMSE and the components of BIAS, DIST and SEQU for WS500 and DIR500 over NC derived from the four NWP models and MME, with lead times of 1–7 days. Generally, the SEQU remains the main source of errors and they rise rapidly with increasing lead times for both WS500 and DIR500. Furthermore, the proportions accounted for by SEQU in total errors are higher than those in 10 m wind forecasts for both WS500 and DIR500. Unlike the 10 m wind forecasts, the insufficiency of the NCEP forecasts at 500 hPa could mainly be attributed to the SEQU. On the other hand, the MME is characterized by lower SEQU, along with higher BIAS and DIST, than all NWP models for the WS500, which is more evident at longer lead times.

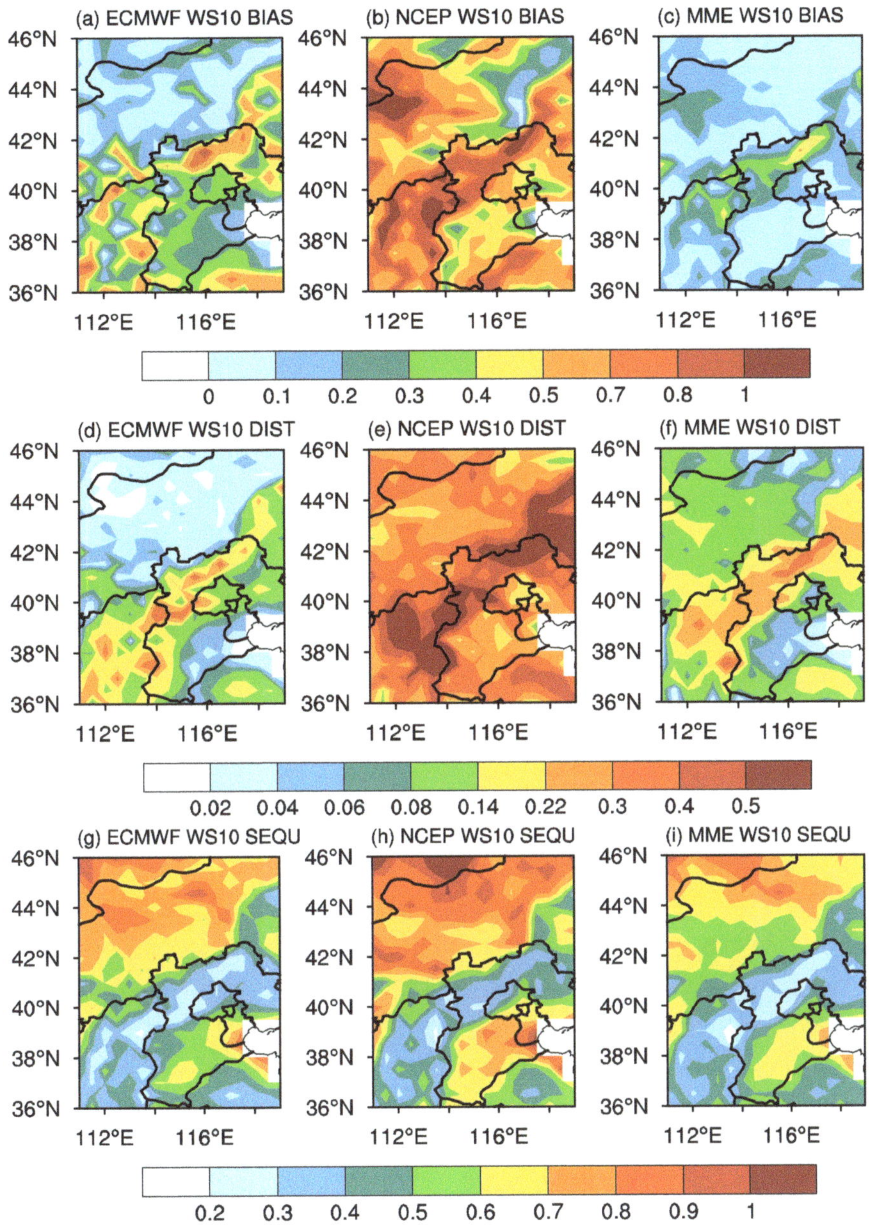

Figure 7. Spatial distributions of decomposed BIAS, DIST and SEQU for WS10 with a lead time of 1 day derived from ECMWF, NCEP and MME.

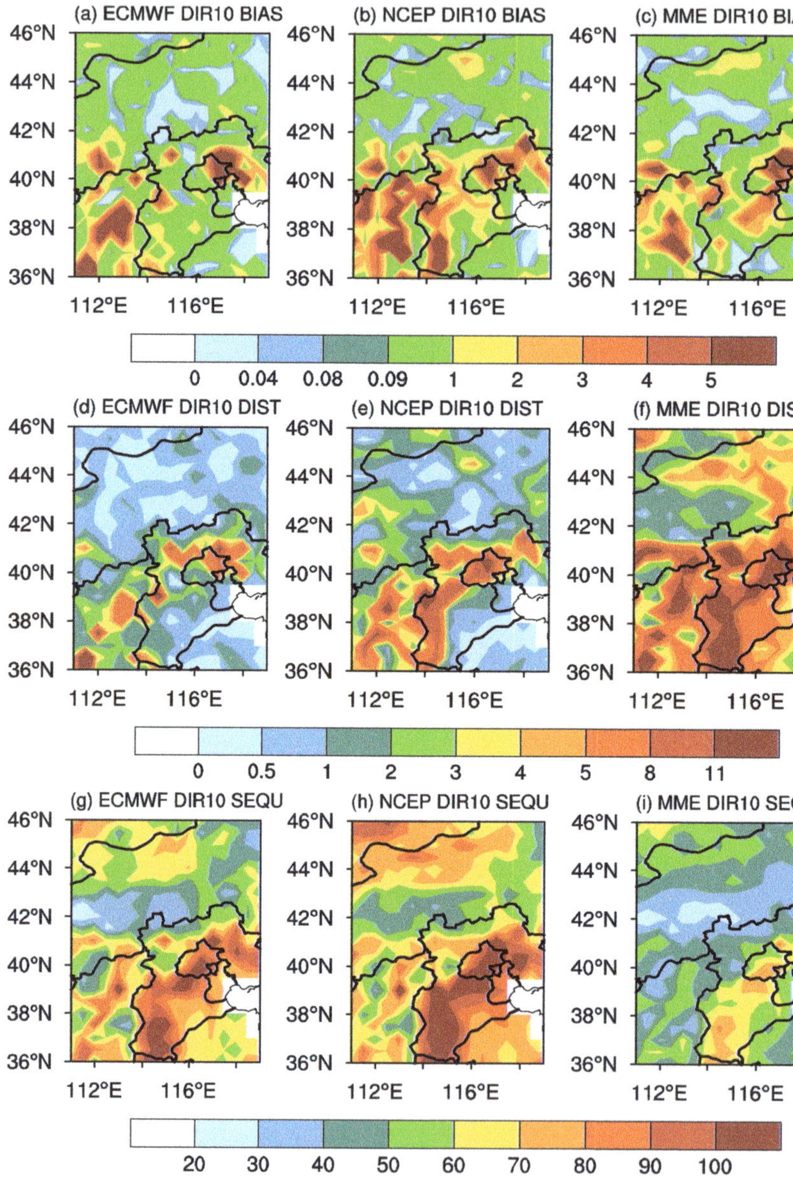

Figure 8. Spatial distributions of decomposed BIAS, DIST and SEQU for DIR10 with a lead time of 1 day derived from ECMWF, NCEP and MME.

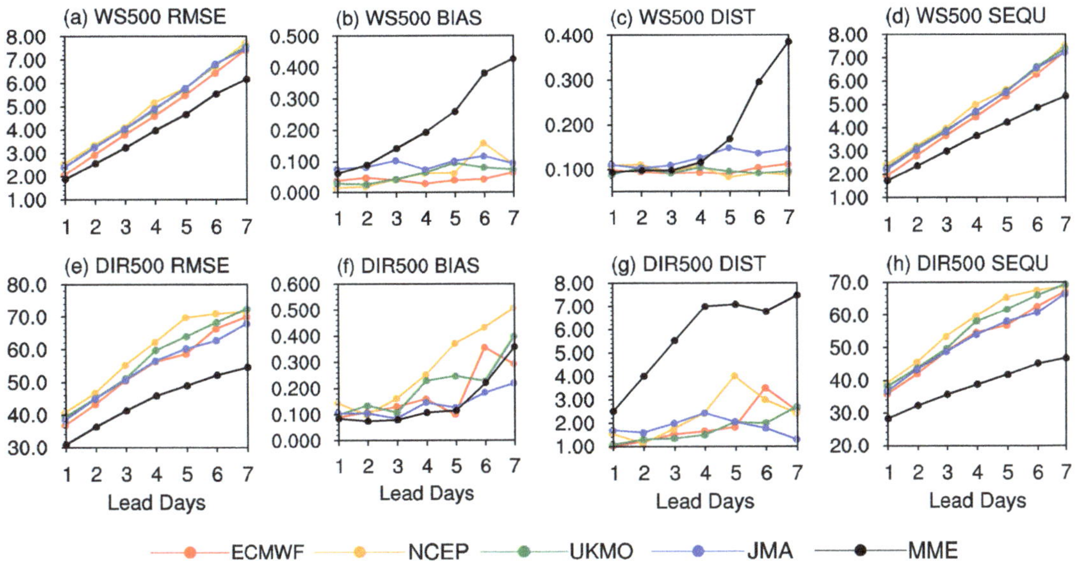

Figure 9. Variations in RMSE, decomposed BIAS, DIST and SEQU for WS500 and DIR500 at lead times of 1–7 days derived from ECMWF, NCEP, UKMO, JMA and MMA, averaged over North China.

Figures 10 and 11 further describe the spatial distributions of BIAS, DIST and SEQU components derived from ECMWF, NCEP and MME over NC for WS500 and DIR500, with the lead time of 1 day taken as an example. In terms of WS500, the SEQUs of NCEP over most areas are greater than 2 m/s, which accounts for the overall insufficiency of the model. Furthermore, the MME shows generally lower SEQUs than the two NWP models, while the BIASs of MME at northern NC are higher than ECMWF and NCEP. For DIR500, the three forecast systems show generally consistent distributions, and the largest SEQUs are mainly distributed at northern NC. Furthermore, MME performs with the lower SEQUs than ECMWF and NCEP for most areas, but there are higher DISTs at northwestern NC in MME than the two models. In addition, MME could not produce overt improvements to ECMWF and NCEP in terms of the BIAS component.

In summary, the main source of wind forecast errors at both 10 m and isobaric surfaces is the SEQU component, which rises rapidly with increasing lead times. The proportions accounted for by SEQU in total errors at isobaric surfaces are higher than that at the 10 m level. The deficiency of NCEP at both 10 m and isobaric surfaces could mainly be attributed to the BIAS and SEQU terms, respectively. Furthermore, the MME tends to perform with lower SEQU than NWP models at both 10 m and isobaric surfaces, which is more distinct for longer lead times. However, the MME shows a slight deficiency in reducing BIAS and DIST. There are even higher DISTs for MME than NWP models, which are not included in detail here and require exploration in future work.

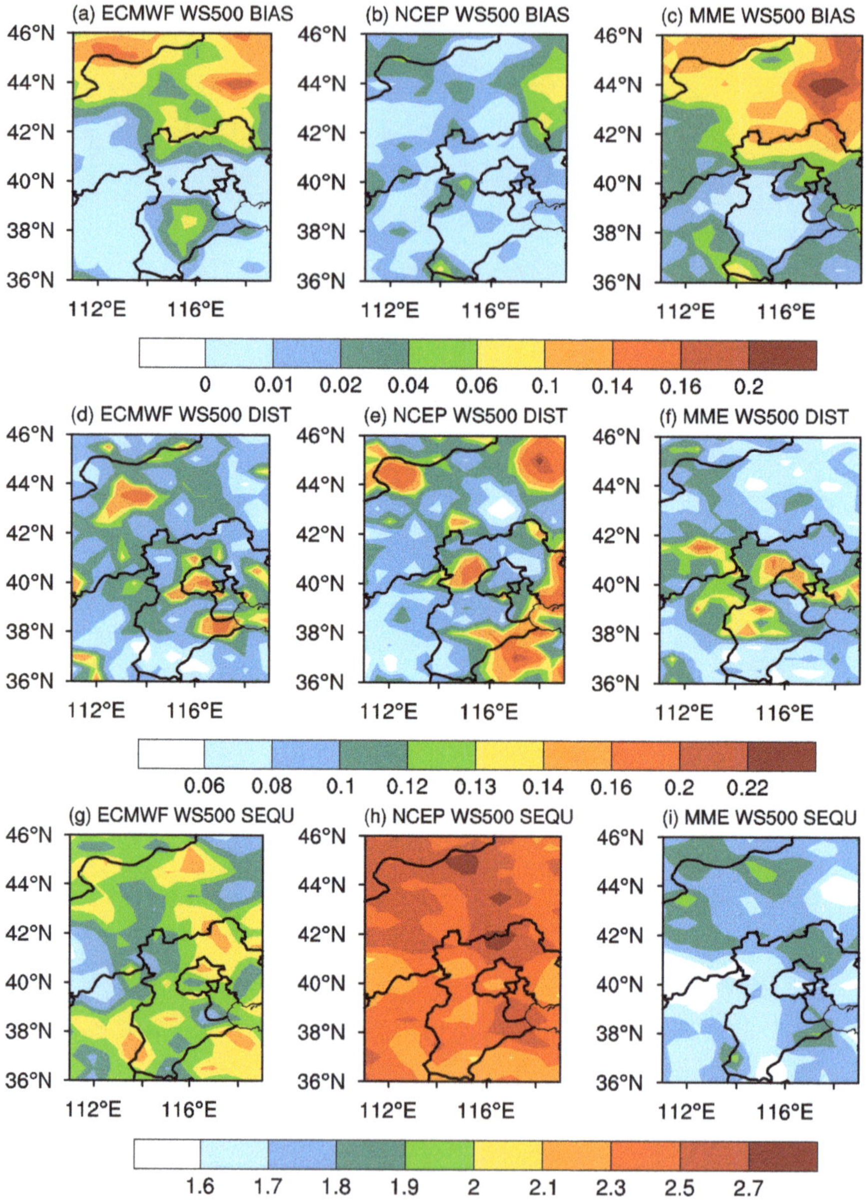

Figure 10. Spatial distributions of decomposed BIAS, DIST and SEQU for WS500 with a lead time of 1 day derived from ECMWF, NCEP and MME.

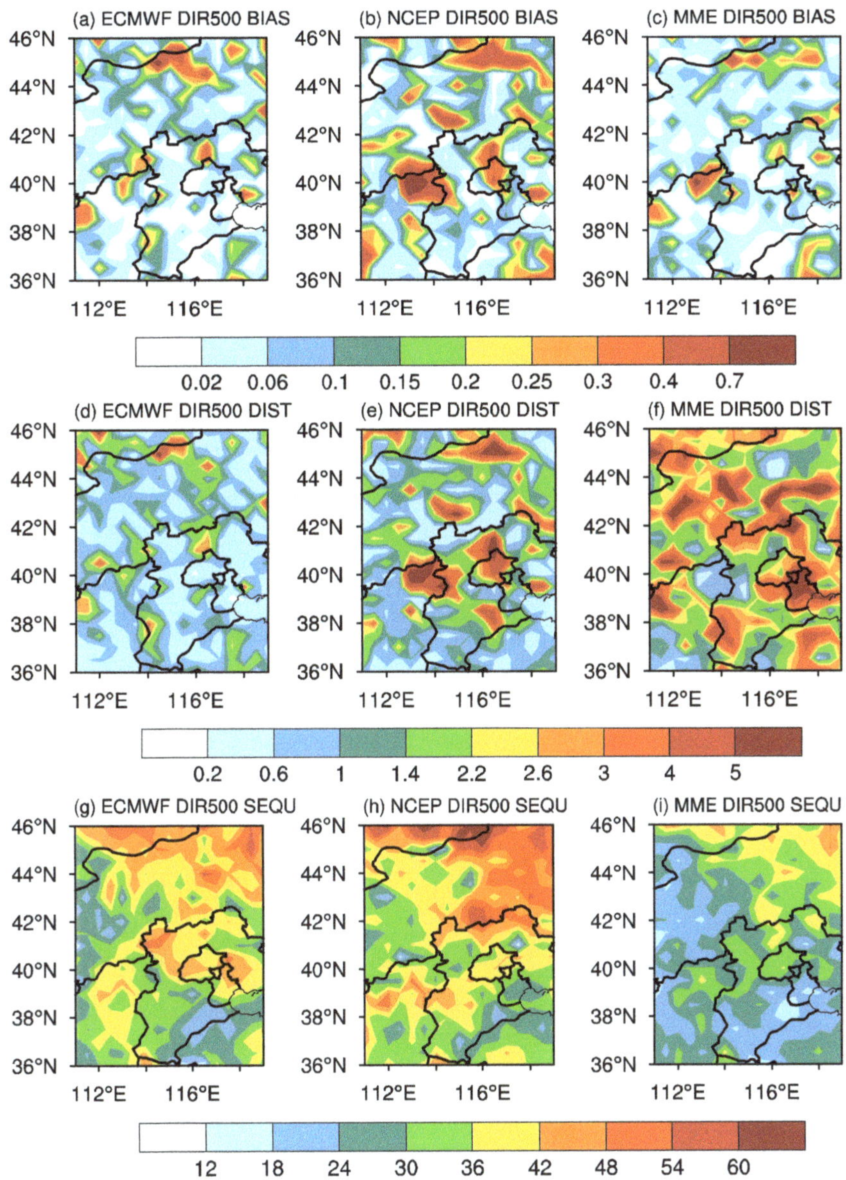

Figure 11. Spatial distributions of decomposed BIAS, DIST and SEQU for DIR500 with a lead time of 1 day derived from ECMWF, NCEP and MME.

4. Conclusions and Discussion

In this study, the wind forecasts of 2020 derived from ECMWF, NCEP, UKMO and JMA over NC for lead times of 1–7 days at 10 m and isobaric surfaces (500 hPa, 700 hPa, 850 hPa and 925 hPa) were evaluated and the straightforward multimodel ensemble mean method (MME) was utilized to improve wind forecast abilities. Furthermore, the error decomposition method was also applied to diagnose the error sources of wind forecasts in NWP models and analyze which aspects of the forecasts were improved by the MME method. Associated results were obtained as follows.

Generally, there was little difference in the performances of the four NWP models in terms of wind direction forecasts, but evident differences occurred in the meridional wind, zonal wind and wind speed forecasts. The ECMWF showed general advantages over the other three NWP models at both 10 m and isobaric surfaces, which were more pronounced at isobaric surfaces. Furthermore, the forecast abilities of MME were superior to ECMWF for U, V, WS and DIR, which were more obvious at higher levels for longer lead times. It is worth noting that multiple forecasts manifested with the consistent trends of increasing (decreasing) RMSE for U, V, WS (DIR) with rising height. In addition, all the NWP models and MME tended to show higher forecast ability at central NC, while they manifested with lower ability at northwestern NC for both ground and isobaric surfaces.

The main source of wind forecast errors at both 10 m and isobaric surfaces was the SEQU component, which rose rapidly with increasing lead times. In addition, the proportions accounted by SEQU in total errors at isobaric surfaces were higher than that at the 10 m level. Furthermore, the deficiency of NCEP at the 10 m and isobaric surfaces could mainly be attributed to the BIAS and SEQU terms, respectively. Furthermore, the MME tended to perform with lower SEQU than NWP models at both 10 m and isobaric surfaces, which was more distinct for longer lead times. However, the MME showed slight deficiency in reducing BIAS and DIST, and there were even higher DISTs for the MME than the NWP models. These results not only provide an important reference for the use of wind NWP results in business departments and scientific research, but also in directing further improvement of NWPs in the future.

Moreover, according to the current study, higher BIASs and DISTs tended to occur at regions with high altitudes for wind forecasts at 10 m, which implied that the BIAS and DIST might be associated with the deficiency of the model in simulating the real terrain [48,49]. Thus, calibration methods incorporating geographic information should also be examined in the future [50,51]. On the other hand, the examined MME method is one of the most basic and straightforward multimodel ensemble methods, which assigns all models with the same role. Considering the deficiency of MME in reducing the BIAS and DIST of wind forecasts, the multimodel ensemble methods based on more complex algorithms assigning different weights for different models, including Kalman filter [52,53], object-based diagnosis [54] and deep learning methods [6,55], are also on the way to be utilized to further improve wind forecast ability. Furthermore, with the development of modern observation channels and technologies, observations are enriched and could be taken into consideration to assess and calibrate the model products in a more realistic way.

Author Contributions: Y.L. and X.Z. contributed to conception and design of the study. H.W., S.Z. and Y.Z. contributed to the analysis. H.Z., D.K. and C.H. organized the database. All authors contributed to manuscript revision, read, and approved the submitted version. All authors have read and agreed to the published version of the manuscript.

Funding: The study was jointly supported by the Collaboration Project of Urumqi Desert Meteorological Institute of China Meteorological Administration "Precipitation forecast based on machine learning", the National Key R&D Program of China (Grant No. 2017YFC1502002), the Basic Research Fund of CAMS (Grant No. 2022Y027), the research project of Jiangsu Meteorological Bureau (Grant No. KQ202209).

Institutional Review Board Statement: Not applicable.

Informed Consent Statement: Not applicable.

Data Availability Statement: The forecast and observation data in this paper are publicly available. The datasets are obtained from the ECMWF archive in https://apps.ecmwf.int/datasets/, 1 August 2022.

Acknowledgments: The authors are grateful to ECMWF, NCEP, UKMO and JMA for their datasets.

Conflicts of Interest: The authors declare that the research was conducted in the absence of any commercial or financial relationships that could be construed as a potential conflict of interest.

References

1. Sloughter, J.M.; Gneiting, T.; Raftery, A.E. Probabilistic wind vector forecasting using ensembles and Bayesian model averaging. *Mon. Weather. Rev.* **2013**, *141*, 2107–2119. [CrossRef]
2. Adeyeye, K.; Ijumba, N.; Colton, J. Exploring the environmental and economic impacts of wind energy: A cost-benefit perspective. *Int. J. Sustain. Dev. World Ecol.* **2020**, *27*, 718–731. [CrossRef]
3. Wen, Y.; Chen, Z. Study on the Estimation of Designed Wind Speed for Jingyue Yangtze River Highway Bridge. *J. Wuhan Univ. Technol. Transp. Sci. Eng.* **2010**, *34*, 306–309.
4. Wynnyk, C.M. Wind analysis in aviation applications. In Proceedings of the 2012 IEEE/AIAA 31st Digital Avionics Systems Conference (DASC), Williamsburg, VA, USA, 14–18 October 2012; pp. 5C2-1–5C2-10.
5. Wilczak, J.; Finley, C.; Freedman, J.; Cline, J.; Bianco, L.; Olson, J.; Djalalova, I.; Sheridan, L.; Ahlstrom, M.; Manobianco, J.; et al. The Wind Forecast Improvement Project (WFIP): A public—Private partnership addressing wind energy forecast needs. *Bull. Am. Meteorol. Soc.* **2015**, *96*, 1699–1718. [CrossRef]
6. Veldkamp, S.; Whan, K.; Dirksen, S. Statistical postprocessing of wind speed forecasts using convolutional neural networks. *Mon. Weather. Rev.* **2021**, *149*, 1141–1152. [CrossRef]
7. Zhu, S.; Remedio, A.R.C.; Sein, D.V.; Sielmann, F.; Ge, F.; Xu, J.; Peng, T.; Jacob, D.; Fraedrich, K.; Zhi, X. Added value of the regionally coupled model ROM in the East Asian summer monsoon modeling. *Theor. Appl. Climatol.* **2020**, *140*, 375–387. [CrossRef]
8. Bauer, P.; Thorpe, A.; Brunet, G. The quiet revolution of numerical weather prediction. *Nature* **2015**, *525*, 47–55. [CrossRef]
9. Zhu, S.; Ge, F.; Sielmann, F.; Pan, M.; Fraedrich, K.; Remedio, A.R.C.; Sein, D.V.; Jacob, D.; Wang, H.; Zhi, X. Seasonal temperature response over the Indochina Peninsula to a worst-case high-emission forcing: A study with the regionally coupled model ROM. *Theor. Appl. Climatol.* **2020**, *142*, 613–622. [CrossRef]
10. Bengtsson, L.; Andrae, U.; Aspelien, T.; Batrak, Y.; Calvo, J.; de Rooy, W.; Gleeson, E.; Hansen-Sass, B.; Homleid, M.; Hortal, M.; et al. The HARMONIE–AROME model configuration in the ALADIN–HIRLAM NWP system. *Mon. Weather. Rev.* **2017**, *145*, 1919–1935. [CrossRef]
11. Zhang, L.; Kim, T.; Yang, T.; Hong, Y.; Zhu, Q. Evaluation of Subseasonal-to-Seasonal (S2S) precipitation forecast from the North American Multi-Model ensemble phase II (NMME-2) over the contiguous US. *J. Hydrol.* **2021**, *603*, 127058. [CrossRef]
12. Lyu, Y.; Zhu, S.; Zhi, X.; Dong, F.; Zhu, C.; Ji, L.; Fan, Y. Subseasonal forecasts of precipitation over Maritime Continent in boreal summer and the sources of predictability. *Front. Earth Sci.* **2022**, *10*, 970791. [CrossRef]
13. Louvet, S.; Sultan, B.; Janicot, S.; Kamsu-Tamo, P.H.; Ndiaye, O. Evaluation of TIGGE precipitation forecasts over West Africa at intraseasonal timescale. *Clim. Dyn.* **2016**, *47*, 31–47. [CrossRef]
14. Lorenz, E.N. Atmospheric predictability as revealed by naturally occurring analogues. *J. Atmos. Sci.* **1969**, *26*, 636–646. [CrossRef]
15. Zhang, H.; Chen, M.; Fan, S. Study on the construction of initial condition perturbations for the regional ensemble prediction system of North China. *Atmosphere* **2019**, *10*, 87. [CrossRef]
16. Schulz, B.; Lerch, S. Machine learning methods for postprocessing ensemble forecasts of wind gusts: A systematic comparison. *Mon. Weather. Rev.* **2022**, *150*, 235–257. [CrossRef]
17. Zhu, Y.; Luo, Y. Precipitation calibration based on the frequency-matching method. *Weather. Forecast.* **2015**, *30*, 1109–1124. [CrossRef]
18. Guo, R.; Yu, H.; Yu, Z.; Tang, J.; Bai, L. Application of the frequency-matching method in the probability forecast of landfalling typhoon rainfall. *Front. Earth Sci.* **2022**, *16*, 52–63. [CrossRef]
19. Hacker, J.P.; Rife, D.L. A practical approach to sequential estimation of systematic error on near-surface mesoscale grids. *Weather. Forecast.* **2007**, *22*, 1257–1273. [CrossRef]
20. Kim, H.M.; Ham, Y.G.; Scaife, A.A. Improvement of initialized decadal predictions over the North Pacific Ocean by systematic anomaly pattern correction. *J. Clim.* **2014**, *27*, 5148–5162. [CrossRef]
21. Lyu, Y.; Zhi, X.; Zhu, S.; Fan, Y.; Pan, M. Statistical calibrations of surface air temperature forecasts over east Asia using pattern projection methods. *Weather. Forecast.* **2021**, *36*, 1661–1674. [CrossRef]
22. Belorid, M.; Kim, K.R.; Cho, C. Bias Correction of short-range ensemble forecasts of daily maximum temperature using decaying average. *Asia Pac. J. Atmos. Sci.* **2020**, *56*, 503–514. [CrossRef]
23. Han, K.; Choi, J.T.; Kim, C. Comparison of statistical post-processing methods for probabilistic wind speed forecasting. *Asia Pac. J. Atmos. Sci.* **2018**, *54*, 91–101. [CrossRef]
24. Krishnamurti, T.N.; Kishtawal, C.M.; LaRow, T.E.; Bachiochi, D.R.; Zhang, Z.; Williford, C.E.; Gadgil, S.; Surendran, S. Improved weather and seasonal climate forecasts from multimodel superensemble. *Science* **1999**, *285*, 1548–1550. [CrossRef]
25. Jun, S.; Kang, N.Y.; Lee, W.; Chung, Y. An alternative multi-model ensemble forecast for tropical cyclone tracks in the western North Pacific. *Atmosphere* **2017**, *8*, 174. [CrossRef]
26. Ji, L.; Zhi, X.; Zhu, S.; Fraedrich, K. Probabilistic precipitation forecasting over East Asia using Bayesian model averaging. *Weather. Forecast.* **2019**, *34*, 377–392. [CrossRef]
27. Zhang, L.; Zhi, X.F. Multimodel consensus forecasting of low temperature and icy weather over central and southern China in early 2008. *J. Trop. Meteorol.* **2015**, *21*, 67–75.
28. Krishnamurti, T.N.; Kishtawal, C.M.; Zhang, Z.; LaRow, T.; Bachiochi, D.; Williford, E.; Gadgil, S.; Surendran, S. Multimodel ensemble forecasts for weather and seasonal climate. *J. Clim.* **2000**, *13*, 4196–4216. [CrossRef]

29. Krishnamurti, T.N.; Kumar, V.; Simon, A.; Bhardwaj, A.; Ghosh, T.; Ross, R. A review of multimodel superensemble forecasting for weather, seasonal climate, and hurricanes. *Rev. Geophys.* **2016**, *54*, 336–377. [CrossRef]
30. Zhi, X.; Qi, H.; Bai, Y.; Lin, C. A comparison of three kinds of multimodel ensemble forecast techniques based on the TIGGE data. *Acta Meteorol. Sin.* **2012**, *26*, 41–51. [CrossRef]
31. Koh, T.Y.; Wang, S.; Bhatt, B.C. A diagnostic suite to assess NWP performance. *J. Geophys. Res. Atmos.* **2012**, *117*, D13. [CrossRef]
32. Gupta, H.V.; Kling, H.; Yilmaz, K.K.; Martinez, G.F. Decomposition of the mean squared error and NSE performance criteria: Implications for improving hydrological modelling. *J. Hydrol.* **2009**, *377*, 80–91. [CrossRef]
33. Zhang, Y.; Ye, A.; Nguyen, P.; Analui, B.; Sorooshian, S.; Hsu, K. New insights into error decomposition for precipitation products. *Geophys. Res. Lett.* **2021**, *48*, e2021GL094092. [CrossRef]
34. Sinha, T.; Sankarasubramanian, A.; Mazrooei, A. Decomposition of sources of errors in monthly to seasonal streamflow forecasts in a rainfall–runoff regime. *J. Hydrometeorol.* **2014**, *15*, 2470–2483. [CrossRef]
35. Mazrooei, A.; Sinha, T.; Sankarasubramanian, A.; Kumar, S.; Peters-Lidard, C.D. Decomposition of sources of errors in seasonal streamflow forecasting over the US Sunbelt. *J. Geophys. Res. Atmos.* **2015**, *120*, 11,809–11,825. [CrossRef]
36. Murphy, A.H. Skill scores based on the mean square error and their relationships to the correlationcoefficient. *Mon. Weather. Rev.* **1988**, *116*, 2417–2424. [CrossRef]
37. Geman, S.; Bienenstock, E.; Doursat, R. Neural networks and the bias/variance dilemma. *Neural Comput.* **1992**, *4*, 1–58. [CrossRef]
38. Hodson, T.O.; Over, T.M.; Foks, S.S. Mean squared error, deconstructed. *J. Adv. Model. Earth Syst.* **2021**, *13*, e2021MS002681. [CrossRef]
39. Zhang, L.; Zhou, T.; Wu, P.; Chen, X. Potential predictability of North China summer drought. *J. Clim.* **2019**, *32*, 7247–7264. [CrossRef]
40. Liu, Y.; Pan, Z.; Zhuang, Q.; Miralles, D.G.; Teuling, A.J.; Zhang, T.; An, P.; Dong, Z.; Zhang, J.; He, D.; et al. Agriculture intensifies soil moisture decline in Northern China. *Sci. Rep.* **2015**, *5*, 11261. [CrossRef]
41. Hersbach, H.; Bell, B.; Berrisford, P.; Hirahara, S.; Horányi, A.; Muñoz-Sabater, J.; Nicolas, J.; Peubey, C.; Radu, R.; Schepers, D.; et al. The ERA5 global reanalysis. *Q. J. R. Meteorol. Soc.* **2020**, *146*, 1999–2049. [CrossRef]
42. Olauson, J. ERA5: The new champion of wind power modelling? *Renew. Energy* **2018**, *126*, 322–331. [CrossRef]
43. Hamill, T.M.; Hagedorn, R.; Whitaker, J.S. Probabilistic forecast calibration using ECMWF and GFS ensemble reforecasts. Part II: Precipitation. *Mon. Weather. Rev.* **2008**, *136*, 2620–2632. [CrossRef]
44. Tarek, M.; Brissette, F.P.; Arsenault, R. Large-scale analysis of global gridded precipitation and temperature datasets for climate change impact studies. *J. Hydrometeorol.* **2020**, *21*, 2623–2640. [CrossRef]
45. Graham, R.M.; Hudson, S.R.; Maturilli, M. Improved performance of ERA5 in Arctic gateway relative to four global atmospheric reanalyses. *Geophys. Res. Lett.* **2019**, *46*, 6138–6147. [CrossRef]
46. Zhang, L.; Ma, X.; Zhu, S.; Guo, Z.; Zhi, X.; Chen, C. Analyses and applications of the precursor signals of a kind of warm sector heavy rainfall over the coast of Guangdong, China. *Atmos. Res.* **2022**, *280*, 106425. [CrossRef]
47. Hagedorn, R.; Buizza, R.; Hamill, T.M.; Leutbecher, M.; Palmer, T.N. Comparing TIGGE multimodel forecasts with reforecast-calibrated ECMWF ensemble forecasts. *Q. J. R. Meteorol. Soc.* **2012**, *138*, 1814–1827. [CrossRef]
48. Bao, X.; Zhang, F.; Sun, J. Diurnal variations of warm-season precipitation east of the Tibetan Plateau over China. *Mon. Weather. Rev.* **2011**, *139*, 2790–2810. [CrossRef]
49. Bromwich, D.H.; Cullather, R.I.; Grumbine, R.W. An assessment of the NCEP operational global spectral model forecasts and analyses for Antarctica during FROST. *Weather. Forecast.* **1999**, *14*, 835–850. [CrossRef]
50. Steinacker, R.; Ratheiser, M.; Bica, B.; Chimani, B.; Dorninger, M.; Gepp, W.; Lotteraner, C.; Schneider, S.; Tschannett, S. A mesoscale data analysis and downscaling method over complex terrain. *Mon. Weather. Rev.* **2006**, *134*, 2758–2771. [CrossRef]
51. Han, L.; Chen, M.; Chen, K.; Chen, H.; Zhang, Y.; Lu, B.; Song, L.; Qin, R. A deep learning method for bias correction of ECMWF 24–240 h forecasts. *Adv. Atmos. Sci.* **2021**, *38*, 1444–1459. [CrossRef]
52. Zhu, S.; Zhi, X.; Ge, F.; Fan, Y.; Zhang, L.; Gao, J. Subseasonal forecast of surface air temperature using superensemble approaches: Experiments over Northeast Asia for 2018. *Weather. Forecast.* **2021**, *36*, 39–51. [CrossRef]
53. He, C.; Zhi, X.; You, Q.; Song, B.; Fraedrich, K. Multi-model ensemble forecasts of tropical cyclones in 2010 and 2011 based on the Kalman Filter method. *Meteorol. Atmos. Phys.* **2015**, *127*, 467–479. [CrossRef]
54. Ji, L.; Zhi, X.; Simmer, C.; Zhu, S.; Ji, Y. Multimodel ensemble forecasts of precipitation based on an object-based diagnostic evaluation. *Mon. Weather. Rev.* **2020**, *148*, 2591–2606. [CrossRef]
55. Peng, T.; Zhi, X.; Ji, Y.; Ji, L.; Tian, Y. Prediction skill of extended range 2-m maximum air temperature probabilistic forecasts using machine learning post-processing methods. *Atmosphere* **2020**, *11*, 823. [CrossRef]

atmosphere

Article

Prediction of a Pilot's Invisible Foe: The Severe Low-Level Wind Shear

Afaq Khattak [1,*], Pak-Wai Chan [2], Feng Chen [1,*] and Haorong Peng [3]

1 The Key Laboratory of Infrastructure Durability and Operation Safety in Airfield of CAAC, Tongji University, 4800 Cao'an Road, Jiading, Shanghai 201804, China
2 Hong Kong Observatory, 134A Nathan Road, Kowloon, Hong Kong, China
3 Shanghai Research Center for Smart Mobility and Road Safety, Shanghai 200092, China
* Correspondence: khattak@tongji.edu.cn (A.K.); fengcheng@tongji.edu.cn (F.C.)

Abstract: Severe low-level wind shear (S-LLWS) in the vicinity of airport runways (25 knots or more) is a growing concern for the safety of civil aviation. By comprehending the causes of S-LLWS events, aviation safety can be enhanced. S-LLWS is a rare occurrence, but it is hazardous for approaching and departing aircraft. This study introduced the self-paced ensemble (SPE) framework and Shapley additive explanations (SHAP) interpretation system for the classification, prediction, and interpretation of LLWS severity. Doppler LiDAR- and PIREPs-based LLWS data from Hong Kong International Airport were obtained, trained, and evaluated to predict LLWS severity. The SPE framework was also compared to state-of-the-art tree-based models, including light gradient boosting machine, adaptive boosting, and classification and regression tree models. The SPE does not require prior data treatment; however, SMOTE-ENN was utilized to treat highly imbalanced LLWS training data for tree-based models. In terms of prediction performance, the SPE framework outperforms all tree-based models. Using SHAP analysis, the SPE was interpreted. It was determined that "runway 25LD", "mean hourly temperature", and "mean wind speed" were the most significant contributors to the occurrence of S-LLWS. The most optimistic projections for the occurrence of S-LLWS events at runway 25LD were during periods of low-to-moderate temperatures and relatively medium-to-high wind speeds. Similarly, the majority of S-LLWS events took place on the runway. Without the need for data augmentation during preprocessing, the SPE framework coupled with the SHAP interpretation system could be utilized effectively for the prediction and interpretation of LLWS severity. This study is an invaluable resource for aviation policymakers and air traffic safety analysts.

Keywords: civil aviation safety; low-level wind shear; pilot reports; machine learning; self-paced ensemble; Shapley additive explanations

Citation: Khattak, A.; Chan, P.-W.; Chen, F.; Peng, H. Prediction of a Pilot's Invisible Foe: The Severe Low-Level Wind Shear. *Atmosphere* **2023**, *14*, 37. https://doi.org/10.3390/atmos14010037

Academic Editors: Duanyang Liu, Hongbin Wang and Shoupeng Zhu

Received: 4 December 2022
Revised: 21 December 2022
Accepted: 21 December 2022
Published: 25 December 2022

1. Introduction

Airline operations are profoundly impacted by weather conditions. Major causes of flight cancellations, delays, and even fatal crashes [1–3] can be traced back to this concern. Wind shear refers to an abrupt shift in the wind's speed or direction in the atmosphere. Particularly during landing and takeoff, aircraft are impacted by low-level wind shear (LLWS), which is present in a lower layer at 1600 feet above ground level (AGL). LLWS is defined by the International Civil Aviation Organization (ICAO) [4] as a 15-knot-or-greater change in wind direction at or below 1600 feet above ground level. It affects the aircraft's lift, and the resulting course deviation could endanger planes taking off or landing [5,6].

Many LLWS events with a magnitude of 25 knots or higher have been registered at airports around the world. Because S-LLWS may have a stronger impact on aircraft operations, timely warnings are crucial. Hong Kong International Airport is one of the most at-risk airports for LLWS (HKIA). It is located in Lantau Island's northern region, which is mountainous, with peaks reaching over 900 m and valleys dropping to 300 m. Lowering the adverse effects of S-LLWS on airport safety and productivity is vital. A reliable LLWS severity prediction approach is crucial for achieving the goal of providing precise and effective wind hazard alerts and ensuring the safety of civil aviation. The development of models for predicting the severity of LLWS closer to airport runways, however, remains among the most challenging areas of research in civil aviation today.

Due to the fact that wind shear exhibits characteristics of meso- and micro-scale meteorological phenomena, such as abrupt changes in speed and direction and a small temporal–spatial scale, predicting wind shear is a difficult endeavor. LLWS events occur in both rainy and non-rainy weather and include phenomena such as frontal gusts and microbursts associated with severe convection, dry microbursts, low-level jets, sea breezes, complex terrain effects, etc [7]. To ensure the safety of civil aircraft, various technologies, including anemometers, terminal Doppler weather radar (TDWR), and Doppler light detection and range (DLDR), have been installed at major airports around the world to detect LLWS (LIDAR). Few airports, including those in Japan, Malaysia, the United States, Germany, France, South Korea, Singapore, and Hong Kong, have LLWS alerting technologies due to high instrument and maintenance costs, a lack of relevant research, and unique local environments [8]. The anemometer-based LLWS alert system and the TDWR have been developed since the 1970s. Their effectiveness for detecting and warning of LLWS in rainy conditions has been demonstrated [9]. The complementary TDWR is also capable of detecting LLWS caused by terrain. However, these technologies are incapable of capturing LLWS events in non-rainy weather [10,11] and are unsuitable for detecting LLWS along the glide path.

Doppler LIDAR [8], a relatively new remote sensing technology, offers a promising alternative for detecting LLWS when the weather is clear. Similarly, certain LLWS events are terrain-induced LLWS phenomena caused by the complex terrain surrounding an airport. Doppler LIDAR technology, which does not depend on humid conditions for detecting LLWS and captures LLWS due to complex terrain near airports, has been developed to address these scenarios. Hong Kong International Airport [12], Nice Côte d'Azur Airport in France [13], Tokyo Haneda International Airport in Japan [14] and Beijing Capital International Airport in China [15] are equipped with the Doppler LIDAR system. It has been added to the TDWR as an augmentation in order to detect and warn of LLWS, even in clear skies. However, the development of a model to predict the severity of LLWS based on Doppler LIDAR observations remains a challenging task that must be addressed. Similarly, all of these LLWS alerting technologies (based on remote sensing and/or on-site measurements) have been proven effective and operational. These technologies send notifications or alerts when LLWS events are detected or observed. However, these hardware-based technologies are incapable of predicting the occurrence of LLWS events and assessing the risk factors that contribute to their occurrence [16].

In the past, numerous numerical modeling techniques, including large-eddy simulations (LES) [17], computational fluid dynamics (CFD) [18] and numerical weather prediction (NWP) [12] have been employed to attempt to predict or simulate wind shear conditions. In general, these studies focused on single or isolated occurrences of reported wind shear events and were conducted on a case-by-case basis. There are insufficient systematic, long-term evaluations of the ability of numerical models to predict the occurrence of LLWS events. These days, machine learning algorithms have gained significant ground. It has become one of the most widely used and beneficial tools in transportation research such as road safety, transportation planning, and pavement analysis [19–22]. However, there is a significant gap in the application of machine learning algorithm in the aviation safety domain, particularity in the prediction and classification of LLWS severity. In this research, efficiently predicting S-LLWS is of interest to us. However, in the data from LiDAR and pilot reports (PIREPs), the S-LLWS class is typically much smaller than the non-severe low-level wind shear (NS-LLWS) class. This creates a data imbalance issue and requires data balancing prior to training and evaluation. Therefore, in contrast to hardware-based technologies and numerical simulation modeling, which efficiently predict LLWS severity while simultaneously dealing with the class imbalance issue, we propose the self-paced ensemble (SPE) framework [23]. This is an ensemble imbalance learning model for dealing with highly imbalanced data. It aims to produce a robust ensemble by the self-paced harmonizing of data hardness via the undersampling method that has been developed. This framework, despite being computationally efficient, has resulted in robust performances in the presence of extremely skewed distributions.

Although machine learning models are efficient in prediction, they do not explicitly demonstrate the relations between input and output factors due to their black-box nature. The interpretation of the model is equally important for appropriately assessing the model's performance. Previously, the machine learning model's results were interpreted using the variable importance analysis methods such as permutation-based importance scores. The variable importance analysis, however, can only provide a ranking of the variables' importance and is unable to explain how each variable individually influences the prediction of the model. Shapley additive explanations (SHAP) analysis, based on the concept of game theory [24], has been utilized in recent studies to quantify each factor's effect on the outcome [25,26]. In this research, we have also employed SHAP, analysis in conjunction with SPE framework, for the assessment of the relative importance of various factors as well as their contributions.

The rest of this paper is organized as follows: The following sections constitute the research methodology, which provides the data description, the details of the proposed SPE framework, a Bayesian optimization strategy for hyperparameter tuning, and the description of the SHAP interpretation system. These are then followed by Section 3, discussing the SPE framework and comparison with other machine learning models, as well as SHAP analysis. Finally, Section 4 summarizes the conclusions and makes additional research recommendations.

2. Materials and Methods

Initially, the LLWS data consisted of LiDAR data and pilot flight reports (PIREPs) obtained from the Hong Kong Observatory (HKO) at HKIA. The details of data extraction from LiDAR and PIREPs are provided in the subsequent section. The extracted data were merged together and preprocessed to separate training–validation and testing datasets into 70% and 30%, respectively. The training dataset was used to develop an SPE framework and tree-based machine learning models, including light gradient boosting machine (LGBM), adaptive boosting (AdaBoost), and classification and regression tree (CART), and the testing dataset was used to evaluate the performance of the developed model. The SPE framework is an ensemble imbalance learning system, which does not require data balancing during the preprocessing phase. In contrast, data balancing was required for the tree-based machine learning models prior to training and validation, which were used to compare the results with the SPE framework. For data balancing, a hybrid synthetic minority oversampling technique—edited nearest neighbor (SMOTE-ENN) treatment was applied to the LLWS training dataset. A portion of the training–validation data were also used to tune model hyperparameters. A Bayesian optimization approach was utilized for the hyperparameter tuning. Afterwards, the SHAP interpretation system was used to evaluate the significance and contribution of various risk factors that generate S-LLWS in the vicinity of airport runways. In addition, factor interaction analysis by SHAP was also conducted. Figure 1 depicts the entire operational paradigm described in this study.

2.1. Study Location

The Hong Kong International Airport (HKIA) is situated on an artificial island called Lantau, which is surrounded on three sides by water. To the south, there are mountains that rise to more than 900 m above sea level. The complex land–sea contrast and intricate orography of HKIA have been the subject of numerous observational and modeling studies, all of which have identified that they are favorable conditions for the occurrence of LLWS [27,28]. As seen in Figure 2, the mountainous area to the south of HKIA amplifies LLWS, disrupting airflow and causing mountain waves, gap effluents, and other disruptions along the HKIA flight paths. The north runway and the south runway are the two runways at HKIA. The directions they orient are 070° and 250°. Eight different arrangements are possible because each runway can be used for takeoffs and landings in either direction. For instance, runway '07LA' stands for a landing ('A' stands for arrival) using the left runway (hence 'L') and a heading angle of 070°. This depicts a plane landing on HKIA's north runway from the west. The same goes for an aircraft taking off from the south runway and heading west—runway 25LD.

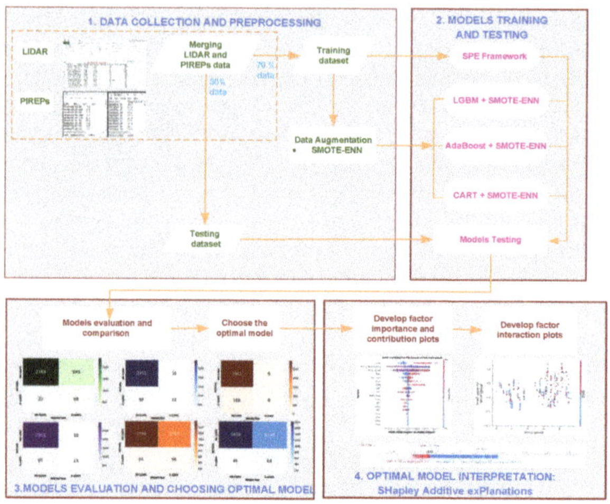

Figure 1. Framework for the prediction and interpretation of LLWS severity in the vicinity of runways.

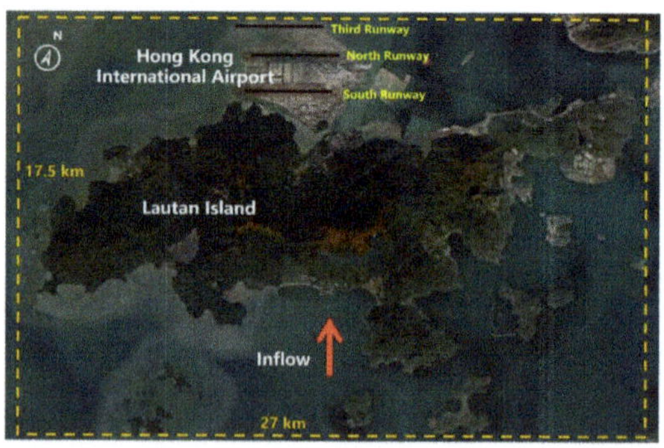

Figure 2. Hong Kong International airport and surrounding terrain.

2.2. Instrument and Data

In this section, the Doppler LiDAR of HKIA and the pilot flight reports (PIREPs) of HKIA inbound and outbound flights, are thoroughly discussed.

2.2.1. Doppler LiDAR at HKIA

In this study, LLWS data gathered from the 2 × long-range Doppler LiDAR at HKIA were analysed. LiDAR operates at an infrared wavelength of approximately 1.5 microns; 100 m is the radial resolution or physical range gate. Maximum radial velocity is roughly 40 m per second. Typically, under ideal weather conditions and in the absence of obstructions such as low clouds, an observation range of 10 or 15 kilometres is achievable. In addition to the standard fixed-elevation scans (plan-position indicator), each LiDAR is configured to conduct "glide-path" scans along take-off and landing flight paths. Coordinating the elevation and azimuth movements of the laser scanner head accomplishes this. Typically, the four possible configurations of the north runway (07LA, 25RA, 07LD, and 25RD) are covered by the north LiDAR, including arrivals (A) and departures (D) directions, and the four possible configurations of the south runway (07RA, 25LA, 07RD and 25LD) are covered by the south LiDAR, including arrivals (A) and departures (D) directions towards the west and the east. The headwind component along each runway configuration (labelled "corridor") can be derived from the "glide-path" scans' radial velocity data. Typically, the scan revisit time for each corridor is roughly one minute, indicating that the temporal resolution or update frequency of the headwind profiles is also roughly one minute.

The LiDAR at HKIA usually operates by a "GLYGA LLWS alerting algorithm" [7]. For each runway corridor, GLYGA receives as input the profile of headwind components gridded with a 100 m interval. The headwind profiles typically extend up to 4–5 NM from the respective runway endpoint, based on scanning range and prevalent atmospheric conditions at the time. Then, a ramp identification procedure is used to identify sudden, consistent changes in the headwind. This is based on the "Peak Spotter" algorithm [29]. First, a profile of velocity increment is quantified by adjusting adjacent data points from the profile of quality-controlled headwind. Next, LLWS "ramps" are identified by sequentially recognizing the velocity increment (i.e., headwind change) within length windows of 400, 800, 1600, and 6400 m. The collection of such "ramps", identified within a single headwind profile, is then ranked using a severity factor [30] that scales with the headwind increment and the inverse cube root of the ramp length. The ramp with the highest severity factor is then used to release an automatic alert when intensity exceeds a predetermined threshold (15 knots) at HKIA.

Mathematically, the quality-controlled headwind profile can be represented as $v(x_k)$, where v is the headwind component at the x_k position, which is the kth data point or range gate along the corresponding glide path. The velocity increment at location x_k can be expressed as $\Delta v(x_k, \lambda) = v(x_k) - v(x_k + \lambda)$ for a given length window (or ramp length), λ. (For a detailed explanation of the ramp identification process at HKIA, please see [7].) The resultant identified ramps, which correspond to a collection of data pairs $(\Delta v, \lambda)$, are then ranked by the severity factor $\mathcal{F}_s$, which is computed using Equation (1).

$$\mathcal{F}_s = \left(\frac{\Delta v}{\lambda}\right)^3 / \Phi_{app} \tag{1}$$

where Φ_{app} denotes the aircrafts' approach speed, which is taken as constant. The $\mathcal{F}_s$ depends primarily on $\Delta v / \lambda$.

2.2.2. HKIA-Based PIREPs

Pilot flight reports (PIREP) of LLWS are an established source for confirming LLWS alerts at HKIA. A PIREP is an abbreviation for pilot reports used in the aviation sector. It is a report that pilots who encounter hazardous weather conditions send to air traffic controllers. Typical PIREPs cover the flight's en route phase and include information on turbulence and aircraft icing. However, the HKIA wind shear PIREPs contain information regarding the timing, location (to the nearest nautical mile), altitude (to the nearest 50 or 100 feet), and velocity (to the nearest 5 knots) of an LLWS event. Pilots can report LLWS events to the air traffic controller at HKIA in two standard ways: by submitting a report form after landing or departure, or by using an on-board radio communication.

2.3. Data Processing

As discussed early, the occurrence of S-LLWS is a substantial risk to inbound and outbound flight safety. Therefore, in order to predict the S-LLWS events, in this study, the occurrence of LLWS severity is defined by the threshold, as shown in Equation (2).

$$\text{LLWS Severity} = \begin{cases} 1: \text{S}-\text{LLWS}, & \text{LLWS} \geq 25 \text{ knots} \\ 0: \text{NS}-\text{LLWS}, & \text{LLWS } 15 - 24 \text{ knots} \end{cases} \quad (2)$$

The original wind shear dataset contains nominal and continuous factors as well as a single target factor, LLWS severity. S-LLWS events represent all LLWS with a magnitude of equal to or greater than 25 knots and are coded as 1, whereas NS-LLWS events have a magnitude between 15 to 24 knots and are coded as 0. S-LLWS events are far less in number than NS-LLWS, but they are an important class for aviation safety. Any ith event in the original dataset can be represented as $(X_i, y_i) = (C_i, N_i, y_i)$, where C_i is the continuous factor, N_i is the nominal factor and y_i is the target factor. As indicated in Table 1, the nominal factors N of the dataset are one-hot encoded. Each nominal value in the dataset is translated into a new column, and the column is assigned a 0 or 1 value. The number of columns is equal to the number of nominal values. For example, an eight-column matrix is created from a nominal factor "Runway" with eight different values (07LA, 07LD, 07RA, 07RD, 25LA, 25LD, 25RA, 25RD). The continuous features of the datasets, on the other hand, are normalized as stated in Equation (3).

$$C_{i,j}^{norm} = \frac{C_{i,j}^{orig} - \min C_j}{\max C_j - \min C_j} \quad (3)$$

where $C_{i,j}^{norm}$ represents the jth normalized continuous factors of the ith instance of the data. $C_{i,j}^{orig}$ represents the original jth continuous factors in the ith instance of the data. The $\min C_j$ and $\max C_j$ represent the minimum and maximum of the jth continuous factor in the original wind shear dataset, respectively.

Table 1. One-hot encoding of categorical factors for the modeling.

Factor	Codes and Description
LLWS Severity	1: If LLWS magnitude is equal to greater than 25 knots, 0: 'Otherwise'
Runways	
07LA	1: If a wind shear event is reported at Runway 07LA, 0: 'Otherwise'
07LD	1: If a wind shear event is reported at Runway 07LD, 0: 'Otherwise'
07RA	1: If a wind shear event is reported at Runway 07RA, 0: 'Otherwise'
25RD	1: If a wind shear event is reported at Runway 25RD, 0: 'Otherwise'
25LA	1: If a wind shear event is reported at Runway 25LA, 0: 'Otherwise'
25LD	1: If a wind shear event is reported at Runway 25LD, 0: 'Otherwise'
25RA	1: If a wind shear event is reported at Runway 25RA, 0: 'Otherwise'
25RD	1: If a wind shear event is reported at Runway 25RD, 0: 'Otherwise'
Location of Occurrence	
1MD	1: If a wind shear event is reported at 1MD from Runway, 0: 'Otherwise'
1MF	1: If a wind shear event is reported at 1MF from Runway, 0: 'Otherwise'
2MD	1: If a wind shear event is reported at 2MD from Runway, 0: 'Otherwise'
2MF	1: If a wind shear event is reported at 2MF from Runway, 0: 'Otherwise'
3MF	1: If a wind shear event is reported at 3MF from Runway, 0: 'Otherwise'
RWY	1: If a wind shear event is reported at Runway, 0: 'Otherwise'
Time of the Day	
Day Time	1: If a wind shear event is reported during daytime, 0: 'Otherwise'
Night Time	1: If a wind shear event is reported during nighttime, 0: 'Otherwise'

Finally, there are 18 factors in the standardized wind shear dataset. The standardized original wind shear dataset consists of the normalized continuous factors (2 factors including hourly temperature and wind speed) as well as one-hot encoded nominal factors (16 factors).

2.4. *Self-Paced Ensemble Framework*

We propose a newly developed SPE framework, which is an ensemble-based imbalance learning framework, to develop a classification and prediction model for S-LLWS using untreated data from LIDAR and PIREPs. Before employing the SPE framework, we present the concepts of hardness harmonization and a self-paced factor.

2.4.1. Hardness Harmonization

All majority class samples are divided into k bins, where k a hyperparameter, based on their hardness values. Each kth bin represents a particular level of hardness. Then, majority class instances are undersampled to create a balanced dataset by maintaining the same total hardness contribution in each bin. Such a method is referred to as "harmonize" in the literature of gradient-based optimization. A similar strategy has been adopted here to harmonize the hardness in the initial iteration. However, hardness harmonization is not utilized in every iteration. The primary reason for this is that the number of trivial samples increases during the training process, as the ensemble classifier gradually conforms to the training set. Consequently, merely harmonizing the hardness contribution leaves a large number of trivial samples. Later iterations of the learning procedure are significantly slowed down by these less informative examples. In lieu of this, "self-paced factors" have been introduced to perform the self-paced harmonization of undersampling.

2.4.2. Self-Paced Factor

In particular, after harmonizing the hardness contribution of each bin, the sample probability of bins with a large population is gradually decreased. The rate of decrease is determined by a self-paced factor (σ). When σ is large, more focus is on the harder samples as opposed to the simple hardness contribution harmonization. In the initial few iterations, the framework focuses primarily on informative borderline samples, and so outliers and noise have little impact on the model's ability to generalize. In later iterations where σ is very large, the framework retains a reasonable proportion of trivial (high confidence) samples as the "skeleton", thereby preventing over-fitting. The detail of SPE framework is shown in Algorithm 1. It is pertinent to mention that the hardness value in each iteration (line 5–6) is updated in order to select those data samples that were most beneficial for the current ensemble. The tangent function (line 8) has been used to control the growth of the self-paced factor. Thus, the self-paced factor is equal to zero in the first iteration and to infinity in the last iteration.

Algorithm 1: Self-Paced Ensemble (SPE) Framework.

1 **Input:** Hardness function ($\hbar$), training dataset $D = \{(x_k, y_k)\}_1^n$, number of bins (k), base classifier (ζ) and number of base classifiers (∂)
2 **Initialize:** $P \Leftarrow$ minority class in training dataset D, $N \Leftarrow$ majority class in training dataset D,
3 Train classifier ζ_0 by using random undersampling of subsets of majority class N_0' and P such that where $|N_0'| = |P|$
4 **for** i = 1 *to* n **do**
5 Ensemble $F_i(x) = \frac{1}{i}\sum_{j=0}^{i-1} \zeta_j(x)$
6 Separate majority class dataset into k bins with respect to $\hbar(x, y, F_i)$: $(b_1, b_2, \ldots, b_\xi)$
7 In the lth bin, the average hardness contribution can be computed as $h_l = \sum_{s \in b_l} h(x_s, y_s, F_i)/|b_l|, \forall\, l = 1, 2, \ldots, k$
8 The self-paced factor is updated as $\sigma = \tan\left(\frac{i\Pi}{2\partial}\right)$
9 The lth bin, non-normalized sampling weight: $p_l = \frac{1}{h_l + \sigma}, \forall\, l = 1, 2, \ldots, k$
10 Undersampling from the lth bin with $\frac{p_l}{\sum_m p_l}|P|$ samples
11 Using newly undersampled data subset, train ζ_i
12 **End**
13 **Return** Final robust ensemble $F(x) = \frac{1}{\partial}\sum_{m=1}^{\partial} \zeta_m(x)$

2.5. Bayesian Optimization for Hyperparameter Tuning

In this study, a Bayesian optimization strategy [31] is employed alongside SPE models and a tree-based machine learning model to determine the optimal hyperparameters. The Bayesian optimization built a probability model for determining the value, which automatically reduces the objective function based on the objective's prior estimation result. Figure 3 is a flowchart of a hybrid Bayesian optimization machine learning approach. Additionally, provided below is the detailed procedure.

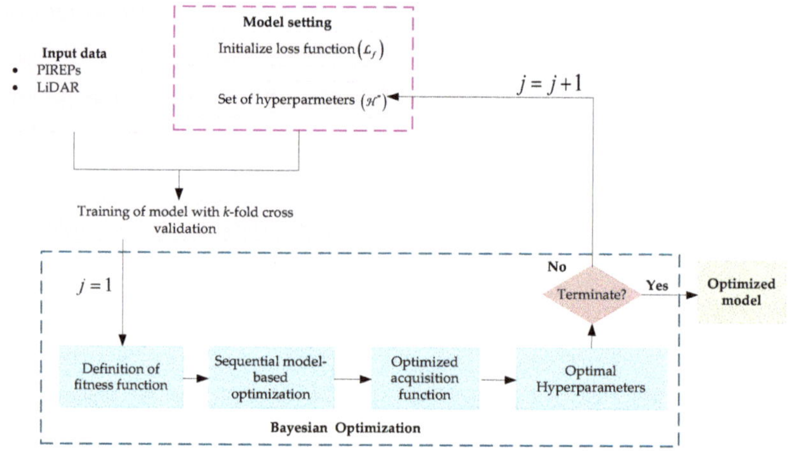

Figure 3. Bayesian approach for hyperparameters tuning.

2.5.1. Initialization

This step involves randomly initializing the appropriate hyperparameter settings (Equation (4)), which can be used to train both the SPE model and machine learning models based on k-fold cross validation. The loss function $\left(L_f\right)$ is additionally initialized.

$$H = \begin{pmatrix} h_{1,1} & h_{1,2} & .. & h_{1,l} \\ h_{2,1} & h_{2,2} & .. & h_{2,l} \\ h_{3,1} & h_{3,2} & .. & h_{3,l} \\ \vdots & \vdots & & \vdots \\ \vdots & \vdots & & \vdots \\ h_{m,1} & h_{m,2} & .. & h_{m,l} \end{pmatrix} \tag{4}$$

2.5.2. Fitness Function

From the initialized values, the solution's random number is generated. Based on the following Equation (5), the fitness function can be used to minimize the objective function.

$$\text{fitness function}\left(\frac{L}{H}\right) = \begin{cases} D(H)\, L < L^* \\ G(H)\, L \geq L^* \end{cases} \tag{5}$$

where L denotes the loss value, $D(H)$ denotes the density estimation, which is based on the loss value during the observations, $G(H)$ is produced by the leftover observations value of loss, and L^* represents the particular quantiles.

2.5.3. Sequential Model-Based Optimization

For fine-tuning the hyperparameters of SPE and tree-based models, sequential model-based optimization is one of the succinct forms of Bayesian optimization. Sequential model-based optimization operates by finding the optimal hyperparameter setting, H^*, by building the Gaussian process, Θ_z, with sampled points which can be obtained by the following Equation (6).

$$H^* = \text{argmin}\Theta_{z-1}(H) \tag{6}$$

Equation (7) can be used to calculate the loss value under ideal hyperparameter settings.

$$L = L_f(H^*) \tag{7}$$

The corresponding L and the H^* settings are stored in the corresponding trails, which can be represented as Ω. These corresponding trails (Ω) are used for parameter settings and evaluations purposes. The Ω update can be determined with the help of following Equation (8).

$$\Omega = \Omega \cup (H^*, L) \tag{8}$$

Finally, build the Gaussian process model, Θ_z, based on updated Ω.

2.5.4. Acquisition Function

The next iteration of the search process is computed using the acquisition function of Bayesian optimization. The maximization of G-Mean, which is the expected improvement in this study, is chosen as an acceptable performance criterion for the SPE model and tree-based machine learning models. Equation (9) can be used to achieve the improvement.

$$D(H) = \max[\{L_{\min} - L(H)\}, 0] \tag{9}$$

2.5.5. Termination

When the termination criteria are satisfied in this step, the best hyperparameters for the SPE model and tree-based machine learning models are obtained.

2.6. Evaluation of Models

In case of binary classification problem, one class is the majority (the negative) and its sample size is represented by n_1; the other class is the minority (the negative) and its sample size is represented by n_2. Let n represent the total size of training dataset, $n = n_1 + n_2$. A binary classifier predicts whether each instance is positive or negative. Therefore, it generates outcomes of four types: true positive (t_p), false positive (f_p), true negative (t_n), and false negative (f_n) (see confusion matrix Figure 4). The confusion matrix provides an in-depth examination of the model's performance when predictions are made for each class. The precision and recall are two exceptionally vital model evaluation metrics. The precision is obtained as the ratio of total number of true positives to the total number of true positives and false positives, whereas recall is the ratio of total number of true positives to the total number of true positives and false negatives. Both precision and recall can be computed from the confusion matrix, as shown by Equations (10) and (11), respectively.

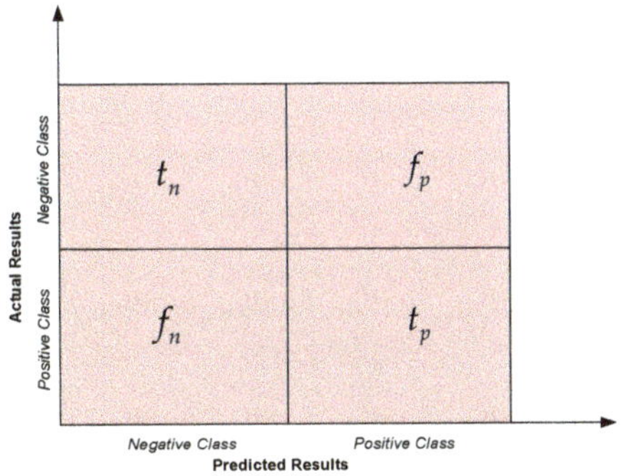

Figure 4. Confusion matrix plot.

However, in ensemble imbalance learning, the imbalanced datasets pose a large challenge to the use of proper metrics for the evaluation of the accuracy in the classification outcomes [32]. The geometric mean (G-Mean), and Matthews' correlation coefficient (MCC) have been used in various studies instead of classification accuracy or F1-score. MCC values should range between −1 and 1. Values closer to +1 represent improved performance. Both MCC and G-Mean have been generally regarded as a balanced measure which can be used even if the classes are of very different sizes. The expressions for the computation of MCC and G-Mean from confusion matrix are given by Equations (12) and (13).

$$Recall = \frac{t_p}{t_p + f_n} \tag{10}$$

$$Precision = \frac{t_p}{t_p + f_p} \tag{11}$$

$$G - Mean = \sqrt{\left(\frac{t_p}{t_p + f_n}\right)\left(\frac{t_n}{f_p + t_n}\right)} \tag{12}$$

$$MCC = \frac{t_p \times t_n - f_p \times f_n}{\sqrt{\left(t_p + f_p\right)\left(t_p + f\right)\left(t_n + f_p\right)\left(t_n + f_n\right)}} \tag{13}$$

2.7. Interpretation of Model by Shapley Additive Explanations (SHAP)

The SHAP analysis relies on a game-theoretical approach to explain the outputs of the ensemble machine learning classifiers. Since machine learning models are "black boxes", the core ideas behind the SHAP analysis involve interpreting these models from both a global and local perspective. The SHAP values, which correspond to the value assigned to each factor in the computation of a machine learning prediction, are estimated. The contribution of each factor is determined and displayed as a Shapley value using Equation (14).

$$\varphi_i = \sum_{\Upsilon \subseteq \Pi\{i\}} \frac{\Upsilon!\left(n - |\Upsilon| - 1\right)!}{n!}\left[\mathrm{E}\left(\Upsilon \cup \{i\}\right) - \mathrm{E}(\Upsilon)\right] \tag{14}$$

where φ_i illustrates the ith factor contribution, Π the set of all factors, Υ is the subset of decision factor, $\mathrm{E}(\Upsilon_i)$ and $\mathrm{E}(\Upsilon)$ illustrate the best model results with and without ith factors, respectively. SHAP analysis yields the outputs of machine learning models through an additive factors imputation strategy, wherein the output model is defined as a linear sum of the input factors (Equation (15)).

$$g(\Psi') = \mu_0 + \sum_{i=1}^{\Lambda} \mu_i \Psi' \quad \Psi' \in \{0,1\}^{\Lambda} \tag{15}$$

It is equal to 1 in cases when a factor is observed, otherwise it is 0. It illustrates that the amount of all input factors, μ_0, represents an outcome without factors (i.e., base value), whereas μ_i shows the Shapley value of factor ith.

3. Results and Discussion

In order to predict the severity of LLWS, this study used an effective and cutting-edge SPE framework along with tree-based machine learning models. Python 3.6.6, a free and open-source programming language, was used in this context. For model training, hyperparameter tuning, performance evaluation, and interpretation, we used the Scikit-learn, sklearn.metrics, HyperOpt, and Shap libraries, as well as Python's sklearn.metrics, imbeans, and sklearn.ensemble. Figure 5 shows how LLWS events are distributed in relation to runway orientation, location of occurrence, and time of day. The box plot of the hourly temperature and wind speed is shown in Figure 6. On the training set for tree-based models, the SMOTE-ENN treatment strategy was used. The training–validation dataset contained 257 instances of S-LLWS and 6908 instances of NS-LLWS prior to treatment. The NS-LLWS instances changed into 6518 and 3069 S-LLWS instances, respectively, after the SMOTE-ENN treatment. The performance evaluation was conducted using a testing dataset, and comparisons were made. The best model is then utilized for SHAP analysis.

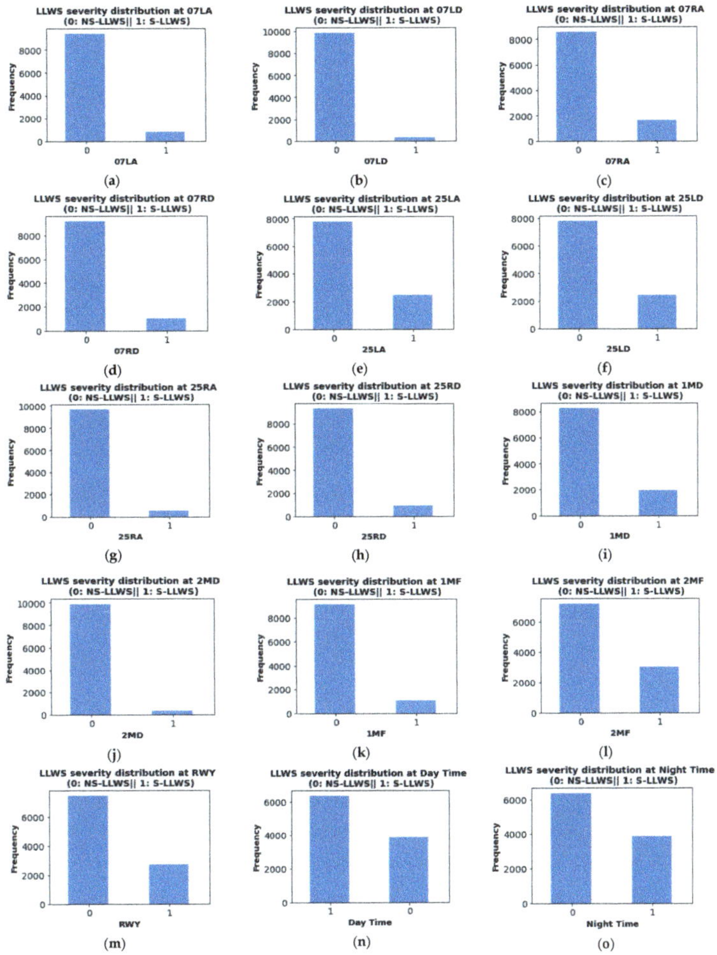

Figure 5. LLWS events distribution: (**a**) Frequency of S-LLWS and NS-LLWS at Runway 07LA, (**b**) Frequency of S-LLWS and NS-LLWS at Runway 07LD, (**c**) Frequency of S-LLWS and NS-LLWS at Runway 07RA, (**d**) Frequency of S-LLWS and NS-LLWS at Runway 07RD, (**e**) Frequency of S-LLWS and NS-LLWS at Runway 25LA, (**f**) Frequency of S-LLWS and NS-LLWS at Runway 25LD, (**g**) Frequency of S-LLWS and NS-LLWS at Runway 25RA, (**h**) Frequency of S-LLWS and NS-LLWS at Runway 25RD, (**i**) Frequency of S-LLWS and NS-LLWS at 1MD from Runway, (**j**) Frequency of S-LLWS and NS-LLWS at 2MD from Runway, (**k**) Frequency of S-LLWS and NS-LLWS at 1MF from Runway, (**l**) Frequency of S-LLWS and NS-LLWS at 2MF from Runway, (**m**) Frequency of S-LLWS and NS-LLWS at RWY, (**n**) Frequency of S-LLWS and NS-LLWS during day time, (**o**) Frequency of S-LLWS and NS-LLWS during night time.

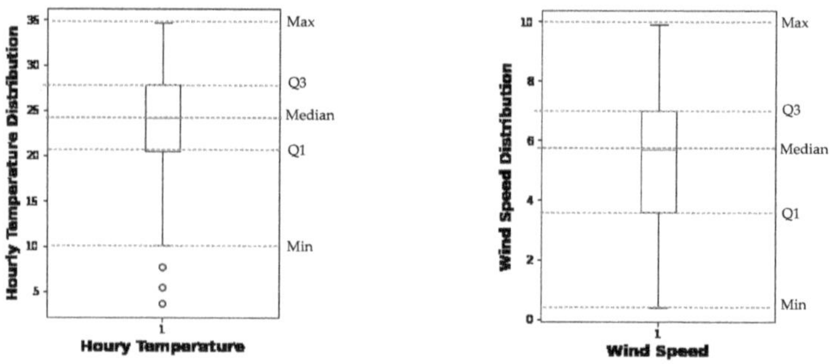

(a) (b)

Figure 6. Box plot: (**a**) hourly temperature distribution (**b**) wind speed distribution.

3.1. Hyperparameter Tuning Using Bayesian Optimization

We used a Bayesian optimization technique that maximized the G-Mean metric to identify the optimal hyperparameters. It is important to note that the SPE framework did not require any prior data treatment, and so imbalanced data were used as input. For tee-based models, both untreated and SMOTE-ENN-treated data were used in the hyperparameter tuning process. Table 2 shows the hyperparameters along with their ranges and optimal values.

Table 2. Machine learning models hyperparameter tuning.

Treatment	Strategy	Hyperparameters	Range	Optimal Values
No treatment	SPE	*n_estimators*	[500, 3000]	833
		max_depth	[0, 10]	7
		learning_rate	[0.001, 0.1]	0.077
	LGBM	*n_estimators*	[500, 3000]	2099
		learning_rate	[0.001, 0.1]	0.043
		max_depth	[0, 10]	5
		lambda_l1	[0.001, 5]	0.39
		lambda_l2	[0.001, 5]	0.22
	AdaBoost	*n_estimators*	[500–3000]	1873
		Learning_rate	[0.01, 1]	0.056
	CART	*min_samples_leaf*	[0.05, 0.1]	0.04
		max_depth	[0, 10]	8
SMOTE-ENN	LGBM	*learning_rate*	[0.001, 0.1]	0.079
		n_estimators	[500, 3000]	2371
		max_depth	[0, 10]	4
		lambda_l1	[0.001, 0.1]	0.57
		lambda_l2	[0.001, 0.1]	0.41
	AdaBoost	*n_estimators*	[500, 3000]	3110
		learning_rate	[0.001, 0.1]	0.093
	CART	*min_samples_leaf*	[0.05, 0.1]	0.03
		max_depth	[0, 10]	8

3.2. Models Performance Assessment and Comparison

The terms S-LLWS and NS-LLWS events were used in this study to designate positive and negative classes of LLWS, respectively. Different performance measures that were derived from the confusion matrix were used to evaluate each model (Figure 7). The recall value and precision values in Table 3 show how well the classifier performed in correctly classifying S-LLWS cases and NS-LLWS cases, respectively. All models were observed to be able to classify NS-LLWS events with high accuracy—more than 95.02%. Given the large number of NS-LLWS cases in the LLWS data, this was expected. The SPE framework utilizing testing data had an 80.13% recall value, compared to all others, each of which had a recall value of between 0.00% and 62.43% regarding the recall values for correctly classifying S-LLWS cases. Figure 7 demonstrates that 88 of the 110 S-LLWS cases in the testing dataset were correctly classified by the SPE framework. After that, CART with SMOTE-ENN-treated data were used, correctly classifying 68 out of 110 S-LLWS while incorrectly classifying the remaining 42. S-LLWS by SPE had a relative classification accuracy rate of 29.41% higher than CART with SMOTE-ENN-treated data. The AdaBoost model with no data treatment did the worst job of correctly classifying S-LLWS. The 110 S-LLWS cases were incorrectly classified in none of them.

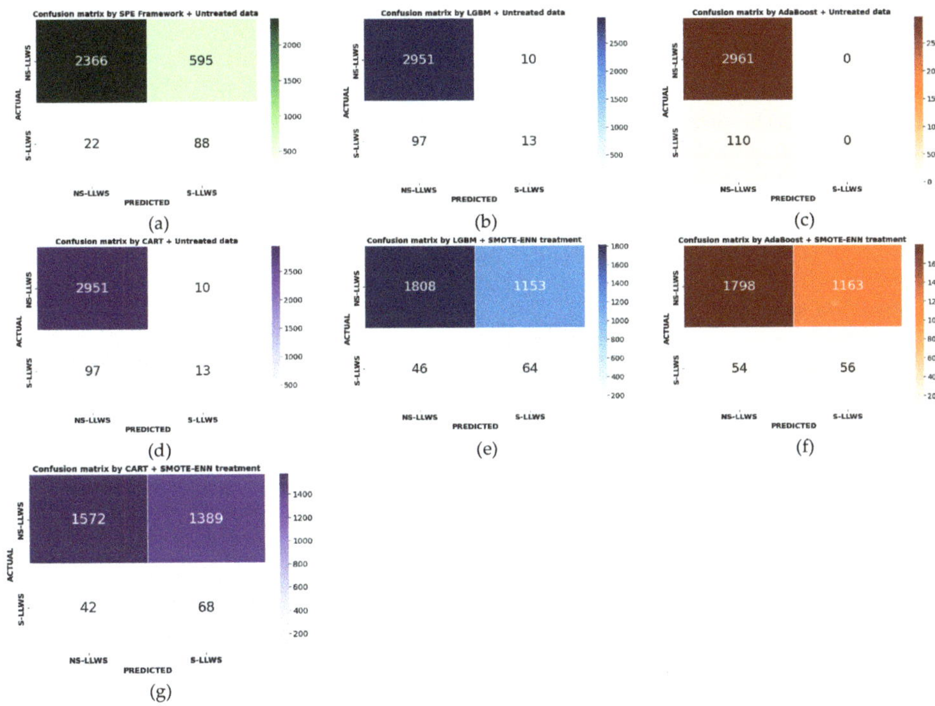

Figure 7. Confusion Matrix: (**a**) SPE framework, (**b**) LGBM without data treatment, (**c**) AdaBoost without data treatment, (**d**) CART without data treatment, (**e**) LGBM with SMOTE-ENN data treatment, (**f**) AdaBoost with SMOTE-ENN data treatment, (**g**) CART with SMOTE-ENN data treatment.

In addition, we have utilized G-Mean and MCC methods in our study. On the testing dataset, the SPE framework demonstrated a higher G-Mean than all other models with treated and untreated data. G-Mean was 0.82 for the SPE framework and 0.59 for LGBM with SMOTE-ENN-treated data. AdaBoost displays the lowest G-Mean value of 0.50 with no treated data. The G-Mean value of the SPE framework was 39.98% greater than that of the LGBM with SMOTE-ENN-treated data. Likewise, comparing MCC values, the SPE framework also outperformed LGBM, AdaBoost, and CART models, with an MCC value of 0.27 indicating the best performance, followed by 0.24 for LGBM. Using G-Mean and MCC as balanced measures of performance, the SPE framework utilizing imbalanced data outperformed the tree-based model SMOTE-ENN that was applied to the balanced data. Consequently, it could be regarded as the optimal model for the interpretation provided by the SHAP analysis, such as the relative importance of factors, their contributions, and their interactions.

Table 3. Performance measure of machine learning models.

Treatment	Model	Class	Precision	Recall	G-Mean	MCC
No treatment	SPE	NS-LLWS	0.99	0.80	0.82	0.27
		S-LLWS	0.13	0.80		
		Average	0.56	0.80		
	LGBM	NS-LLWS	0.97	1.00	0.55	0.24
		S-LLWS	0.57	0.12		
		Average	0.77	0.56		
	AdaBoost	NS-LLWS	0.96	1.00	0.50	0.00
		S-LLWS	0.00	0.00		
		Average	0.48	0.56		
	CART	NS-LLWS	0.97	1.00	0.55	0.23
		S-LLWS	0.57	0.12		
		Average	0.77	0.56		
SMOTE-ENN	LGBM	NS-LLWS	0.98	0.61	0.59	0.07
		S-LLWS	0.05	0.58		
		Average	0.51	0.60		
	AdaBoost	NS-LLWS	0.97	0.61	0.58	0.04
		S-LLWS	0.05	0.51		
		Average	0.51	0.56		
	CART	NS-LLWS	0.97	0.53	0.57	0.05
		S-LLWS	0.05	0.62		
		Average	0.51	0.57		

3.3. *Self-Paced Ensemble Framework Interpretation by SHAP*

3.3.1. Global Factor Interpretation

Numerous techniques can be employed to determine the relative significance of factors. However, factor contribution is distinct from factor significance. The contribution of a factor indicates which factor has the greatest influence on a model's performance. In addition to identifying relevant factors, the factor contributions provide a rational explanation for the observed results. This study investigated the significance of each factor and its contribution using SHAP analysis. Figure 8a depicts, initially, the factor importance of the input factors, indicating the overarching influence of the factors on the predictions. It is the mean of the absolute Shapley values for the entire training dataset. The average absolute SHAP value of 0.185 indicates that, of all the features, "Runway 25LD" is the most vulnerable to S-LLWS occurrences. The average absolute SHAP values for "hourly temperature" and "wind speed" are 0.145 and 0.135, respectively, making them the second and third most influential factors.

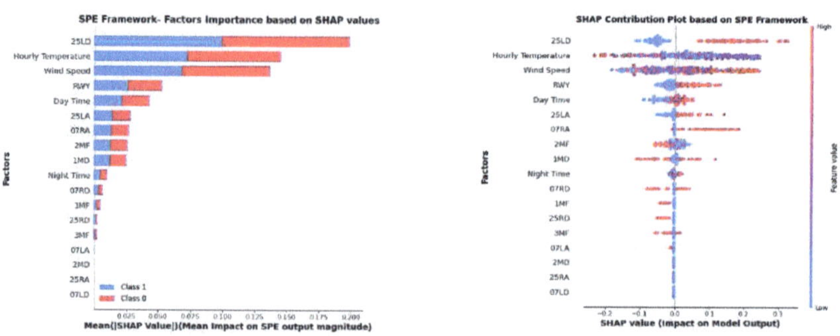

(a) (b)

Figure 8. Global Factor Interpretation; (**a**) Factor Importance plot; (**b**) Factor Contribution plot.

Figure 8b is a SHAP contribution plot of the factors, illustrating the distribution of SHAP values for each factor and the corresponding impact patterns. It is also known as the SHAP bee swarm plot. The horizontal axis of this plot represents the SHAP value, while the vertical axis contains all of the factors in the LLWS dataset. Each point on the plot represents a single SHAP value for a given prediction. Red indicates a higher value for a factor, while blue indicates a lower value. Based on the distribution of the red and blue dots, we can derive a general sense of the impact of factors' directionality. Some valuable insights can be drawn from the plots for the top three factors.

The runway 25LD factor, denoted by red dots, is coded as 1. All the red dots fall to the right of the vertical reference line, indicating the likelihood of the occurrence of S-LLWS over runway 25LD. Blue dots fall to the left of the vertical line, indicating the occurrence of NS-LLWS over other runways of HKIA. The previous studies [33,34] indicated that hourly prevailing wind directions such as east, south-east, south, and south-west were found to cause a higher risk of S-LLWS. This indicates that at 25LD, an S-LLWS event could be more likely to happen under the easterly, southeasterly, southerly, and southwesterly directions.

In the case of the hourly temperature factor, most of the purple dots fall to the right of the vertical line, while most of the blue dots and red dots fall to the left of the vertical line. This illustrates that S-LLWS is most likely to occur at low-to-moderate hourly temperatures, while a few high temperatures are more likely to cause the occurrence of NS-LLWS. The reason for this might be a temperature inversion [35–37], which is an alteration in the troposphere's typical temperature lapse rate, i.e., the reduction in temperature with altitude. On chilly, clear nights, this phenomenon typically occurs close to the ground, where the air immediately above the ground rapidly cools and becomes much colder than the layer of air higher up. As a result, the densely packed lower-level cold air is trapped by the layer of warm air. This may result in severe turbulence and, subsequently, S-LLWS.

Moderate-to-high values of wind speed mostly caused the occurrence of S-LLWS and vice versa. The findings are also consistent with previous HKIA research [33,38–41]. As for the occurrence of LLWS, however, wind speed variation is more significant than mean wind speed. Due to the fact that the average duration of an LLWS event confronted by an aircraft is somewhere between a few seconds and several minutes, the hourly mean wind speed cannot offer an accurate indication of LLWS. Therefore, more sophisticated data about wind conditions, such as a 1 min mean turbulence intensity, is necessary to enhance the performance of the models.

3.3.2. Local Factor Interpretation

Figure 9 depicts the SHAP explanatory force chart for two instances, randomly selected from the actual estimate results. The base value (0.656) on the plot represents the mean optimal SPE framework model estimations for the training dataset. The NS-LLWS condition occurs when the SPE framework output value is less than the base value. The S-LLWS condition occurs when the output value of the SPE Framework exceeds the base value. The blue arrows represent the magnitude of the influence of an input factor on the probability of NS-LLWS events. The influence of input factors on the occurrence of S-LLWS is highlighted by red arrows. The amount of space a factor occupies on each arrow demonstrates the size of its effect.

Consider two LLWS severity cases, one from S-LLWS and the other from NS-LLWS, which were correctly classified with estimated values of 1.03 and 0.52, respectively (see Figure 9). The value for S-LLWS is greater than the base value (0.656). Similarly, the value for NS-LLWS is less than 0.656. Figure 9a depicts an S-LLWS event that occurred when runway 25LD = 1, wind speed = 2.2 m/s, and hourly temperature = 17.9C. This is shown by the red arrows pointing to the right. The size of the "Runway 25LD" arrow is larger than the "Wind Speed" and "Hourly Temperature" arrows. This shows that "Runway 25LD" is a stronger predictor of S-LLWS in this case than "Wind Speed" and "Hourly Temperature." In contrast, for the same instance, "Day Time = 0", as represented by the blue arrow pointing to the left, indicates nighttime and depicts the likelihood of the occurrence of NS-LLWS.

Similarly, in Figure 9b, for another instance correctly classified as NS-LLWS, "1MD = 1", "Wind Speed = 6.9 m/s", and "pointing" the blue arrows, pointing to the left, are more likely to result in the occurrence of NS-LLWS. It demonstrates that, 1 nautical mile away from the end of the runway, an NS-LLWS event occurred.

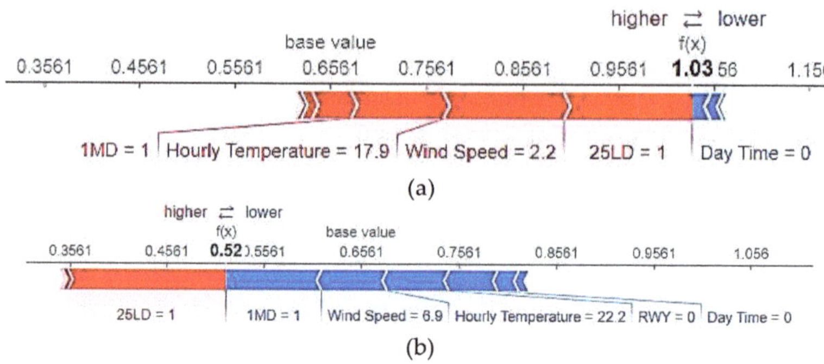

Figure 9. SHAP Explanatory Force Pot: (**a**) Plot for an instance value equals to 1.03; (**b**) Plot for an instance value equals to 0.52.

3.3.3. Factor Interaction Analysis

The SHAP interaction plots are examined to identify how the input factors, used to evaluate the SPE framework, interact with one another in terms of their contributions (see Figure 10). The interaction analysis of the top four influential factors, i.e., runway 25LD, hourly temperature, wind speed, and RWY (horizontal location of LLWS occurrence), is provided. Other factors' interactions, however, could be examined as well. The red and blue scatter plots in Figure 10a depict the variability in the runway 25LD and 25LD SHAP values. When the hourly temperature is low to moderate, the SHAP value for runway 25LD is higher. This means that most of the S-LLWS occurs in the vicinity of runway 25LD when the hourly temperature ranges from low to moderate. The temperature inversion on Hong Kong's Lantau Island could also be contributing to this scenario.

Figure 10b depicts the distribution of wind speed at runway 25LD. Wind speed points greater than 5 m/s have a higher SHAP value, indicating the likelihood of an S-LLWS event. Figure 10c illustrates that most of the S-LLWS events occurred "on the runway." The PIREPs reported S-LLWS when aircraft were making their final approach or just when they became airborne after takeoff.

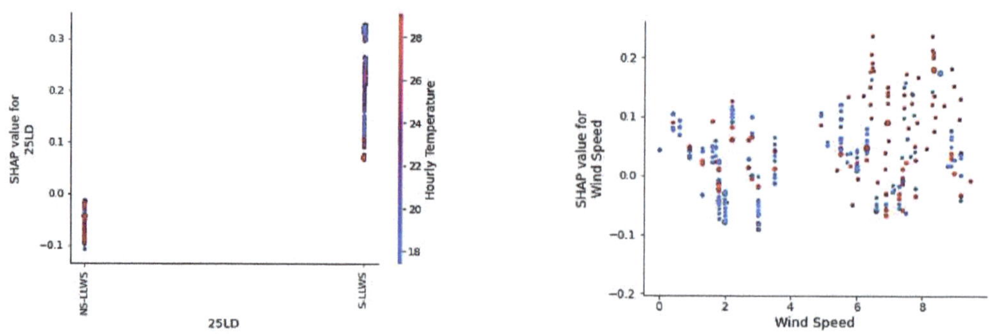

(a) (b)

Figure 10. *Cont.*

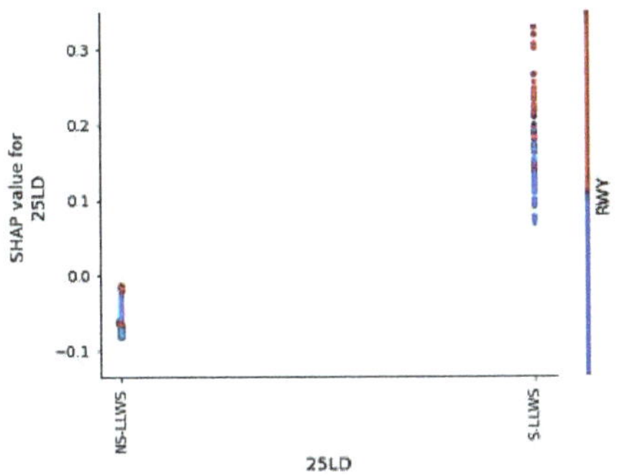

(c)

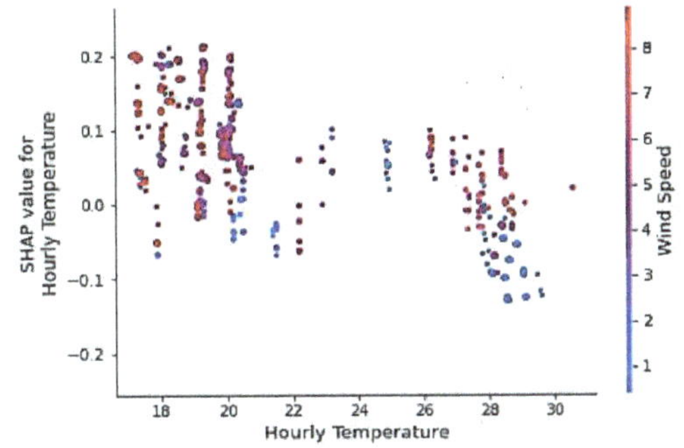

(d)

Figure 10. SHAP Interaction Plots: (**a**) Interaction of Runway 25LD and Hourly Temperature; (**b**) Interaction of Wind Speed and Runway 25LD; (**c**) Interaction of Runway 25LD and RWY (location of LLWS occurrence); (**d**) Interaction of Hourly Temperature and Wind Speed.

Figure 10d shows that the optimum conditions for the occurrence of S-LLWS were lower than average hourly temperatures and medium-to-high wind speeds. The points representing that scenario fall to the left of the plot and above the SHAP 0.00 reference line. However, to obtain a clear threshold, it may be necessary to know the altitude at which LLWS happen in addition to the parameters that are already known.

4. Conclusions and Recommendations

In this research, a novel SPE framework for the prediction and imbalance classification of LLWS severity has been proposed and compared with tree-based machine learning models, using both treated and untreated HKIA-based LLWS data from LiDAR and PIREPs. The SHAP interpretation system was also used to identify key risk factors and quantify their effects on the occurrence of S-LLWS. In this study, the SPE framework was trained and evaluated using untreated data, whereas both untreated and treated data were used to train the LGBM, AdaBoost, and CART machine learning models. The SMOTE-ENN technique was used as a treatment technique for highly imbalanced LLWS data. In terms of precision, recall, G-Mean, and MCC, the experimental results demonstrated that the SPE framework, based on the untreated data, outperforms all other tree-based models. The newly introduced SPE framework model offers a viable option for modeling and predicting LLWS severity based on imbalanced LLWS data.

Machine learning models, on the other hand, are regularly chastised for their lack of transparency and interpretability. Despite the fact that machine learning models are more adaptable and efficient than statistical approaches, their widespread recognition in the engineering domain continues to be a challenge. To tackle the SPE framework's interpretability issue, the SHAP interpretation system was used to evaluate the SPE's output in order to identify major risk factors and assess their impact on the severity of the LLWS. The results of the SHAP interpretation system can be used to rank the risk factor's overall significance. It can also be used to look into the individual and interaction effects of risk factors (for instance, how specific effects alter in response to changes in the risk factor's value). The analysis revealed that runway 25LD, hourly temperature, wind speed, and RWY (location of LLWS occurrence) were the top four most significant factors in predicting LLWS severity. The optimistic projections for the occurrence of S-LLWS events were low-to-medium temperatures at runway 25LD with relatively moderate-to-high wind speeds. Likewise, most of the S-LLWS events happen on the runway.

This research outlines a strategy that can be used to conduct a large-scale analysis of LLWS in aviation and serves as a useful tool for aviation policymakers and air traffic safety researchers. This paper discussed the SPE framework using highly imbalanced

LLWS data and the SHAP interpretation system. Additional research could be conducted by combining a number of other machine learning techniques with a number of additional risk factors, including monthly variation, location of occurrence of LLWS above ground level, etc. Future research could be expanded by employing additional techniques for augmenting data to deal with highly imbalanced LLWS data.

Author Contributions: Conceptualization, A.K.; Data curation, P.-W.C.; Formal analysis, A.K.; Funding acquisition, P.-W.C. and F.C.; Investigation, H.P.; Methodology, A.K., P.-W.C. and H.P.; Project administration, F.C.; Resources, P.-W.C.; Supervision, F.C.; Validation, F.C. and H.P.; Visualization, A.K.; Writing—original draft, A.K.; Writing—review and editing, H.P. All authors have read and agreed to the published version of the manuscript.

Funding: This research was supported by National Natural Science Foundation of China (U1733113), Shanghai Municipal Science and Technology Major Project (2021SHZDZX0100), Research Fund for International Young Scientists (RFIS) of National Natural Science Foundation of China (NSFC) (Grant No. 52250410351) and National Foreign Expert Project (Grant No. QN2022133001L).

Institutional Review Board Statement: Not applicable.

Informed Consent Statement: Not applicable.

Data Availability Statement: Not applicable.

Acknowledgments: We are thankful to the Hong Kong Observatory at Hong Kong International Airport for providing us PIREPs and LiDAR data.

Conflicts of Interest: The authors declare no conflict of interest.

References

1. Borsky, S.; Unterberger, C. Bad weather and flight delays: The impact of sudden and slow onset weather events. *Econ. Transp.* **2019**, *18*, 10–26. [CrossRef]
2. Choi, S.; Kim, Y.J.; Briceno, S.; Mavris, D. Prediction of weather-induced airline delays based on machine learning algorithms. In Proceedings of the 2016 IEEE/AIAA 35th Digital Avionics Systems Conference (DASC), Sacramento, CA, USA, 25–29 September 2016; IEEE: New York, NY, USA; pp. 1–6.
3. Gultepe, I.; Sharman, R.; Williams, P.D.; Zhou, B.; Ellrod, G.; Minnis, P.; Trier, S.; Griffin, S.; Yum, S.; Gharabaghi, B.; et al. A review of high impact weather for aviation meteorology. *Pure Appl. Geophys.* **2019**, *176*, 1869–1921. [CrossRef]
4. Airport Council International. *World Airport Traffic Forecast 2017–2040 Airport Council International, Montréal*. 2017. Available online: https://issuu.com/aciworld/docs/aci_annualreport2017 (accessed on 21 February 2022).
5. Fichtl, G.H.; Camp, D.W.; Frost, W. Sources of low-level wind shear around airports. *J. Aircr.* **1977**, *14*, 5–14. [CrossRef]
6. Stratton, D.A.; Stengel, R.F. Probabilistic reasoning for intelligent wind shear avoidance. *J. Guid. Control. Dyn.* **1992**, *15*, 247–254. [CrossRef]
7. Shun, C.; Chan, P. Applications of an infrared Doppler LiDAR in detection of wind shear. *J. Atmos. Ocean. Technol.* **2008**, *25*, 637–655. [CrossRef]
8. Thobois, L.; Cariou, J.P.; Gultepe, I. Review of LiDAR-based applications for aviation weather. *Pure Appl. Geophys.* **2019**, *176*, 1959–1976. [CrossRef]
9. Hallowell, R.G.; Cho, J.Y. Wind-shear system cost-benefit analysis. *Linc. Lab. J.* **2010**, *18*, 47–68.
10. Lau, S.Y.; Shun, C.M. Terrain-induced wind shear during the passage of Typhoon Utor near Hong Kong in July 2001. In Proceedings of the Tenth Conference on Mountain Meteorology and MAP Meeting, Park City, UT, USA, 16–21 June 2002; pp. 433–436, *in preprints*.
11. Hon, K.K.; Chan, P.W. Improving LiDAR Windshear Detection Efficiency by Removal of "Gentle Ramps". *Atmosphere* **2021**, *12*, 1539. [CrossRef]
12. Chan, P.W.; Hon, K.K. Performance of super high resolution numerical weather prediction model in forecasting terrain-disrupted airflow at the Hong Kong International Airport: Case studies. *Meteorol. Appl.* **2016**, *23*, 101–114. [CrossRef]
13. Boilley, A.; Mahfouf, J.F. Wind shear over the Nice Côte d'Azur airport: Case Study. *Nat. Hazards Earth Syst. Sci.* **2013**, *13*, 2223–2238. [CrossRef]
14. Matayoshi, N.; Iijima, T.; Yamamoto, K.; Fujita, E. Development of Airport Low-level Wind Information (ALWIN). In Proceedings of the 16th AIAA Aviation Technology, Integration, and Operations Conference, Washington, DC, USA, 13–17 June 2016; p. 4362.
15. Zhang, H.; Wu, S.; Wang, Q.; Liu, B.; Yin, B.; Zhai, X. Airport low-level wind shear LiDAR observation at Beijing Capital International Airport. *Infrared Phys. Technol.* **2019**, *96*, 113–122. [CrossRef]
16. Hon, K.-K. Predicting low-level wind shear using 200-m-resolution NWP at the Hong Kong International Airport. *J. Appl. Meteorol. Climatol.* **2020**, *59*, 193–206. [CrossRef]

17. Keck, R.E.; Mikkelsen, R.; Troldborg, N.; de Maré, M.; Hansen, K.S. Synthetic atmospheric turbulence and wind shear in large eddy simulations of wind turbine wakes. *Wind Energy* **2014**, *17*, 1247–1267. [CrossRef]
18. Lei, L.; Chan, P.W.; Li-Jie, Z.; Hui, M. Numerical simulation of terrain-induced vortex/wave shedding at the Hong Kong International Airport. *Meteorol. Z.* **2013**, *22*, 317–327. [CrossRef]
19. Jiang, L.; Xie, Y.; Wen, X.; Ren, T. Modeling highly imbalanced crash severity data by ensemble methods and global sensitivity analysis. *J. Transp. Saf. Secur.* **2022**, *14*, 562–584. [CrossRef]
20. Guo, R.; Fu, D.; Sollazzo, G. An ensemble learning model for asphalt pavement performance prediction based on gradient boosting decision tree. *Int. J. Pavement Eng.* **2021**, *23*, 3633–3646. [CrossRef]
21. Feng, D.C.; Wang, W.J.; Mangalathu, S.; Taciroglu, E. Interpretable XGBoost-SHAP machine-learning model for shear strength prediction of squat RC walls. *J. Struct. Eng.* **2021**, *147*, 04021173. [CrossRef]
22. Zhang, S.; Khattak, A.; Matara, C.M.; Hussain, A.; Farooq, A. Hybrid feature selection-based machine learning Classification system for the prediction of injury severity in single and multiple-vehicle accidents. *PLoS ONE* **2022**, *17*, e0262941. [CrossRef]
23. Khattak, A.; Almujibah, H.; Elamary, A.; Matara, C.M. Interpretable Dynamic Ensemble Selection Approach for the Prediction of Road Traffic Injury Severity: A Case Study of Pakistan's National Highway N-5. *Sustainability* **2022**, *14*, 12340. [CrossRef]
24. Liu, Z.; Cao, W.; Gao, Z.; Bian, J.; Chen, H.; Chang, Y.; Liu, T.Y. Self-paced ensemble for highly imbalanced massive data classification. In Proceedings of the 2020 IEEE 36th International Conference on Data Engineering (ICDE), Dallas, TX, USA, 20–24 April 2020; IEEE: New York, NY, USA; pp. 841–852.
25. Lundberg, S.M.; Lee, S.-I. A unified approach to interpreting model predictions. *Adv. Neural Inf. Process. Syst.* **2017**, *30*.
26. Mangalathu, S.; Hwang, S.H.; Jeon, J.S. Failure mode and effects analysis of RC members based on machine-learning-based SHapley Additive exPlanations (SHAP) approach. *Eng. Struct.* **2020**, *219*, 110927. [CrossRef]
27. Dong, S.; Khattak, A.; Ullah, I.; Zhou, J.; Hussain, A. Predicting and analyzing road traffic injury severity using boosting-based ensemble learning models with SHAPley Additive exPlanations. *Int. J. Environ. Res. Public Health* **2022**, *19*, 2925. [CrossRef] [PubMed]
28. Szeto, K.C.; Chan, P.W. High resolution numerical modelling of wind shear episodes at the Hong Kong International Airport. In Proceedings of the 12th Conference on Aviation, Range, and Aerospace Meteorology, Atlanta, GA, USA, 29 January–2 February 2006.
29. Hon, K.K.; Chan, P.W. Historical analysis (2001–2019) of low-level wind shear at the Hong Kong International Airport. *Meteorol. Appl.* **2022**, *29*, e2063. [CrossRef]
30. Jones, J.G.; Haynes, A. *A Peakspotter Program Applied to the Analysis of Increments in Turbulence Velocity*; RAE: Bedford, VA, USA, 1984.
31. Woodfield, A.A.; Woods, J.F. *Worldwide Experience of Wind Shear during 1981–1982*; Royal Aircraft Establishment: Bedford, VA, USA, 1983.
32. Wu, J.; Chen, X.Y.; Zhang, H.; Ziong, I.-D.; Lei, H.; Deng, S.-H. Hyperparameter optimization for machine learning mod-els based on Bayesian optimization. *J. Electron. Sci. Technol.* **2019**, *17*, 26–40.
33. Zhu, Q. On the performance of Matthews correlation coefficient (MCC) for imbalanced dataset. *Pattern Recognit. Lett.* **2020**, *136*, 71–80. [CrossRef]
34. Chen, F.; Peng, H.; Chan, P.W.; Ma, X.; Zeng, X. Assessing the risk of windshear occurrence at HKIA using rare-event logistic regression. *Meteorol. Appl.* **2020**, *27*, e1962. [CrossRef]
35. Chan, P.W. Case study of a special event of low-level windshear and turbulence at the Hong Kong International Airport. *Atmos. Sci. Lett.* **2022**, e1143. [CrossRef]
36. Stocker, J.; Johnson, K.; Forsyth, E.; Smith, S.; Gray, S.; Carruthers, D.; Chan, P.W. Derivation of High-Resolution Meteorological Parameters for Use in Airport Wind Shear Now-Casting Applications. *Atmosphere* **2022**, *13*, 328. [CrossRef]
37. Gernowo, R.; Subagio, A.; Adi, K.; Widodo, A.P.; Widodo, C.E.; Putranto, A.B. Atmospheric dynamics and early warning system low level windshear for airport runway hazard mitigations. In *Journal of Physics: Conference Series*; IOP Publishing: Bristol, UK, 2021; Volume 1943, p. 012029.
38. Leonidov, V.I.; Semenets, V.V. Analysis of methods for wind shear detection in area of airports by data of atmosphere acoustic sounding. *Telecommun. Radio Eng.* **2018**, *77*, 363–372. [CrossRef]
39. Chen, F.; Peng, H.; Chan, P.W.; Zeng, X. Low-level wind effects on the glide paths of the North Runway of HKIA: A wind tunnel study. *Build. Environ.* **2019**, *164*, 106337. [CrossRef]
40. Chen, F.; Peng, H.; Chan, P.W.; Zeng, X. Wind tunnel testing of the effect of terrain on the wind characteristics of airport glide paths. *J. Wind Eng. Ind. Aerodyn.* **2020**, *203*, 104253. [CrossRef]
41. Chen, F.; Peng, H.; Chan, P.W.; Huang, Y.; Hon, K.K. Identification and analysis of terrain-induced low-level wind shear at Hong Kong International Airport based on WRF–LES combining method. *Meteorol. Atmos. Phys.* **2022**, *134*, 60. [CrossRef]

atmosphere

Article

Prediction and Interpretation of Low-Level Wind Shear Criticality Based on Its Altitude above Runway Level: Application of Bayesian Optimization–Ensemble Learning Classifiers and SHapley Additive exPlanations

Afaq Khattak [1,*], Pak-Wai Chan [2], Feng Chen [1] and Haorong Peng [3]

1 The Key Laboratory of Infrastructure Durability and Operation Safety in Airfield of CAAC, Tongji University, 4800 Cao'an Road, Jiading, Shanghai 201804, China
2 Hong Kong Observatory, 134A Nathan Road, Kowloon, Hong Kong, China
3 Shanghai Research Center for Smart Mobility and Road Safety, Shanghai 200092, China
* Correspondence: khattak@tongji.edu.cn

Abstract: Low-level wind shear (LLWS) is a rare occurrence and yet poses a major hazard to the safety of aircraft. LLWS event occurrence within 800 feet of the runway level are dangerous to approaching and departing aircraft and must be accurately predicted. In this study, first the Bayesian Optimization–Ensemble Learning Classifiers (BO-ELCs) including Adaptive Boosting, Light Gradient Boosting Machine, Categorical Boosting, Extreme Gradient Boosting, and Random Forest were trained and tested using a dataset of 234 LLWS events extracted from pilot flight reports (PIREPS) and weather reports at Hong Kong International Airport. Afterward, the SHapley Additive exPlanations (SHAP) algorithm was utilized to interpret the best BO-ELC. Based on the testing set, the results revealed that the Bayesian Optimization–Random Forest Classifier outperformed the other BO-ELCs in accuracy (0.714), F1-score (0.713), AUC-ROC (0.76), and AUR-PRC (0.75). The SHAP analysis found that the hourly temperature, wind speed, and runway 07LA were the top three crucial factors. A high hourly temperature and a moderate-to-high wind speed made Runway 07LA vulnerable to the occurrence of critical LLWS events. This research was a first attempt to forecast the criticality of LLWS in airport runway vicinities and will assist civil aviation airport authorities in making timely flight operation decisions.

Keywords: low-level wind shear; ensemble learning classifiers; Bayesian optimization; SHapley Additive exPlanations

Citation: Khattak, A.; Chan, P.-W.; Chen, F.; Peng, H. Prediction and Interpretation of Low-Level Wind Shear Criticality Based on Its Altitude above Runway Level: Application of Bayesian Optimization–Ensemble Learning Classifiers and SHapley Additive exPlanations. *Atmosphere* **2022**, *13*, 2102. https://doi.org/10.3390/atmos13122102

Academic Editors: Duanyang Liu, Hongbin Wang and Shoupeng Zhu

Received: 8 November 2022
Accepted: 12 December 2022
Published: 15 December 2022

1. Introduction

Globally, the civil aviation industry has grown rapidly in the last decade as a consequence of enhanced economic development. Passenger traffic worldwide surpassed 8.8 billion in 2018 and is expected to triple to 10 billion by 2037. It is projected to grow at a 3.7% annual rate in the long run and reach 19.7 billion by 2040 [1]. Although there is a boom in the aviation industry worldwide, weather is one of the key factors that has a major impact on overall airline operations. It is a significant contributor to flight cancellations, delays, and—in the worst-case scenario—accidents. Wind shear is an aviation term that refers to a sudden, abrupt change in wind direction or speed, whereas low-level wind shear (LLWS) refers to wind shear that occurs below 1600 feet (500 m) above ground level (AGL). Low-level jet streams, frontal systems, low-level temperature inversions, and LLWS are closely associated, more specifically with the unique wind-shear conditions created by man-made structures such as the distribution of various buildings, terrain roughness, and natural obstructions such as mountains and water/land interfaces, among other factors, at and around a particular airport [2].

1.1. Low-Level Wind Shear: Pilots' Invisible Enemy

The cockpit remains extremely active during the landing phase, and the captain and co-pilot must make a number of quick decisions to complete their landing checklist. However, poor weather conditions, complex terrain, and the presence of buildings near the airport will increase turbulence along the glide path. Therefore, the occurrence of LLWS below 800 feet above the ground is regarded as the most critical phenomenon for both approaching and departing aircraft. The pilot must contend with violent updrafts and downdrafts as well as abrupt changes in the aircraft's horizontal and vertical movement while completing the landing checklist. As depicted in Figure 1, this critical condition may lead to a missed approach, landing short of the runway (loss of lift), or deviation from the actual flight path during final approach. Basically, there are two detrimental and potentially hazardous effects of LLWS on approaching aircraft: perturbation of the glide path and deviancy of the approach speed from the established (set) value [3]. As a result, the pilot may perceive additional pressure during the approach phase when the engine power is low and the airspeed is close to stall speed due to unexpected changes in wind direction or speed. This effect of declining and raising headwind shear on an aircraft during an approach is depicted in Figure 2 (assuming no pilot intervention is being used); both scenarios utilize a conventional instrument landing system with a 3-degree glide path and final approach speed (v_a). In the first scenario (Figure 2a), the approaching aircraft is subjected to a declining headwind with the headwind speed (v_{hw}). As the aircraft approaches the ground, its airspeed (the aircraft's speed in relation to the surrounding air flow) declines, thereby lowering lift and tending to result in a greater descent angle due to the transient force imbalance. In this scenario, the aircraft may possibly land short of the runway. The second case (Figure 2b) assumes a rising headwind on the same glide path and slope (3 degrees). As a result, the aircraft's airspeed increases in relation to the surrounding air flow, thereby generating more lift and resulting in a flatter angle of descent or even a climb. In this scenario, landing may possibly be aborted and a go-around initiated.

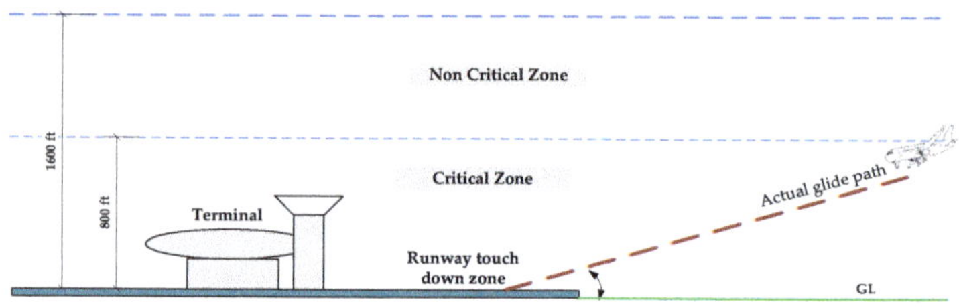

Figure 1. Occurrence location of LLWS events in the vicinity of an airport runway.

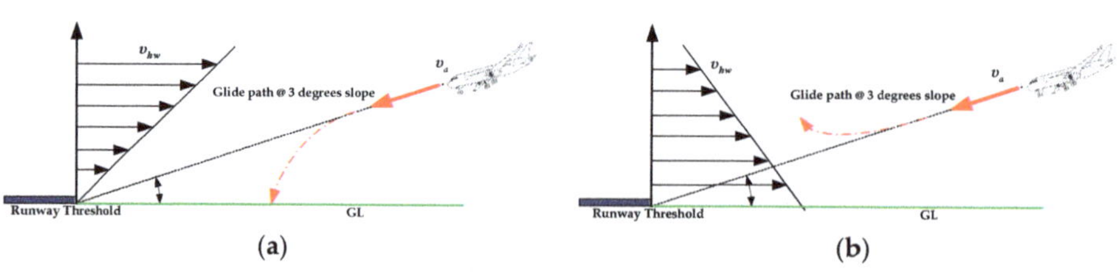

Figure 2. Detrimental and potentially hazardous effects of LLWS on approaching aircraft: (**a**) declining headwind during final approach; (**b**) rising headwind during final approach.

1.2. Low-Level Wind Shear Detection Technologies

Airports worldwide have profited significantly from the availability of meticulous, high-resolution technologies for remote sensing including Terminal Doppler Weather Radar (TDWR) and Doppler Light Detection and Ranging (LiDAR) [4–6]. By far, the most extensively used approaches for detecting wind shear are TDWR, ground-based anemometer networks, and wind profilers. This approach has been shown to be effective in alerting LLWS since the mid-1990s, most notably during the passage of tropical cyclones and thunderstorms. When the weather is clear, the TDWR system does not offer accurate wind information. Certain LLWS incidents, on the other hand, are connected to airflow that reaches the airport from rugged terrain. To deal with these scenarios, a new method of detection that is not dependent on humid conditions must be developed. The LiDAR system has been added as a booster to the TDWR in order to identify and warn of LLWS in clear skies. When the air is clear, Doppler LiDAR can detect return signals from aerosols and offer precise Doppler wind measurements. To ensure the safety of civil aircraft, TDWRs and LiDARs have been extensively installed at major airports worldwide. However, only a few airports worldwide, including Japan, Malaysia, Germany, France, Korea, Singapore, and Hong Kong, possess these LLWS alerting system technologies due to high maintenance and equipment costs, a lack of pertinent research, and specific geographical characteristics [7]. Additionally, these LLWS alerting technologies, which are based on remote sensing and/or on-site measurements, have been demonstrated to be successful and operational. When an LLWS event is detected or observed, these detection- or observation-based technologies send notifications. However, these detection- or observation-based technologies cannot predict when the next LLWS event will occur or what risk factors contribute to its occurrence as well as the criticality [8].

Extreme weather conditions such as microbursts and sea breezes, as well as the geographic surroundings of an airport, which include complex topography and structures, both contribute to wind shear events. Over 70% of pilot flight reports illustrated terrain-induced wind shear at Hong Kong International Airport (HKIA, International Civil Aviation Organization (ICAO) code: VHHH) [9]. Several researchers used analytical and simulation techniques to assess the impact of LLWS, such as Lei et al. [10], who employed a computational fluid dynamics (CFD) model to simulate the shedding of vortices from the mountains near HKIA. It was observed that accurately modeling this shedding had a considerable impact on forecasting terrain-induced wind shear at airports. Using data from TDWRs and LiDARs, a high-resolution aviation model (AVM) [11] was developed to evaluate the occurrence of terrain-induced wind shear at HKIA. The model was proven to accurately simulate terrain-induced wind shear, including microbursts caused by Lantau Island's mountains. During the winter, when wind shear occurs over the runway owing to turbulence generated by neighboring hills, Shimoyama et al. [12] researched the turbulence over Japan's Shonai Airport. According to the models, terrain attributes may have a considerable impact on the amount of turbulence encountered along flight paths, implying that aircraft safety may be influenced by wind direction. Furthermore, it was demonstrated that the turbulence induced by terrain features may be predicted using this modeling method depending on the degree to which the findings match the turbulence measured using a Doppler radar.

1.3. Ensemble Learning Classifiers and Interpretation

In comparison to prior hardware-based techniques and numerical and simulation models, we proposed in this study to use Bayesian Optimization–Ensemble Learning Classifiers (BO-ELCs) to predict the criticality of LLWS events. ELCs have been applied in a number of fields, including health care modeling, transportation and traffic safety, finance, and economics [13–20]. However, there is a significant gap in the literature regarding the use of ELCs in the civil aviation safety domain. In the past, Liu et al. [21] developed a neural-network-based prediction model for the assessment of wind fields along the glide path near HKIA using LiDAR data. It was quite effective at predicting wind shear. However, one could argue that neural network models are difficult to comprehend since their structures or weights include only a limited amount of information about the estimated function [22,23]. On the other hand, decision-tree-based machine learning models are easy to understand and their outcomes can be easily explained. The models empower predictive models with high accuracy and stability. The predictions of ELCs do not, however, explicitly and clearly demonstrate the relationship between changes in input and output variables, in contrast to statistical or empirical models. The interpretation of the model is equally important for appropriately assessing the model's performance. Previously, the ELC results were interpreted using the variable-importance analysis technique. The variable-importance analysis methodologies, however, can only provide a ranking of the variables' importance and are unable to explain how each variable individually influences the prediction of model. The SHapley Additive exPlanations (SHAP) algorithm, which is based on the concept of game theory [24], has been utilized in recent studies to quantify each variable's effect on the outcomes and to provide information about the strength and direction of each variable's influence on each individual sample [25–30]. Civil aviation safety researchers should take advantage of this opportunity because understanding the complex interactions between several risk factors that determine the criticality of LLWS is crucial for aviation and meteorological applications.

1.4. Research Process

The purpose of this research was to develop a model for predicting the criticality of LLWS events in the vicinity of an airport runway and then to interpret the results via SHAP analysis. There were four stages to the research procedure. Before constructing and comparing the ELCs, the hyperparameters were adjusted via Bayesian Optimization (BO), which is one of the machine learning hyperparameter tuning techniques [31]. The reasons for which we chose the BO technique in contrast to the Grid Search CV [32] and Random Search CV [33] techniques were its ability to lower the time needed to obtain an optimal set of hyperparameters and its better generalized performance on the test instances. The Bayesian-optimized models were subsequently compared to evaluate their performance. A SHAP analysis was then employed in both the individual and global interpretation of risk factors. It investigated the significance of risk factors and their interactions. This research is expected to fill a gap in the literature on ensemble learning applications in civil aviation safety.

2. Materials and Methods

In this study, five state-of-the-art ELCs, namely the Light Gradient Boosting Machine (LGBM) [34], Random Forest (RF) [35], the Extreme Gradient Boosting Machine (XG-Boost) [36], Categorical Boosting (CatBoost) [37], and Adaptive Boosting (AdaBoost) [38] optimized via BO were used to predict the criticality of LLWS in the vicinity of runways at HKIA. The data for modeling, which was extracted from Pilot Flight Reports (PIREPs) and Hong Kong Observatory (HKO) weather reports, included the LLWS magnitude and altitude experienced by the pilots of approaching or departing aircrafts, runway used by approaching and departing aircraft, wind direction, time of the day, mean hourly temperature, and mean wind speed. Based on these input data, the BO-ELC models were developed for the prediction of LLWS criticality. The hyperparameters such as n_estimators, learning rate, num_leaves, reg_lambda, reg_alpha, max_depth were a few hyperparameters of the ELC models that were considered for optimization via the BO technique. Using the well-tuned ELCs, a performance assessment was conducted to obtain the necessary performance measures and assess the best model.

Afterward, using the best ELC, the Shapley additives values were computed to characterize the influence of each factor on the final inference of LLWS criticality. The best model was assessed from global and local perspectives using the SHAP model. The SHAP algorithm is basically a local explainability model but can be employed to construct a global explanation after taking the average of all of the local instances that illustrate macro-level details. The global interpretations based on SHAP were consistent with the local explanations. Figure 3 depicts the entire operational paradigm proposed in this research.

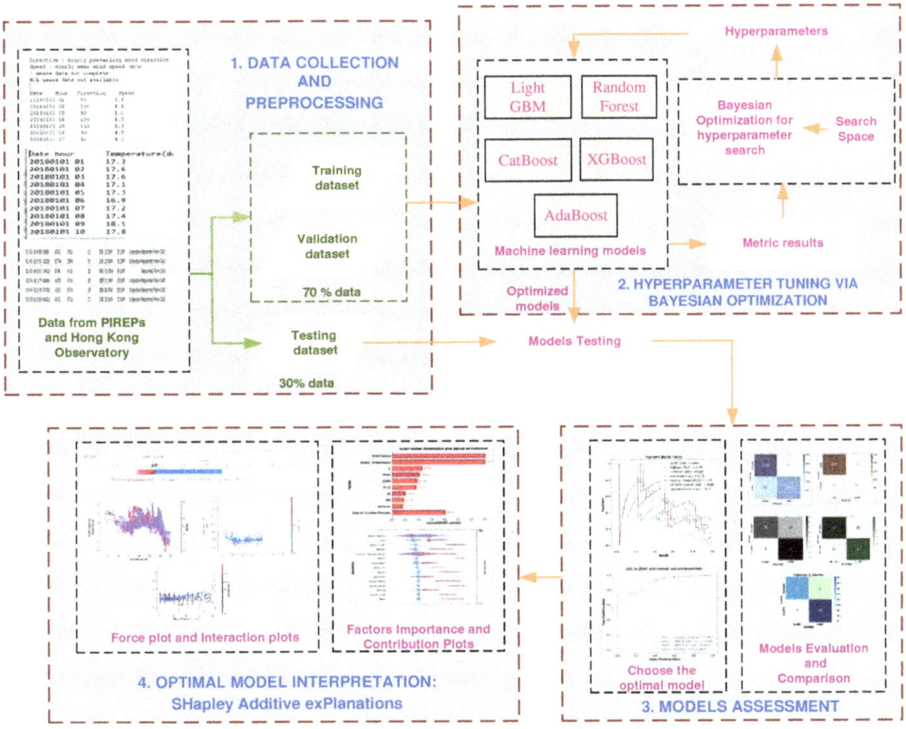

Figure 3. Framework for the prediction and interpretation of LLWS criticality at the vicinity of runways at HKIA.

2.1. Study Location

HKIA is located on an artificial Lantau island surrounded on three sides by open sea water with mountains to the south that reach elevations of over 900 m above sea level. Numerous observational and modeling studies have shown that HKIA's intricate orography and complex land–sea contrast are conducive to the occurrence of LLWS [39]. It is one of the most susceptible airports to wind shear in the world. Significant LLWS events occur once every 400 to 500 flights [40]. As shown in Figure 4, the mountainous terrain to the south of HKIA amplifies LLWS, which disrupts airflow and generates mountain waves, gap discharges, and other disturbances along the HKIA flight paths. Two runways exist at HKIA: the North Runway and the South Runway. They are oriented in the 070° and 250° directions, respectively. Due to the fact that each runway can be used for takeoffs and landings in either direction, there are a total of eight possible configurations. For example, runway '07LA' denotes landing ('A' refers to arrival) with a heading angle of 070° (shortened to '07') using the left runway (hence 'L'); this shows aircraft landing on North Runway from the western side of HKIA. Likewise, an aircraft departing from the South Runway in the west would use runway 25LD.

2.2. Data Processing from PIREPs and Hong Kong Observatory

In aviation, pilot reports are abbreviated as PIREPs. Pilots who encounter hazardous weather conditions report them to air traffic controllers. The traditional PIREPs typically include information on turbulence and aircraft icing and cover the flight's en-route phase. However, information about the timing, location (to the nearest nautical mile), speed (to the nearest 5 knots), and altitude (to the nearest 50 or 100 feet) of an LLWS event is encapsulated in the HKIA wind shear PIREPs. The positive or negative signs show a gain or loss, respectively, in headwind. Pilots can submit a report form after landing or taking off, or they can use on-board radio communication to communicate LLWS events to the air traffic controller at HKIA.

Figure 4. Hong Kong International Airport and surrounding terrain.

A substantial amount of wind shear data was essential to develop the prediction models. To this end, 243 LLWS data were obtained from the PIREPs and HKO weather reports. The PIREPs were used to ascertain the LLWS height and intensity as well as the runway that arriving and departing aircraft used. As depicted in Figure 3, the occurrence location of LLWS was divided into two zones—the critical zone and the non-critical zone—based on the PIREPs. All LLWS events occurring within 800 ft of the surface were deemed critical, whereas all others were deemed non-critical. An LLWS event was therefore a binary factor with two possible outcomes as indicated by Equation (1):

$$\text{LLWS} = \begin{cases} 1\ \text{C}-\text{LLWS}, & \text{if LLWS is 800ft above runway level} \\ 0\ \text{NC}-\text{LLWS}, & \text{otherwise} \end{cases} \quad (1)$$

The combined wind shear data from HKO weather reports and PIREPs contained both nominal factors (such as runway orientation, wind direction, month of the year, and time of the day) and continuous factors (such and mean hourly temperature and mean wind speed). Any i-th LLWS event in the original dataset could be represented as $(X_i, y_i) = (C_i, N_i, y_i)$, where C_i is the continuous factors, N_i is the nominal factors, and y_i is the target factors. The nominal factors N of the dataset were one-hot encoded as shown in Table 1. Each nominal value in the dataset was translated into a new column, and the column was assigned a 0 or 1 value. The number of columns was equal to the number of nominal values. For example, an eight-column matrix was created from a nominal factor "Runway" with six different values (07LA, 07RA, 07RD, 25LA, 25LD, and 25RA). The continuous factors of the wind shear datasets, on the other hand, were normalized as stated in Equation (2):

$$C_{i,j}^{norm} = \frac{C_{i,j}^{orig} - \min C_j}{max\ C_j - \min C_j} \quad (2)$$

where $C_{i,j}^{norm}$ represents the j-th normalized continuous factors of the i-th instance of the data, and $C_{i,j}^{orig}$ represents the original j-th continuous factors in the i-th instance of the data. The min C_j and max C_j represent the minimum and maximum of the j-th continuous factor in the combined wind shear dataset, respectively. Finally, there were 22 dependent factors in the standardized wind shear dataset; i.e., normalized continuous factors (2 × factors) and one-hot encoded nominal factors (20 × factors).

Table 1. One-hot encoding of categorical factors for the modeling.

Factor	Codes and Description
Runway Orientation	
07LA	1: If a wind shear event is reported at Runway 07LA, 0: Otherwise
07RA	1: If a wind shear event is reported at Runway 07RA, 0: Otherwise
25RA	1: If a wind shear event is reported at Runway 25RA, 0: Otherwise
25LA	1: If a wind shear event is reported at Runway 25LA, 0: Otherwise
25LD	1: If a wind shear event is reported at Runway 25LD, 0: Otherwise
07RD	1: If a wind shear event is reported at Runway 07LD, 0: Otherwise
Wind Direction	
N	1: If wind direction is North, 0: Otherwise
NE	1: If wind direction is North-East, 0: Otherwise
E	1: If wind direction is East 0: Otherwise
SE	1: If wind direction is South-East, 0: Otherwise
S	1: If wind direction is South, 0: Otherwise
SW	1: If wind direction is South-West, 0: Otherwise
W	1: If wind direction is West, 0: Otherwise
NW	1: If wind direction is North-West, 0: Otherwise
Season of the Year	
Winter	1: If a wind shear event occurs in Winter, 0: Otherwise
Spring	1: If a wind shear event occurs in Spring, 0: Otherwise
Summer	1: If a wind shear event occurs in Summer, 0: Otherwise
Autumn	1: If a wind shear event occurs in Autumn, 0: Otherwise
Time of the Day	
Day Time	1:If a wind shear event occurs during day time, 0: Otherwise
Night Time	1:If a wind shear event occurs during night time, 0: Otherwise

2.3. Hybrid Bayesian Optimization–Ensemble Learning Classifier (BO-ELC)

In this work, BO was utilized in conjunction with ELCs to train and tune the ELCs and find the optimal hyperparameters. The BO assembled a probability model for finding the value that automatically diminished the objective function based on the precedent estimation outcome of the objective. Figure 5 shows the flowchart of the hybrid BO-ELC approach. The step-by-step procedure for ELC optimization via BO is also given below.

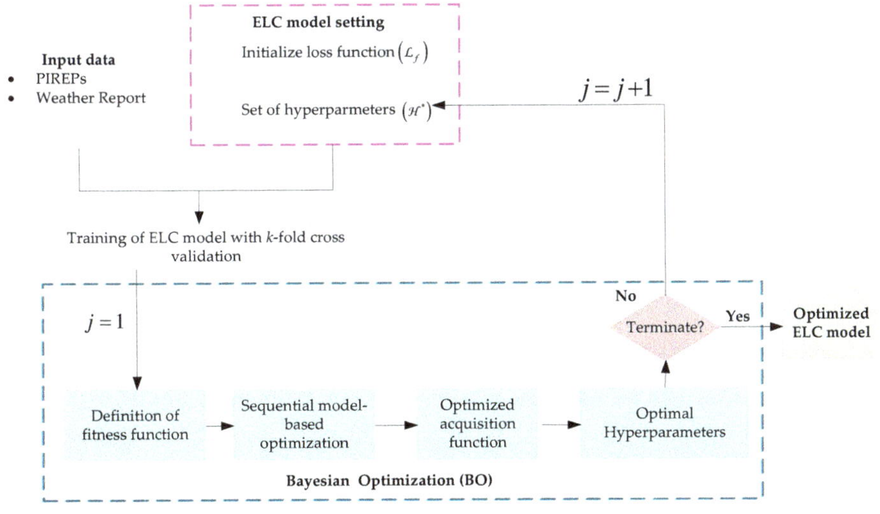

Figure 5. Hybrid BO-ELC approach for the prediction of LLWS criticality.

2.3.1. Initialization

In this step, the appropriate hyperparameters settings were initialized randomly (Equation (3)), which could be used to train the ELCs based on k-fold cross validation. In addition, a loss function $\left(L_f\right)$, which was the black-box function and that was required to be optimized, was also initialized. The aim was to determine the optimal set of hyperparameters that globally optimized the loss function $\left(L_f\right)$.

$$H = \begin{pmatrix} h_{1,1} & h_{1,2} & .. & h_{1,l} \\ h_{2,1} & h_{2,2} & .. & h_{2,l} \\ h_{3,1} & h_{3,2} & .. & h_{3,l} \\ \vdots & \vdots & & \vdots \\ \vdots & \vdots & & \vdots \\ h_{m,1} & h_{m,2} & .. & h_{m,l} \end{pmatrix} \tag{3}$$

2.3.2. Fitness Function

The random number of the solution was generated from the initialized values. The *fitness function* was used to minimize the objective function based on the following Equation (4):

$$\textit{fitness function}\left(\frac{L}{H}\right) = \begin{cases} D(H)\,L < L^* \\ G(H)\;\; L \geq L^* \end{cases} \tag{4}$$

where L denotes the loss value; $D(H)$ denotes the density estimation, which was based on the loss value during the observations; $G(H)$ is produced by the leftover observations value of loss, and L^* represents the particular quantiles.

2.3.3. Sequential Model-Based Optimization

Sequential model-based optimization was one of the concise forms of BO used to tune the hyperparameters of the ELCs. Sequential model-based optimization operates by finding the optimal hyperparameter setting H^* by building the Gaussian process Θ_z with a sampled point, which can be obtained using the following Equation (5):

$$H^* = \text{argmin}\Theta_{z-1}(H) \tag{5}$$

The loss value can be determined under the optimal hyperparameter setting by using Equation (6):

$$L = L_f(H^*) \tag{6}$$

The corresponding L and the H^* setting were stored in the corresponding trails, which can be represented as Ω. These corresponding trails (Ω) were used for parameter settings and evaluation purposes. The Ω update could be determined with the help of the following Equation (7):

$$\Omega = \Omega \cup (H^*, L) \tag{7}$$

Finally, we built the Gaussian process Θ_z model based on the updated Ω.

2.3.4. Acquisition Function

The acquisition function of BO was employed to compute the next iteration in the search process. In this study, the expected improvement was chosen as an acceptable performance criterion of the ELCs, which was the maximization of AUC-ROC. The improvement could be obtained with the help of L by using Equation (8):

$$D(H) = \max(L_{\min} - L(H), 0) \tag{8}$$

2.3.5. Termination

In this step, the optimal hyperparameters were obtained for the ELCs with the help of the BO.

2.4. Evaluation of BO-ELCs

In EL modeling, performance assessment of the classifiers is a vital task. When a classification problem requires checking or visualizing the performance, the area under the receiver operating characteristics curve (AUC-ROC) and the area under the Precision and Recall curve (AUC-PRC) can be used. Both the AUC-ROC and AUC-PRC were used as performance metrics for the assessment of the classification models' performances. In the case of the ROC, the AUC-ROC ranged from 0 (fully incorrect) to 1 (perfectly classified).

In addition, we also used a confusion matrix, which provided an in-depth examination of the model's performance when predictions were made for each class. For the binary classification problem, one class was the majority (the negative) and its sample size was represented by n_1; the other class was the minority (the negative) and its sample size was represented by n_2. Let n represent the total size of the training data set ($n = n_1 + n_2$). The binary classifier predicted whether each instance was positive or negative. Therefore, it generated four types of outcomes: true positive T^ρ, false positive F^ρ, true negative T^n, and false negative F^n (see Figure 6). The Accuracy, Recall, Precision, and F1-score were extracted from the confusion matrix and are given as Equations (9)–(12).

$$Classification\ Accuracy = \frac{T^\rho + T^\eta}{T^\rho + F^\eta + T^\eta + F^\rho} \tag{9}$$

$$Precision = \frac{T^\rho}{T^\rho + F^\rho} \tag{10}$$

$$Recall = \frac{T^\rho}{T^\rho + F^\eta} \tag{11}$$

$$F1 - Score = \frac{T^\rho}{T^\rho + \frac{1}{2}(F^\eta + F^\rho)} \tag{12}$$

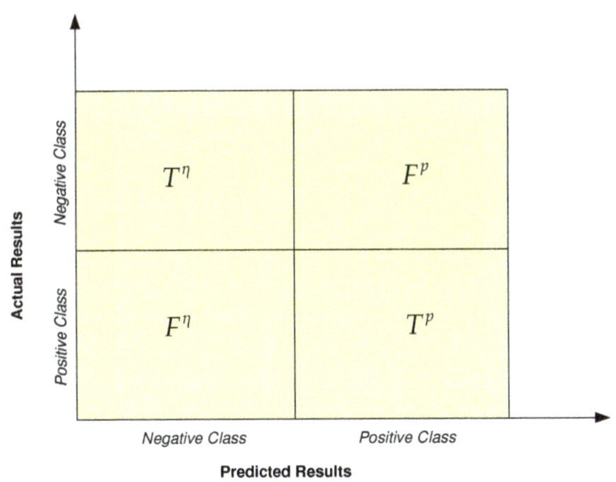

Figure 6. Confusion matrix plot.

On the basis of the recall and precision extracted from the confusion matrix, we could also plot the precision–recall curve and calculate area under the curve (AUC-PRC).

2.5. BO-ELC Interpretation Using Shapley Additive exPlanations (SHAP)

The SHAP analysis is based on a game-theory mechanism for interpretation of ensemble learning models. The fundamental concept behind the SHAP tool is to compute the marginal contribution of factors to the ELC output and then a "black box model" is interpreted from both the global and local perspectives. During the training or testing of the ELCs, a prediction value was computed for each instance, and the SHAP value corresponded to the value assigned to each factors in the instance. The contribution of each factors denoted by the Shapley value was computed using Equation (13):

$$\varphi_i = \sum_{\Upsilon \subseteq \Pi\{i\}} \frac{\Upsilon!(n - |\Upsilon| - 1)!}{n!} [f(\Upsilon \cup \{i\}) - f(\Upsilon)] \tag{13}$$

where φ_i indicates the contribution of the i-th factor; Π represents the set of all factors; Υ represents the subset of the given predicted factor; and $f(\Upsilon_i)$ and $f(\Upsilon)$ represent the model results with and without i-th factors, respectively. The SHAP analysis tool produced interpretable ELCs via an additive factors imputation strategy in which the output model was defined as a linear sum of the input factors (Equation (14)):

$$g(z\prime) = \varphi_0 + \sum_{i=1}^{\Lambda} \varphi_i z\prime \tag{14}$$

where $z\prime \in \{0,1\}^{\Lambda}$ when a factor is observed = 1, otherwise = 0; Λ denotes the number of input factors; φ_0 is the base values (i.e., the predicted outcome without factors); and φ_i denotes the Shapley value of the i-th factor. The SHAP model was used in this study for the interpretation of Bayesian-optimization (BO)-ELC; the important factors that are likely to cause critical LLWS were assessed. The SHAP tool performed a factor-interaction analysis as well.

3. Results and Discussion

To evaluate the capability of five BO-ELCs to predict LLWS criticality, the combined PIREPs and HKO weather reports were separated into training and testing sets at a 7:3 ratio. Using Bayesian Optimization and 10-fold cross-validation, the hyperparameters for each ELC were tuned to obtain the optimal set of hyperparameters. Each tuned ELC was then evaluated using unseen instances from the testing set. In addition, the performance of the BO-ELCs on the testing set was compared to determine the best BO-ELC model. Finally, the game-theory-based SHapley Additive exPlanation mechanism was implemented using the best BO-ELC model to provide explanations for the prediction of LLWS events. Based on the 234 PIREPs, 96 (39.51%) of the LLWS events occurred over runway 07LA, 13 (5.34%) occurred over runway 07RA, 34 (13.99%) occurred over runway 07RD, 8 (2.30%) occurred over runway 25LA, 67 (27.57%) occurred over runway 25RA, and 25 (10.28%) occurred over runway 25LD. In the winter season (January, February, and December), 26 LLWS events occurred out of a total of 234 wind shear events; 139 LLWS events occurred in spring 2016 (March, April, and May); 53 occurred in summer (June, July, and August); and 25 occurred in autumn (September, October, and November). The PIREPs also illustrated that 61.9% of the LLWS events occurred during day time (07:00 AM–07:00 PM) and 38.1% during night time. At the time of the LLWS event occurrence, the HKO weather reports illustrated north-bound wind flows 1.73% of time, northeast-bound 7.35% of time, east-bound 51.9% of the time, southeast-bound 10.3% of the time, south-bound 7.7% of the time, southwest-bound 12.1% of the time, west-bound 4.7% of the time, and northwest-bound 3.8% of the time. The HKO weather reports also provided the hourly temperature and wind speed at the time of the LLWS occurrences. Figure 7 shows the distribution bar plots of LLWS events with respect to the runway orientation, seasons of the year, wind direction, and time of day or night. The figure also contains box plots of the hourly temperature and wind speed that show the maximum, minimum, Q1, Q3, and median values.

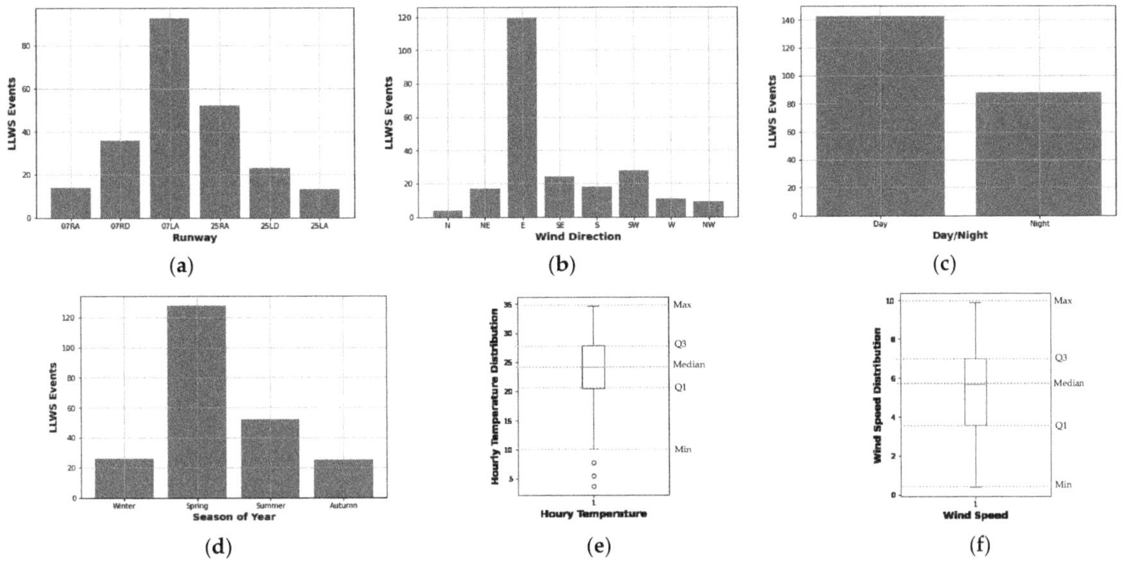

Figure 7. (**a**) LLWS events with respect to runway; (**b**) LLWS events with respect to wind direction; (**c**) LLWS events with respect to day/night; (**d**) LLWS events with respect to season of year; (**e**) box plot of hourly temperature; (**f**) box plot of wind speed.

3.1. Hyperparameter Tuning Using Bayesian Optimization

Table 2 shows the optimal hyperparameters along with their ranges and optimal values that were obtained via the hybrid BO-ELC approach. Each ELC model with the optimal hyperparameters was then used for the performance evaluation.

Table 2. Hyperparameter tunings of ELCs.

Algorithm	Hyperparameters	Range	Optimal Values
LGBM	{(*n*_estimators), (num_leaves), (learning rate), (reg_lambda), (reg_alpha}	{(100–1500), (30–100), (0.001–0.2), (1.1–1.5), (1.1–1.5)}	{900, 38, 0.07, 1.24, 1.18}
CatBoost	{(*n*_estimators), (max_depth), (learning rate)}	{(200–1500), (2–15), (0.001–0.2)}	{727, 5, 0.1}
AdaBoost	{(n_estimators), (learning rate)}	{(100–1500), (0.001–0.2)}	{871, 0.08}
RF	{(n_estimators), (max_depth)}	{(50–1000), (2–15)}	{1041, 7}
XGBoost	{(*n*_estimators), (num_leaves), (learning rate), (reg_lambda), (reg_alpha}	{(100–1500), (30–100), (0.001–0.2), (1.1–1.5), (1.1–1.5)}	{1105, 46, 0.05, 1.41, 1.27}

3.2. Performance Assessment of BO-ELCs

To assess the performances of the BO-ELCs, the ROC curves were plotted and the AUC-ROC was calculated for each ensemble classifier. The AUC-ROC curves were used to provide a basis for the comparison between each classifier. Figure 8 demonstrates that all models utilized showed strong predictive values. All the developed classifiers showed AUC-ROC values greater than 0.50. The most accurate classifier among all of the classifiers was the BO-Random Forest model, which had an AUC-ROC of 0.759. The worst AUC-ROC was shown by BO-AdaBoost, which was equal to 0.687.

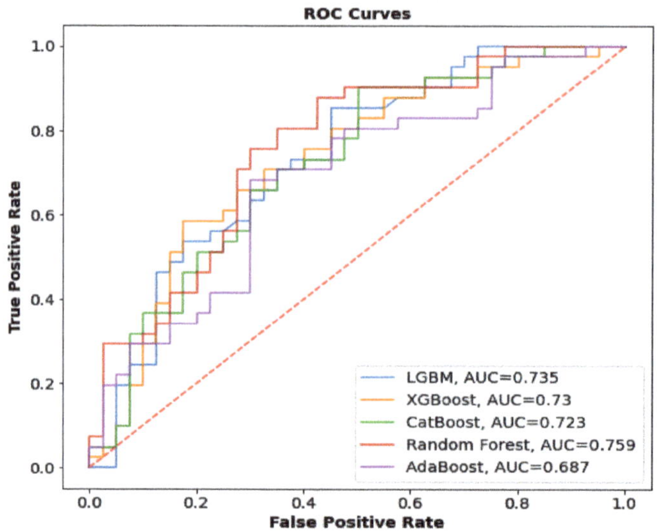

Figure 8. Combined ROC curve for all Bayesian-optimized ELC models.

Although the AUC-ROC is a helpful metric for determining the overall accuracy of a binary prediction model, it does not provide class-specific accuracy (predicted accuracy of NC-LLWS vs. predicted accuracy of C-LLWS). To illustrate the accuracy of both predictions, a confusion matrix for each classifier was generated, and several performance indicators including accuracy, precision, recall, and the F1-score, were extracted. Table 3 reports the comparison results among five BO-ELCs and Figure 9 illustrates the AUC-PRC when employing the testing dataset. The BO-Random Forest Classifier showed the highest Accuracy, Precision, Recall and F1-score (Accuracy = 0.714, Precision = 0.724, Recall = 0.710, F1-score = 0.713, AUC-PRC = 0.75). The CatBoost model was the second-best model (Accuracy = 0.681, Precision = 0.674, Recall = 0.689, F1-score = 0.686, and AUC-PRC = 0.69). The XGBoost model had the worst prediction performance among all the classifier (Accuracy = 0.652, Precision = 0.664, Recall = 0.652, F1-score = 0.566, and AUC-PRC = 0.68). Based on the results of the AUC-ROC, the performance indicators that were extracted from the confusion matrix, and AUC-PRC, the BO-Random Forest classifier had a better predicted LLWS criticality performance and could be used for the SHAP analysis interpretation.

Table 3. Performance measures of BO-ELCs.

BO-ELC	Performance Metrics			
	Accuracy	Precision	Recall	F1-Score
LGBM	0.672	0.681	0.672	0.676
AdaBoost	0.681	0.673	0.661	0.663
Random Forest	0.714	0.724	0.710	0.713
CatBoost	0.681	0.674	0.689	0.686
XGBoost	0.652	0.664	0.652	0.656

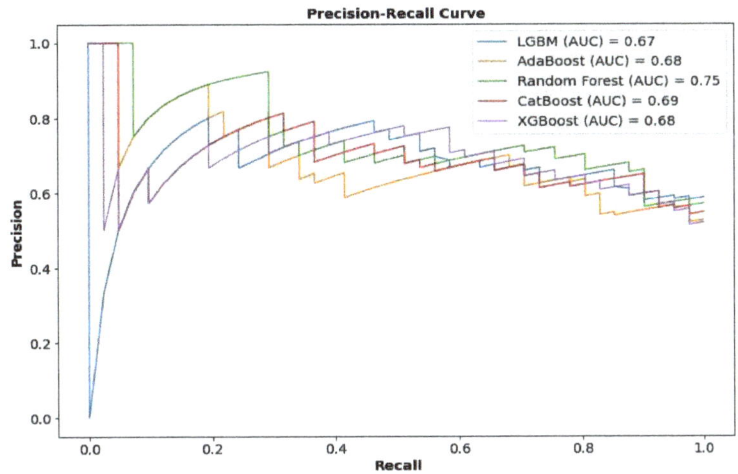

Figure 9. Combined precision–recall curve (PRC) for all Bayesian-optimized ELC models.

3.3. Sensitivity Analysis

Developing a concise LLWS criticality prediction model is essential because more precise models may capture the relationship between LLWS criticality and risk factors better. The capability to interpret the classifier modeling results is equally essential. This section describes how the SHAP method was implemented to interpret the BO-Random Forest classifier results and the BO-CatBoost classifier results to estimate the impact of the top three individual risk factors and their interactions.

3.3.1. Global Factors' Importance and Contribution

For the global factors' importance and contribution analysis, we used the BO-Random Forest classifier, which was the best model in our case, followed by the BO-CatBoost classifier, which was the second-best model. In using these two optimal models with accurate LLWS criticality predictions, there was strong merit in investigating which factors were the most important and quantifying how these factors contributed to the final predictions. To explore the impact of each factor on the final prediction, the SHAP values were used. It is worth mentioning that factor importance is not the same as factor contribution. Factor importance indicates which factors have the greatest impact on a model's performance. The factor contributions not only identify relevant factors, but they also provide a logical explanation for the observed outcome (NC-LLWS or C-LLWS). This study determined the importance of each factor and its contribution to the model estimate using the top two BO-ELCs with better accuracies. Figure 10a illustrates the SHAP global importance scores for the factors used in the BO-Random Forest classifier. However, the outcome did not indicate the proportionate contribution of each factor to the likelihood of an LLWS criticality. It showed that the most important factor that caused the occurrence of C-LLWS was the hourly temperature, which had a mean SHAP value of +0.98, followed by the mean wind speed with a mean SHAP value of +0.64 and Runway 07LA (+0.41). Figure 10b illustrates the SHAP global importance scores for the factors using the CatBoost model. The results revealed that the most important factor that caused C-LLWS was the hourly temperature (+0.82) followed by wind speed (+0.49) and Runway 07LA (+0.38). The sequences of the factor importance in the case of both the BO-Random Forest classifier and the BO-CatBoost classifier were consistent, while there was a slight difference in their SHAP values.

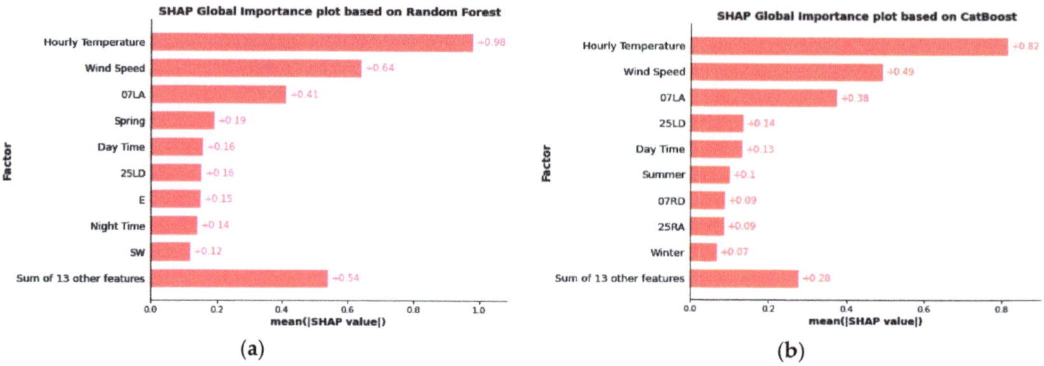

Figure 10. SHAP global importance plots: (**a**) Random Forest; (**b**) CatBoost.

Similarly, a SHAP contribution evaluation was conducted to conduct a more in-depth examination of the Random Forest and CatBoost models using SHAP beeswarm plots (Figure 11). We created a quantitative value from the SHAP contribution plots that combined the Shaply values and expressed the classifier contributions of factors. The input factors were arranged on the vertical axis in order of increasing influence, starting with the most influential. The SHAP value is shown on the horizontal axis, and the significance of the factor is shown by the color scale, (blue to pinkish-red for low significance to high significance). The SHAP beeswarm plots of the Random Forest and CatBoost models illustrate that most of the moderate-to-high mean wind speeds resulted in the occurrence of C-LLWS events, which are represented by the pinkish-red color toward the right side of the vertical reference line with positive SHAP values (Figure 11a,b). The blue color toward the left of the vertical reference line indicates the occurrence of NC-LLWS events due to a low mean wind speed. Similarly, in the case of the mean hourly temperature, a high temperature (represented in red) is shown to the right of the vertical reference line with a positive SHAP value and blue to the left of the vertical reference line. It shows that high temperatures resulted in C-LLWS events while low temperatures were more likely to cause NC-LLWS events. The same was the case for wind speed and Runway 07LA.

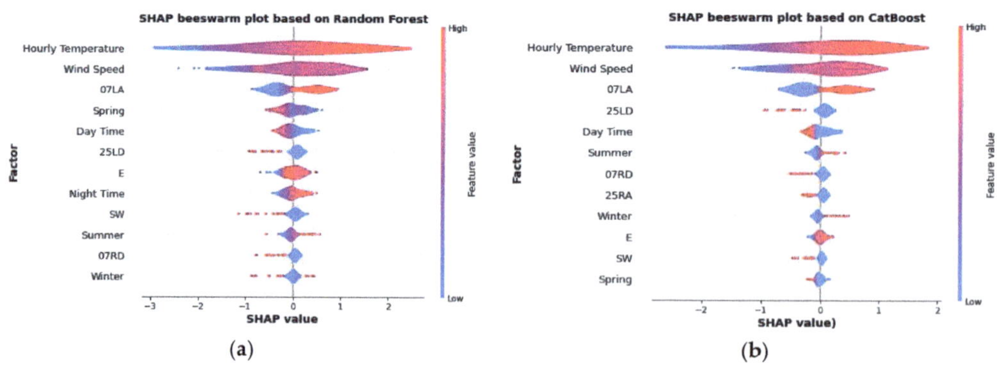

Figure 11. SHAP beeswarm plots: (**a**) Random Forest; (**b**) CatBoost.

3.3.2. Factor Dependence and Interaction

There was no obvious correlation between the changes in the factor value and the changes in the SHAP value in the factor global importance and contribution (beeswarm) plot. Figure 12 supplements the contribution plot by providing more information about how the SHAP values varied with the eigenvalues and by displaying the individual interpretation outcomes for the three major factors. The SHAP dependence and interaction plots were examined to ascertain the extent to which the input variables used to evaluate the Random Forest classifier interacted in terms of their contributions (see Figure 12). The SHAP dependence plot is a scatter plot that demonstrates the effect a single factor had on the predictions made by the classifier, which in our case was the Random Forest model. The SHAP interaction plot shows the effect of two factors on the models' predictions.

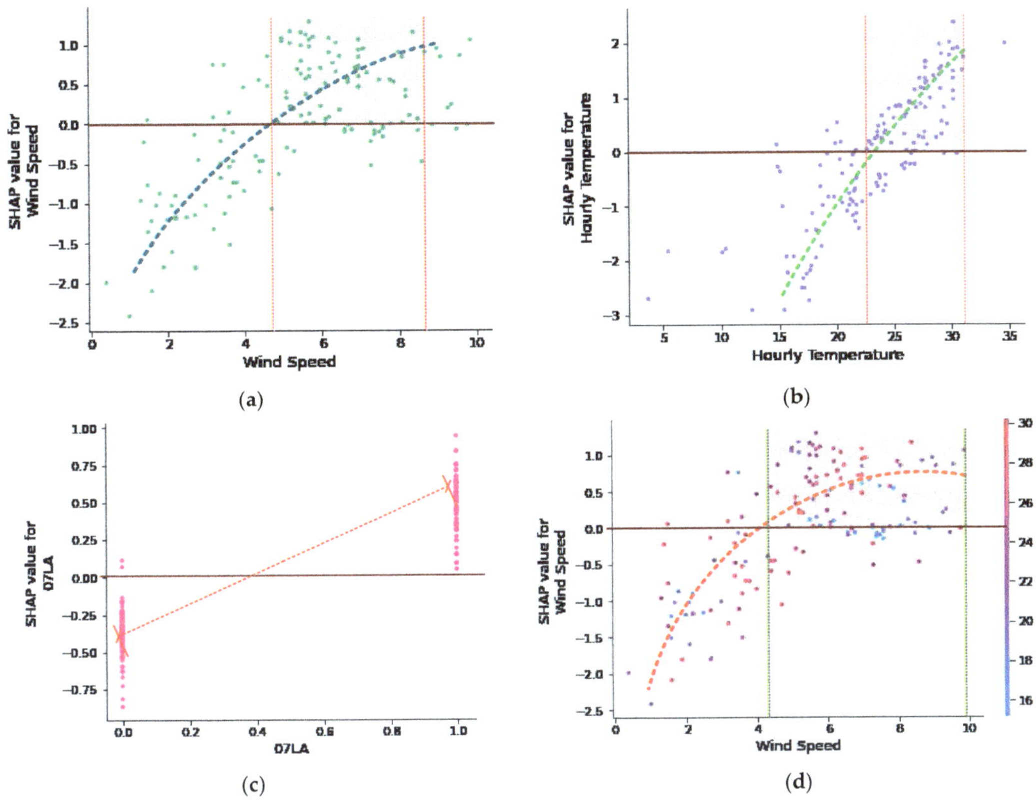

Figure 12. (**a**) SHAP wind speed dependence plot; (**b**) SHAP hourly temperature dependence plot output; (**c**) SHAP Runway 07LA dependence plot; (**d**) SHAP interaction plot of wind speed and hourly temperature.

The dependence and interaction plot examines the top three influential factors; namely, hourly temperature, wind speed, and Runway 07LA. Other factor interactions, however, could be explored as well. Figure 12a depicts the effect of wind speed on the models' predictions. The points with high density fell above the SHAP 0.0 reference line and at wind speeds of more than 4.4 m/s up to 8 m/s had a positive impact on the prediction of LLWS, which meant that wind speeds higher than 4.4 m/s were more likely to cause C-LLWS events. The results were consistent with the findings of previous studies [41,42]. However, it is pertinent to mention that for the occurrence of C-LLWS events, the variation in wind speed is more important than the mean wind speed. The duration of a C-LLWS event that might be encountered by an aircraft is generally within a few seconds to a few minutes. Therefore, the hourly mean wind speed can hardly provide an accurate indication of LLWS criticality. Therefore, more refined data on wind conditions such as a 1 min mean in turbulence intensity may be required to improve the models' accuracies.

Figure 12b depicts the effect of hourly temperature on the models' predictions. The points with high density fell above the SHAP 0.0 reference line at an hourly temperature of 23 °C to 31 °C, which had a positive impact on the predictions of LLWS. This illustrated that C-LLWS events were more likely to occur at temperatures between 23 °C and 31 °C. The SHAP value for Runway 07LA when labeled '1' was higher than reference 0.0 and lower when labeled as '0' (Figure 12c). This illustrated that Runway 07LA was highly vulnerable to the occurrence of C-LLWS events. This also showed that C-LLWS events were more likely to occur under the easterly, southeasterly, southerly, and southwesterly winds, which was also consistent with the previous findings [5,43,44]. Pilots should be cautious when making a "final approach" to Runway 07LA. Figure 12d demonstrates the interaction of the wind speed and hourly temperature and their combined influence on the BO-Random Forest classifier prediction. When the wind speed ranged from 4.2 m/s to 9.8 m/s and the hourly temperature ranged from 24.8 °C to 29.5 °C, high density points formed in the shaded area above the SHAP 0.0 reference line. Within these ranges, C-LLWS events were more likely to occur.

3.3.3. Local Factor Interpretation

Figure 13 shows the SHAP explanatory force chart for two randomly selected cases from the actual estimations. The base value (0.045) on the graph represents the mean of the BO-Random Forest classifier estimations for the training data set. The NC-LLWS condition occurred when the outcome value of classifier was less than the classifier's base value. C-LLWS events occurred when the classifier's output value exceeded the base value. The blue arrows illustrate the magnitude of an input factor's effect on the likelihood of an NC-LLWS occurrence. The likelihood of occurrence of a C-LLWS event was influenced by input factors as indicated by the red arrows. Each arrow's area occupied by a factor reflects the magnitude of that factor's effect. Consider two instances of the BO-Random Forest classifier that were correctly classified as C-LLWS and NC-LLWS from the training dataset. The two instances depicted in Figure 13 correctly classified as NC-LLWS and C-LLWS had estimated values of −2.91 and 3.62, respectively. For the first randomly selected instance (Figure 13a), when the wind speed was equal to 3.4 m/s with a moderate hourly temperature equal to 23.8 °C, an NC-LLWS occurred. This figure also illustrates that seasons other than spring can have occurrences of NC-LLWS events. The spring season designated as 0 highlighted that for this randomly selected instance, the spring season did not contribute to the occurrence of NC-LLWS events. Contrary to this (Figure 13b), the combination of a moderate temperature equal to 21.6 °C with a high wind speed and spring season, a C-LLWS event occurred. However, the autumn season did not contribute to the occurrence of C-LLWS events. Similarly, for this very instance, Runway 07LA contributed to the occurrence of a C-LLWS event. In a similar fashion, we could randomly select other correctly classified instances for their local interpretation.

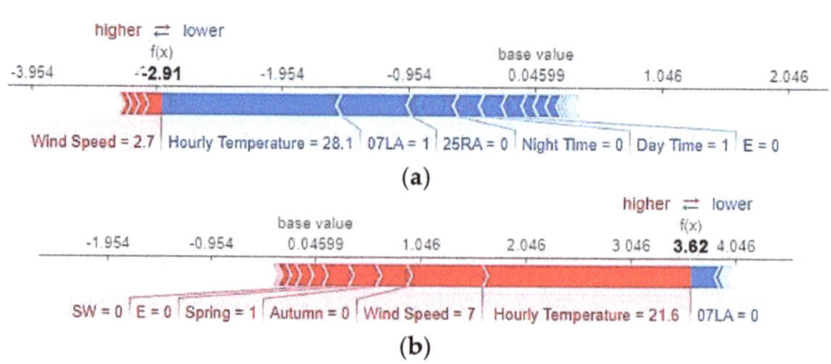

Figure 13. SHAP explanatory force plots: (**a**) plot of case with a value equal to 0.291; (**b**) plot of case with a value equal to 3.62.

4. Conclusions and Recommendations

This study presented the application of five state-of-the-art BO-ELCs in the prediction of the occurrence of an LLWS criticality. Six factors, including the hourly temperature, wind speed, runway orientation, wind direction, time of day, season, and the height of the LLWS as a binary target factor from the PIREPs and HKO weather reports, were used as inputs. Under the comprehensive evaluation criteria, all models achieved a high prediction accuracy. Nevertheless, ensemble learning algorithms are frequently criticized for their lack of interpretability and transparency. Despite the fact that engineering-domain models are more flexible and frequently more accurate than traditional predictive statistical techniques, this has an effect on their widespread acceptability. In this study, the model with the best prediction was interpreted using the SHAP algorithm, and the influence of the top three factors on the occurrence of an LLWS criticality was demonstrated. Based on the study, the following conclusions were drawn:

- In the testing dataset, the BO-Random Forest classifier had the best overall performance of all BO-ELCs investigated in this study with an AUC-ROC of 0.759 and accuracy, precision, recall, F1-score, and AUC-PRC values of 0.714, 0.724, 0.710, 0.713, and 0.75, respectively.
- The performance of each individual BO-ELC varied marginally. Despite the fact that XGBoost's AUC-ROC was 0.73, its accuracy, recall, precision, F1-score, and AUC-PRC values were 0.652, 0.656, 0.664, 0.656, and 0.68, respectively.
- The AdaBoost and LGBM models demonstrated the lowest AUC-ROC (0.687) and AUC-PRC (0.67) scores, respectively.
- SHAP demonstrated efficacy in interpreting the optimal model's outcome (BO-Random Forest classifier). In terms of the factor influence, the SHAP analysis revealed that the hourly temperature is the most influential factor followed by the wind speed and runway 07LA.
- When the wind speed was moderate to high (>4.2 m/s) and the temperature was moderate to high (>24.5 °C), aircrafts on a final approach to Runway 07LA were more likely to experience critical LLWS.

The technique proposed in this research work can be utilized to undertake a large-scale investigation of wind shear and can serve as a useful resource for aviation authorities and researchers who are concerned with aviation safety. In addition, this paper focused exclusively on the prediction of LLWS criticality as computed by using five BO-ELC classifiers (CatBoost, XGBoost, LGBM, RF, and AdaBoost) in aggregation with the SHAP model. This study was limited to the application of machine learning models. Future studies might be undertaken by combining a number of other BO-ELCs such as a stacking ensemble as well as neural network models with a range of additional risk factors such as inter-annual changes in wind shear events and their spatial distributions.

Author Contributions: Conceptualization, A.K. and P.-W.C.; Data curation, P.-W.C.; Formal analysis, A.K.; Funding acquisition, P.-W.C. and F.C.; Investigation, A.K., F.C., and H.P.; Methodology, A.K.; Project administration, F.C.; Resources, H.P.; Validation, F.C.; Writing—original draft, A.K.; Writing—review and editing, H.P. All authors have read and agreed to the published version of the manuscript.

Funding: This research was supported by the National Natural Science Foundation of China (U1733113), the Shanghai Municipal Science and Technology Major Project (2021SHZDZX0100), the Research Fund for International Young Scientists (RFIS) of the National Natural Science Foundation of China (NSFC) (Grant No. 52250410351), and the National Foreign Expert Project (Grant No. QN2022133001L).

Institutional Review Board Statement: Not applicable.

Informed Consent Statement: Not applicable.

Data Availability Statement: Not applicable.

Acknowledgments: We are thankful to the Hong Kong Observatory at Hong Kong International Airport for providing us with the pilot reports and weather-related data.

Conflicts of Interest: The authors declare no conflict of interest.

References

1. ACI. *Annual World Airport Traffic Report*; ACI: Montreal, QC, Canada, 2018.
2. Fichtl, G.H.; Camp, D.W.; Frost, W. Sources of low-level wind shear around airports. *J. Aircr.* **1977**, *14*, 5–14. [CrossRef]
3. Bretschneider, L.; Hankers, R.; Schönhals, S.; Heimann, J.M.; Lampert, A. Wind Shear of Low-Level Jets and Their Influence on Manned and Unmanned Fixed-Wing Aircraft during Landing Approach. *Atmosphere* **2021**, *13*, 35. [CrossRef]
4. Michelson, M.; Shrader, W.; Wieler, J. Terminal Doppler weather radar. *Microw. J.* **1990**, *33*, 139.
5. Shun, C.; Chan, P. Applications of an infrared Doppler LiDAR in detection of wind shear. *J. Atmos. Ocean. Technol.* **2008**, *25*, 637–655. [CrossRef]
6. Li, L.; Shao, A.; Zhang, K.; Ding, N.; Chan, P.-W. Low-level wind shear characteristics and LiDAR-based alerting at Lanzhou Zhongchuan International Airport, China. *J. Meteorol. Res.* **2020**, *34*, 633–645. [CrossRef]
7. Thobois, L.; Cariou, J.P.; Gultepe, I. Review of LiDAR-based applications for aviation weather. *Pure Appl. Geophys.* **2019**, *176*, 1959–1976. [CrossRef]
8. Hon, K.-K. Predicting low-level wind shear using 200-m-resolution NWP at the Hong Kong International Airport. *J. Appl. Meteorol. Climatol.* **2020**, *59*, 193–206. [CrossRef]
9. Chan, P. A tail strike event of an aircraft due to terrain-induced wind shear at the Hong Kong International Airport. *Meteorol. Appl.* **2014**, *21*, 504–511. [CrossRef]
10. Lei, L.; Chan, P.; Li-Jie, Z.; Hui, M. Numerical simulation of terrain-induced vortex/wave shedding at the Ho;ng Kong International Airport. *Meteorol. Z.* **2013**, *22*, 317–327. [CrossRef]
11. Chan, P.; Hon, K. Observation and numerical simulation of terrain-induced wind shear at the Hong Kong International Airport in a planetary boundary layer without temperature inversions. *Adv. Meteorol.* **2016**, *2016*, 1454513. [CrossRef]
12. Shimoyama, K.; Nakanomyo, H.; Obayashi, S. Airport terrain-induced turbulence simulations integrated with weather prediction data. *Trans. Jpn. Soc. Aeronaut. Space Sci.* **2013**, *56*, 286–292. [CrossRef]
13. Casado-Sanz, N.; Guirao, B.; Lara Galera, A.; Attard, M. Investigating the risk factors associated with the severity of the pedestrians injured on Spanish crosstown roads. *Sustainability* **2019**, *11*, 5194. [CrossRef]
14. Shaik, M.E.; Islam, M.M.; Hossain, Q.S. A review on neural network techniques for the prediction of road traffic accident severity. *Asian Transp. Stud.* **2021**, *7*, 100040. [CrossRef]
15. Mujalli, R.O.; de Oña, J. Injury severity models for motor vehicle accidents: A review. In *Proceedings of the Institution of Civil Engineers-Transport*; Thomas Telford Ltd.: London, UK, 2013; pp. 255–270.
16. Yuan, C.; Li, Y.; Huang, H.; Wang, S.; Sun, Z.; Li, Y. Using traffic flow characteristics to predict real-time conflict risk: A novel method for trajectory data analysis. *Anal. Methods Accid. Res.* **2022**, *35*, 100217. [CrossRef]
17. Zhao, Y.; Deng, W. Prediction in Traffic Accident Duration Based on Heterogeneous Ensemble Learning. *Appl. Artif. Intell.* **2022**, 1–24. [CrossRef]
18. Zhang, S.; Khattak, A.; Matara, C.M.; Hussain, A.; Farooq, A. Hybrid feature selection-based machine learning Classification system for the prediction of injury severity in single and multiple-vehicle accidents. *PLoS ONE* **2022**, *17*, e0262941. [CrossRef]
19. Goodman, S.N.; Goel, S.; Cullen, M.R. Machine learning, health disparities, and causal reasoning. *Ann. Intern. Med.* **2018**, *169*, 883–884. [CrossRef]
20. Guo, R.; Fu, D.; Sollazzo, G. An ensemble learning model for asphalt pavement performance prediction based on gradient boosting decision tree. *Int. J. Pavement Eng.* **2022**, *23*, 3633–3646. [CrossRef]
21. Liu, J.N.; Kwong, K.; Chan, P.W. Chaotic oscillatory-based neural network for wind shear and turbulence forecast with LiDAR data. *IEEE Trans. Syst. Man Cybern. Part C* **2012**, *42*, 1412–1423. [CrossRef]
22. Guidotti, R.; Monreale, A.; Ruggieri, S.; Pedreschi, D.; Turini, F.; Giannotti, F. Local rule-based explanations of black box decision systems. *arXiv* **2018**, arXiv:1805.10820.
23. Rudin, C. Stop explaining black box machine learning models for high stakes decisions and use interpretable models instead. *Nat. Mach. Intell.* **2019**, *1*, 206–215. [CrossRef] [PubMed]
24. Lundberg, S.M.; Lee, S.-I. A unified approach to interpreting model predictions. *Adv. Neural Inf. Process. Syst.* **2017**, *30*.
25. Merrick, L.; Taly, A. The explanation game: Explaining machine learning models using Shapley values. In Proceedings of the International Cross-Domain Conference for Machine Learning and Knowledge Extraction, Dublin, Ireland, 25–28 August 2020; pp. 17–38.
26. Dong, S.; Khattak, A.; Ullah, I.; Zhou, J.; Hussain, A. Predicting and analyzing road traffic injury severity using boosting-based ensemble learning models with SHAPley Additive exPlanations. *Int. J. Environ. Res. Public Health* **2022**, *19*, 2925. [CrossRef] [PubMed]
27. Mangalathu, S.; Hwang, S.H.; Jeon, J.S. Failure mode and effects analysis of RC members based on machine-learning-based SHapley Additive exPlanations (SHAP) approach. *Eng. Struct.* **2020**, *219*, 110927. [CrossRef]

28. Khattak, A.; Almujibah, H.; Elamary, A.; Matara, C.M. Interpretable Dynamic Ensemble Selection Approach for the Prediction of Road Traffic Injury Severity: A Case Study of Pakistan's National Highway N-5. *Sustainability* **2022**, *14*, 12340. [CrossRef]
29. Feng, D.C.; Wang, W.J.; Mangalathu, S.; Taciroglu, E. Interpretable XGBoost-SHAP machine-learning model for shear strength prediction of squat RC walls. *J. Struct. Eng.* **2021**, *147*, 04021173. [CrossRef]
30. García, M.V.; Aznarte, J.L. Shapley additive explanations for NO_2 forecasting. *Ecol. Inform.* **2020**, *56*, 101039. [CrossRef]
31. Wu, J.; Chen, X.Y.; Zhang, H.; Ziong, I.-D.; Lei, H.; Deng, S.-H. Hyperparameter optimization for machine learning models based on Bayesian optimization. *J. Electron. Sci. Technol.* **2019**, *17*, 26–40.
32. Shekar, B.; Dagnew, G. Grid search-based hyperparameter tuning and classification of microarray cancer data. In Proceedings of the 2019 2nd International conference on advanced computational and communication paradigms (ICACCP), Sikkim, India, 25–28 February 2019; pp. 1–8.
33. Kadam, V.J.; Jadhav, S.M. Performance analysis of hyperparameter optimization methods for machine learning with small and medium sized medical datasets. *J. Discret. Math. Sci. Cryptogr.* **2020**, *23*, 115–123. [CrossRef]
34. Ke, G.; Meng, Q.; Finley, T.; Wang, T.; Chen, W.; Ma, W.; Ye, Q.; Liu, T.Y. LightGBM: A highly efficient gradient boosting decision tree. *Adv. Neural Inf. Process. Syst.* **2017**, *30*.
35. Parmar, A.; Katariya, R.; Patel, V. A review on random forest: An ensemble classifier. In Proceedings of the International Conference on Intelligent Data Communication Technologies and Internet of Things, Coimbatore, India, 7–8 August 2018; pp. 758–763.
36. Chen, T.; Guestrin, C. Xgboost: A scalable tree boosting system. In Proceedings of the 22nd ACM SIGKDD International Conference on Knowledge Discovery and Data Mining, San Francisco, CA, USA, 13–17 August 2016; pp. 785–794.
37. Hancock, J.T.; Khoshgoftaar, T.M. CatBoost for big data: An interdisciplinary review. *J. BIG Data* **2020**, *7*, 1–45. [CrossRef] [PubMed]
38. An, T.-K.; Kim, M.-H. A new diverse AdaBoost classifier. In Proceedings of the 2010 International conference on artificial intelligence and computational intelligence, Sanya, China, 23–24 October 2010; pp. 359–363.
39. Chan, P.W. A significant wind shear event leading to aircraft diversion at the Hong Kong international airport. *Meteorol. Appl.* **2012**, *19*, 10–16. [CrossRef]
40. Chan, P. Severe wind shear at Hong Kong International airport: Climatology and case studies. *Meteorol. Appl.* **2017**, *24*, 397–403. [CrossRef]
41. Chen, F.; Peng, H.; Chan, P.W.; Ma, X.; Zeng, X. Assessing the risk of wind shear occurrence at HKIA using rare-event logistic regression. *Meteorol. Appl.* **2020**, *27*, e1962. [CrossRef]
42. Chan, P.W.; Hon, K.K. Observations and numerical simulations of sea breezes at Hong Kong International Airport. *Weather* 2022. [CrossRef]
43. Chen, F.; Peng, H.; Chan, P.-w.; Zeng, X. Low-level wind effects on the glide paths of the North Runway of HKIA: A wind tunnel study. *Build. Environ.* **2019**, *164*, 106337. [CrossRef]
44. Chen, F.; Peng, H.; Chan, P.-w.; Zeng, X. Wind tunnel testing of the effect of terrain on the wind characteristics of airport glide paths. *J. Wind. Eng. Ind. Aerodyn.* **2020**, *203*, 104253. [CrossRef]

atmosphere

MDPI

Article

Prediction of Aircraft Go-Around during Wind Shear Using the Dynamic Ensemble Selection Framework and Pilot Reports

Afaq Khattak [1,*], Pak-Wai Chan [2], Feng Chen [1,*] and Haorong Peng [3]

1 The Key Laboratory of Infrastructure Durability and Operation Safety in Airfield of CAAC, Tongji University, 4800 Cao'an Road, Jiading, Shanghai 201804, China
2 Hong Kong Observatory, 134A Nathan Road, Kowloon, Hong Kong, China
3 Shanghai Research Center for Smart Mobility and Road Safety, Shanghai 200092, China
* Correspondence: khattak@tongji.edu.cn (A.K.); fengchen@tongji.edu.cn (F.C.)

Abstract: Pilots typically implement the go-around protocol to avoid landings that are hazardous due to wind shear, runway excursions, or unstable approaches. Despite its rarity, it is essential for safety. First, in this study, we present three Dynamic Ensemble Selection (DES) frameworks: Meta-Learning for Dynamic Ensemble Selection (META-DES), Dynamic Ensemble Selection Performance (DES-P), and K-Nearest Oracle Elimination (KNORAE), with homogeneous and heterogeneous pools of machine learning classifiers as base estimators for the prediction of aircraft go-around in wind shear (WS) events. When generating a prediction, the DES approach automatically selects the subset of machine learning classifiers which is most probable to perform well for each new test instance to be classified, thereby making it more effective and adaptable. In terms of Precision (86%), Recall (83%), and F1-Score (84%), the META-DES model employing a pool of Random Forest (RF) classifiers outperforms other models. Environmental and situational factors are subsequently assessed using SHapley Additive exPlanations (SHAP). The wind shear magnitude, corridor, time of day, and WS altitude had the greatest effect on SHAP estimation. When a strong tailwind was present at low altitude, runways 07R and 07C were highly susceptible to go-arounds. The proposed META-DES with a pool of RF classifiers and SHAP for predicting aircraft go-around in WS events may be of interest to researchers in the field of air traffic safety.

Keywords: wind shear; go-around; machine learning; dynamic ensemble selection; SHapley Additive exPlanations

Citation: Khattak, A.; Chan, P.-W.; Chen, F.; Peng, H. Prediction of Aircraft Go-Around during Wind Shear Using the Dynamic Ensemble Selection Framework and Pilot Reports. *Atmosphere* **2022**, *13*, 2104. https://doi.org/10.3390/atmos13122104

Academic Editors: Duanyang Liu, Hongbin Wang and Shoupeng Zhu

Received: 25 November 2022
Accepted: 12 December 2022
Published: 15 December 2022

1. Introduction

An abrupt change in wind direction or speed of at least 14 knots and below 1600 feet (500 m) above runway level is referred to as wind shear (WS) in the aviation industry [1]. This could be the result of environmental conditions such as a thunderstorm, gust, or sea breeze, or it could be the result of the airport's proximity to complex terrain, such as mountains or man-made structures. The occurrence of wind shear is regarded as one of the most dangerous phenomena for approaching and departing aircrafts [2].

During the landing phase, the flight deck remains highly engaged, and the pilots must make a number of split-second decisions to complete their landing checklist. However, adverse weather conditions such as wind shear, mountainous terrain, and the presence of buildings close to the airport could increase turbulence along the glide path. While completing the landing checklist, the pilot must contend with violent updrafts and downdrafts and abrupt changes in the aircraft's horizontal and vertical movement. As shown in Figure 1, the head wind shear or tail wind shear may result in landing short of the runway (loss of lift) or deviating from the actual flight path during the final approach. Consequently, pilots initiate a go-around procedure. Despite that this protocol is implemented to prevent unsafe landings, their complicated maneuvering procedures and limited available time can raise additional safety concerns, particularly in wind shear events. As a result of this operational anomaly, air traffic controllers have a greater workload, and noise levels have massively increased [3,4]. Additionally, the airport throughput and punctuality of flights are negatively impacted [5,6]. Majority of go-arounds are performed at low altitudes and low speeds, necessitating immediate adjustments to the aircraft's altitude, thrust, and flight path to avoid collisions with nearby air traffic.

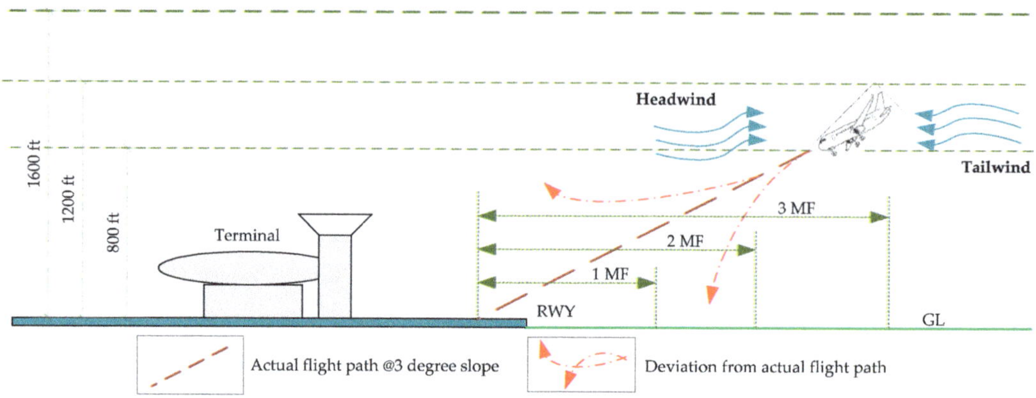

Figure 1. Occurrence location of WS events in the vicinity of the airport runway.

Since wind shear plays a major role in the execution of go-around protocols, airports around the world have benefited greatly from the availability of precise remote sensing technologies, including Terminal Doppler Weather Radar (TDWR) and Doppler Light Detection and Range (LiDAR), to timely detect WS events [7–9]. Researchers in the past have used a wide range of approaches to predict go-around based on various parameters as well as contributing factors, including the environment, such as wind speed, visibility, and pressure, etc., unstable approach and a change in runway configuration, as well as physiological conditions associated with the pilot and air traffic controller, as shown in Table 1.

While these studies have shed light on the many factors that can lead to a go-around, none of them have examined the role that wind shear plays in this phenomenon. There is a significant gap in the literature about the prediction of go-around under wind shear conditions. The occurrence of go-around due to wind shear is usually a rare event, however, predicting its occurrence under wind shear conditions is of utmost importance. Therefore, the goal of this research is to quantify the factors that contribute to the occurrence of go-around triggered by wind shear and situational factors, such as time of day, season of the year, and flight and aircraft type. In this study, our study location is Hong Kong International Airport (HKIA) and we used HKIA-based pilot report (PIREPs) data. We then employed dynamic ensemble learning strategies to classify go-around and approaches of aircrafts. In many practical situations, ensemble learning has outperformed a single machine learning approach [19–22]. Stacking, bagging, and boosting are the three main ideas of ensemble learning, which encapsulates the techniques and strategies of model blending. The fundamental aim of ensemble learning is to pool the efficacy of several classification models into a single conclusion. A dataset with many factors or characteristics for each instance constitutes a binary classification problem. One of the considerations is the decision label, which should be categorical and reveal to which group each instance belongs. The goal of classification strategies is to build classification models that can predict and classify the dependent label for the given sample. The two most common kinds of classification schemes are dynamic and static. A comparison of ensemble and classification model selection techniques for static and dynamic classification approaches is depicted in Figure 2 [23,24]. The primary difference between static and dynamic classification approaches is whether all the test samples are predicted with the same classifier. Similar to how classifier selection differs from ensemble classifier selection, a single classifier model can be comprised of several base classifiers that are employed to predict a test sample, leading to a wide number of classification techniques that rely on their unique combination. In most cases, the performance of a static classification strategy is inferior to that of a dynamic one, as various classification models excel in various settings.

Table 1. Literature on various factors contributing to the occurrence of aircraft go-around.

Serial No.	Parameters	Contributing Factors	Model Employed	Literature
1.	Environment	Visibility, wind speed, and localizer deviation significantly impacted go-around.	Flight simulation of Airbus A330-200 and Boeing 737-800	[10]
		Visibility, wind speed, and pressure significantly impacted go-around.	Categorical Boosting	[11]
		Thunderstorms and winds exceeding 29 mph significantly impacted go-around	Statistical model	[12]
2.	Pilot and air traffic controller	Unpleasant psychological condition compromised pilot decision-making and cognitive performance that resulted in go-around	Neuro-economics brain imaging protocol	[13]
		Anomalies in pilot flying performance, including flight path deviations and visual scanning behaviors caused go-around	Flight simulator test	[14]
		Situational unawareness by air traffic controllers caused go-around	Path analysis and bootstrap	[15]
		Age and experience of air traffic controllers contributed to go-around	Flight simulator test	[16]
		Pilot and controller experiences and mental states	Surveys and interviews	[6]
3.	Unstable approach/runway configuration	Quantification of aircraft deviation at final approach	Sparse Variation Gaussian process	[17]
		Approach stability, departure air traffic, flight spacing, departure traffic, and ceiling contributed to go-around	Principal component analysis	[18]

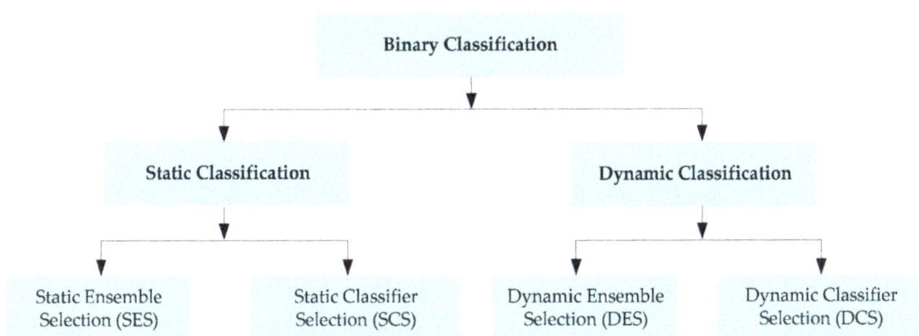

Figure 2. Types of binary classification.

For this research, we used three DES models, including Meta-Learning for Dynamic Ensemble Selection (META-DES) [25], K-Nearest Oracle Elimination (KNORAE) [26], and Dynamic Ensemble Selection Performance (DES-P) [27], whose input is the pools of homogenous and heterogeneous classification algorithms. The pools of homogenous and homogenous classification algorithms are highlighted in Table 2. Afterward, SHAP analysis interpreted the results of the optimal DES model and illustrated important factors contributing to go-around under WS conditions.

Machine learning models are typically black boxes, so their predictions may not make the connection between input and output changes crystal clear. The interpretation of the model is equally important for an insight of the model's performance. Factor analysis methods, such as permutation-based importance scores, were previously employed to decipher the outcomes of machine learning studies. However, the factor importance analysis can only rank the significance of the factors, and it does not comprehend how each factor affects the model's prediction on its own. SHapley Additive exPlanations (SHAP) analysis, inspired by game theory [33], has been used in recent studies to quantitatively assess the relative importance of each contributing factor [34–36]. Use of SHAP with machine learning models allows for the interpretation of the relative contributions and the importance of different factors [37–40].

Table 2. Pools of various classification algorithms for the study.

Ensemble	Pools of Algorithm	Reference
Homogenous	Random Forest (RF)	[28]
	Extremely Randomized Tree (ERT)	[29]
	Bagging Multi-Layer Perceptron (BMLP)	[30]
Heterogeneous	K-Nearest Neighbor (KNN)	[27]
	Support Vector Machine (SVM)	[31]
	Binary Logistic Regression (BLR)	[32]

Our findings would aid pilots, flight attendants, air traffic controllers, and policymakers in estimating when a go-around is requisite. Second, identifying mitigation strategies to reduce aircraft go-around and, more generally, the circumstances that lend credence to them, which may be deemed anomalous and inherently unappealing, can be aided by quantifying the contributing factors of go-around occurrences. It is possible to reduce the need for go-around by implementing mitigation strategies such as adjustment of protocols, enhancing pilot education, and revamping hardware.

The remainder of this paper is structured as follows. Section 2 illustrates the research methodology and discusses our sources of data, DES models, and the SHAP interpretation strategy. Section 3 details the DES models' performance as a comparison as well as the SHAP analysis results. Section 4 encompasses the conclusion of our study and recommendations.

2. Methodology

In this study, we first analyzed the pilot reports (PIREPs) of Hong Kong International Airport (HKIA) to determine the factors that most likely contributed to the go-around. A PIREP is an abbreviation for pilot reports used in civil aviation. The pilots who encounter hazardous weather conditions and go-around are sent to air traffic controllers. The factors that can influence go-around include weather conditions such as wind shear conditions (wind shear magnitude, altitude, and horizontal location of wind shear from the runway as well as its causes), precipitation (rainfall), aircraft and flight (wide or narrow-body aircraft, international or domestic flight), landing runway, and temporally specific factors such as the season of the year and time of the day (daytime/nighttime).

Secondly, we built DES models with different pools of homogenous and heterogeneous classifiers as base estimators to predict aircraft go-around in case of WS events. Based on the model with the best performance, lastly, we estimated the importance and contributions of various factors to go-around occurrence using the SHAP interpretation approach. Figure 3 depicts the whole operational paradigm proposed in this study.

2.1. Study Location

The HKIA is located on an artificial Lantau Island on the southeastern coast of mainland China in a subtropical zone. The tropical cyclones and southwest monsoon are two typical convective weather conditions that occur in Hong Kong. In addition to bringing thunderstorms and showers to the region, the convective weather interrupts air traffic. Due to these reasons, Hong Kong International Airport (HKIA) is among the airports most susceptible to WS in the vicinity of the runway. Numerous observational and modeling studies have shown that HKIA's intricate orography and complex land–sea contrast are also conducive to the occurrence of WS [41]. Significant WS events occur once every 400 to 500 flights. From the opening of HKIA in 1998 until 2015, 97.70% of reports illustrated 15–25 knots of WS [42].

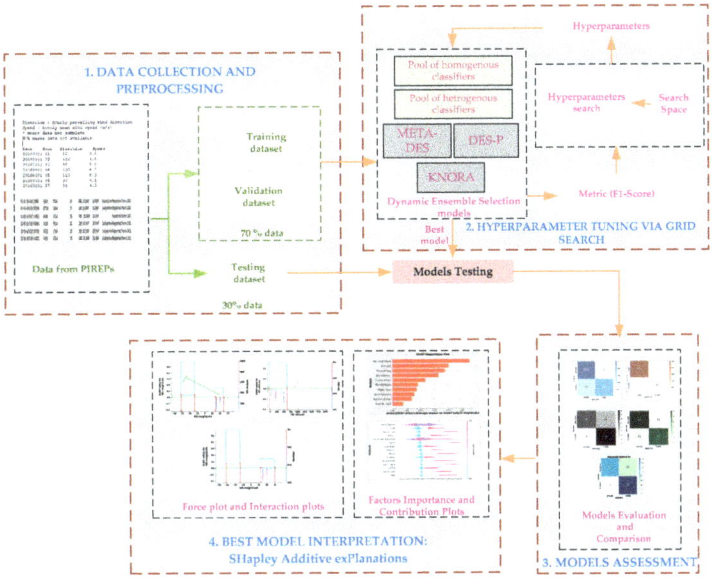

Figure 3. Proposed framework of our study.

Figure 4 shows that HKIA is surrounded on three sides by open sea water and mountains to the south, which reaches elevations of over 900 m above sea level. This complex terrain surrounding HKIA also contributes to terrain-induced WS. The mountainous terrain to the south of HKIA amplifies WS, disrupting airflow and generating mountain waves, gap discharge, and other disturbances along the HKIA flight paths. Three runway corridors exist at HKIA: the North Runway (Northern Corridor), the Central Runway (Central Corridor), and the South Runway (Southern Corridor). The Northern Corridor is a newly constructed runway, and therefore the previous Northern Corridor is now designated as the Central Corridor. They are oriented in the 070° and 250° directions. Since each runway can be used for takeoffs and landings in either direction, there are a total of twelve possible configurations. For example, runway '07LA' denotes landing ('A' refers to arrival) with a heading angle of 070° (shortened to '07') using the left runway (hence 'L'). This shows aircraft landing on the Northern Corridor from the western side of HKIA. Likewise, an aircraft departing the Southern Corridor in the west would use runway 25LD.

Figure 4. Hong Kong International Airport and surrounding terrain.

2.2. Data Processing from PIREPs

As stated earlier, pilot reports are abbreviated as PIREPs in aviation. When pilots encounter hazardous weather, they notify air traffic controllers. Traditionally, PIREPs include information about turbulence, aircraft icing, and the flight route phase. However, because HKIA is vulnerable to WS, information about the occurrence of WS is explicitly provided, including the occurrence date and time, the horizontal location of WS from the runway threshold (nearest nautical mile), WS magnitude (nearest 5 knots), vertical location or altitude of WS (to the nearest 50 or 100 ft), type of aircraft, and flight number. In addition, if an aircraft performs a go-around during WS caused by a sea breeze or gust front, the pilot reports go-around in the HKIA-based PIREPs, as indicated in Table 3. Note that in Table 3, the positive or negative sign associated with the magnitude of WS indicates a headwind and tailwind, respectively. Moreover, pilots at HKIA can submit PIREPs after landing or use on-board radio communication to relay pertinent information to the air traffic controller.

Table 3. Extracted environmental and situational factors from HKIA-based PIREPs.

Date	Time	Runway	Flight Type	Aircraft Type	WS Magnitude	WS H-Location	WS Altitude	PPT	Go-Around	Cause of WS
2021-01-16	6:17 AM	07RA	CX495	A35K	−20 knots	3-NM	900 ft	No	No	See breeze
2021-01-21	3:18 PM	25LA	5Y4511	B744	15 knots	2-NM	500 ft	Yes	No	See breeze
—	—	—	—	—	—	—	—	—	—	—
—	—	—	—	—	—	—	—	—	—	—
2021-03-29	10:12 PM	07CA	CX8178	B77W	25 knots	RWY	50 ft	No	Yes	Gust front
—	—	—	—	—	—	—	—	—	—	—
—	—	—	—	—	—	—	—	—	—	—
2021-09-21	3:58 AM	07RA	PO980	B748	20 knots	2-NM	200 ft	No	Yes	Gust front

A total of 1731 instances of WS events were illustrated by PIREPs from 2017 to 2021, including both departing and approaching flights. However, out of 1731 instances, 1388 (80.18%) instances were reported by approaching flights and 343 (19.81%) by departing flights. In this study, we dealt with the causes of go-around during WS events, and therefore, the information reported by approaching flights was retained while that from departing flights was discarded from the dataset. Furthermore, the dataset was preprocessed to deal with the missing values and other irrelevant information. After carefully cleaning redundant and erroneous information, the finalized dataset was obtained with 872 instances in which go-around was observed 196 times. In addition, to develop a binary classification problem, all the go-around events (being the minority class) were labeled as '1', while all the approaches (being the majority class) were labeled as "0". A detailed description of all the factors is shown in Table 4. The summary statistics of all the factors from HKIA-based PIREPs are provided in Table 5.

Table 4. Environmental and situational factors' description and coding.

Factors	Descriptions	Type of Data	Coding
Go-around	Go-around/approach	Discrete	'Go-around = 1', 'Approach = 0'
Vehicle-Specific	Airline Flight Type	Discrete	'International flight = 1', 'Others = 0'
	Aircraft Type	Discrete	'Wide-body = 1', 'Others = 0'
Runway-Specific	Corridor	Discrete	'07C = 0', '07R = 1', '25C=2', '25L = 3'
Environment-Specific	WS magnitude	Continuous	-
	WS H-Location	Discrete	'At RWY = 0', '1-NM = 1', '2-NM = 2', '3-NM = 3'
	WS altitude	Continuous	-
	Cause of WS	Discrete	'Gust Front = 0', 'Sea Breeze = 1,
	Precipitation	Discrete	'Yes = 1', 'No = 0'
Temporal-Specific	Time of day	Discrete	'Day = 1', 'Night=0'
	Seasons	Discrete	'Winter = 0', 'Spring = 1', 'Summer = 2', 'Autumn = 3'

Table 5. Descriptive statistics of various environmental and situational factors.

Factors	Descriptions	Statistics			
		Mean	**St. dev**	**Min**	**Max**
Vehicle-Specific	Airline Flight Type	0.554	0.497	0	1
	Aircraft Type	0.741	0.434	0	1
Runway-Specific	Orientation	0.897	1.002	0	3
Environment-Specific	WS Magnitude (−/+)	17.17/−19.23	3.86/4.85	−15/15	−40/45
	WS H-Location	1.473	0.896	0	3
	WS V-Location (ft)	335.52	304.723	15	2000
	Cause of WS	0.457	0.492	0	1
	Precipitation	0.530	0.497	0	1
Temporal-Specific	Time of day	0.623	0.482	0	1
	Seasons	1.551	0.865	0	3

2.3. *Dynamic Ensemble Selection (DES) Algorithms*

As stated before, we proposed three DES models to develop a reliable classification and prediction model for aircraft go-around and approach during WS events. The DES models are Meta-Learning for Dynamic Ensemble Selection (META-DES), K-Nearest Oracle Elimination (KNORAE), and Dynamic Ensemble Selection Performance (DES-P). The DES modeling process flowchart is depicted in Figure 5.

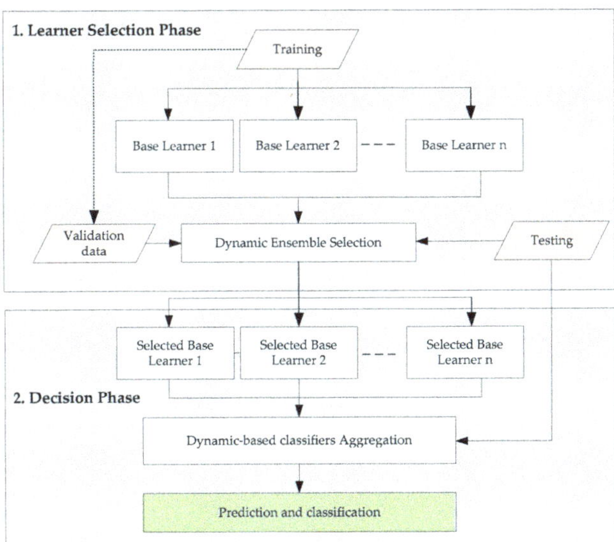

Figure 5. Dynamic Ensemble Selection process.

2.3.1. META-DES

The objective of the META-DES algorithm [25] is to determine if the selected classification model from a pool of latent classification models is able to classify the given test data. This meta-problem can primarily be tackled in two steps.

Finding the meta-features for each classification model in the pool is the first step. There are four types of meta-features: (a) posterior likelihood/probability for each target label, (b) overall local accuracy (OLA) of the classification model in the region of competence, and (c) the neighbor's hard classification (NHC) (a vector of 'n' is generated, where 'n' is the number of training instances in the region of competence). The value of the vector is set to 1 if the classification model correctly classifies the instance within its area of competence; otherwise, it is set to 0. (d) The confidence of the classifier (the orthogonal distance between the input instance and the classifiers' decision boundary).

Step two is to determine, using meta-features, whether a particular classification algorithm is capable of producing precise predictions for a given set of test instances. As a result, the ensemble of classifiers for the given test data consisted of every classification algorithm selected by meta-classification models.

2.3.2. KNORAE

For any given set of test data, the KNORAE algorithm will find the subset of classification models that correctly classifies all K-Nearest Neighbors. The classification of the test data is then given to the ensemble of these chosen classification algorithms and open to voting (the KNORAE algorithm uses the majority voting rule for prediction). In other words, the algorithm gets rid of classification models that incorrectly classify nearby data [26]. The algorithm stops prioritizing nearest neighbors and looks for a classification model that can correctly label all training instances that are close to the test data if it cannot find a classification algorithm that can do so.

2.3.3. DES-P

By contrasting the effectiveness of each classification algorithm to that of a random classification algorithm, this DES procedure eliminates the inefficient ones. For a given number of classes in a training dataset, the efficacy of the random classification algorithm is $1/C$ (see the explanation in [27]). The dynamic selection of classification models is carried out by comparing the performance of the classification algorithm to that of a random classification algorithm in the neighborhood defined by the test data. For the provided test data, the classification algorithm can be added to the ensemble if its performance is better than a random classification algorithm. If no classification algorithm is picked, all the algorithms in the pool will be used on the given test data.

2.4. Pool of Classifiers

The following pool of classifiers was used for the DES algorithms: homogeneous ensembles such as Random Forest (RF), Extremely Randomized Tree (ERT), and Bagging Multi-Layer Perceptron (BMLP), and heterogeneous ensembles consisting of pooling of Support Vector Machine (SVM), K-Nearest Neighbor (KNN), and Binary Logistic Regression (BLR) classifiers.

2.5. Performance Evaluation

The Recall, Precision, and F1-scores were used to analyze the performance of the DES models in classifying the aircraft's go-around and approach during WS events. For each diagnostic label, the performance indicators were independently evaluated. For a complete understanding of all performance metrics, below is a list of terms.

TP (True Positive): The total number of predictions that correctly identified instances of "go-around" as "go-around." TN (True Negative): The number of predictions that correctly identified "approach" as "approach." FP (False Positive): The total number of instances in which "approach" was incorrectly predicted as "go-around." FN (False Negative) is the total amount of predictions that incorrectly classified "go-around" as "approach." The following is an explanation of the evaluation metrics:

Recall

Recall for a single class 'i' is the ratio between the TP to the sum of the TP and FN in the confusion matrix for that class. It can be calculated by using Equation (1):

$$Recall_i = \frac{TP}{TP + FN} \tag{1}$$

The overall Recall is the average of the Recall of each class, which is given by Equation (2):

$$Recall = \frac{1}{L}\sum_{i=1}^{L} Recall_i \quad (2)$$

Precision

Precision for a single class 'i' is the ratio between the *TN* to the sum of the *TN* and *FP* in the confusion matrix for that class. It can be calculated by using Equation (3):

$$Precision_i = \frac{TN}{TN + FP} \quad (3)$$

The overall Recall is the average of the Recall of each class, which is given by Equation (4):

$$Precision = \frac{1}{L}\sum_{i=1}^{L} Precision_i \quad (4)$$

F1-Score

The F1-Score is a metric that considers both the Precision and the Recall of the test instances to compute the score. It can be interpreted as a weighted mean of the Recall and Precision. It can be calculated for class 'i' by using Equation (5):

$$F1\text{-}Score_i = \frac{2[(precision_i)(recall_i)]}{precision_i + recall_i} \quad (5)$$

The overall F1-Score is the average of the F1-Score of each class, which is given by Equation (6):

$$F1\text{-}Score = \frac{1}{L}\sum_{i=1}^{L}\left(\frac{2[(precision_i)(recall_i)]}{precision_i + recall_i}\right) \quad (6)$$

2.6. Dynamic Ensemble Selection Interpretation by SHapley Additive exPlanations (SHAP)

The SHAP analysis is based on a game theory approach for the explanation of the machine learning-ensemble classifiers' outputs. As machine learning models are "black-box", therefore, when interpreting these models, both a global and local perspective are the core ideas behind the SHAP analysis. The SHAP values were estimated, which correspond to the value given to each factor in the instance when a machine learning prediction was computed. Equation (7) is used to calculate the contribution of each factor, which is shown as the Shapley value:

$$\varphi_i = \sum_{\gamma \subseteq \Pi\{i\}} \frac{\gamma!(n - |\gamma| - 1)!}{n!}[\mathrm{E}(\gamma \cup \{i\}) - \mathrm{E}(\gamma)] \quad (7)$$

where φ_i illustrates the ith factor contribution, Π is the set of all factors, γ is the subset of the decision factors, and $\mathrm{E}(\gamma_i)$ and $\mathrm{E}(\gamma)$ illustrate the best model results with and without ith factors, respectively. SHAP analysis basically results in interpretable DES models through an additive factors imputation strategy, wherein the output model is defined as a linear sum of the input factors (Equation (8)):

$$g(\Psi') = \Delta_0 + \sum_{i=1}^{\Lambda} \Delta_i \Psi' \qquad \Psi' \in \{0,1\}^{\Lambda} \quad (8)$$

It is equal to 1 in case when a factor is observed, otherwise it is 0. It illustrates the amount of all input factors, where Δ_0 represents an outcome without factors (i.e., base value), and Δ_i shows the Shapley value of the ith factor.

In this study, the SHAP analysis was employed for the interpretation of the proposed DES model, i.e., the global importance and contribution of factors that are likely to cause aircraft go-around as well as the interactions of factors.

3. Results and Discussion

To predict the occurrence of go-around in WS conditions, the DES models with different pools of base estimators were employed by using HKIA-based PIREPs. Figure 6 shows the frequency distributions of the factors from the PIREPs. To assess the potential correlations between the factors of the PIREPs, we performed Pearson correlation analysis. Statistically, Figure 7 illustrates that the absolute value of Pearson's correlation coefficient is between 1 and -1. Although we have observed a Pearson correlation coefficient value of –0.63 for causes of WS and PPT, the correlation is moderate, and we will not exclude them for subsequent modeling. Both the factors are environmental-specific and their inclusion in the model may have a significant impact. For the analysis, we used the Python sklearn.metrics, imbeans, and sklearn.ensemble, Scikit-learn, and SHAP libraries.

3.1. Data Partitioning

The dataset of 872 go-arounds and approaches under WS conditions that was extracted from HKIA-based PIREPS and used for DES modeling has been split into primarily two sets, which are known as the training validation set and the test set. Seventy percent of the data was used for training validation, while thirty percent of the data was used for actual testing. The training validation set had a total of 468 and 143 records, respectively, for the number of approaches and the go-around events. The testing set included a total of 209 approaches and 53 records of the go-around attempts.

3.2. Grid Search Strategy for Hyperparameter Tuning

Using Stratified 10-Fold Cross-Validation, the training validation set was evaluated. The training validation set was split into 10 equal-sized folds. Utilizing stratified sampling, each fold retained a proportional amount of each label. The Stratified 10-Fold Cross-Validation strategy was chosen because it maintains a proportional representation of each label. The DES model was initially trained with nine folds, and then its F1-Score performance was evaluated with the final fold. This procedure was repeated ten times until all available folds (those that comprised the training set in the initial fold) comprised the validation set. The average F1-Score of each 10 folds was then determined.

Grid Search [43] is one of the most frequently employed hyperparameter tuning techniques for machine learning approaches. Through using the Grid Search technique, the feasible set (search space) of hyperparameters was pre-determined, and the model's best hyperparameters were chosen based on their performance in cross-validation. For our studies, the model's hyperparameters were determined by the set of hyperparameters that maximized the overall F1-Score (mean F1-Score across all folds). The F1-Score was chosen as the performance indicator because it combines the recall and precision of diagnostic labels. Table 6 shows the optimal values of the hyperparameter of the models.

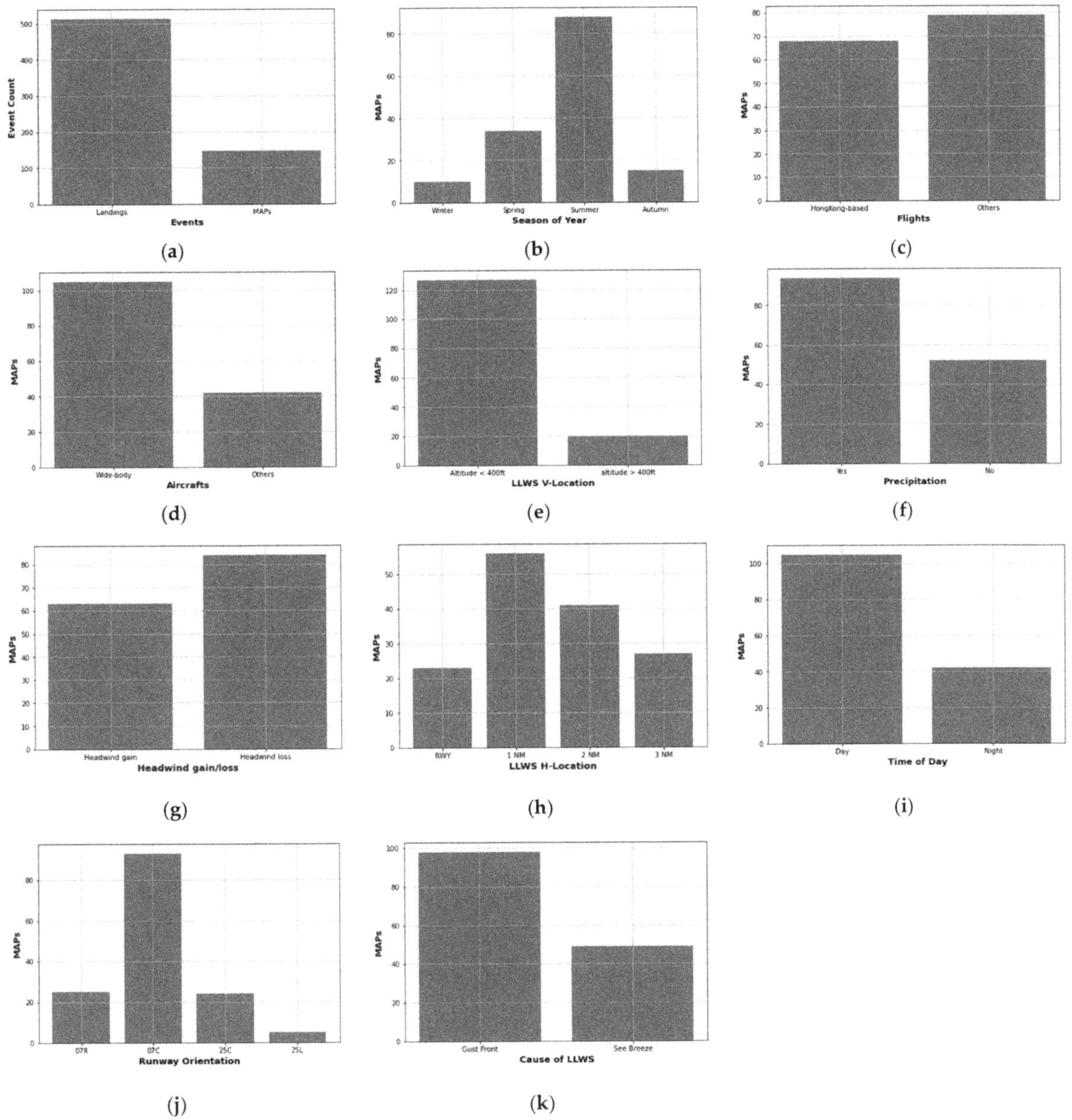

Figure 6. Distribution of go-around with respect to environmental and situational factors (**a**) Distribution of Landing (approaches) and MAPs (Go-around); (**b**) Distribution of Go-around in different season of years; (**c**) Distribution of Go-around with respect to type of flight; (**d**) Distribution of Go-around with respect to type of aircraft; (**e**) Distribution of Go-around with respect to altitude (V-Location) of the wind shear; (**f**) Distribution of Go-around with respect to precipitation (**g**) Distribution of Go-around with respect to wind shear magnitude; (**h**) Distribution of Go-around with respect to wind shear horizontal (H)-location; (**i**) Distribution of Go-around with respect to time of the day; (**j**) Distribution of Go-around with respect to corridor/runway orientation; (**k**) Distribution of Go-around with respect to wind shear causes.

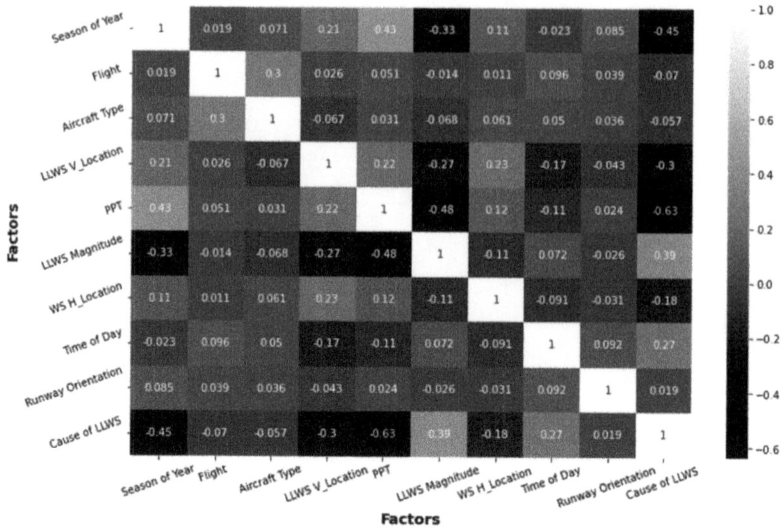

Figure 7. Pearson's correlation matrix of the explanatory factors.

Table 6. Optimal hyperparameter values of the models.

Model	Hyperparameter	Space	Optimal Value
RF	Number of trees	[100, 500, 1000, 1500, 2000, 2500, 3000]	2500
	Max depth of tree	[3, 5, 7, 9, 11, 13, 15]	11
BMPL	Number of estimators	[200, 400, 600, 800, 1000]	500
	Batch size	[50, 100, 150, 200, 250, 300]	200
	Epoch size	[50, 100, 150, 200]	110
ERT	Number of trees	[100, 500, 1000, 1500, 2000, 2500, 3000]	2000
	Max depth of tree	[3, 5, 7, 9, 11, 13, 15]	11
SVM	C	[0.1, 1.0, 100]	100
	Gamma	[1.0, 0.1, 0.01, 0.001, 0.0001]	0.01

3.3. DES Models' Performance Assessment and Comparison

As was previously mentioned, the positive and negative classes were referred to as approach and go-around, respectively. The Precision, Recall, and F1-Score performance metrics were extracted from the confusion matrices of each DES algorithm and used to evaluate all models. Homogeneous and heterogeneous pools of classification algorithms were used as the base estimators (Tables 7–10). META-DES produced a higher performance measure for DES algorithms using RF classifiers as base estimators with Precision (86%), Recall (83%), and F1-Score (84%) (Table 7). KNORAE-RF, the second-best DES model when used with the RF classifier, produced an F1-Score of 82%, a Precision value of 82%, and a Recall value of 82%. Similar to this, DES-P-BMLP produced higher performance measures, with Precision (78%), Recall (75%), and F1-Score (77%), in the case of DES algorithms with BMLP (Table 8). When using the ERT classifier with other DES algorithms, the META-DES performed well (Table 9). It displayed a Precision of 78%, a Recall of 76%, and an F1-Score of 77%. Furthermore, the META-DES with the pool of heterogeneous classifiers (SVM+KNN+BLR) performed well as compared to DES-P and KNORAE (Table 10). It showed a Precision of 78%, a Recall of 76%, and an F1-Score of 77%. Overall, it was found that the META-DES-RF model performed better than the other DES models and could be used in conjunction with SHAP analysis to determine the relative importance of different factors as well as their contributions.

Table 7. Comparison of performance measures of DES algorithms based on the pool of RF.

Approach			Performance Measures				
			Predicted		Precision	Recall	F1-Score
			Approach	Go-Around			
KNORAE-RF	Actual	Approach	193	16	0.82	0.82	0.82
		Go-around	15	38			
DES-P-RF	Actual	Approach	182	27	0.75	0.68	0.71
		Go-around	30	23			
META-DES-RF	Actual	Approach	195	14	0.86	0.83	0.84
		Go-around	16	37			

Table 8. Comparison of performance measures of DES based on the pool of BMLP.

Approach		Class	Performance Measures				
			Predicted		Precision	Recall	F1-Score
			Approach	Go-Around			
KNORAE-BMLP	Actual	Approach	195	15	0.77	0.75	0.76
		Go-around	22	31			
DES-P-BMLP	Actual	Approach	182	27	0.78	0.75	0.77
		Go-around	23	30			
META-DES-BMLP	Actual	Approach	195	15	0.73	0.73	0.73
		Go-around	23	30			

Table 9. Comparison of performance measures of DES based on the pool of ERT.

Approach		Class	Performance Measures				
			Predicted		Precision	Recall	F1-Score
			Approach	Go-Around			
KNORAE- ERT	Actual	Approach	185	24	0.76	0.73	0.75
		Go-around	24	29			
DES-P-ERT	Actual	Approach	184	25	0.75	0.72	0.74
		Go-around	25	28			
META-DES-ERT	Actual	Approach	188	21	0.78	0.76	0.77
		Go-around	21	32			

Table 10. Comparison of performance measures of DES based on the pool of heterogeneous classifiers.

Approach		Class	Performance Measures				
			Predicted		Precision	Recall	F1-Score
			Approach	Go-Around			
KNORAE-(SVM+KNN+BLR)	Actual	Approach	172	37	0.71	0.72	0.72
		Go-around	21	32			
DES-P-(SVM+KNN+BLR)	Actual	Approach	168	41	0.72	0.70	0.71
		Go-around	23	30			
META-DES-(SVM+KNN+BLR)	Actual	Approach	188	21	0.78	0.76	0.77
		Go-around	21	32			

3.4. Sensitivity Analysis

It is vital to develop an evident go-around prediction model because more accurate models might effectively capture the association between go-around and various environmental and situational factors. The ability to comprehend the optimal META-DES-RF model is immensely valuable. The SHAP method was used in this section to interpret the best META-DES-RF results and calculate the combined effect of each individual risk factor.

3.4.1. Global Factors' Importance and Contribution

We utilized the META-DES-RF model for the factors' importance and contribution analysis due to its superior go-around prediction compared to other models. There is a compelling case for determining which factors are most crucial and for quantifying their contributions to the final predictions. It is important to note that factor contribution and factor importance are two different concepts. The importance of a factor reveals which variables have the biggest effects on a model's performance. The factor contributions not only point out important factors but also give a logical justification for the observed result, in our case "go-around" and "approach."

The SHAP global importance scores for the factors used in the META-DES-RF are shown in Figure 8a. The result does not, however, show how much each factor contributed to the likelihood of a go-around happening. It demonstrates that WS magnitude, with a mean SHAP value of +0.257, was the most significant factor that contributed to the occurrence of go-arounds, followed by corridor, with a mean SHAP value of +0.190, time of day (+0.190), and WS altitude (+0.160). Similar to this, a SHAP contribution evaluation was carried out to examine the META-DES-RF model in greater detail using SHAP beeswarm plots (Figure 8b). From the SHAP contribution plots, which combined the Shapely values and expressed the contributions of the various factors to the META-DES-RF model, we were able to derive a quantitative value. On the vertical axis, the input factors are arranged from most influential to least influential in order of increasing influence. The horizontal axis displays the SHAP value, and the color scale, which ranges from blue to red for low significance to high significance, displays the factor's significance.

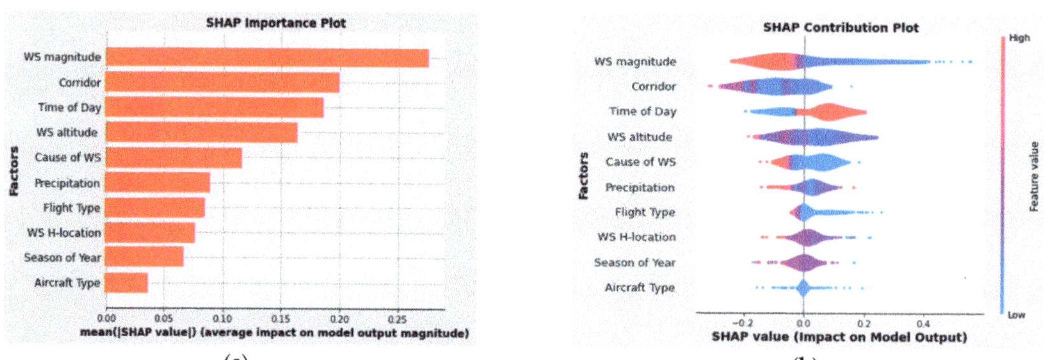

(a) (b)

Figure 8. SHAP global interpretation: (**a**) SHAP importance plot and (**b**) SHAP beeswarm plot.

The META-DES-RF model's SHAP beeswarm plot showed that majority of the tailwind led to the commencement of the aircraft go-around. The cause may be that in strong tailwinds, an aircraft's airspeed—the speed of the aircraft relative to the airflow around it—does not significantly decrease as it approaches the ground, and with a high airspeed, an aircraft may not be able to land at the designated touchdown location. Pilots increase the throttle to go around, try again, or ask for a different runway to ensure safety. The outcome is also in line with earlier research [44]. The second important factor was the corridor's orientation. Runways 07C and 07R were more likely to initiate go-arounds when there was wind shear. Runways 07C and 07R should not be used for landings during WS events because go-arounds have become a safety concern. The third crucial factor was the time of day. Although we could not pinpoint any prior research on the effect of the time of day on the go-around, our data nonetheless revealed that majority of the go-around happened during the day (07:00 AM to 19:00 PM).

The fourth crucial factor was WS altitude. Figure 8b illustrates how WS events that took place at lower altitudes were held responsible for the high number of go-arounds. This is also consistent with a previous study [45]. The cockpit remains incredibly active during the landing phase, and the captain and co-pilot must make a number of quick decisions to wrap up their landing checklist. However, the best course of action is to abort the landing and perform a go-around when an unexpected WS happens very close to the runway. As a result, majority of go-arounds happened when the aircraft ran into WS close to the ground.

3.4.2. Factor Dependence and Interaction

In the factor importance and contribution (beeswarm) plots, there was no evidence of a correlation between the alteration in the factor value and the change in the SHAP value. The interpretation results for the factors are shown in Figure 9, which also adds more relevant information about how the SHAP values varied with the eigenvalues to the contribution plot. To assess the extent to which the critical environment factors used to evaluate the META-DES-RF interacted in terms of their contributions, the SHAP interaction plots were examined.

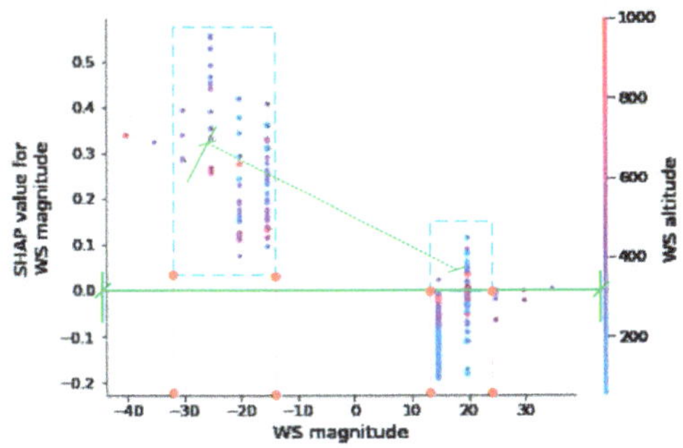

(a)

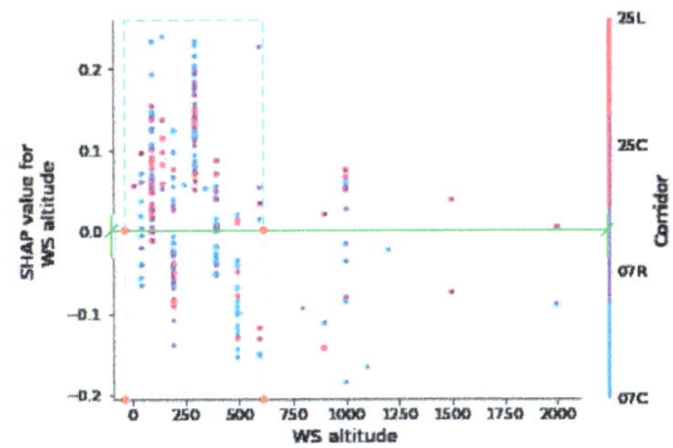

(b)

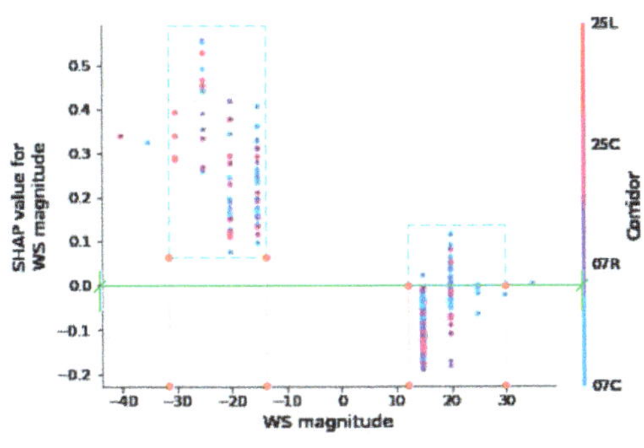

(c)

Figure 9. (**a**) SHAP WS magnitude vs. WS altitude plot. (**b**) SHAP WS altitude vs. Corridor plot. (**c**) SHAP WS magnitude vs. Corridor plot.

Figure 9a shows how the models' predictions were impacted by the WS magnitude and WS altitude. The go-around phenomenon is heavily influenced by the points that are above the SHAP 0.00 green reference line. Thus, it is evident that the points with magnitudes of −14 to −32 knots are above the SHAP 0.00 green reference line. Most of the points have labels in blue and purple, which indicate low altitude between 0 and 600 feet. It shows that strong tailwinds at low altitudes play a greater role in the occurrence of go-arounds. Figure 9b depicts how the WS altitude and Corridor influenced the model predictions. It is apparent that the points with high density that fall between WS altitudes of 0 and 600 feet are located above the SHAP 0.00 green reference line. Majority of the points have blue and purple labels, which denote corridors 07C and 07R. It demonstrates that runways 07C and 07R are highly susceptible to the occurrence of WS at low altitude, thereby increasing the likelihood of a go-around.

Figure 9c illustrates the effect of the WS magnitude and Corridor on model predictions. Clearly, the dense points that fall between WS altitudes of −14 and −32 knots are located above the SHAP 0.00 green reference line. A significant proportion of the points is marked with blue and purple labels, denoting corridors 07C and 07R. It reveals that runways 07C and 07R are particularly prone to the occurrence of WS at −14 to −32 knots (tailwind condition), as well as the low altitude of WS, thereby boosting the likelihood of a go-around.

4. Conclusions and Recommendations

In this study, a Dynamic Ensemble Selection model was used with a pool of homogeneous (Random Forest, Extremely Randomized Tree, and Bagging Multilayer Perceptron) and heterogeneous (Support Vector Machine, K-Nearest Neighbor, and Binary Logistic Regression) classifiers to predict the occurrence of go-arounds using the Hong Kong International Airport-based Pilot Reports from 2018 to 2021. The META-DES-RF model outperformed all the other models in terms of the Precision value, the Recall value, and the F1-Score. As a result, the META-DES framework that has been proposed presents a novel approach to modeling and forecasting aircraft go-around in WS conditions.

The lack of inclusivity and interpretability of machine learning models has been widely criticized. Although these approaches are often more flexible and reliable than traditional statistical models, this hinders their widespread adoption for prediction. Therefore, in this study, the results of META-DES-RF were evaluated, and both key risk factors and their impact on the occurrence of go-around were analyzed using the SHAP strategy to deal with the problem of interpretability introduced by META-DES-RF.

The top four crucial risk factors that enhance the probability of the occurrence of go-around under WS events were WS magnitude, Corridor, time of day, and WS altitude. The SHAP analysis revealed that there was a strong interaction among WS magnitude, WS altitude, and Corridor. It has been observed that runways 07C and 07R of HKIA were more prone to the occurrence of go-around. These go-around events occurred when strong tailwinds of −14 to −32 knots occurred within 600 ft above the runway level.

The novel method used in this research could be applied to a comprehensive investigation of how WS events have affected air traffic operations. It is a helpful tool for experts in air traffic safety and decision-makers in the aviation industry. In this study, SHAP analysis and dynamic ensemble classifiers were only used to predict the aircraft go-around under WS events. Future research initiatives may employ additional DES algorithms with various pools of classification models and risk factors. Doppler LiDAR data could also be combined with PIREPs in future research to evaluate a wide range of other parameters, including the impact of pressure, the direction of the wind, and others.

Author Contributions: Conceptualization, F.C.; data curation, P.-W.C.; formal analysis, A.K.; funding acquisition, A.K.; investigation, P.-W.C.; methodology, A.K. and P.-W.C.; project administration, A.K. and H.P.; resources, H.P.; software, F.C.; validation, F.C. All authors have read and agreed to the published version of the manuscript.

Funding: This research was supported by the National Natural Science Foundation of China (U1733113), the Shanghai Municipal Science and Technology Major Project (2021SHZDZX0100), the Research Fund for International Young Scientists (RFIS) of the National Natural Science Foundation of China (NSFC) (Grant No. 52250410351), and the National Foreign Expert Project (Grant No. QN2022133001L).

Institutional Review Board Statement: Not applicable.

Informed Consent Statement: Not applicable.

Data Availability Statement: Not applicable.

Acknowledgments: We are thankful to the Hong Kong Observatory at Hong Kong International Airport for providing us Pilot Report data of go-around events.

Conflicts of Interest: The authors declare no conflict of interest.

References

1. Airport Council International. *World Airport Traffic Forecast 2017–2040 Airport Council International*; Airport Council International: Montréal, QC, USA, 2017.
2. Fichtl, G.H.; Camp, D.W.; Frost, W. Sources of low-level wind shear around airports. *J. Aircr.* **1977**, *14*, 5–14. [CrossRef]
3. Metzger, U.; Parasuraman, R. Automation in Future Air Traffic Management: Effects of Decision Aid Reliability on Controller Performance and Mental Workload. In *Decision Making in Aviation*; Routledge: London, UK, 2017; pp. 345–360.
4. Prats, X.; Puig, V.; Quevedo, J.; Nejjari, F. Multi-objective optimization for aircraft departure trajectories minimizing noise annoyance. *Transp. Res. Part C Emerg. Technol.* **2010**, *18*, 975–989. [CrossRef]
5. Shortle, J.; Sherry, L. A Model for Investigating the Interaction between Go-Arounds and Runway Throughput. In Proceedings of the 2013 Aviation Technology, Integration, and Operations Conference, Los Angeles, CA, USA, 12–14 August 2013; p. 4235.
6. Blajev, T.; Curtis, W. *Go-around Decision-Making and Execution Project: Final Report to Flight Safety Foundation*; Flight Safety Foundation: Alexandria, VA, USA, 2017.
7. Michelson, M.; Shrader, W.; Wieler, J. Terminal Doppler weather radar. *Microw. J.* **1990**, *33*, 139.
8. Shun, C.; Chan, P. Applications of an infrared Doppler LiDAR in detection of wind shear. *J. Atmos. Ocean. Technol.* **2008**, *25*, 637–655. [CrossRef]
9. Li, L.; Shao, A.; Zhang, K.; Ding, N.; Chan, P.-W. Low-level wind shear characteristics and LiDAR-based alerting at Lanzhou Zhongchuan International Airport, China. *J. Meteorol. Res.* **2020**, *34*, 633–645. [CrossRef]
10. Zaal, P.; Campbell, A.; Schroeder, J.A.; Shah, S. Validation of Proposed Go-Around Criteria under Various Environmental Conditions. In Proceedings of the AIAA Aviation 2019 Forum, Dallas, TX, USA, 17–21 June 2019; p. 2993.
11. Chou, C.S.; Tien, A.; Bateman, H. A Machine Learning Application for Predicting and Alerting Missed Approaches for Airport Management. In Proceedings of the2021 IEEE/AIAA 40th Digital Avionics Systems Conference (DASC), San Antonio, TX, USA, 3–7 October 2021; pp. 1–9.
12. Donavalli, B.; Mattingly, S.P.; Massidda, A. Impact of Weather Factors on Go-Around Frequency (No. 17-03934). In Proceedings of the Transportation Research Board 96th Annual Meeting, Washington, DC, USA, 8–12 January 2017.
13. Causse, M.; Dehais, F.; Péran, P.; Sabatini, U.; Pastor, J. The effects of emotion on pilot decision-making: A neuroergonomic approach to aviation safety. *Transp. Res. Part C Emerg. Technol.* **2013**, *33*, 272–281. [CrossRef]
14. Dehais, F.; Behrend, J.; Peysakhovich, V.; Causse, M.; Wickens, C.D. Pilot flying and pilot monitoring's aircraft state awareness during go-around execution in aviation: A behavioral and eye tracking study. *Int. J. Aerosp. Psychol.* **2017**, *27*, 15–28. [CrossRef]
15. Jou, R.C.; Kuo, C.W.; Tang, M.L. A study of job stress and turnover tendency among air traffic controllers: The mediating effects of job satisfaction. *Transp. Res. Part E Logist. Transp. Rev.* **2013**, *57*, 95–104. [CrossRef]
16. Kennedy, Q.; Taylor, J.L.; Reade, G.; Yesavage, J.A. Age and expertise effects in aviation decision making and flight control in a flight simulator. *Aviat. Space Environ. Med.* **2010**, *81*, 489–497. [CrossRef]
17. Singh, N.P.; Goh, S.K.; Alam, S. Real-time unstable approach detection using sparse variational gaussian process. In Proceedings of the 2020 International Conference on Artificial Intelligence and Data Analytics for Air Transportation (AIDA-AT), Singapore, 3–4 February 2020; pp. 1–10.
18. Dai, L.; Liu, Y.; Hansen, M. Modeling go-around occurrence using principal component logistic regression. *Transp. Res. Part C Emerg. Technol.* **2021**, *129*, 103262. [CrossRef]
19. Dong, S.; Khattak, A.; Ullah, I.; Zhou, J.; Hussain, A. Predicting and analyzing road traffic injury severity using boosting-based ensemble learning models with SHAPley Additive exPlanations. *Int. J. Environ. Res. Public Health* **2022**, *19*, 2925. [CrossRef]
20. Guo, R.; Fu, D.; Sollazzo, G. An ensemble learning model for asphalt pavement performance prediction based on gradient boosting decision tree. *Int. J. Pavement Eng.* **2021**, *23*, 3633–3646. [CrossRef]
21. Feng, D.C.; Wang, W.J.; Mangalathu, S.; Taciroglu, E. Interpretable XGBoost-SHAP machine-learning model for shear strength prediction of squat RC walls. *J. Struct. Eng.* **2021**, *147*, 04021173. [CrossRef]

22. Zhang, S.; Khattak, A.; Matara, C.M.; Hussain, A.; Farooq, A. Hybrid feature selection-based machine learning Classification system for the prediction of injury severity in single and multiple-vehicle accidents. *PLoS ONE* **2022**, *17*, e0262941. [CrossRef]
23. Khattak, A.; Almujibah, H.; Elamary, A.; Matara, C.M. Interpretable Dynamic Ensemble Selection Approach for the Prediction of Road Traffic Injury Severity: A Case Study of Pakistan's National Highway N-5. *Sustainability* **2022**, *14*, 12340. [CrossRef]
24. Ko, A.H.; Sabourin, R.; Britto, A.S., Jr. From dynamic classifier selection to dynamic ensemble selection. *Pattern Recognit.* **2008**, *41*, 1718–1731. [CrossRef]
25. Cruz, R.M.; Sabourin, R.; Cavalcanti, G.D.; Ren, T.I. META-DES: A dynamic ensemble selection framework using meta-learning. *Pattern Recognit.* **2015**, *48*, 1925–1935. [CrossRef]
26. Walmsley, F.N.; Cavalcanti, G.D.; Sabourin, R.; Cruz, R.M. An investigation into the effects of label noise on Dynamic Selection algorithms. *Inf. Fusion* **2022**, *80*, 104–120. [CrossRef]
27. Cruz, R.M.; Zakane, H.H.; Sabourin, R.; Cavalcanti, G.D. Dynamic ensemble selection vs. KNN: Why and when dynamic selection obtains higher classification performance? In Proceedings of the 2017 Seventh International Conference on Image Processing Theory, Tools and Applications (IPTA), Montreal, QC, Canada, 28 November–1 December 2017; pp. 1–6.
28. Zheng, J.; Liu, Y.; Ge, Z. Dynamic ensemble selection based improved random forests for fault classification in industrial processes. *IFAC J. Syst. Control.* **2022**, *20*, 100189. [CrossRef]
29. Li, X.; Zhang, K.; Niu, J.; Liu, L. A machine learning-based dynamic ensemble selection algorithm for microwave retrieval of surface soil freeze/thaw: A Case Study Across China. *GI Sci. Remote Sens.* **2022**, *59*, 1550–1569. [CrossRef]
30. Zhao, Z.; Xu, S.; Kang, B.H.; Kabir, M.M.; Liu, Y.; Wasinger, R. Investigation and improvement of multi-layer perceptron neural networks for credit scoring. *Expert Syst. Appl.* **2015**, *42*, 3508–3516. [CrossRef]
31. Niu, P.; Wei, W. Classification of hyperspectral remote sensing images with dynamic support vector machine ensemble. *J. Comput. Appl.* **2010**, *30*, 1590. [CrossRef]
32. Potha, N.; Stamatatos, E. Dynamic ensemble selection for author verification. In *European Conference on Information Retrieval*; Springer: Cham, Switzerland, 2019; pp. 102–115.
33. Lundberg, S.M.; Lee, S.-I. A unified approach to interpreting model predictions. *Adv. Neural Inf. Process. Syst.* **2017**, *30*.
34. Mangalathu, S.; Hwang, S.H.; Jeon, J.S. Failure mode and effects analysis of RC members based on machine-learning-based SHapley Additive exPlanations (SHAP) approach. *Eng. Struct.* **2020**, *219*, 110927. [CrossRef]
35. Ndichu, S.; Kim, S.; Ozawa, S.; Ban, T.; Takahashi, T.; Inoue, D. Detecting Web-Based Attacks with SHAP and Tree Ensemble Machine Learning Methods. *Appl. Sci.* **2021**, *12*, 60. [CrossRef]
36. Wang, D.; Thunéll, S.; Lindberg, U.; Jiang, L.; Trygg, J.; Tysklind, M. Towards better process management in wastewater treatment plants: Process analytics based on SHAP values for tree-based machine learning methods. *J. Environ. Manag.* **2022**, *301*, 113941. [CrossRef]
37. Scavuzzo, C.M.; Scavuzzo, J.M.; Campero, M.N.; Anegagrie, M.; Aramendia, A.A.; Benito, A.; Periago, V. Feature importance: Opening a soil-transmitted helminth machine learning model via SHAP. *Infect. Dis. Model.* **2022**, *7*, 262–276. [CrossRef]
38. Alkadhim, H.A.; Amin, M.N.; Ahmad, W.; Khan, K.; Nazar, S.; Faraz, M.I.; Imran, M. Evaluating the Strength and Impact of Raw Ingredients of Cement Mortar Incorporating Waste Glass Powder Using Machine Learning and SHapley Additive ExPlanations (SHAP) Methods. *Materials* **2022**, *15*, 7344. [CrossRef]
39. Li, X.; Zhao, Y.; Zhang, D.; Kuang, L.; Huang, H.; Chen, W.; Fu, X.; Wu, Y.; Li, T.; Zhang, J.; et al. Development of an interpretable machine learning model associated with heavy metals' exposure to identify coronary heart disease among US adults via SHAP: Findings of the US NHANES from 2003 to 2018. *Chemosphere* **2023**, *311*, 137039. [CrossRef]
40. Jabeur, S.B.; Mefteh-Wali, S.; Viviani, J.L. Forecasting gold price with the XGBoost algorithm and SHAP interaction values. *Ann. Oper. Res.* **2021**, 1–21. [CrossRef]
41. Chan, P.W.; Hon, K.K. Observations and numerical simulations of sea breezes at Hong Kong International Airport. *Weather* **2022**. [CrossRef]
42. Hon, K.-K. Predicting low-level wind shear using 200-m-resolution NWP at the Hong Kong International Airport. *J. Appl. Meteorol. Climatol.* **2020**, *59*, 193–206. [CrossRef]
43. Purushotham, S.; Tripathy, B.K. Evaluation of Classifier Models Using Stratified Tenfold Cross Validation Techniques. In *International Conference on Computing and Communication Systems*; Springer: Berlin/Heidelberg, Germany, 2011; pp. 680–690.
44. Chan, P.W. A significant wind shear event leading to aircraft diversion at the Hong Kong international airport. *Meteorol. Appl.* **2012**, *19*, 10–16. [CrossRef]
45. Chen, F.; Peng, H.; Chan, P.W.; Ma, X.; Zeng, X. Assessing the risk of wind shear occurrence at HKIA using rare-event logistic regression. *Meteorol. Appl.* **2020**, *27*, e1962. [CrossRef]

atmosphere

Article

Spatiotemporal Patterns of Sea Ice Cover in the Marginal Seas of East Asia

Lei Zhang [1,2], Guoyu Ren [3,4], Mei Xu [1,2], Fanchao Meng [5], Rongwei Liao [6,*], Duanyang Liu [7], Minyan Wang [8] and Dan Jia [9]

1 Tianjin Meteorological Information Centre, Tianjin 300074, China
2 Tianjin Key Laboratory for Oceanic Meteorology, Tianjin 300074, China
3 Department of Atmospheric Science, School of Environmental Science, China University of Geosciences, Wuhan 430074, China
4 Laboratory for Climate Studies, National Climate Center, China Meteorological Administration, Beijing 100081, China
5 Tianjin Climate Center, Tianjin 300074, China
6 State Key Laboratory of Severe Weather, Chinese Academy of Meteorological Sciences, Beijing 100081, China
7 Key Laboratory of Transportation Meteorology of China Meteorological Administration, Nanjing Joint Institute for Atmospheric Sciences, Nanjing 210041, China
8 CMA Earth System Modeling and Prediction Centre, Beijing 100081, China
9 Binhai New Area Meteorological Bureau, Tianjin 300450, China
* Correspondence: liaorw@cma.gov.cn

Abstract: Using multisource sea ice fusion data, the spatiotemporal characteristics of sea ice cover were analyzed for the marginal seas of East Asia for the period 2005–2021. The results show that there were obvious differences in the beginning and end dates of the sea ice in the different sea areas. The northern Sea of Japan had the longest ice period, and Laizhou Bay and Bohai Bay in the Bohai Sea had the shortest ice period. The time when the largest sea ice extent appeared was relatively stable and mostly concentrated in late January to mid-February. There were obvious spatial differences in the duration of the sea ice cover in the marginal seas of East Asia. The duration of the sea ice cover gradually decreased from high latitude to low latitude and from nearshore to open seas. The annual average duration of the sea ice cover was more than 100 days in most of the Sea of Japan and approximately 20 days in most of Laizhou Bay and Bohai Bay. The melting speed was significantly faster than the freezing speed in the Bohai Sea and Yellow Sea, resulting in asymmetric changes in the daily sea ice extent in the two seas. The increasing trends in the maximum sea ice extent and total sea ice extent were 0.912×10^5 km^2/10 yr and 0.722×10^7 km^2/10 yr, respectively, from 2005 to 2013, both of which passed the significance test at the 0.05 level.

Keywords: climate change; climatology; sea ice; marginal sea; East Asia; observation

Citation: Zhang, L.; Ren, G.; Xu, M.; Meng, F.; Liao, R.; Liu, D.; Wang, M.; Jia, D. Spatiotemporal Patterns of Sea Ice Cover in the Marginal Seas of East Asia. *Atmosphere* **2023**, *14*, 207. https://doi.org/10.3390/atmos14020207

Academic Editor: Christof Lüpkes

Received: 3 December 2022
Revised: 10 January 2023
Accepted: 12 January 2023
Published: 19 January 2023

1. Introduction

The marginal seas of East Asia include the Bohai Sea, the Yellow Sea and the Sea of Japan. There are frequent economic and trade activities among the coastal countries in this sea area, and sea ice disaster is one of the major marine disasters in winter. The huge destructive power of sea ice seriously threatens marine transportation, offshore oil exploration, offshore engineering construction and marine fishery [1–3]. Under the background of global climate change, there are obvious interannual and interdecadal variations in sea ice in the marginal seas of East Asia in winter [4–11]. The research on the characteristics of intra-annual, interannual and long-term changes in the sea ice of the marginal seas of East Asia has important theoretical and practical significance. It will help to accurately and comprehensively reveal the temporal and spatial patterns of sea ice formation, development and ablation, and it will be useful for understanding the mechanism and impact of climate change in the northern region, improving the prediction and prevention of sea ice disaster in the marginal seas of East Asia.

The trend in the extent of sea ice in the marginal seas of East Asia has received extensive attention. Using a variety of satellite data, Wang et al. [12] analyzed the temporal and spatial distribution of sea ice in the Bohai Sea from 1996 to 2011, where the maximum value of the annual maximum sea ice extent exceeded 3.0×10^4 km^2, and the minimum value of the annual maximum sea ice extent was less than 1.0×10^4 km^2. Using moderate resolution imaging spectroradiometer (MODIS) sea ice data, Ouyang et al. [13] found that the average maximum sea ice extent in the Bohai Sea was $2.3 \times 10^4 \pm 0.8 \times 10^4$ km^2 during the period 2000–2016. Using the reconstructed daily sea ice extent data, Yu et al. [14] showed that the trends in the annual maximum sea ice extent and the annual average sea ice extent were $-0.33 \pm 0.18\%$ yr^{-1} and $-0.51 \pm 0.16\%$ yr^{-1} from 1958 to 2015 in the Bohai Sea. Based on the daily sea ice data of the National Snow and Ice Data Center (NSIDC) in the United States, Liu et al. [15] showed that the period with the largest sea ice extent in the Bohai Sea and Yellow Sea was from late January to late February each year. The interannual oscillation of the ice extent was obvious, showing a trend of first increasing and then decreasing over 2007–2018, and there was a significant negative correlation with the coastal temperature of the Bohai Sea and Yellow Sea during the same period. According to Ken and Takuya [16], the sea ice in the Sea of Japan is mainly located in the Tatar Strait in the northwest of the Sea of Japan and the coast of Siberia. The changes in the sea ice in the Sea of Japan are not consistent with global climate change, which may be affected by local climatic factors. The change in the sea ice extent in the marginal seas of East Asia may be closely related to the warming of Northeast Asia.

At present, studies on sea ice mainly use satellite remote data, including visible, infrared, passive microwave and active microwave remote sensing [8,17,18]. Visible remote sensing has a high resolution, but it cannot be applied for monitoring at night. In addition, clouds have a great influence on visible and infrared remote sensing [19]. The microwave data are not limited by day and night, are less affected by clouds and fog, and have high spatiotemporal continuity, but passive microwave remote sensing currently has a low resolution [20–22]. Active microwave remote sensing has a high resolution but has shortcomings such as a long revisit time and high cost [23]. Therefore, the visible, infrared, passive microwave, and active microwave sensors for monitoring sea ice have their own advantages and disadvantages. The multisource fusion data combine the advantages of the above sensor data to form high-resolution sea ice cover fusion data [24–26]. In the marginal seas of East Asia, high-resolution fusion sea ice data are used to conduct sea ice cover research, and the high-resolution continuous monitoring of sea ice in the marginal seas of East Asia can be realized, which can clearly and continuously display the variation characteristics of sea ice in the marginal seas of East Asia. At present, research on sea ice cover has yet to be carried out in the marginal seas of East Asia using high-resolution fusion sea ice data.

Based on the high-resolution daily sea ice cover data from 2005 to 2021, which was developed using multisource data, this study analyzed the temporal and spatial characteristics of the sea ice cover and sea ice extent in the marginal seas of East Asia and estimated the trends in the sea ice extent in the sea areas over the last 17 years.

2. Materials and Methods

The IMS (interactive multisensory snow and ice mapping system) sea ice cover data used in this study were from the National Ice Center (NIC) [27,28]. The spatial resolution of the IMS data is 4 km, and the time resolution is 1 day. The NIC sea ice products have an ideal spatiotemporal resolution and high confidence relative to other single-source sea ice extent data [29]. The NIC sea ice product contains daily sea ice cover data from March 2004 to June 2021, with a total of only 8 missing days.

The value of the sea ice extent (SIE) in this study was the sum of the extent of all ice pixels. The annual SIE was the sea ice cover from November of the previous year to May of the current year. For example, the SIE from November 2006 to May 2007 was defined as the annual SIE in 2007. For the missing sea ice data, the values of the previous day and the next day were used to fill in by means of linear interpolation. The maximum value of the SIE in the marginal seas of East Asia in a year is defined as the maximum sea ice extent (MSIE) in that year, and the total sea ice extent (TSIE) in that year is represented by the daily cumulative amount of the SIE. In the marginal seas of East Asia, there are only SIE values from November to May of the following year. The trend in the SIE was estimated using the least squares method, and the significance level of the linear trend was judged by the *t*-test method.

The study area (117–142.3° E, 35–52° N) was the Bohai Sea, the central and northern Yellow Sea, and the Sea of Japan (Figure 1). The six areas where sea ice frequently occurs include Liaodong Bay in the northern Bohai Sea, Bohai Bay in the western Bohai Sea, Laizhou Bay in the southern Bohai Sea, Korea Bay in the northern Yellow Sea, Peter the Great Bay in the western Sea of Japan, and the northern Sea of Japan. The latitudes of the northern Sea of Japan and the Laizhou Bay are the highest and the lowest, respectively.

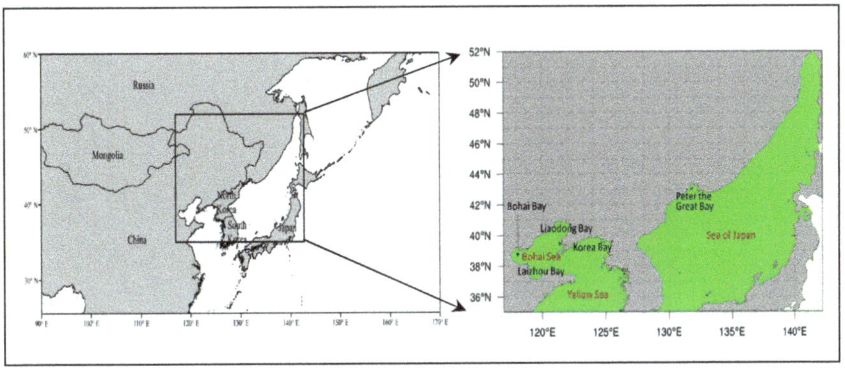

Figure 1. The study region.

3. Results

3.1. Temporal Characteristics of Sea Ice Extent

Table 1 shows the multiyear average first ice date and last ice date of the sea ice in the major sea areas of the marginal seas of East Asia from 2005 to 2021, as well as the occurrence time of the MSIE. There were obvious differences in the first and last ice dates of the sea ice in the different sea areas of the marginal seas of East Asia, but the time of the MSIE in each sea area was concentrated from late January to mid-February. As the latitude of the northern Sea of Japan is the highest, the average freezing time was the earliest in this area, which was 13 November. The average melting time of the sea ice was the latest in the northern Sea of Japan, which was May 16. The ice period lasted for approximately 184 days in the northern Sea of Japan, which was significantly longer than the other sea areas. Liaodong Bay, Korea Bay and Peter the Great Bay had similar first ice dates, with sea ice appearing in mid-to-late December, but the ice period in the Peter the Great Bay was longer than that of the other sea areas. The freezing times of Bohai Bay and Laizhou Bay were the latest, the melting time was the earliest, and the ice period was the shortest of all of the areas.

Table 1. The statistics for the first ice date, last ice date, the date of the MSIE and the ice period in the different regions of the marginal seas of East Asia from 2005 to 2021.

Region	First Ice Date	Last Ice Date	The Date of MSIE	Ice Period and Standard Deviation (Days)
Liaodong Bay	8 December	12 March	8 February	94 ± 19
Bohai Bay	19 December	6 March	29 January	77 ± 30
Laizhou Bay	25 December	2 March	28 January	67 ± 34
Korea Bay	10 December	10 March	9 February	90 ± 20
Peter the Great Bay	9 December	11 April	13 February	123 ± 27
Northern Sea of Japan	13 November	16 May	10 February	184 ± 7

MSIE: maximum sea ice extent.

3.2. Spatial Characteristics of Sea Ice Extent

Figure 2 shows the spatial distribution of the annual average duration of sea ice cover in the marginal seas of East Asia from 2005 to 2021. The area with the longest duration of sea ice cover was the northern Sea of Japan. The annual average duration of the sea ice cover was more than 100 days in most of the Sea of Japan. The areas with the shortest duration of sea ice cover were Laizhou Bay and Bohai Bay. The annual average duration of the sea ice cover was approximately 20 days in most of Laizhou Bay and Bohai Bay. The duration of the sea ice cover reached 60 days in most of Liaodong Bay and Korea Bay. The duration of the sea ice cover in the nearshore waters was significantly longer than that in the open sea, and it gradually decreased from the coast to the open sea. The main reasons for the spatial differences of the sea ice were that the latitude of the northern Sea of Japan is higher, and the temperature is lower in winter, while the latitude of the Bohai Sea and Yellow Sea is lower, and the temperature is relatively higher in winter.

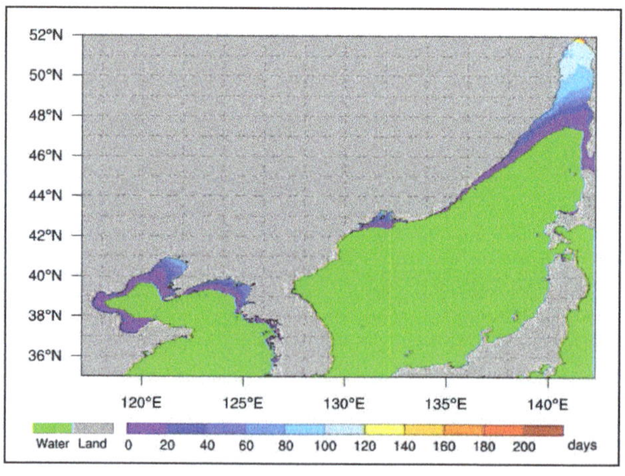

Figure 2. Annual average duration of sea ice cover in the marginal seas of East Asia from 2005 to 2021.

Large-scale sea ice appeared in 2013, and this year was selected as the typical year to analyze the intraseasonal variation characteristics of the sea ice in the whole marginal seas of East Asia (Figure 3). In the starting stage (Figure 3a), the sea ice first appeared in the northern Sea of Japan. In the increasing stage (Figure 3b), the sea ice appeared in the Bohai Sea, Yellow Sea and Peter the Great Bay, and the sea ice extent in the northern Sea of Japan continued to increase. In the peak stage (Figure 3c), the northwest coast of the Sea of Japan was covered by long and narrow sea ice, and there was a large amount of sea ice in Peter the Great Bay and on the coast of the Bohai Sea and northern Yellow Sea. In the decreasing stage (Figure 3d), the sea ice in the Yellow Sea and the Bohai Sea was mainly concentrated in Liaodong Bay and Korea Bay, and the sea ice in the Sea of Japan was distributed in the northern Sea of Japan and Peter the Great Bay. In the disappearing stage (Figure 3e), the extent of the sea ice in the areas with sea ice continuously decreased, and this was mainly limited to the northern Sea of Japan.

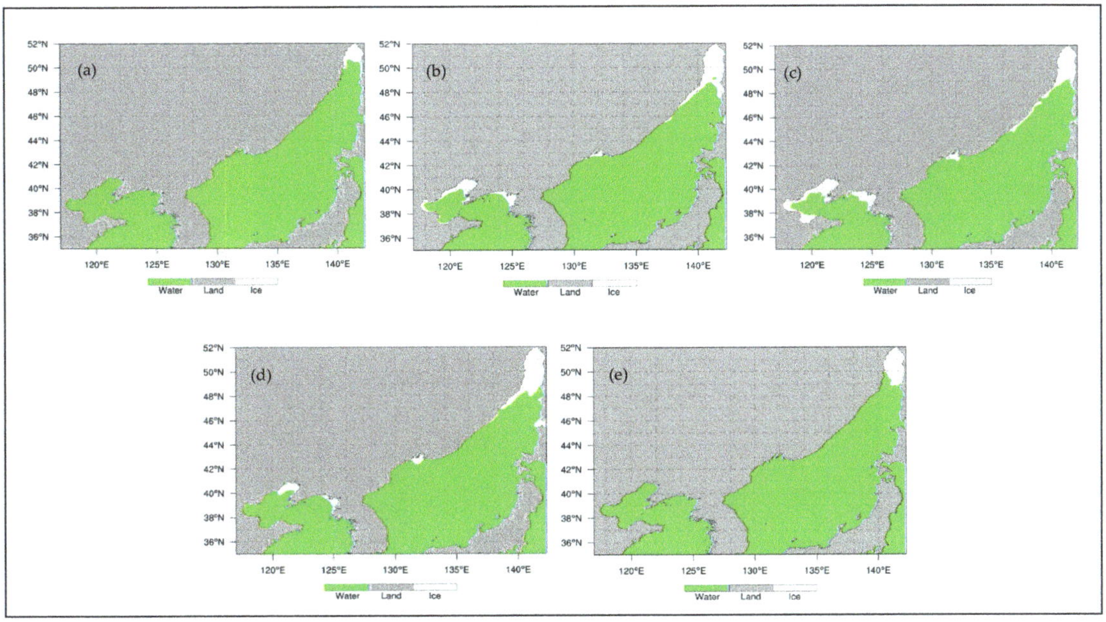

Figure 3. Spatial distribution of sea ice extents in the different stages of a start (**a**), increase (**b**), peak (**c**), decrease (**d**) and disappearance (**e**) in the marginal seas of East Asia.

3.3. Variation Characteristics of Sea Ice Extent

Figure 4 shows the intraseasonal variation of the daily mean sea ice extent in the Bohai Sea, Yellow Sea, Peter the Great Bay and the northern Sea of Japan from 2005 to 2021. The maximum sea ice extent occurred from late January to mid-February, when the sea ice extent in the seas reached the maximum value in one year. The melting speed of the sea ice was significantly faster than the freezing speed, resulting in the asymmetry of the daily sea ice extent variations in the Bohai Sea and Yellow Sea (Figure 4a). The freezing and melting times of the sea ice were similar the Peter the Great Bay and the northern Sea of Japan. The ice period in the northern Sea of Japan was significantly longer than that of the Bohai Sea, Yellow Sea and Peter the Great Bay.

Figure 5a shows the interannual fluctuations of the MSIE in the marginal seas of East Asia from 2005 to 2021. The interannual fluctuations of the MSIE were large, with a maximum value of 17.4×10^4 km^2 in 2016 and a minimum value of 7.6×10^4 km^2 in 2007. Figure 5b shows the changes in the total sea ice extent of the marginal seas of East Asia over the past 17 years. The maximum value of 11.7×10^6 km^2 occurred in 2013, and the minimum value of 5.42×10^6 km^2 occurred in 2006 and 2007. In the past 17 years, the minimum value was less than half of the maximum value.

From 2005 to 2021, the trends in the MSIE (Figure 5a) and the total sea ice extent (Figure 5b) in the marginal seas of East Asia were not significant, but there were significant differences in the trends in each of the 9 years before and after 2013. This shows the change characteristic of increasing first and then decreasing over the past 17 years. From 2005 to 2013, the MSIE (Figure 5a) and the total sea ice extent (Figure 5b) both showed increasing trends, which were 0.912×10^5 km^2/10 yr and 0.722×10^7 km^2/10 yr, respectively, and both passed the significance test at the 0.05 level. From 2013 to 2021, the MSIE (Figure 5a) and the total sea ice extent (Figure 5b) both showed decreasing trends of -0.409×10^5 km^2/10 yr and -0.370×10^7 km^2/10 yr, respectively, but neither passed the significance test at the 0.05 level. From Table 2, it can be seen that the total sea ice extent for each month from 2005 to 2013 showed an increasing trend, and the increase was most obvious in January and February, with the trends of 0.286×10^7 km^2/10 yr and 0.221×10^7 km^2/10 yr, both of which passed the significance test at the 0.05 level. The total sea ice extent in each month from 2013 to 2021 showed a decreasing trend, with the most obvious decrease in January at -0.124×10^7 km^2/10 yr, which passed the significance test at the 0.1 level.

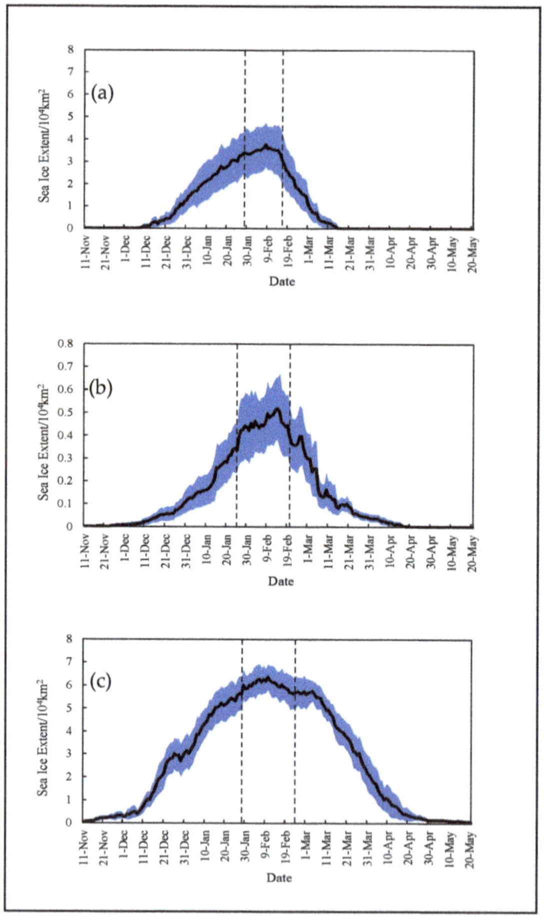

Figure 4. Daily mean sea ice extent (black, solid line) and standard deviation (blue area) in the Bohai Sea and Yellow Sea (**a**), Peter the Great Bay (**b**) and the northern Sea of Japan (**c**) from 2005 to 2021.

Table 2. Trends in the total sea ice extent in the marginal seas of East Asia (unit: 10^7 km^2/10 yr).

	Dec	Jan	Feb	Mar	Apr	Year
2005–2021	0.015	0.039	0.012	−0.005	−0.002	−0.058
2005–2013	0.098 **	0.286 **	0.221 **	0.058	0.056 **	0.722 **
2013–2021	−0.049	−0.124 *	−0.101	−0.058	−0.032	−0.370 *

* Passed the significance test at the 0.1 level; ** passed the significance test at the 0.05 level.

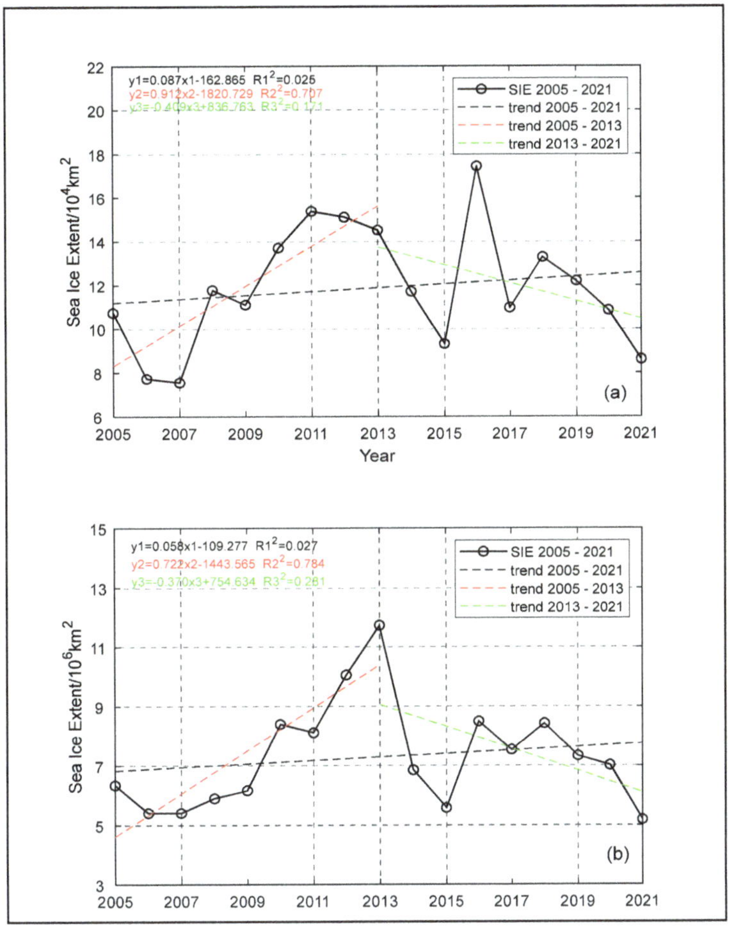

Figure 5. Annual maximum sea ice extent (**a**) and total sea ice extent (**b**) in the marginal seas of East Asia from 2005 to 2021.

Figure 6 shows the annual-pentad profile of the sea ice extent in the marginal seas of East Asia from 2005 to 2021. The years 2008, 2012, 2016 and 2018 can be classified as the unimodal type, with the peaks appearing from the 1st pentad in February to the 3rd pentad in February. There were two relatively obvious peaks in 2010, which were characterized by the sea ice extent increasing first, then slightly decreasing, and then increasing once again, with peaks appearing on the 4th pentad in January and the 3rd and 4th pentads in February, respectively. There was no obvious peak in the sea ice extent in 2006 and 2007. The sea ice extent of the marginal seas of East Asia was always relatively large from the 6th pentad in January to the 1st pentad in March, resulting in the total sea ice extent in 2013 being the maximum over the past 17 years.

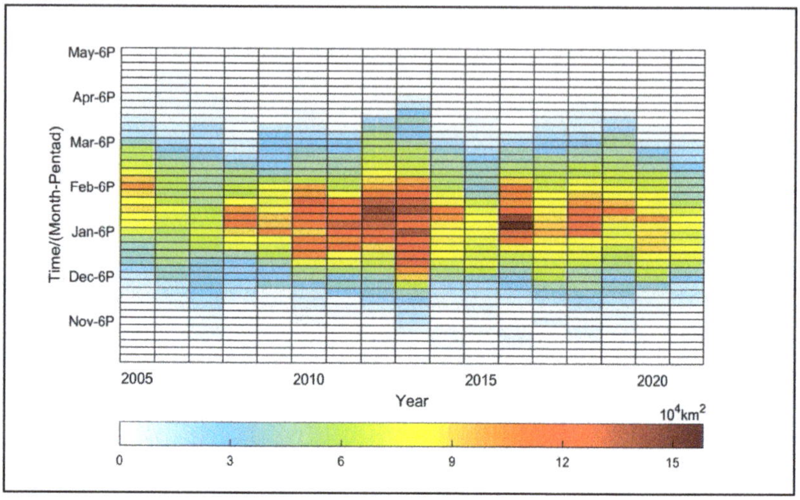

Figure 6. Sea ice extent of the marginal seas of East Asia in 2005–2021 (unit: 10^4 km^2).

4. Discussion

The research shows that the longest duration of sea ice cover in the marginal seas of East Asia was in the northern coast of the Sea of Japan, and the duration of the sea ice cover in the coastal waters of all regions was significantly longer than that in the open sea. The phenomenon that the duration of the sea ice decreased from the seashore to the deep sea in Liaodong Bay was also pointed out by Wang et al. [12] and Baoleerqimuge [30]. The trend in the sea ice extent in the Bohai Sea and Yellow Sea showed to increase first and then decrease from 2005 to 2021. Liu et al. [15] found that the sea ice extent increased at first and then reduced in the Yellow Sea and Bohai Sea in 2007–2018. The results are similar to previous studies [12,15,30].

The different studies used different kinds of data in determining the first and last ice dates of the sea ice in the marginal seas of East Asia. Due to the differences in resolution, temporal and spatial coverage, cloud cover and data processing methods among the different studies, the results of the first and last ice dates in the different studies are, to a certain extent, different [12,24,26,31].

Influenced by atmospheric and marine environments, ice conditions have obvious interannual and interdecadal variabilities in the marginal seas of East Asia [32]. The interannual variability of sea ice in this area is obviously related to atmospheric factors, such as the intensity of the East Asia winter monsoon and the atmospheric circulation at 500 hPa [33]. The formation and evolution of the sea ice are not only affected by atmospheric conditions but also restricted by factors such as ocean circulation conditions, sea temperature and salinity structure [6]. The marginal seas of East Asia are relatively closed, and the changes in the sea ice are also controlled by local climatic factors [16].

The long-term change in the sea ice extent in the marginal seas of East Asia may be related to the overall trend of climate warming in East Asia and North China [34]. However, the time scale of the data in this study was short, and climate warming in Northeast Asia slowed down during the analysis period [35], which may be the main reason for the insignificant long-term trends of some sea ice indicators. Obtaining a longer series of high-resolution sea ice extent data and discussing the response mechanism of the long-term changes and variations in the sea ice extent in the marginal seas of East Asia to climate warming and natural climate variability are work that needs to be strengthened in the future.

5. Conclusions

Using high-resolution IMS multisource sea ice fusion data, the climatological characteristics and the trends in the sea ice cover in the marginal seas of East Asia were analyzed from 2005 to 2021, and the following conclusions were obtained.

(1) There were obvious differences in the first and last ice dates of the sea ice in the different sea areas of the marginal seas of East Asia, but the date of the maximum sea ice extent was relatively concentrated from 2005 to 2021. As the latitude of the northern Sea of Japan is the highest, the earliest freezing time, the latest melting time and the longest ice period were in this area. Laizhou Bay and Bohai Bay had the latest freezing time, the earliest melting time and the shortest ice period. The occurrence time of the largest sea ice extent was relatively stable in the marginal seas of East Asia and mostly concentrated in late January to mid-February.

(2) There were obvious spatial differences in the duration of the sea ice cover in the marginal seas of East Asia. The duration of the sea ice cover gradually decreased from the high latitudes to low latitudes and from near-shore waters to open seas in the marginal seas of East Asia over the past 17 years. The area with the longest duration of sea ice cover was the northern Sea of Japan. The areas with the shortest duration of sea ice cover were Laizhou Bay and Bohai Bay. The annual average duration of the sea ice cover was more than 100 days in most of the Sea of Japan and approximately 20 days in most of Laizhou Bay and Bohai Bay, reaching 60 days in most of Liaodong Bay and Korea Bay. The duration of the sea ice cover in the coastal waters was significantly longer than that in the open seas, and there were high-value bands of duration of sea ice cover along the coastline.

(3) The melting speed was significantly faster than the freezing speed in Bohai Sea and Yellow Sea, resulting in the asymmetric changes in the daily sea ice extent in the Bohai Sea and Yellow Sea. According to the number of occurrences of sea ice extent peaks, the style of the sea ice extents could be divided into single peak, double peak and stable.

(4) Over the past 17 years, the maximum sea ice extent in the marginal seas of East Asia reached a maximum value of 17.4×10^4 km^2 in 2016 and a minimum value of 7.6×10^4 km^2 in 2007. The sea ice extent of the marginal seas of East Asia from the 6th pentad in January to the 1st pentad in March was relatively large, resulting in the maximum of the total sea ice extent in 2013. The maximum sea ice extent and the total sea ice extent over the past 17 years had obvious differences in the change trend for the first and second halves of the period, and they all showed the change characteristics of first increasing and then decreasing. From 2005 to 2013, the trends in the maximum sea ice extent and the total sea ice extent were, respectively, 0.912×10^5 km^2/10 yr and 0.722×10^7 km^2/10 yr, which passed the significance test at the 0.05 level.

Author Contributions: L.Z. analyzed the data, made the figures and wrote the paper; G.R., M.X., F.M., R.L. and D.L. provided scientific advice and helped improve the paper; M.W. and D.J. revised the language and figures. All authors have read and agreed to the published version of the manuscript.

Funding: This study was funded by the Collaborative Innovation Project of Meteorological Science and Technology in the Bohai Rim (QYXM202101), Research on Key Technologies of Sea Visibility Grid Prediction Based on Multisource Fusion and Machine Learning in the Yellow Sea and Bohai Sea (CXFZ2022J028), the National Key Research and Development Program (2018YFA0605603) and the Transverse Item-Research Project of the Chinese Academy of Meteorological Sciences (Grant No. IN_JS_2022031).

Institutional Review Board Statement: Not applicable.

Informed Consent Statement: Not applicable.

Data Availability Statement: The data used in this study are publicly available via the National Snow and Ice Data Center https://nsidc.org/data/g02156 (Recently accessed date: 30 March 2022).

Acknowledgments: The authors acknowledge, with deep gratitude, the contribution of the anonymous reviewers whose time and effort in providing comments has improved this paper substantially.

Conflicts of Interest: The authors declare that there are no conflict of interest regarding the publication of this paper.

References

1. Wang, X.Y.; Yuan, B.K.; Shang, J.; Ge, L. Disaster of sea ice and defense Countermeasures in the Bohai Sea and the Yellow Sea. *Coast. Eng.* **2011**, *30*, 46–55.
2. Gu, W.; Liu, C.Y.; Yuan, S.; Li, N.; Chao, J.L.; Li, L.T.; Xu, Y. Spatial distribution characteristics of sea-ice-hazard risk in Bohai, China. *Ann. Glaciol.* **2013**, *54*, 73–79. [CrossRef]
3. Zhang, D.Y.; Xu, N.; Yue, Q.J.; Liu, D. Sea Ice Problems in Bohai Bay Oil and Gas Exploitation. *J. Coastal. Res.* **2015**, *73*, 676–680. [CrossRef]
4. IPCC. Summary for policymakers. In *Climate Change 2021: The Physical Science Basis*; Masson-Delmotte, V., Zhai, P., Pirani, A., Connors, S.L., Eds.; Working Group I Contribution to the Sixth Assessment Report of the Intergovernmental Panel on Climate Change; Cambridge University Press: Cambridge, UK, 2021; pp. 1–31.
5. Park, K.A.; Kim, K.; Cornillon, P.C.; Chung, J.Y. Relationship between satellite-observed cold water along the Primorye coast and sea ice in the East Sea (the Sea of Japan). *Geophys. Res. Lett.* **2006**, *33*, L10602. [CrossRef]
6. Liu, Y.; Liu, Q.Z.; Sui, J.P.; Song, J.X.; Bai, S. The respond of ice for the Bohai Sea and the Huanghai Sea with the general circulation and the climate change in winters. *Acta Oceanol. Sin.* **2013**, *35*, 18–27.
7. Nihashi, S.; Ohshima, K.I.; Saitoh, S.I. Sea-ice production in the northern Japan Sea. *Deep. Sea Res.* **2017**, *127*, 65–76. [CrossRef]
8. Yuan, S.; Liu, C.Y.; Liu, X.Q.; Chen, Y.; Zhang, Y.J. Research advances in remote sensing monitoring of sea ice in the Bohai Sea. *Earth Sci. Inform.* **2021**, *14*, 1729–1743. [CrossRef]
9. Zhang, L.; Xu, B.; Liao, Z.H.; Shi, C.X.; Xu, M.; Sun, M.L. Development experiment of a global multi-source sea ice concentration fusion dataset. *Acta Meteorol. Sin.* **2021**, *79*, 703–715.
10. Matveeva, T.A.; Semenov, V.A. Regional Features of the Arctic Sea Ice Area Changes in 2000–2019 versus 1979–1999 Periods. *Atmosphere* **2022**, *13*, 1434. [CrossRef]
11. Jiao, Y.; Zhang, Y.Q.; Hu, P. Can Arctic Sea Ice Influence the Extremely Cold Days and Nights in Winter over the Tibetan Plateau. *Atmosphere* **2022**, *13*, 246. [CrossRef]
12. Wang, M.; Wu, S.L.; Zheng, W.; Zhao, C.H.; Liu, C. Temporal-Spatial Distribution of Bohai Sea Sea Ice in Long-Time Series and Its Correlation with Air Temperature. *Meteor. Mon.* **2016**, *42*, 1237–1244.
13. Ouyang, L.X.; Hui, F.M.; Zhu, L.X.; Cheng, X.; Cheng, B.; Shokr, M.; Zhao, J.C.; Ding, M.H.; Zeng, T. The spatiotemporal patterns of sea ice in the Bohai Sea during the winter seasons of 2000–2016. *Int. J. Digit. Earth.* **2019**, *12*, 893–909. [CrossRef]
14. Yan, Y.; Uotila, P.; Huang, K.Y.; Gu, W. Variability of sea ice area in the Bohai Sea from 1958 to 2015. *Sci. Total. Environ.* **2020**, *709*, 136164. [CrossRef] [PubMed]
15. Liu, C.; Cao, X.C.; Yin, Z.H.; Geng, D.P. Climatic characteristics of sea ice extent and its response to meteorological factors in the Yellow Sea and the Bohai Sea during 2007–2018. *Mar. Sci. Bull.* **2019**, *38*, 173–178.
16. Ken, I.; Takuya, I. Millennial-scale fluctuations in seasonal sea-ice and deep-water formation in the Japan Sea during the late Quaternary. *Palaeogeogr. Palaeoclimatol. Palaeoecol.* **2007**, *247*, 131–143.
17. Wang, R.F.; Huang, D.; Zhang, X.; Wei, P. Combined pattern matching and feature tracking for Bohai Sea ice drift detection using Gaofen-4 imagery. *Int. J. Remote Sens.* **2020**, *41*, 7486–7508. [CrossRef]
18. Lee, H.S.; Lee, K.S. Capability of geostationary satellite imagery for sea ice monitoring in the Bohai and Yellow seas. *J. Mar. Sci. Technol.* **2016**, *24*, 1129–1135.
19. Shi, W.; Wang, M.H. Sea ice properties in the Bohai Sea measured by MODIS-Aqua: 2. Study of sea ice seasonal and interannual variability. *J. Mar. Syst.* **2012**, *95*, 32–40. [CrossRef]
20. Comiso, J.C.; Nishio, F. Trends in the sea ice cover using enhanced and compatible AMSR-E, SSM/I, and SMMR data. *J. Geophys. Res.* **2008**, *113*, C02S07. [CrossRef]
21. Cavalieri, D.J.; Parkinson, C.L. Arctic sea ice variability and trends, 1979–2010. *Cryosphere* **2012**, *6*, 881–889. [CrossRef]
22. Lavergne, T.; Sørensen, A.M.; Kern, S.; Tonboe, R.; Notz, D.; Aaboe, S.; Bell, L.; Dybkjær, G.; Eastwood, S.; Gabarro, C.; et al. Version 2 of the EUMETSAT OSI SAF and ESA CCI sea-ice concentration climate data records. *Cryosphere* **2019**, *13*, 49–78. [CrossRef]
23. Su, H.; Ji, B.W.; Wang, Y.P. Sea Ice Extent Detection in the Bohai Sea Using Sentinel-3 OLCI Data. *Remote Sens.* **2019**, *11*, 2436. [CrossRef]
24. Xie, Y.; Koch, S.; Mcginley, J.; Albers, S.; Bieringer, P.E.; Wolfson, M.; Chan, M. A space-time multiscale analysis system: A sequential variational analysis approach. *Mon. Weather. Rev.* **2011**, *139*, 1224–1240. [CrossRef]
25. Shen, Y.; Zhao, P.; Pan, Y.; Yu, J.J. A high spatiotemporal gauge-satellite merged precipitation analysis over China. *J. Geophys. Res. Atmos.* **2014**, *119*, 3063–3075. [CrossRef]
26. Shi, C.X.; Pan, Y.; Gu, J.X.; Xu, B.; Han, S.; Zhu, Z.; Zhang, L.; Sun, S.; Jiang, Z.J. A review of multi-source meteorological data fusion products. *Acta. Meteor. Sin.* **2019**, *77*, 774–783.
27. Ramsay, B.H. The interactive multisensor snow and ice mapping system. *Hydrol. Process.* **1998**, *12*, 1537–1546. [CrossRef]
28. Helfrich, S.; McNamara, D.; Ramsay, B.; Baldwin, T.; Kasheta, T. Enhancements to, and forthcoming developments in the interactive multisensor snow and ice mapping system (IMS). *Hydrol. Process.* **2007**, *21*, 1576–1586. [CrossRef]

29. Meier, W.N.; Fetterer, F.; Stewart, J.S.; Helfrich, S. How do sea-ice concentrations from operational data compare with passive microwave estimates? Implications for improved model evaluations and forecasting. *Ann. Glaciol.* **2015**, *56*, 332–340. [CrossRef]
30. Baoleerqimuge. *The Long Term Change Characteristics of China Offshore Sea Surface Temperature and Its Effect on Coastal Climate*; Chinese Academy of Meteorological Sciences: Beijing, China, 2014; pp. 1–99.
31. Liu, Y.; Wu, H.D. Sea ice in the Bohai sea and the northern yellow sea. *Mar. Forecast* **2017**, *34*, 94–101.
32. Xiao, W.X. *Characteristics of Sea Ice in Northeast Asia and its Reasons*; Chinese Academy of Meteorological Sciences: Beijing, China, 2017; pp. 1–69.
33. Tang, M.N.; Hong, J.L.; Liu, Y.; Sui, J.P.; Zhao, Q. Statistical Analysis of Climatic factors impacting on the Bohai Sea ice. *Mar. Sci. Bull.* **2015**, *34*, 152–157.
34. Wen, K.M.; Ren, G.Y.; Li, J.; Zhang, A.Y.; Ren, Y.Y.; Sun, X.B.; Zhou, Y.Q. Recent surface air temperature change over mainland China based on an urbanization-bias adjusted dataset. *J. Clim.* **2019**, *32*, 2691–2705. [CrossRef]
35. Sun, X.B.; Ren, G.Y.; Ren, Y.Y.; Fang, Y.H.; Liu, Y.L.; Xue, X.Y.; Zhang, P.F. A remarkable climate warming hiatus over Northeast China since 1998. *Theor. Appl. Climatol.* **2018**, *133*, 579–594. [CrossRef]

atmosphere

MDPI

Article

Potential Effect of Air Pollution on the Urban Traffic Vitality: A Case Study of Nanjing, China

Yang Cao [1], Hao Wu [2,*], Hongbin Wang [2], Duanyang Liu [2] and Shuqi Yan [2]

1 Jiangsu Province City Layout Design Research Institute, Nanjing 210036, China
2 Key Laboratory of Transportation Meteorology of China Meteorological Administration, Nanjing Joint Institute for Atmospheric Sciences, Nanjing 210041, China
* Correspondence: haowu@cma.gov.cn

Abstract: Studies on the vitality of urban residents' daily commuting and air pollution are scarce. Based on the cell phone mobile signaling data, urban air quality observation data, and urban transportation infrastructure environment data of Nanjing in 2019, and through the panel regression model and the standard deviation ellipse analysis (SDE) to measure the impact of air pollution on residents' daily traffic vitality, we construct the survey panel matrix data system with streets as spatial units. Through SDE and panel regression model analysis, we measured the restraining effect of air pollution on the traffic vitality. The scope of the traffic vitality area SDE was found to shrink as the air quality index (AQI) increases. The study found three main characteristics: (1) Under different transportation models and different location conditions, there are obvious differences in traffic vitality. The entire city presents a trend of "northeast-southwest" axial expansion in the spatial pattern of the traffic vitality. Compared with the urban core area, the traffic vitality of residents in the north-south areas of Nanjing's periphery has declined significantly. (2) The inhibitory effect of air pollution on public traffic vitality and self-driving traffic vitality are differences. Approximately one-tenth of traffic activities may be inhibited by air pollution. The weakening of traffic vitality greatly reduces the city's ability to attract and gather people, materials, and resources. (3) The inhibitory effect of air pollution on traffic vitality is heterogeneous under different transportation infrastructure environments. The higher the public transportation station density and public transportation frequency of the street, the more obvious the suppression effect of air pollution. The higher the parking density, station accessibility, road intersections density, and transportation facility diversity, the lower the suppression effect of air pollution. This study elucidates the relationship among air pollution, the transportation infrastructure environment, and the traffic vitality, and provides significant guidelines for optimizing the organization of elements in the transportation infrastructure environment, thereby mitigating the inhibitory effect of air pollution on traffic vitality.

Keywords: air pollution; traffic vitality; built environment; spatial correlation; spatial lag model; phone signaling data; air quality; behavioral habits; activity density; population distribution; land use mix

Citation: Cao, Y.; Wu, H.; Wang, H.; Liu, D.; Yan, S. Potential Effect of Air Pollution on the Urban Traffic Vitality: A Case Study of Nanjing, China. *Atmosphere* **2022**, *13*, 1592. https://doi.org/10.3390/atmos13101592

Academic Editor: Ashok Luhar

Received: 25 August 2022
Accepted: 26 September 2022
Published: 29 September 2022

1. Introduction

The traffic vitality of residents in the city directly reflects the attractiveness and diversity of the city, which is embodied in the intensity and type of resident activities in the urban space [1–3]. Improving the traffic vitality is conducive to improving the quality of life of residents and is significant to the sustainable development of the city [4,5]. In urban transportation planning, unanimous attention has been paid to the traffic vitality, particularly the impact of the built environment on the traffic vitality [6,7]. However, researchers have found it difficult to accurately measure the traffic vitality. Past studies are mostly based on observational surveys in small-scale spaces (such as streets) or use short-term (such as 1 week) resident activity logs for discussion. Describing commuting vitality distribution and change characteristics for large-scale spaces (such as the entire city) and long-term periods (such as spanning several weeks) is difficult [8,9]. Existing studies have identified the spatial displacement of residents over a period of time as traffic travel activities [10]. The frequency and density of traffic trips represent traffic vitality. We identify the people whose location changes in a relatively short period of time as the traffic travel crowd. We take the street as a unit to measure the density of traffic people per unit area as traffic vitality. Therefore, daily traffic vitality refers to the daily traffic density of residents in the street unit. Generally speaking, traffic behavior mainly refers to the traffic travel activities of individual residents, while traffic vitality is used to describe the frequency and density of group traffic travel at the medium and macro spatial level (such as streets and larger spatial scales). Following the continuous development of information and communications technology (ICT) in recent years, mobile phones record the activities and travel behaviors of residents in all aspects of daily life. These geographically attributed location big data are real-time, continuous, and accurate among other characteristics, making them a suitable data source for the study of residents' transportation activities [11]. Past studies collected mobile phone signaling data [12,13], LBS positioning data [14], GPS data [15,16], and social media sign-in data (such as Weibo and Twitter) [17–20] among others to measure residents' commuting activities. The specific spatial dimensions involve three levels of streets, cities, and regions; the time dimension is accurate to the hour as the unit; and the intensity of commuting activities and temporal and spatial characteristics are recorded [21–24].

Atmospheric particles have multiple impacts on human health and the environment. Studies have shown that PM10 adversely affects human health and increases mortality; while fine particulate PM2.5 and ultrafine particles are at higher risk [24–26]. At the same time, there is a correlation between air pollution and traffic vitality. Atmospheric particles can absorb and scatter solar radiation, and particulate matter greatly reduces the visibility of surface traffic [27]. In addition, studies have shown that there are positive and negative effects between transportation infrastructure and air pollution. On the one hand, with the continuous improvement of transportation infrastructure, the transportation structure is increasingly optimized. A new generation of transportation infrastructure facilitates low-energy, low-polluting modes of transportation. A developed urban transportation system can significantly reduce traffic congestion, shorten the waiting time caused by road congestion, and increase the frequency of traffic travel. In addition, good transportation infrastructure can also reduce the level of exhaust emissions when vehicles are idling, thereby improving the overall level of urban haze pollution. On the other hand, from the perspective of the utilization of transportation infrastructure, the improvement of transportation infrastructure conditions will increase the scale of urban transportation. The improvement of transportation infrastructure will increase the daily commuting frequency and vitality of residents to a certain extent, which is conducive to the formation of more commuting needs and longer commuting distances [28].

In the past three decades, China's urbanization level has developed rapidly, and the urban population agglomeration also enhanced the traffic vitality. However, industrialization and the rapid development of motor vehicles have caused increasingly severe air pollution problems. Smog, PM2.5, and air quality are common in government and media reports and have become some of the main limitations of the sustainable development of cities [29]. Traffic sources are significant to urban air quality as they could impact the air quality along highways and streets. For example, black carbon PM is emitted primarily by traffic sources and affected the air pollutants' distribution [30–32]. Presently, urban geographers mainly focus on the temporal and spatial evolution characteristics and driving factors of air pollution, the impact of air pollution on the physical and mental health of residents, the differences between residents in different regions and different socioeconomic backgrounds, and the governance of air pollution areas [33]. At the same time, certain scholars focus on urban residents' exposure to air pollution, and use GPS, portable environmental monitors, and other tools to determine the impact of air pollution on residents' travel (such as travel trajectories, traffic ways) [34,35]. These studies have shown that following the popularization of health awareness, air quality affects residents' activities in urban spaces (whether to go out for activities, travel modes, and choice of location and time of activities), and thereby affecting residents' traffic vitality.

Overall, current empirical research on the impact of air pollution on the traffic vitality is lacking. Owing to the development of information technology in recent years, the trajectory of residents' activities can be more accurately determined. A small number of studies have collected micro-blog sign-in data and used the city as a spatial unit to reveal that air pollution inhibits residents' activities (sign-in scale) [36], thereby affecting their happiness (emotions expressed in the sign-in text) [37]. In addition, certain scholars have focused on the impact of air pollution on specific types of activities, such as determining whether air pollution will inhibit residents from eating out based on changes in the number of online reviews on food and beverages [38]. However, these studies did not clearly distinguish and quantify related factors such as air pollution, built environment, and traffic vitality. Past studies have not analyzed the differential impact of air pollution on traffic activity under different transportation infrastructure environments. On the one hand, the studies that consider cities as the research unit disregard the spatial imbalance of air pollution and mobility in large cities [39]. On the other hand, current researchers have difficulty measuring the impact mechanism of built environment factors such as density, mixing degree, and location (distance) on air pollution's inhibition of traffic travel activities.

The differences among the inhibitory effects of air pollution on the traffic vitality vary for cities at different stages of social and economic development [36]. On the other hand, the built environment significantly influences residents' activities. The inhibitory effect of air pollution on the traffic vitality may be heterogeneous in the built environment. For example, although air pollution generally reduces residents' willingness to go out, in places where the built environment is relatively attractive, the reduction in residents' willingness to travel may be relatively low. To better measure the impact of built environment elements on the traffic vitality in an air-polluted environment, an analysis of the micro- and medium-scale space within the city is required [37]. In addition, to control the impact of hidden variable factors that may exist in the observation on traffic travel, it is necessary to obtain long-term panel data to improve the accuracy of model effect analysis [38]. Residents with different socioeconomic backgrounds live in different environments, and different residents have different perceptions of air pollution and outdoor activity habits [39]; this disparity exacerbates environmental injustice [40].

In recent years, the technical methods for measuring residents' daily traffic travel have constantly been updated. Following the popularization of mobile communication technology, local and international researchers use mobile phone signaling data to perform applied research on the urban center system, job–resident relationship analysis, and urban overall planning evaluation, and implement the urban system [41–43]. Mobile phone signaling data mainly extract the information exchange data and time stamp information between the base station and the mobile phone terminal in the mobile communication system to determine the spatial location and status of the phone user [44]. The identification of residents' travel behavior based on the travel laws of phone users has the advantages of presenting a large sample size, balanced sampling distribution, high spatial accuracy, and robust data timeliness. Scholars measure the built environment based on data sources such as points of interest (POI) and Open Street Maps. The specific influencing factors include urban structure, population density, street density, land use mix, and location [45,46]. These studies show that big data with geographic location attributes, such as mobile phone signaling check-in data, can reflect the daily commuting activities of residents in real time [47].

Based on the location data of mobile phone signaling, the urban air quality observation data, and the urban built environment data in Nanjing, 2019, this study comprehensively constructs the panel survey matrix data with the street as the spatial unit and the day as the time unit. The impact of air pollution on residents' daily traffic vitality is quantified through the panel regression model and standard deviation ellipse (SDE) analysis. According to existing research, people have expectations about the time spent on different types of travel activities. Generally speaking, in the urban road environment, the walking speed is 4–5 km/h, the bicycle speed is 15–20 km/h, the car speed is 60–80 km/h, the bus speed is 40–80 km/h, and the subway speed is 60–120 km/h [48]. Residents choose the exact traffic travel mode, and within a fixed travel time period, they will get a stable traffic travel range. Studies have shown that people's choice of long-distance transportation activities is affected by many factors, including atmospheric environment and urban built environment [49]. Different from the necessary activities within the 15 min daily life circle of residents, the traffic vitality pays more attention to the traffic travel behavior with a speed of more than 20 km/h and a travel time of more than 15 min. This paper takes 15 min as the minimum time unit and calculates the average displacement distance of mobile phone signaling location data within the urban area of Nanjing as the minimum activity radius of residents' travel. On this basis, this paper uses the standard deviation ellipse analysis model to measure the core range and direction distribution of large-scale residents' daily traffic travel. This study elucidates the relationships among air pollution, the built environment, and the traffic vitality through the statistical analysis of air pollution panel data at the street level. It also provides a theoretical basis for city managers to optimize the built environment to reduce the inhibitory effect of air pollution on the traffic vitality.

2. Data and Methods

2.1. Data Collection

2.1.1. Mobile Phone Signaling Data

This study regards the street as the basic spatial unit and the day as the basic time unit for investigation. Specifically, mobile phone signaling data are used to measure the intensity of residents' commuting connections to reflect the traffic vitality [37–39]. In 2019, we obtained 60 days of mobile phone signaling data for 2 consecutive months from Chinese mobile operators in Nanjing, Jiangsu Province (Figure 1). On the one hand, the data provider had deleted private information such as the resident's name, age, work unit, and residential address from the original data submitted. On the other hand, we focused on analyzing the change characteristics of large-scale residents' traffic travel location and did not undertake in-depth analysis of individual residents' information. We were not involved in the description of individual private information of residents. We have supplemented the characteristics of Nanjing's social environment in this section and the details are as follows: Nanjing is the capital city of Jiangsu Province, with a land area of 6587.02 km^2. The urban area is 868.28 km^2. In 2019, the population of Nanjing was 9.282 million. There are 11 subway lines, 8395 buses, and more than 3 million cars in Nanjing. According to the data in the report "2019 Nanjing Environmental Status Bulletin", the main source of air pollutants in Nanjing is industrial production, not motor vehicle emissions [45]. Therefore, the built environment factors have a greater impact on the air quality in Nanjing. According to the Bulletin on traffic transportation in Nanjing in 2016, the residents transported by private cars take a percentage of 11.89% and those transported by public ways were 26.86% [46].

First, the user's residence and work place, and commuters were identified based on the staying time and recurrence rate of mobile phone users in different urban spaces. We determined the residents' movement trajectories based on cell phone signaling locations and timestamps. In addition, we judged land use attributes based on the information of points of interest on the network electronic map. Combining residents' travel trajectories and land use attributes, we can determine the types of residents' daily activities. Specifically, we collected cell phone signaling data between 22:00 at night and 5:00 a.m. the next day. Residents were generally sleeping during this time period, and we then confirmed the place of residence. Additionally, we collected cell phone signaling data between 10 a.m. and 5 p.m. We combined it with industrial land, commercial land, public service land, etc., to determine where residents work and the activity types. In addition, we took 15 min as the minimum time unit to further measure the movement range of the resident's mobile phone signaling position during this period. We determined the residents' traffic pattern.

Second, considering limitations such as the sample representativeness of mobile phone signaling location data (significant difference between the signaling check-in frequency of active commuter users and inactive users), data cleaning, and integration were carried out through methods reported in past studies [32,39]. Specifically, cell phone signaling records outside the scope of Nanjing were excluded. For multiple check-ins at a fixed location over a long period, this study records the value of traffic vitality as 1. Furthermore, according to the land development attributes of the user's place of residence, this study divides the urban functional space into seven types of labels, including workplaces, residences, public transportation, public services, leisure and entertainment, and tourist destinations [36].

Based on mobile phone signaling data, the working population in Nanjing accounts for 43.2% of the total population. The commuter population was compared with the most recent census data in Nanjing, and the consistency test was carried out in the city street unit. The two data results were linearly correlated, and the model results passed the 99% confidence interval test, with an R^2 value of 0.86, which indicated strong correlation. The results show that there is no significant difference between the two sets of data and predicting the commuter population based on mobile phone signaling data is reliable. Regardless of air pollution, residents have to commute to work; however, transportation for other activities is greatly affected by residents' subjective wishes. Therefore, this study focuses on the activities of commuters in five other functional spaces other than the work place and residence. These five activity tags accounted for 62.45% of all check-in records. The data reflect that in addition to the daily commute to work, more than half of the residents access urban public services, leisure, tourism, and other functions through daily commuting [22,36]. Finally, this study obtained more than 5.63 million mobile phone signaling records in the five functional spaces mentioned above. This study calculated the sign-in density of residents' activities in Nanjing, comprising 95 streets of 11 jurisdictions in 2019. The average daily check-in density was 17.9 times per 10 km^2.

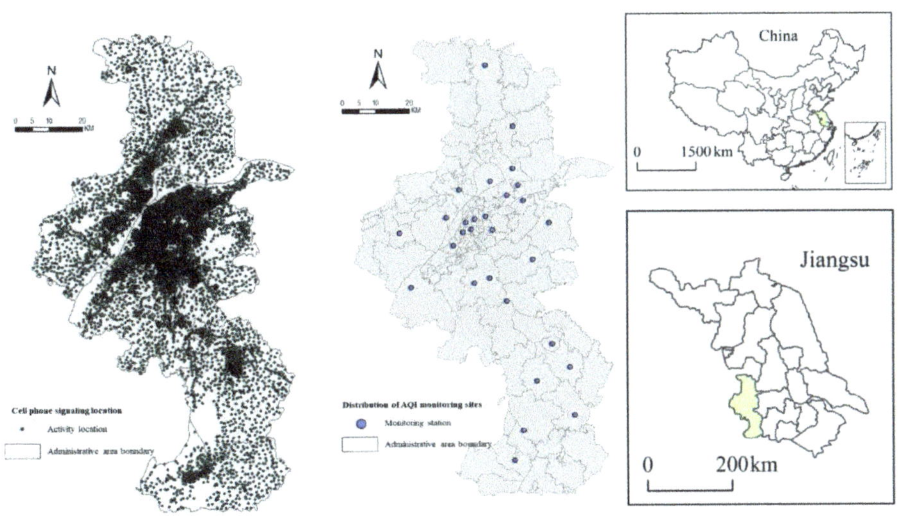

(a) (b) (c)

Figure 1. Distributions of mobile phone sign-in locations (**a**), and AQI monitoring stations (**b**) and the case site location (**c**).

2.1.2. Air Pollution and Meteorological Data

Owing to the concern regarding air pollution, each city has set up air monitoring stations, and declared the city's AQI and the concentrations of 6 types of pollutants ($PM_{2.5}$, PM_{10}, CO, NO_2, O_3, SO_2) in real time. The standard of the AQI is the Technical Regulation on Ambient Air Quality Index of China, which is a national standard. It can be found on the website of the Ministry of Ecology and Environment of the People's Republic of China.

The data acquisition time was from 20 October 2019 to 20 December 2019, for 60 consecutive days, including 17 non-working days and 43 working days. This time period was the alternating period of autumn and winter, and the AQI changes obviously, which can support the research on the influence of AQI concentration changes on traffic vitality in this paper. This paper focuses on the general impact of air pollution on the urban traffic vitality. Therefore, in terms of data collection time, we avoided traditional holidays such as National Day, Labor Day, and Spring Festival, and tried our best to avoid the unstable impact of the high traffic density during holidays on the research results. Research data were obtained from the daily update data of the China Environment Ministry, which contain 29 air monitoring stations in Nanjing. This study continuously collected the daily AQI and 6 types of pollutant concentration data through the 29 air monitoring stations in 2019. We selected 342 observation days, which basically covered the meteorological observation data of a whole year. We obtained daily real-time meteorological observation data through the website of the Ministry of Environment of China (http://www.cnemc.cn/, accessed on 20 December 2019). We obtained multiple sets of meteorological observation data on each observation day, averaging 5–6 times a day, accumulatively obtained 1890 wind speed, temperature, and precipitation observation data, and correspondingly obtained 1890 AQI data. Studies have shown a significant high correlation between the AQI and the concentration of six types of pollutants [32]. Therefore, this study used the AQI data of each air monitoring station to represent the air pollution in the streets. According to Chinese national standards, air quality is divided into six levels: excellent (AQI $\leq$ 50), good (50 < AQI $\leq$ 100), light pollution (100 < AQI $\leq$ 150), moderate pollution (150 < AQI $\leq$ 200). As shown in Table 1, the average daily AQI of Nanjing in 2019 was 61.75 and the air quality was mainly excellent and moderately polluted. In addition, significant differences were observed among the AQI values of each monitoring station (F = 7.982, *p*-value = 0.000), indicating significant differences in air quality in various streets of Nanjing. According to our previous work [50–52], the stable boundary layer with less precipitation and high air pollutants emissions works together to make winter the most polluted season [53]. In the spring and summer, the enhanced convective activity of the atmosphere and increased precipitation created optimal weather conditions for diffusion, which was also beneficial for the removal of air pollutants. The dense population, large number of motor vehicles, industrial production, and human activities in the urban area also result in the urban area being more polluted than the suburbs [54,55].

Table 1. Variable definitions and summary statistics.

Variable		Description	Observation	Min	Max	Mean	Variance
Explanatory variables (air quality)	AQI	Daily Air Quality Index	1890	20.08	61.75	27.44	161.19
Explanatory variables (transportation facilities)	Parking density	$\text{paking density} = \frac{\text{parking numbers}}{\text{Street area }(\text{km}^2)}$	132	0.49	15.47	15.27	82.88
	Station accessibility	$\text{station walkability} = \frac{\text{walking distance}}{\text{Street area }(\text{km}^2)}$	132	0.06	14.73	17.84	81.74
	Metro/Bus station density	$\text{station density} = \frac{\text{subway/bus stations}}{\text{Street area }(\text{km}^2)}$	132	0.00	0.71	1.17	6.26
	Transportation facility diversity	$X_i = \frac{\text{Area of Type i transportaion facility }(\text{km}^2)}{\text{Total land area }(\text{km}^2)}$ $P_i = \frac{X_i}{\sum_{i=1}^{n} X_i}$ $\text{Facility Diversity Index} = \frac{(-1)\left(\sum_{i=1}^{n} P_i \ln(P_i)\right)}{\ln(n)},\ n = 6$	132	0.27	0.68	0.09	0.83
	Road intersection density	$\text{Road intersection dense} = \frac{\text{street road intersections}}{\text{Street area }(\text{km}^2)}$	132	0.00	12.74	14.82	63.79

Table 1. *Cont.*

Variable		Description	Observation	Min	Max	Mean	Variance
Dependent variable	Traffic vitallity intensity	Activity intensity $= \frac{\text{Daily signaling check-in}}{\text{Street area } (10\ \text{km}^2)}$	49,831	0.00	17.83	50.58	3014.00
Control variable (weather)	Precipitation	Daily precipitation, dummy variable: 1 = heavy rain and above, 0 = other	1890	0	0.02	0.12	1
	Wind speed	Daily average wind speed (m/s)	1890	0.45	1.72	0.76	6.82
	Temperature	Daily average temperature (°C)	1890	6.94	22.65	5.32	32.76
Control variable (date attributes)	Date	Non-working days, dummy variables: 1 = non-working days, 0 = working days	342	0	0.27	0.38	1

Note: Transportation facility diversity index are closely related to resident traffic activities, including bus stops, subway stations, bicycle stops, car parking spaces, bus parking stations, gas stations, etc.

Studies have shown that meteorological conditions affect the choice of residents' activities [35]. The China Meteorological Administration updates the data daily. This study collected data on the daily average wind speed, average temperature, and rainfall in 2019 released by 29 meteorological monitoring stations in Nanjing. This necessitates control of the impact of meteorological and environmental factors when measuring the air pollution impact on traffic vitality. We regarded heavy and extra heavy rain (precipitation $\geq$60 mm/d) are as extreme weather conditions in cities. This situation is likely to cause urban waterlogging and affect residents' traffic activities [39] and was regarded as a dummy variable in the model in this study (Table 1). Based on the monitoring data of each meteorological station, we take the average recorded data of the meteorological stations in the street as the street meteorological measurement result. Based on the results of descriptive statistical analysis, we can find that there are significant differences in temperature, humidity and wind speed among streets (F = 33.72, p-value = 0.000; F = 175.83, p-value = 0.000), but no significant differences in rainfall between streets (F = 1.548, p-value = 0.119).

2.1.3. Transportation Infrastructure Elements Data

This study considers the streets as the survey unit, and measures the impact of the built environment on the vitality of residents' transportation through five variables [6,7,9,13]: parking density, subway station density, road intersection density, transportation facility diversity, and spatial location (Table 1). We use streets as spatial units to measure indicators such as traffic stops, road intersections and AQI. We take the average AQI within the street as the street's AQI value. In addition, we calculate the average number of road intersections per unit area within the street. In short, the three indicators of transportation infrastructure, air quality, and traffic vitality are all statistically calculated on the street as the spatial unit. The population data were obtained from the database of the seventh national census in 2020. The data on subway/bus/parking stations and road intersections were obtained from the facility points of interest and road network database in Baidu electronic map. The calculation of the transportation facility diversity index was mainly based on the land use classification data of Nanjing in 2020. The location distance specifically refers to the straight-line distance from the city's geometric center to the administrative center. Xi refers to the ratio of the land use area of category i in a street to the land area of the street. This indicator is used to measure the degree of transportation facility mixing, which indirectly reflects the intensity of transportation facility and the maturity of block development. The type of transportation facility directly affects the types of residents' traffic activities. The higher the degree of transportation facility mixing, the richer the types of residents' traffic activities, and the higher the density of residents' activities, the stronger willingness of residents to travel.

2.2. Research Methods

2.2.1. Spatial Panel Regression Model

The study involved 60 days of survey panel data on 95 streets. To better reveal the restraining effect of air pollution on the traffic vitality, we selected the fixed effects model for controlling the unobservable time and individual effects [42]. The basic panel regression model is expressed as follows:

$$Y_{it} = \gamma_0 + \gamma_1 X_{it}^a + \gamma_2 X_{it}^w + \gamma_3 X_t^d + \eta_t + u_i + \varepsilon_{it} \quad (1)$$

where the dependent variable Y_{it} is the activity intensity of street i on day t, reflecting the traffic vitality. X_{it}^a refers to the AQI of street i on day t, and encompasses wind speed, temperature, and precipitation. X_{it}^w refers to the variable group of the meteorological conditions of the i street on the t day, including meteorological data such as wind speed, temperature, and precipitation. X_t^d refers to a dummy variable that distinguishes between working days and non-working days. Studies show a significant difference between the residential activity spaces of working and non-working days (including rest days and holidays) [18,22]. Therefore, this study controlled for the working and non-working day factors, and then analyzed the impact of air pollution on the traffic vitality. η_t is a time-fixed-effect variable. u_i is the street fixed effect variable. This variable is a dummy variable for streets that controls for unobservable differences in streets that do not change over time. ε_{it} is the error term. γ_0 is the variable coefficient adjustment variable, which represents the influence weight of each variable.

Furthermore, we explore the impact of air pollution on the traffic vitality and the heterogeneity of different transportation infrastructure environments, and based on the panel regression model (Equation (1)), interaction terms between built environment variables and air pollution variables were introduced and expressed as follows:

$$Y_{it} = \gamma_0 + \gamma_1 X_{it}^a + \gamma_2 X_{it}^w + \gamma_3 X_t^d + \gamma_4 X_i^b X_{it}^a + \eta_t + u_i + \varepsilon_{it} \quad (2)$$

where X_i^b refers to the streets' built environment variable, including population density, subway station density, road intersection density, land use mix, and location variables. The traditional panel data regression model does not consider the spatial autocorrelation of the elements, which may cause errors in the regression coefficients.

2.2.2. Spatial Lag Model (SLM) and the Spatial Error Model (SEM)

Studies show an obvious spatial dependence in the traffic vitality space, and there may be a significant correlation between the intensity of residents' activities in adjacent spatial units [3]. Moran's I test identified a spatial autocorrelation ($p < 0.05$) of street traffic vitality. Therefore, we used GeoDa software to obtain the spatial matrix that measures the neighborhood relationship between streets and introduced the spatial lag model (SLM) and the spatial error model (SEM) to measure the impact of air pollution on the traffic vitality and the heterogeneity in the built environment. SLM and SEM are two spatial econometric regression models commonly used to calculate spatial correlations [43,44]. SLM is expressed as follows:

$$Y_{it} = vKY_{it} + \gamma X + \varepsilon \quad (3)$$

where K is an exogenous spatial weight matrix (150 × 150) reflecting the neighboring relationships among spatial units. KY_{it} refers to a lagging dependent variable that reflects spatial autocorrelation. v is a parameter of spatial dependence, and its absolute value represents the strength of spatial relevance. γ is the parameter vector. X is the independent variable measured by the model. ε refers to the error term that satisfies the spatial autocorrelation. In contrast with SLM which measures the spatial correlation through the spatial lag of the dependent variable, SEM measures the spatial correlation in the error term, and is expressed as follows:

$$Y_{it} = \gamma X + \beta W\varepsilon + u \quad (4)$$

where u is the error term, and its distribution conforms to a normal distribution with a mean value of 0 and a fixed variance. β is the spatial dependency parameter for filtering the spatial relevance of the error term.

2.2.3. Standard Deviational Ellipse Model

The standard deviational ellipse (SDE) model was used to describe the directionality and range characteristics of the spatial distribution of urban geographic elements [24]. It has been widely used in activity space analysis [26]. In this study, the SDE tool in ArcGIS software was used to visualize the characteristics and variations of residents' traffic vitality space under different air quality levels, and then quantitatively analyze the impact of air pollution factors on the traffic travel space vitality.

3. Results

3.1. *Spatial Change Characteristics of Traffic Vitality under Air Pollution*

This paper measures the trajectory position coordinates of residents' traffic travel through mobile phone signaling data. Based on the mobile phone signaling data, this paper uses the standard deviation ellipse model for spatial fitting, and then obtains the core area and distribution direction of residents' traffic vitality. Population data comes from the seventh national census, and the statistical unit is the street. We verified and corrected the distribution of the permanent population in the census through mobile phone signaling data. Generally speaking, 10:00 p.m. to 6:00 a.m. is the time for residents to rest and sleep, which can reflect the actual spatial location of residents. Based on the census data, we use the distribution of mobile phone signaling data during sleep as a correction index to finely measure the spatial distribution of the daily urban population. Therefore, the area and direction of the urban population distribution ellipse will change slightly on different survey days. To visualize the spatial changes in traffic vitality under the air pollution environment, this study separately calculates the residents' average traffic vitality under the air quality of each street in four levels: excellent ($AQI \leq 50$), good ($50 < AQI \leq 100$), light pollution ($100 < AQI \leq 150$), and moderate pollution ($150 < AQI \leq 200$). The vitality distribution of traffic space is shown in Figure 2. Moreover, based on the data from the 7th census of Nanjing City and the land use survey data in 2019, this study presents a map of the spatial distribution characteristics of the population and facilities. It further reflects the relationship between the traffic vitality (red circle) and the distribution of population (green circle) and facilities (blue circle) (Figure 2 and Table 2).

Table 2. Standard deviation ellipse features of urban vibrancy under different air quality levels (i.e., $AQI \leq 50$, $50 < AQI \leq 100$, $100 < AQI \leq 150$, and $150 < AQI \leq 200$) and its comparison with the SDE of population and facilities.

	Traffic Pattern	Longitude of Center Point (°)	Latitude of Center Point (°)	Long Axis Radius (km)	Short Axis Radius (km)	Direction (°)	Area (km^2)
Traffic vitality ellipse (excellent)	Public traffic	118.37	31.17	35.29	19.59	30.73	548.97
	Self-driving	118.23	30.06	31.29	18.32	28.65	482.32
Traffic vitality ellipse (good)	Public traffic	118.36	31.12	30.62	18.13	29.22	477.15
	Self-driving	118.13	28.33	28.53	17.42	27.51	433.56
Traffic vitality ellipse (light pollution)	Public traffic	118.22	30.42	21.75	17.86	38.82	392.25
	Self-driving	118.19	29.57	20.33	16.92	37.65	388.78
Traffic vitality ellipse (moderate pollution)	Public traffic	118.28	31.11	13.39	13.72	42.68	332.23
	Self-driving	118.26	32.19	12.76	13.65	40.72	297.84
Population ellipse	-	118.31	31.17	33.57	21.23	37.21	528.13
Facilities ellipse	-	118.32	31.19	27.50	21.17	32.29	526.82

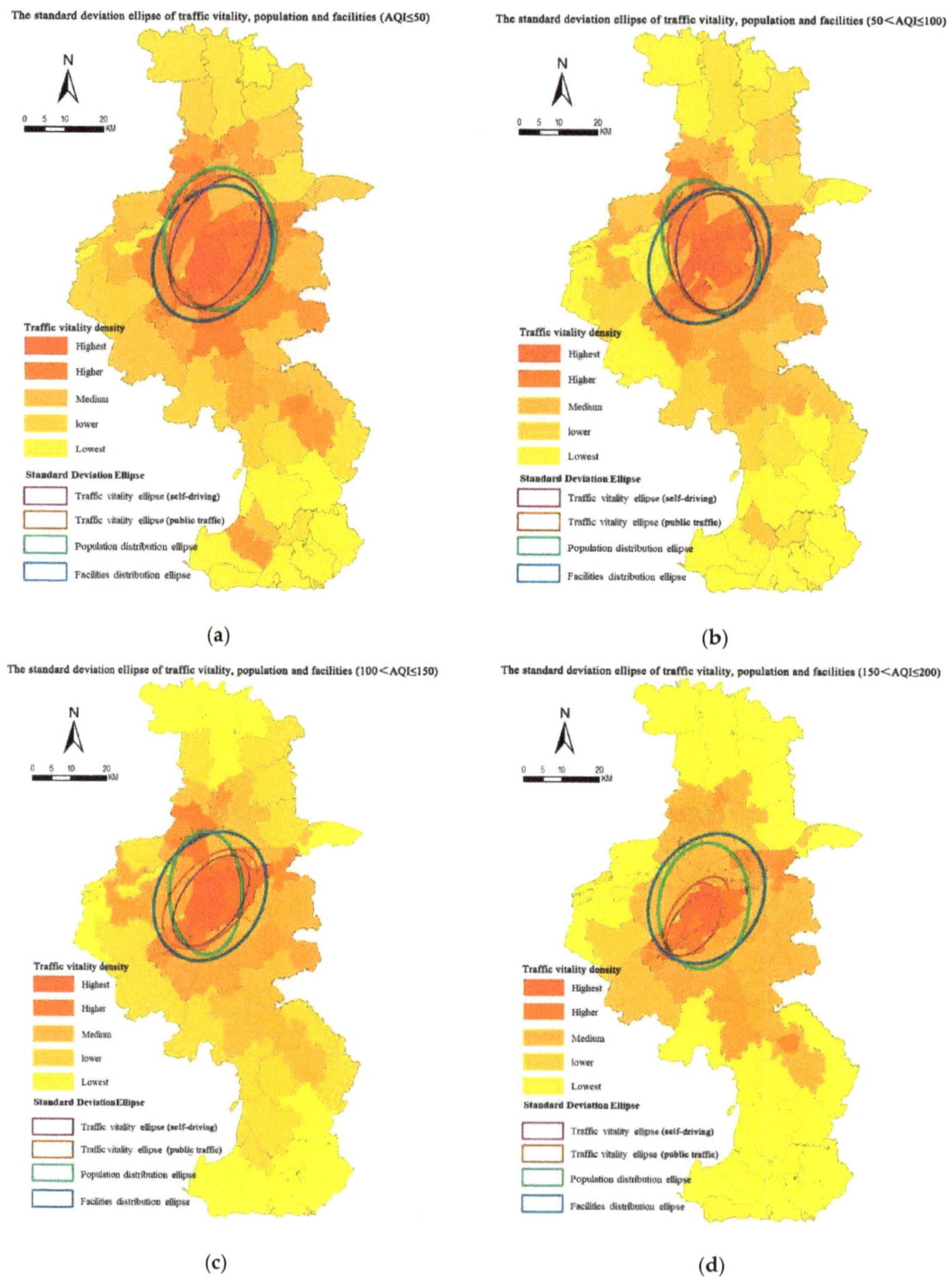

Figure 2. Standard deviation ellipse (SDE) features of traffic vitality under different air quality levels (i.e., (**a**) AQI ≤ 50, (**b**) 50 < AQI ≤ 100, (**c**) 100 < AQI ≤ 150, and (**d**) 150 < AQI ≤ 200) in Nanjing.

Under good air quality, Nanjing's traffic vitality is obviously consistent with the spatial direction of population and distribution of facilities. This confirms that mobile phone signaling check-in data can accurately reflect the spatial distribution of residents' commuting travel activities [19–22]. The traffic vitality space in Nanjing presents the spatial characteristics of "one core–multiple centers–multiple nodes", and areas with a high degree of travel activity are located along the Yangtze River. The core of residents' activities is located in the main urban area of Nanjing, and there are many hot spots for travel activities in the periphery of the urban area. The entire city presents a trend of "northeast–southwest" axial expansion in the spatial pattern of the traffic vitality. The "multi-node" refers to high-density area nodes with multiple traffic trips within the city.

The model results show that the influence of air pollution degree on residents' willingness to travel is not a simple linear relationship, but a fluctuating relationship, which drops sharply upon reaching a certain level. When the air quality is moderately polluted, the latitude and longitude of the SDE center point and the long and short axes change significantly. The direction of traffic vitality SDE changed from "Southwest-Northeast" to "East-West". This indicates that as the risk of air pollution exposure increased, although the intensity of traffic vitality in the urban core area declined, it remained the main choice for residents' travel activities. Compared with the urban core area, the traffic vitality of residents in the north-south areas of Nanjing's periphery has declined significantly.

In addition, the model results show that air pollution significantly compresses the size of the active area for residents' traffic and travel, and changes the distribution direction of the active area. When air quality is excellent, the SDE area of traffic vitality between the population and the distribution of facilities, indicating that residents are more willing to go to the outer areas of the city to carry out activities. As the risk of exposure to air pollution increases, when air quality is good, the SDE area for traffic vitality decreases to 71.3% of the excellent air quality. When air quality is light pollution, the SDE area for traffic vitality decreases to 47.8% of the excellent air quality. As the air quality transformed into moderate pollution, the area of the SDE for traffic vitality is significantly reduced to 34.2% of the area when the air quality is excellent. The SDE area of traffic vitality is significantly lower than the SDE area of population distribution, showing significant spatial shrinkage.

3.2. Inhibitory Effect of Air Pollution on Traffic Vitality

Table 3 shows the results of the panel regression model of the impact of air pollution on traffic vitality. Model 1 is a basic panel regression model. We used Model 1 as the base control model, Model 2 represented SLM panel regression models, and model 3 represented SEM panel regression models.

(1) Air pollution significantly inhibits the traffic vitality. After controlling the weather conditions, whether it is working day or not, and the spatial correlation conditions, AQI shows a significant negative correlation with traffic vitality. The weakening of traffic vitality greatly reduces the city's ability to attract and gather people, materials, and resources. Based on models 4 and model 5, each increase of 10 unit in the AQI value reduces the public traffic vitality intensity by approximately 0.46 times/10 km^2. Therefore, as the air quality decreases by one level (AQI increases by 50 units), the activity intensity decreases by 2.3 times/10 km^2. Based on models 6, each increase of 10 unit in the AQI value reduces the public traffic vitality intensity by approximately 0.52 times/10 km^2. Therefore, as the air quality decreases by one level (AQI increases by 50 units), the activity intensity decreases by 2.6 times/10 km^2. The daily average of 19.7 times/10 km^2 of mobile phone signaling sign-in intensity in Nanjing 2019. A change in the air quality in Nanjing from good to light pollution causes public traffic vitality to drop by 11.67% and self-driving traffic vitality to drop by 13.19%. Therefore, approximately one-tenth of traffic travel activities may be inhibited by air pollution. However, a shift on the air quality in Nanjing from good to moderate pollution (AQI increase by 100 units) inhibits nearly 23.35% of traffic travel.

Table 3. Results of the basic panel regression and spatial panel regression models.

Variable	Model 1		Model 2		Model 3		Model 4 (Bus)		Model 5 (Subway)		Model 6 (Self-Driving)	
	Coef.	**S.E.**	**Coef.**	**S.E.**	**Coef.**	**S.E.**	**Coef.**	**S.E.**	**Coef.**	**S.E.**	**Coef.**	**S.E.**
Date (Non-working day = ref.)	0.761 *	0.453	0.792 *	0.383	0.862 *	0.408	0.738 *	0.373	0.738 *	0.398	0.743 *	0.477
AQI	−0.152 ***	0.024	−0.087 ***	0.024	−0.097 ***	0.024	−0.271 ***	0.063	−0.248 ***	0.063	−0.233 ***	0.063
Wind speed	−1.063 ***	0.142	−0.093 ***	0.142	−1.103 ***	0.152	−0.498 ***	0.142	−0.472 ***	0.142	−0.518 ***	0.15
Precipitation (Non-torrential rain = ref.)	−2.017 ***	0.351	−2.018 ***	0.35	−2.112 ***	0.375	−2.075 ***	0.35	−1.931 ***	0.35	−2.032 ***	0.371
Temperature	0.937 ***	0.126	0.664 ***	0.126	0.746 ***	0.135	0.672 ***	0.126	0.663 ***	0.126	0.724 ***	0.134
(Temperature)^2	−0.023 ***	0.019	−0.017 ***	0.019	−0.009 ***	0.019	−0.013 ***	0.019	−0.013 **	0.019	−0.010 **	0.019
AQI* Parking density	-	-	-	-	-	-	-	-	-	-	−0.001 **	0.000
AQI* Public transportation frequency							0.325 **	0.043	0.215 ***	0.037	-	-
AQI* Station accessibility	-	-	-	-	-	-	0.267 **	0.053	0.229 ***	0.053	-	-
AQI* Transportation facility diversity	-	-	-	-	-	-	−0.003 ***	0.000	−0.004 ***	0.000	−0.003 ***	0.001
AQI* Metro/Bus station density	-	-	-	-	-	-	0.024 **	0.007	0.025 **	0.006	-	-
AQI* Road intersection density	-	-	-	-	-	-	0.002 **	0.000	0.001 *	0.000	0.001 *	0.000
ρ	-	-	0.082 ***	0.008	0.075 ***	0.008	-	-	0.052 ***	0.007	0.069 ***	0.007
Location fixed effect							Control					
Street fixation effect							Control					
Season fixed effects							Control					
N							49,831					
					Model fit							
Adjusted R^2	0.076		0.076		0.082		0.274		0.218		0.235	
Log-likelihood	−276,912.2		−261,782.9		−252,838.2		−273,091.1		−292,071.3		−266,721.5	
AIC	438,912.2		462,295.9		462,276.5		463,212.2		473,118.6		456,115.1	

Note: * $p < 0.05$; ** $p < 0.01$; *** $p < 0.001$.

(2) At the same time, the model results also show that meteorological conditions have a significant impact on traffic vitality. Nanjing has a subtropical monsoon climate. The mild and humid climatic conditions are conducive to the transportation of residents, and eliminate the inapplicability of the model results caused by the drastic changes in outdoor temperature in different cities [44]. The regression model results show that there is a negative correlation between meteorological conditions (temperature, humidity, etc.) and traffic vitality (public traffic pattern, self-driving traffic pattern, etc.). The increase in temperature and wind speed, and heavy rain lead to a significant reduction in residents' activities, thereby reducing the traffic vitality. In addition, as the temperature and wind speed rises, the decreasing trend of traffic vitality in public traffic pattern gradually decreased, while the trend of decreasing traffic vitality in self-driving traffic pattern did not change significantly.

(3) Under different traffic patterns, the traffic vitality intensity between adjacent streets also shows spatially correlated. In general, the fit of the SLM and SEM panel regression models shows relatively accurate. As shown in Models 1–3, compared with the results of traditional regression models, the significance and influence trends of variables in the SLM and SEM models are more stable. In the SEM and SLM panel regression models, the spatial dependence parameter (ρ) value was positively significant, thereby indicating a significant positive correlation between the adjacent streets' traffic vitality. The model result also indirectly indicates that there is spatial heterogeneity in the inhibitory effect of air pollution on traffic vitality. In different streets, the differences in transportation infrastructure conditions will also indirectly affect the traffic vitality under different traffic patterns.

3.3. Impact of Air Pollution on Traffic Activity Is Heterogeneous in Different Transportation Infrastructure Environments

Models 4–6 further introduce interaction terms between transportation infrastructure environments and air pollution variables. Owing to the accuracy of model fitting, models 4–6 exhibit greater improvement than models 1–3, and the influence coefficients of the 6 interaction variables are all significant. Model 1, Model 2, and Model 3 analyze the correlation between air quality-related variables and traffic vitality, respectively. Among them, Model 2 focuses on the influence analysis of "rain or not" factor on traffic vitality. Model 3 focuses on the influence of "whether it is a working day" factor on traffic vitality. Model 4, Model 5 and Model 6 respectively measure the interaction impact between AQI index and transportation infrastructure environment factors on traffic vitality. Transportation infrastructure environment factors include parking density, public transportation frequency, station accessibility, road intersection density, transportation facility diversity, and subway station density. The model specifically analyzes whether there is a significant correlation between the interaction terms of different indicators and traffic vitality. Model 4–6 respectively measure the cross-influence effects of transportation infrastructure and air pollution level on urban traffic vitality under different traffic patterns of bus, subway and self-driving, etc. It proves that the inhibitory effect of air pollution on traffic vitality shows significant heterogeneity in different transportation infrastructure environments.

(1) The higher the public transportation station density and public transportation frequency of the street, the more obvious the suppression effect of air pollution. Generally speaking, the public transportation station agglomeration is conducive to promoting the accessibility of station and connection of urban elements, thereby enhancing the traffic vitality [36]. The results show that densely distributed public stations areas are more sensitive to air pollution, thus strengthening the inhibitory effect of air pollution on traffic vitality. The research results of the models are consistent with the past studies stating that air pollution has a significant inhibitory effect on traffic vitality [39]. Air pollution has a strong inhibitory effect on the vitality of traffic in densely distributed public stations areas. This inhibitory effect adversely affects the sustainable development of urban society and economy.

(2) The higher the parking density, station accessibility, road intersections density, and transportation facility diversity, the lower the suppression effect of air pollution. Compared

with public traffic such as buses and subways, the impact of air pollution on self-driving traffic pattern is relatively low. The higher the parking density in the street, the more the opportunities for residents to drive by themselves to reach their destination and carry out related activities. To a certain extent, the inhibitory effect of air pollution on traffic vitality is alleviated. When the density of road intersections is high, the road system in the street presents the spatial morphological characteristics of wide lanes, network interweaving, and small scale. Studies have shown that a road system with small scale, small road width, and network intertwined can help increase diversified traffic travel [26,31,37]. In addition, the higher the degree of transportation facility diversity, the more abundant the types of transportation provided. In addition, the diversification of activity types is considered to be an important aspect of improving the traffic vitality [32,35]. This study shows that the diversification of transportation facility types in built-up areas has alleviated the inhibitory effect of air pollution on traffic vitality.

(3) The inhibitory effect of air pollution on traffic vitality presents significant space heterogeneity. There is spatial heterogeneity in the inhibitory effect of air pollution on traffic vitality, which will weaken the enthusiasm of residents to use urban infrastructure and hinder the optimization and adjustment of urban functional spatial structure. If this air pollution situation continues to increase and does not improve for a long time, the disadvantage of the suburb's lack of attractiveness to the population will be further magnified. The model results are consistent with the distribution characteristics of traffic vitality space in the central urban streets and suburban streets under different air quality levels in the previous section. Under the increasingly serious of air pollution, the decline in traffic vitality in the suburbs of Nanjing is significantly higher than that of the central urban area. This will lead to the lack of suburban traffic vitality, which will further affect the layout strategy of suburban infrastructure. Therefore, urban management departments should pay more attention to the supervision of air quality in suburban areas, so as to alleviate the inhibitory effect of air pollution on traffic vitality as much as possible.

4. Discussion

This study integrated mobile phone signaling data, weather, and air pollution among other multi-source data, and conducted panel statistical analysis at the street level. The research results have enriched the empirical research on the micro-spatial scale of traffic vitality in the city interior area, and will elucidate the relationship among air pollution, the built environment, and traffic vitality. In terms of the research content, this study measured the inhibitory effect of air pollution on traffic vitality and verified the heterogeneity impact under different transportation infrastructure environments. Considering China's current urban development stage of rapid urbanization, industrialization, and motorization, the urban management departments should strengthen pollution emission management to effectively reduce regional air pollution. In contrast, urban construction departments should strengthen the construction of urban rail transit systems with high coverage, high carrying capacity, and high operating frequency, optimize the layout of land use, fortify the construction of supporting facilities, and alleviate the inhibitory effect of air pollution on residents' transportation.

This study is limited in the following three aspects, which should be addressed in future. First, collecting the residents' activity data from multiple dimensions, combining the residents' daily life trajectories with the built environment attributes, and refining the relationship between different traffic modes and air pollution are necessary [23–25]. Second, in future research, we will try to integrate different types of data to enrich the measurement models and methods of traffic vitality. Third, people with different socioeconomic backgrounds have different views on transportation choices and air pollution tolerance. We need to study the differences in the inhibitory effects of air pollution on different population types to better address environmental equity issues [32,33]. Presently, certain theoretical studies have shown that air pollution has differentiated inhibitory effects and behavioral feedback on different types of people, but empirical research is required to verify the mechanism and quantitative feedback.

5. Conclusions

This study mainly obtains the location data of mobile phone signaling in Nanjing 2019, using the street as the space unit to calculate the traffic vitality in each street. Combining the air quality index, daily climate state, traffic commuting pattern and the characteristics of the street transportation infrastructure environment, we focus on exploring the inhibitory effect of air pollution on traffic vitality and the heterogeneous impact of different transportation infrastructure environments. The study found three main characteristics:

(1) There are obvious differences among traffic vitality under different air quality levels. Specifically, in different transportation models and different location conditions, there are obvious differences in traffic vitality. The traffic vitality space in Nanjing presents the spatial characteristics of "one core-multiple centers-multiple nodes", and areas with a high degree of travel activity are located along the Yangtze River. The core of res-idents' activities is located in the main urban area of Nanjing, and there are many hot spots for travel activities in the periphery of the urban area. The entire city presents a trend of "northeast-southwest" axial expansion in the spatial pattern of the traffic vitality. The "Multi-node" refers to high-density area nodes with multiple traffic trips within the city. This indicates that as the risk of air pollution exposure increased, although the in-tensity of traffic vitality in the urban core area declined, it remained the main choice for residents' travel activities. Compared with the urban core area, the traffic vitality of residents in the north-south areas of Nanjing's periphery has declined significantly.

(2) The inhibitory effect of air pollution on public traffic vitality and self-driving traffic vitality are differences. Specifically, each increase of 10 unit in the AQI value reduces the public traffic vitality intensity by approximately 0.46 times/10 km^2. Therefore, as the air quality decreases by one level (AQI increases by 50 units), the activity intensity decreases by 2.3 times/10 km^2. In addition, each increase of 10 unit in the AQI value reduces the public traffic vitality intensity by approximately 0.52 times/10 km^2. Therefore, as the air quality decreases by one level (AQI increases by 50 units), the activity intensity decreases by 2.6 times/10 km^2. Therefore, approximately one-tenth of traffic activities may be inhibited by air pollution. The weakening of traffic vitality greatly reduces the city's ability to attract and gather people, materials, and resources.

(3) The inhibitory effect of air pollution on traffic vitality is heterogeneous under different transportation infrastructure environments. The higher the public transportation station density and public transportation frequency of the street, the more obvious the suppression effect of air pollution. The higher the parking density, station accessibility, road intersections density, and transportation facility diversity, the lower the suppression effect of air pollution. Com-pared with public traffic such as buses and subways, the impact of air pollution on self-driving traffic pattern is relatively low. There is spatial heterogeneity in the inhibitory effect of air pollution on traffic vitality, which will weaken the enthusiasm of residents to use urban infrastructure and hinder the optimization and adjustment of urban functional spatial structure.

This paper introduces the Spatial Lag Model and Spatial Error Model to further investigate the impact of air pollution on traffic vitality and its heterogeneity in the built environment after controlling for spatial dependencies. The spatial matrix model of street adjacent relationship was constructed with the help of ArcGIS software. Specifically, the SEM model and the SLM model integrate the spatial element information, and integrate the spatial dimension effects such as the agglomeration and diffusion of elements in the space into the traditional regression model. This model is beneficial to analyze the interaction effect of spatial elements in the influence mechanism. However, the model used in this paper still needs to be further optimized, and it is necessary to study the inhibitory effect of air pollution on different populations and different types of activities. The optimized model can better respond to research on international environmental justice issues. People with different socioeconomic attributes, genders, and types of activities have different tolerances for air pollution. Under different circumstances, there are still large differences in the traffic vitality of residents, which needs to be further explored by empirical research.

The results of this study help to clarify the relationship between air pollution, the built environment, and the vitality of urban transportation. In general, air pollution significantly affects the vitality of urban traffic. Under different built environment conditions, the effect of air pollution on urban traffic vitality also showed significant differences. Therefore, the government needs to work with environmental protection departments, natural resource management departments, transportation management departments, and other departments to work together to improve the urban environment. On the one hand, environmental protection departments should strengthen the prevention and control of air pollution, including reducing pollutant emissions, and strictly controlling new production capacity in high-energy-consuming and high-polluting industries. Relevant departments need to vigorously promote clean production, speed up the adjustment of the energy structure, and increase the supply of clean energy such as natural gas and coal-to-methane. At the same time, the environmental protection department needs to strengthen the supervision and management of industrial projects, strengthen the constraints of energy conservation and environmental protection indicators, and shall not approve the construction of projects that have not passed the environmental assessment. On the other hand, the urban construction department should appropriately increase the scale of commercial land in the built-up area and increase the density of the road network and subway stations. In the urban area, the proportion of industrial land will be reduced, and the mixed degree of land use will be improved, thereby enhancing the traffic vitality of residents in the built-up area.

Author Contributions: Conceptualization, H.W. (Hongbin Wang) and D.L.; Data curation, H.W. (Hao Wu) and Y.C.; Formal analysis, S.Y.; Methodology, H.W. (Hao Wu); Visualization, D.L., Y.C. and S.Y.; Writing—original draft, H.W. (Hao Wu); Writing—review and editing, H.W. (Hongbin Wang). All authors have read and agreed to the published version of the manuscript.

Funding: This research was funded by the Basic Research Fund of CAMS (Grant No. 2022Y023), the 333 Project of Jiangsu Province (BRA2018420), the Jiangsu Innovative and Enterpreneurial Talent Programme (JSSCBS20221645), and the Beijing foundation of NJIAS (BJG202209).

Institutional Review Board Statement: Not applicable.

Informed Consent Statement: Not applicable.

Data Availability Statement: Not applicable.

Conflicts of Interest: The authors declare no conflict of interest.

References

1. Wu, J.; Ta, N.; Song, Y.; Lin, J.; Chai, Y. Urban form breeds neighborhood vibrancy: A case study using a GPS-based activity survey in suburban Beijing. *Cities* **2018**, *74*, 100–108. [CrossRef]
2. Mehta, V. Lively streets: Determining environmental characteristics to support social behavior. *J. Plan. Educ. Res.* **2007**, *27*, 165–187. [CrossRef]
3. Wu, C.; Ye, X.; Ren, F.; Du, Q. Check-in behavior and spatio-temporal vibrancy: An exploratory analysis in Shenzhen, China. *Cities* **2018**, *77*, 104–116. [CrossRef]
4. Glaeser, E. Cities, productivity, and quality of life. *Science* **2011**, *333*, 592–594. [CrossRef]
5. Lynch, K. *The Image of the City*; The MIT Press: Boston, MA, USA, 1960.
6. Jacbos, J. *The Death and Life of Great American Cities*; Random House: New York, NY, USA, 1961.
7. Gehl, J. *Life between Buildings: Using Public Space*; Island Press: Washington, DC, USA, 2011.
8. Chai, Y.; Ta, N.; Ma, J. The socio-spatial dimension of behavior analysis: Frontiers and progress in Chinese behavior geography. *J. Geogr. Sci.* **2016**, *26*, 1243–1260. [CrossRef]
9. Bo, W.; Becky, L.; Feng, Z. Urban geography research in the e-society: A perspective from human activity. *Geogr. Res.* **2018**, *37*, 2075–2086.
10. Shen, Y.; Chai, Y.; Guo, W. Day-to-day variability in activity-travel behavior based on GPS data: A case study in suburbs of Beijing. *Geogr. Res.* **2013**, *32*, 701–710.

11. Yuan, Y.; Raubal, M.; Liu, Y. Correlating mobile phone usage and travel behaviour—A case study of Harbin, China. *Comput. Environ. Urban Syst.* **2012**, *36*, 118–130. [CrossRef]
12. Long, Y.; Huang, C. Does block size matter? The impact of urban design on economic vitality for Chinese cities. *Environ. Plan. B Urban Anal. City Sci.* **2019**, *46*, 406–422. [CrossRef]
13. Luo, S.; Zhen, F. How to measure public space vitality based on mobile phone data: An empirical analysis of Nanjing's parks. *Geogr. Res.* **2019**, *38*, 1594–1608.
14. Zhong, W.; Wang, D.; Xie, D.; Yan, L. Dynamic characteristics of Shanghai's population distribution using cell phone signaling data. *Geogr. Res.* **2017**, *36*, 972–984.
15. Niu, X.; Wu, W.; Li, M. Influence of built environment on street vitality and its spatiotemporal characteristics based on LBS positioning data. *Urban Plan. Int.* **2019**, *34*, 28–37. [CrossRef]
16. Zhou, S.; Hao, X.; Liu, L. Validation of spatial decay law caused by urban commercial center's mutual attraction in polycentric city: Spatio-temporal data mining of floating cars' GPS data in Shenzhen. *Acta Geogr. Sin.* **2015**, *69*, 1810–1820.
17. Wang, B.; Zhen, F.; Wei, Z.; Guo, S.; Chen, T. A theoretical framework and methodology for urban activity spatial structure in e-society: Empirical evidence for Nanjing city, China. *Chin. Geogr. Sci.* **2015**, *25*, 672–683. [CrossRef]
18. Zhen, F.; Cao, Y.; Qin, X.; Wang, B. Delineation of an urban agglomeration boundary based on Sina Weibo microblog 'check-in'data: A case study of the Yangtze River Delta. *Cities* **2017**, *60*, 180–191. [CrossRef]
19. Malleson, N.; Birkin, M. Analysis of crime patters through the integration of an agent-based model and a population microsimulation. *Comput. Environ. Urban Syst.* **2012**, *36*, 551–561. [CrossRef]
20. Wang, B.; Zhen, F.; Zhang, H. The dynamic changes of urban space-time activity and activity zoning based on check-in data in Sina Web. *Sci. Geogr. Sin.* **2015**, *35*, 151–160.
21. Liu, Y.; Liu, X.; Gao, S.; Gong, L.; Kang, C.; Zhi, Y.; Chi, G.; Shi, L. Social sensing: A new approach to understanding our socioeconomic environments. *Ann. Assoc. Am. Geogr.* **2015**, *105*, 512–530. [CrossRef]
22. Zhen, F.; Wang, B. Rethinking human geography in the age of big data. *Geogr. Res.* **2015**, *34*, 803–811.
23. Na, T.; Zeng, Y.; Zhu, Q.; Wu, J. Relationship between built environment and urban vitality in Shanghai downtown area based on big data. *Sci. Geogr. Sin.* **2020**, *40*, 60–68.
24. Qin, X.; Zhen, F.; Zhu, S.; Xi, G. Spatial Pattern of Catering Industry in Nanjing Urban Area Based on the Degree of Public Praise from Internet: A Case Study of Dianping.com. *Sci. Geogr. Sin.* **2014**, *34*, 810–817.
25. Jin, X.; Long, Y.; Sun, W.; Lu, Y.; Yang, X.; Tang, J. Evaluating cities' vitality and identifying ghost cities with emerging geographical data. *Cities* **2017**, *63*, 98–109. [CrossRef]
26. Zhen, F.; Wang, B.; Qin, X. *Urban Studies and Innovation in Urban Planning Method Based on Big Data*; China Architecture & Building Press: Beijing, China, 2015.
27. Yue, Y.; Zhuang, Y.; Yeh, A.G.O.; Xie, J.Y.; Ma, C.L.; Li, Q.Q. Measurements of POI-based mixed use and their relationships with neighborhood vibrancy. *Int. J. Geogr. Inf. Sci.* **2016**, *31*, 658–675. [CrossRef]
28. Zheng, S.; Kahn, M.E. Understanding China's urban pollution dynamics. *J. Econ. Lit.* **2013**, *51*, 731–772. [CrossRef]
29. Wang, Z.; Liang, L.; Wang, X. Spatio-temporal evolution patterns and influencing factors of PM2.5 in Chinese urban agglomerations. *Acta Geogr. Sin.* **2019**, *74*, 2614–2630.
30. Bohemen, H.; Laak, W. The Influence of Road Infrastructure and Traffic on Soil, Water, and Air Quality. *Environ. Manag.* **2003**, *31*, 0050–0068. [CrossRef]
31. Ma, J.; Chai, Y.; Fu, T. Progress of research on the health impact of people's space-time behavior and environmental pollution exposure. *Prog. Geogr.* **2017**, *36*, 1260–1269.
32. Ma, J.; Tao, Y.; Kwan, M.P.; Chai, Y. Assessing mobility-based real-time air pollution exposure in space and time using smart sensors and GPS trajectories in Beijing. *Ann. Am. Assoc. Geogr.* **2020**, *110*, 434–448. [CrossRef]
33. Cao, X.; Min, J.; Huang, X. The influence of rainfall and air pollution on public bicycle usage: A case study of Xi'an. *Hum. Geogr.* **2019**, *34*, 151–158.
34. Invernizzi, G.; Ruprecht, A.; Mazza, R.; De Marco, C.; Močnik, G.; Sioutas, C.; Westerdahl, D. Measurement of black carbon concentration as an indicator of air quality benefits of traffic restriction policies within the ecopass zone in Milan, Italy. *Atmos. Environ.* **2011**, *45*, 3522–3527. [CrossRef]
35. Zheng, S.; Wang, J.; Sun, C.; Zhang, X.; Kahn, M.E. Air pollution lowers Chinese urbanites' expressed happiness on social media. *Nat. Hum. Behav.* **2019**, *3*, 237–243. [CrossRef] [PubMed]
36. Sun, C.; Zheng, S.; Wang, J.; Kahn, M.E. Does clean air increase the demand for the consumer city? Evidence from Beijing. *J. Reg. Sci.* **2019**, *59*, 409–434. [CrossRef]
37. Brauer, M.; Hoek, G.; Van Vliet, P.; Meliefste, K.; Fischer, P.H.; Wijga, A.; Koopman, L.P.; Neijens, H.J.; Gerritsen, J.; Kerkhof, M.; et al. Air Pollution from Traffic and the Development of Respiratory Infections and Asthmatic and Allergic Symptoms in Children. *Am. J. Respir. Crit. Care Med.* **2002**, *166*, 1092–1098. [CrossRef] [PubMed]
38. Zhou, L.; Zhou, C.; Yang, F.; Che, L.; Wang, B.; Sun, D. Spatio-temporal evolution and the influencing factors of PM2.5 in China between 2000 and 2011. *Acta Geogr. Sin.* **2017**, *72*, 2079–2092.
39. Chen, Y.; Ebenstein, A.; Greenstone, M.; He, G.; Zhou, M. Evidence on the impact of sustained exposure to air pollution on life expectancy from China's Huai River policy. *Proc. Natl. Acad. Sci. USA* **2013**, *110*, 12936–12941. [CrossRef]

40. Bevc, C.A.; Marshall, B.K.; Picou, J.S. Environmental justice and toxic exposure: Toward a spatial model of physical health and psychological well-being. *Soc. Sci. Res.* **2007**, *36*, 48–67. [CrossRef]
41. Yan, L.; Duarte, F.; Wang, D.; Zheng, S.; Ratti, C. Exploring the effect of air pollution on social activity in China using geotagged social media check-in data. *Cities* **2019**, *91*, 116–125. [CrossRef]
42. Yi, J.; Wang, N.; Qian, J.; Ma, T.; Du, Y.Y.; Pei, T.; Zhou, C.; Tu, W.; Liu, Z.; Wang, H.; et al. Spatio-temporal responses of urban road traffic and human activities in an extreme rainfall event using big data. *Acta Geogr. Sin.* **2020**, *75*, 497–508.
43. Shuting, J.; Bo, L.; Yongchun, Y.; Peiji, S.; Kunbo, S.; Fuwen, D. The distribution of cities in China and its influencing factors. *Geogr. Res.* **2015**, *34*, 1352–1366.
44. Yuanyuan, M.; Jiajun, D. Study on spatial-temporal patterns of robbery and snatch: A case study of Pudong new area, Shanghai. *Hum. Geogr.* **2014**, *29*, 49–54.
45. Griliches, Z.; Hausman, J.A. Errors in variables in panel data. *J. Econom.* **1986**, *31*, 93–118. [CrossRef]
46. Anselin, L. *Spatial Econometrics: Methods and Models*; Kluwer Academic Publisher: Boston, MA, USA, 2013.
47. Yang, L.; Wang, B.; Zhou, J.; Wang, X. Walking accessibility and property prices. *Transp. Res. Part D* **2018**, *62*, 551–562. [CrossRef]
48. Nanjing Ecological Environment Bureau. Bulletin on Environmental Conditions in Nanjing in 2019. 2020. Available online: http://hbj.nanjing.gov.cn/njshjbhj/202006/P020211101345930912841.pdf (accessed on 1 May 2020).
49. Nanjing Municipal Bureau of Transportation. Nanjing Transportation Industry Development Statistical Bulletin in 2016. 2017. Available online: https://www.renrendoc.com/paper/210006449.html (accessed on 1 June 2017).
50. Wu, H.; Wang, T.; Riemer, N.; Chen, P.; Li, M.; Li, S. Urban heat island impacted by fine particles in Nanjing, China. *Sci. Rep.* **2017**, *7*, 11422. [CrossRef] [PubMed]
51. Wu, H.; Wang, T.; Wang, Q.G.; Riemer, N.; Cao, Y.; Liu, C.; Ma, C.; Xie, X. Relieved air pollution enhanced urban heat island intensity in the Yangtze River Delta, China. *Aerosol Air Qual. Res.* **2019**, *19*, 2683–2696. [CrossRef]
52. Wu, H.; Wang, T.; Wang, Q.; Cao, Y.; Qu, Y.; Nie, D. Radiative effects and chemical compositions of fine particles modulating urban heat island in Nanjing, China. *Atmos. Environ.* **2021**, *247*, 118201. [CrossRef]
53. Tong, Y.; Yin, Y.; Qian, L.; An, J. Analysis of the characteristics of hazy phenomena in Nanjing area. *China Environ. Sci.* **2007**, *27*, 584–588.
54. Ma, W.L.; Qi, H.; Li, Y.F.; Liu, L.Y.; Sun, D.Z.; Wang, D.G.; Zhang, Z.; Tian, C.G.; Shen, J.M. Seasonal and spatial variations of air concentrations of polycyclic aromatic hydrocarbons in northeastern Chinese urban region. *Bull. Environ. Contam. Toxicol.* **2011**, *86*, 43–49. [CrossRef]
55. Eiguren-Fernandez, A.; Miguel, A.H.; Froines, J.R.; Thurairatnam, S.; Avol, E.L. Seasonal and spatial variation of polycyclic aromatic hydrocarbons in vapor-phase and PM2.5 in Southern California urban and rural communities. *Aerosol Sci. Technol.* **2004**, *38*, 447–455. [CrossRef]

atmosphere

MDPI

Article

Attention-Based BiLSTM Model for Pavement Temperature Prediction of Asphalt Pavement in Winter

Shumin Bai [1,2], Wenchen Yang [2,3,*], Meng Zhang [4], Duanyang Liu [5,6], Wei Li [2,3] and Linyi Zhou [5,6]

1 School of Transportation Engineering, Kunming University of Science and Technology, Kunming 650500, China
2 National Engineering Laboratory for Surface Transportation Weather Impacts Prevention, Broadvision Engineering Consultants Co., Ltd., Kunming 650200, China
3 Yunnan Key Laboratory of Digital Communications, Kunming 650103, China
4 Yunnan Yunling Highway Science and Technology Co., Ltd., Kunming 650011, China
5 Key Laboratory of Transportation Meteorology, CMA, Nanjing 210008, China
6 Nanjing Joint Institute for Atmospheric Sciences, Nanjing 210008, China
* Correspondence: tongjiywc@163.com

Abstract: Pavement temperature is the main factor determining road icing, and accurate and timely pavement temperature prediction is of significant importance to regional traffic safety management and preventive maintenance. The prediction of pavement temperature at the micro-scale has been a challenge to be tackled. To solve this problem, a bidirectional extended short-term memory network model based on the attention mechanism (Att-BiLSTM) was proposed to improve the prediction performance by using the time series features of pavement temperature and meteorological factors. Pavement temperature data and climatic data were collected from a road weather station in Yunnan, China. The results show that the *MAE*, *MSE*, and *MAPE* of the proposed Att-BiLSTM model were 0.330, 0.339, and 10.1%, respectively, which were better than the other baseline models. It was shown that 93.4% of the predicted values had an error less than 1 °C, and 82.1% had an error less than 0.5 °C, indicating that the proposed Att-BiLSTM model enables significant performance improvement. In addition, this paper quantified and analyzed the effects of parameters such as the size of the sliding window, the number of hidden layer neurons, and the optimizer on the performance of the prediction model.

Keywords: transportation meteorology; pavement temperature prediction; deep learning; BiLSTM; attention mechanisms; winter icing

Citation: Bai, S.; Yang, W.; Zhang, M.; Liu, D.; Li, W.; Zhou, L. Attention-Based BiLSTM Model for Pavement Temperature Prediction of Asphalt Pavement in Winter. *Atmosphere* **2022**, *13*, 1524. https://doi.org/10.3390/atmos13091524

Academic Editors: Da-Lin Zhang and Begoña Artíñano

Received: 4 August 2022
Accepted: 15 September 2022
Published: 18 September 2022

1. Introduction

Road surface conditions have a significant impact on the safe operation of vehicles [1,2]. Especially in winter, rain and snow tend to cause the road surface to freeze, which can significantly reduce the friction coefficient of asphalt pavement and create poor road pavement driving conditions, which can cause serious traffic accidents. In winter, pavement temperature is a significant factor determining road icing, and the accurate prediction of pavement temperature can provide guidance for preventive and proactive pavement maintenance and improve service levels [3,4]. For example, real-time data from pavement condition monitoring systems can be used to predict future pavement temperatures and salt dangerous road sections before they are at risk of icing up, preventing the risk before it happens.

Pavement temperature prediction is a nonstationary time series prediction problem, and traditional methods usually only rely on a previous moment of observation for prediction, such as the Markov model and the autoregressive moving average model. These methods cannot consider the thermal inertia of a pavement, so the accuracy of the models is poor. With the rise in machine learning models, especially recurrent neural networks

(RNNs), RNNs have made a major breakthrough in time series forecasting because RNNs can find and model higher-order nonlinear relationships in time series. Although researchers have applied RNNs and other deep learning models to pavement temperature prediction, the meteorological factors affecting pavement temperature are less considered in the models, so the influence of meteorological elements on pavement temperature is not well modeled.

In this paper, a bi-directional long short-term memory (BiLSTM) model based on an attention mechanism was proposed that is practical and implementable. The BiLSTM can effectively solve the gradient disappearance and gradient explosion problems and is used to capture the forward and reverse information of the sequence more completely [5]. Attentional mechanisms were used to precisely identify the most important features. The proposed model has the ability to accurately predict pavement temperatures using historical pavement temperature data and can provide support for preventive maintenance.

2. Literature Review

Most of the existing studies solved the pavement temperature prediction problem in different ways, using both physical and statistical models.

Physical models are used to predict pavement temperatures by solving partial differential equations for heat transfer. For example, Sass developed a surface energy equation in 1992 to predict pavement temperatures over a 3 h period [6]. Voldborg developed a forecasting model that can generate short-term indicators such as air temperature, humidity, and road surface temperature for each of the more than 200 road weather stations in Denmark [7]. Meng developed a refined numerical model for the prediction of pavement parameters, taking into account the influence of pavement factors and basic urban properties, and the results showed that solar radiation correction factors, asphalt depth, and asphalt thermal conductivity are important parameters for the simulation of road interface temperatures [8]. Chen J et al. proposed an innovative time-varying function to predict pavement temperature in relation to solar radiation and air temperature [9]. However, the physical model is complex to model and requires a large number of difficult-to-collect parameters as input. At the same time, as Karsistoa's results show, errors can be significant when physical variations are complex [10].

In contrast to physical models, statistical models do not require analytical deviations and numerical calculations to estimate pavement temperatures, but rather statistical analysis based on historical data to obtain a reasonable predictive model. Statistical models are divided into linear and nonlinear models depending on the relationship between the influencing factors and the pavement temperature. For example, Park et al. developed a linear regression model for estimating the minimum surface temperature of a pavement based on the ambient air temperature [11]. Asefzadeh et al. developed separate models for predicting daily average pavement temperatures for different seasons and daily maximum and minimum pavement temperatures for different asphalt layer depths [12]. Kršmanc et al. adjusted the input parameters and different time intervals to predict pavement temperatures based on stepwise linear regression analysis [13]. Zapata et al. developed a medium-depth pavement temperature prediction model and conducted a sensitivity analysis on the influencing factors, and found that there is a nonlinear relationship between the influencing factors and the pavement temperature [14]. In contrast to the linear regression model, nonlinear regression models both typically involve more complex equations and better capture the nonlinear relationship between pavement temperature and the influencing factors, which makes it the classical model in this field.

With the rise in machine learning models, many promising methods have been widely used to model pavement temperatures. Yang et al. used K-Nearest Neighbors to explore the variation in pavement temperature on different road sections [15]. Molavi et al. evaluated the performance of different machine learning models for the prediction of asphalt pavement temperatures under average, minimum, and maximum daily temperatures [16]. Milad et al. proposed an asphalt pavement temperature prediction model through deep learning techniques and suggested that future researchers should integrate loss-balancing algorithms into multi-task learning to improve the efficiency of difficult tasks. Meanwhile, future studies of predicted pavement temperatures should consider the effects of factors such as air temperature, wind speed, and relative humidity [17]. Li et al. proposed that the prediction of pavement surface temperature should not be a single value, but a probability distribution. They developed a prediction model for evaluating the probability distribution of pavement surface temperature in winter [18].

3. Objective and General Outline

3.1. Objective

The present study aimed to proposes an attention-based BiLSTM model for the pavement temperature prediction of asphalt pavement in winter. The BiLSTM was used to completely capture the forward and reverse information of pavement temperature sequences and meteorological feature sequences, and the attention mechanism was used to accurately identify the most important features and improve feature utilization to further improve the performance of pavement temperature prediction. In addition, this study analyzed the effects of the size of the sliding window, the number of hidden layer neurons, the optimizer, and the training epoch on the prediction accuracy.

3.2. General Outline

Figure 1 provides the general outline of the research, which consisted of three main steps. (1) The first step was the collection and preprocessing of winter pavement temperature data and meteorological element data. In order to collect accurate data, several road weather stations were installed and checked regularly to ensure the stations worked well. To further improve the data quality, data preprocessing was conducted. (2) The second step determined the model inputs by Spearman correlation coefficients. The input feature matrix had a significant impact on the model, and important variable extraction was performed in order to capture the influence of meteorological factors on pavement temperature and pavement temperature time series characteristics. (3) The third step established the optimal pavement temperature prediction model by adjusting the model hyperparameters to predict the future pavement temperature. The hyperparameters controlled the performance of the model, and in order to obtain the optimal model, the optimal values of parameters such as the size of the sliding window, the number of hidden layer neurons, and the optimizer were obtained through the experiment. Finally, the established attention-based BiLSTM model was used for pavement temperature prediction to further support preventive maintenance.

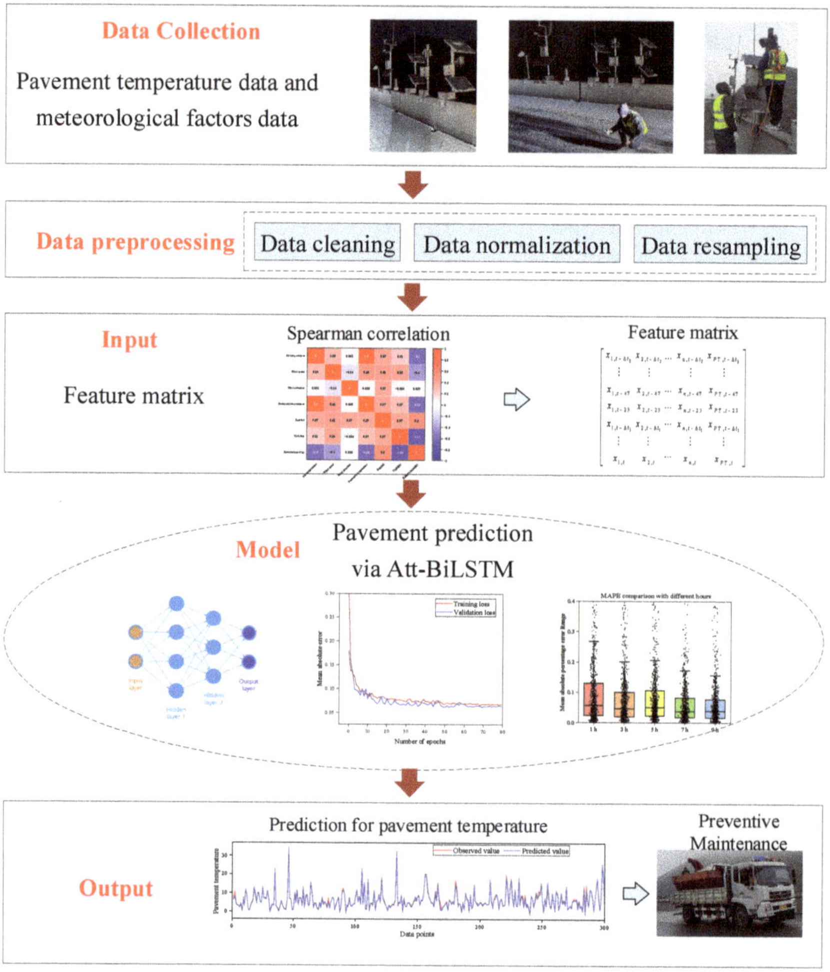

Figure 1. General outline of the research methodology.

4. Data

4.1. Data Description

Compared to northern China, southern China receives less snow. When the temperature is low, thin ice is easily created on the road. Thin ice, being smooth and transparent, prevents drivers from observing it and slowing down in advance. Most drivers in Yunnan lack experience in driving in ice and snow, and when they find that the vehicle is out of control, they cannot handle it rationally. As a result, thin ice causes casualties in Yunnan every year.

The observing station is on the Niujiagou bridge on the Maliuwan–Zhaotong line (103°76′13″, 27°74′04″) in the Wumeng Mountains. The raw data are real-time data collected by VAISALA automatic road weather stations every minute, including the main meteorological factors such as pavement temperature, air temperature, humidity, wind speed, and rainfall. An example of the raw data is shown in Table 1.

Table 1. The raw data format of VAISALA road weather station.

Time	Visibility/m	Temperature/°C	Related Humidity/%	Rainfall/mm	Wind Speed/m/s	Wind Direction/°	Pavement Temperature °C
1 December 2019 0:00	20,000	3.7	92	0	2.3	54	3.7
1 December 2019 0:01	20,000	3.7	93	0	2	32	3.7
…	…	…	…	…	…	…	…

The total data collection period included two time periods from November 2019 to March 2020 and November 2020 to March 2021. Figure 2 shows the time sequence distribution diagrams. After eliminating duplicate, missing, and abnormal data, the data were resampled at the interval of one hour. Thereby, 4344 samples remained for modeling. Figure 2 shows the time sequence distribution diagrams for November 2019 to March 2020. The descriptive statistics of the climatic data and pavement temperature data are presented in Table 2. The results of the analysis show that the average winter air temperature is about 5.4 degrees Celsius, and the lowest pavement temperature is about −4.1 degrees Celsius. At the same time, the average relative humidity is 83.7% due to the high vegetation cover. Figure 2 shows that the air and pavement temperatures are specifically cyclical.

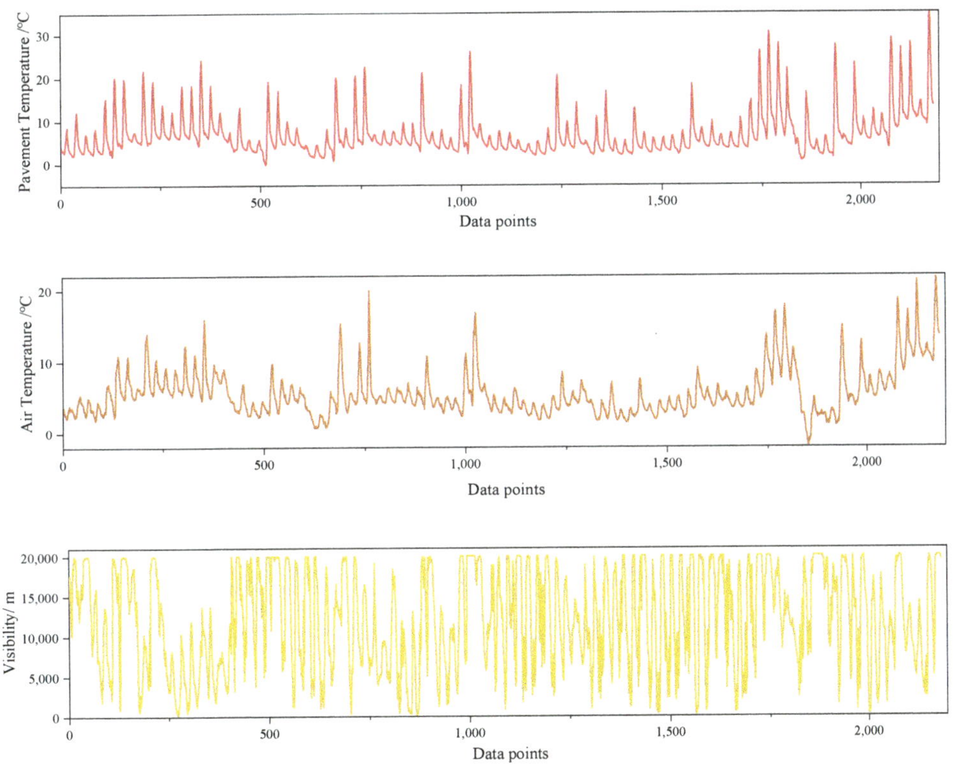

Figure 2. *Cont.*

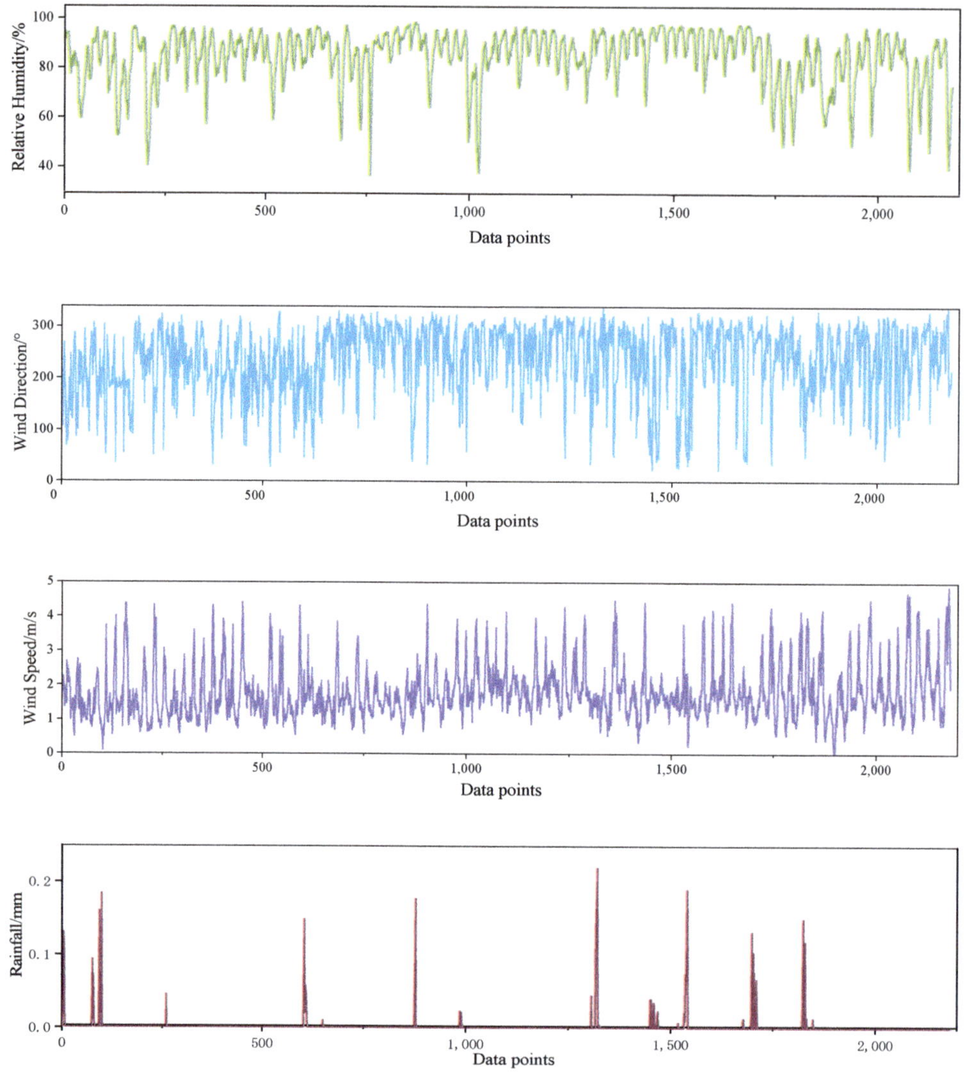

Figure 2. Distribution of each measured variable. Pavement temperature (**first**), air temperature (**second**), visibility (**third**), relative humidity (**fourth**), wind direction (**fifth**), wind speed (**sixth**) and rainfall (**last**).

Table 2. Descriptive statistics of the climatic data and pavement temperature data.

Variable	Mean	St. Dev	Max	Min
Visibility	11,110.0	5994.4	20,000.0	112.8
Air temperature	5.4	3.7	22.3	−2.1
Humidity	83.7	11.9	98.9	34.6
Rainfall	0.0	0.0	0.2	0.0
Wind speed	1.8	0.8	5.0	0.0
Wind direction	218.2	75.2	351.3	10.1
Pavement temperature	6.5	5.3	35.9	−4.1

4.2. Data Preprocessing

There are many uncontrollable factors in the road weather station data collection process, especially unexpected factors such as vehicle movement and equipment failure, which can lead to missing values and noisy data in the raw data. Therefore, the data processing flow, as shown in Figure 3, was designed, which will be discussed later.

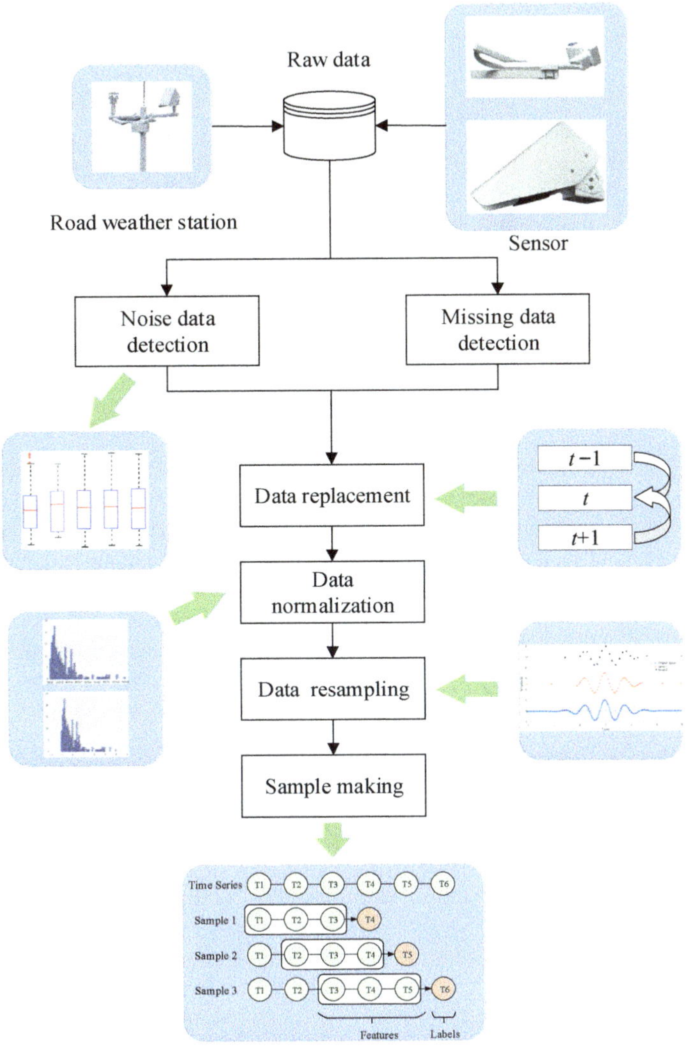

Figure 3. The data processing flow.

4.2.1. Data Cleaning and Replacement

The three-sigma guidelines were used to identify noisy data, which were considered noisy if the absolute value of the difference between the value and the mean was greater than three times its standard deviation.

$$|X_i - \overline{X}| > 3\sigma_X \quad (1)$$

where X_i is the observed value of the feature; $\overline{X}$ is the mean value of the feature; σ_X is the standard deviation of the feature.

In this way, noisy data could be detected, and missing values could be found directly from the data. After completing the noisy data and missing data detection, we removed them and filled in the proper data. Due to the high frequency of data collection, we used the average value for filling. The calculation formula is as shown in Equation (2).

$$X_i' = \frac{X_{i-1} + X_{i+1}}{2} \tag{2}$$

where X_i' is the value calculated by the averaging method.

4.2.2. Data Normalization and Resampling

In order to improve the training speed of the model and reduce the impact of different magnitudes between different features on the complexity, the z-score normalization was chosen to linearly transform the original data. The calculation formula is as shown in Equation (3).

$$\hat{X}_i = \frac{X_i - \overline{X}}{\sigma_X} \tag{3}$$

where $\hat{X}_i$ is the normalized data.

It was considered that the model could not provide a reference for the prevention of pavement icing if the prediction time interval was too small. Therefore, the minute-Scale data set was resampled at 1 h intervals to form a new data set.

4.2.3. Generating Samples Making

In this paper, the prediction of pavement temperature was considered as a time series problem, which means that the model used the sliding window approach to construct supervised learning samples. As shown in Figure 4, green represents ordinary time series data, time series framed by black lines are used as features, and orange time series represent labels. The process of the sliding window approach is as follows:

(1) Suppose the number of the sliding window is set as Δt, which means the model uses the data in the previous times $[t - \Delta t, t)$ to predict the pavement temperature in time t.
(2) If the number of time series data is d, a total of $d - \Delta t$ samples can be constructed.

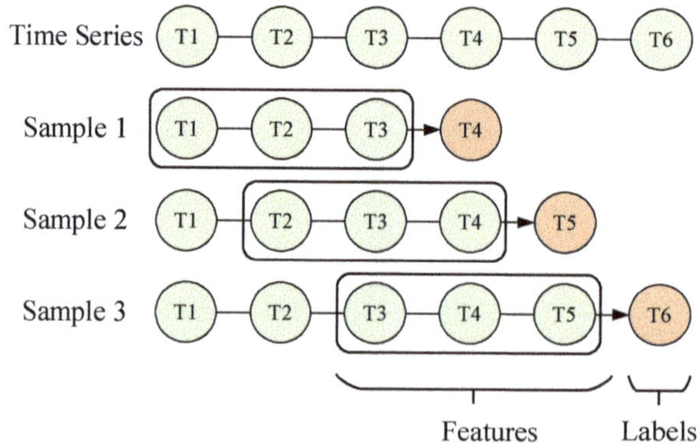

Figure 4. Sliding window approach.

5. Attention-Based BiLSTM Modeling

In this paper, an attention-based BiLSTM deep learning model for pavement temperature prediction was proposed. The model consists of five parts: an input layer, BiLSTM layer, attention layer, fully connected layer, and output layer. Figure 5 illustrates the overall architecture of the Att-BiLSTM model. The BiLSTM layer is capable of extracting features from the front and back directions of the pavement temperature time series data. After that, the important features are further extracted using the attention layer to form a new feature vector. The attention mechanism was introduced mainly to optimize the LSTM structure to compensate for its lack of ability to give different levels of attention to features over multiple time steps. Finally, the attention layer is followed by the fully connected (FC) layers, which are regression layers used to make predictions. Each module will be described in detail in the following subsections.

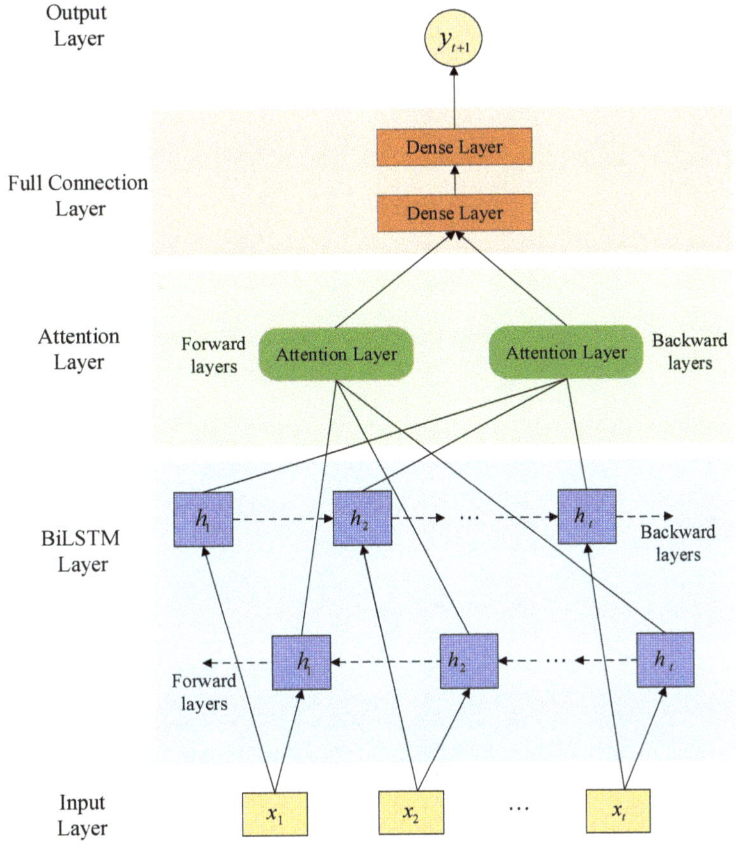

Figure 5. Framework of the proposed model.

5.1. BiLSTM

Pavement temperatures are affected by the cumulative effect of pavement temperatures at multiple historical moments. When extracting temporal features, the influence of pavement temperature at multiple historical moments should be considered. Recurrent neural networks (RNNs) are a classical architecture for time sequence data prediction, proposed by Hopfield [19]. The advantage of RNNs is the use of output as feedback in RNNs compared to traditional artificial neural networks, which makes RNNs more effective in learning time-dependence [20]. However, when handling problems with long-term dependencies, RNNs may fail to converge. In order to solve this problem, Hochreiter and Schmidhuber proposed a long-short-term memory (LSTM) neural network, which introduces memory cells to deal with long-term dependencies [21]. The LSTM neural network is shown in Figure 6.

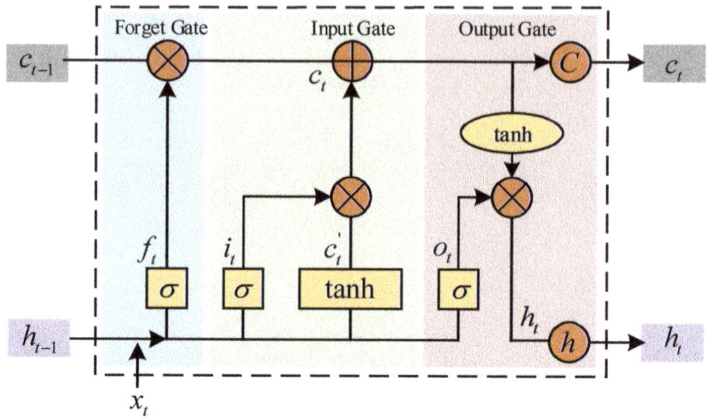

Figure 6. Long short-term memory neural network.

With the help of the LSTM neural network, the temporal characteristics of the actual values at the predicted target moment can be extracted from the actual pavement temperature sequence at the target window moment and only mapped to the actual pavement temperature at the target prediction moment, enabling prediction on a time series scale. The LSTM network is calculated as follows:

$$f_t = \sigma(W_f \cdot [h_{t-1}, x_t] + b_f) \tag{4}$$

$$i_t = \sigma(W_i \cdot [h_{t-1}, x_t] + b_i) \tag{5}$$

$$c_t' = \tanh(W_C \cdot [h_{t-1}, x_t] + b_C) \tag{6}$$

$$c_t = f_t \cdot c_{t-1} + i_t \cdot c_t' \tag{7}$$

$$o_t = \sigma(W_o \cdot [h_{t-1}, x_t] + b_o) \tag{8}$$

$$h_t = o_t \cdot \tanh(c_t) \tag{9}$$

where t is the moment; x_t is the current moment input; f_t is forget gate; i_t is input gate; c_t' is a temporary cell state; c_t is a cell state; o_t is output gate; h_t is the output of the hidden layer at the current moment; c_{t-1} is the state of the cell at the previous moment; h_{t-1} is the output of the hidden layer at the previous moment; W_f, W_i, W_C, W_o are the weights to be learned, respectively; b_f, b_i, b_C, b_o are the offsets to be learned; σ is the *sigmoid* activation function.

Although LSTM overcomes the limitations of RNNs, it can still only process sequence information from the past and cannot utilize front sequence information. Huang et al. [22] proposed a bidirectional LSTM (BiLSTM) including forward and backward LSTM layers, as shown in Figure 7. A BiLSTM is able to integrate and process information from both the front and rear, capturing road temperature and associated parameter time information more effectively.

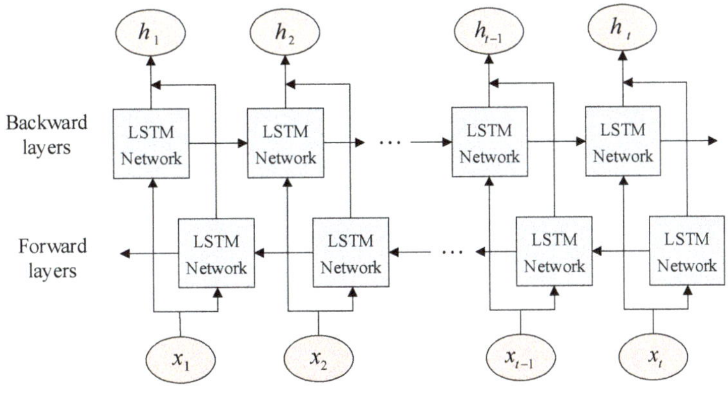

Figure 7. Bidirectional long short-term memory neural network.

5.2. Attention Mechanism

The attention mechanism is a distribution mechanism inspired by the human brain. The human brain focuses on the area that needs to be focused, reducing or even not giving attention to other areas to obtain more important detailed information. In other words, the attention mechanism gives higher weights to relevant parts while minimizing irrelevant parts by giving them lower weights, thus improving the accuracy of the model. The attention mechanism structure is shown in Figure 8.

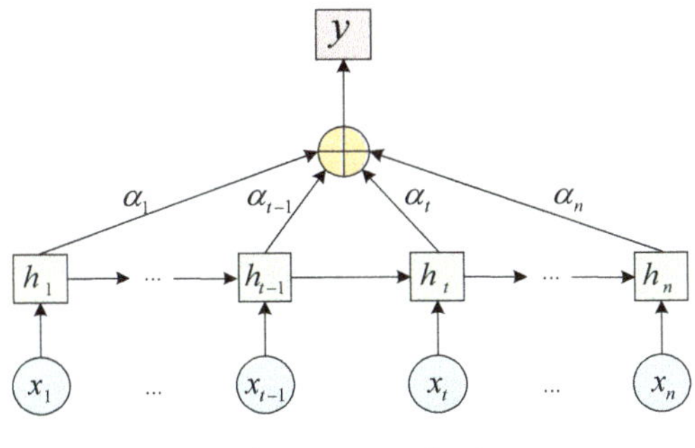

Figure 8. Structure of attention mechanism.

Here, $x_t(t \in [1, n])$ denotes the input to the BiLSTM network, $h_t(t \in [1, n])$ is the output of the hidden layer obtained by BiLSTM for each input, $\alpha_t(t \in [1, n])$ is the output of the attention mechanism for the BiLSTM hidden layer attention probability distribution, and y is the output value of the BiLSTM with the introduction of the attention mechanism.

5.3. Evaluation Metric

Several common performance metrics are used to evaluate the performance of the model: mean absolute error (*MAE*), mean square error (*MSE*), and mean absolute percentage error (*MAPE*), which are calculated using (10)–(12).

$$MAE = \frac{\sum_{i=1}^{N} |y_i - \hat{y}_i|}{N} \tag{10}$$

$$MSE = \frac{\sum_{n=1}^{N} (y_i - \hat{y}_i)^2}{N} \tag{11}$$

$$MAPE = \frac{100\%}{N} \sum_{i=1}^{N} \left| \frac{y_i - \hat{y}_i}{y_i} \right| \tag{12}$$

where y_i represents the observed pavement temperature and $\hat{y}_i$ represents the predicted pavement temperature.

6. Results and Discussion

6.1. Selection of Important Characteristic Variables

We aimed to reduce the complexity of the model input and improve the accuracy of the prediction model. The characteristic variables with significant correlation with the predicted target value of the pavement temperature were used as the input variables of the pavement temperature prediction model. Considering the possible nonlinear correlation between meteorological characteristics and pavement temperature, a Spearman correlation analysis was performed for the six meteorological characteristics variables as well as pavement temperature. The Spearman's rank correlation coefficient method is often used to analyze the closeness of a relationship between two variables and is calculated as:

$$r = \frac{\sum_{i=1}^{n}(f_i - \overline{f_i})(f_j - \overline{f_j})}{\sqrt{\sum_{i=1}^{n}(f_i - \overline{f_i})^2 \sum_{i=1}^{n}(f_j - \overline{f_j})^2}} \quad (13)$$

where $\overline{f_i}$ and $\overline{f_j}$ are the mean of the experimental values f_i and f_j, respectively, r is the Spearman rank correlation coefficient, and the closer $|r|$ is to 1, the higher the degree of linear correlation between f_i and f_j. The results are shown in Figure 9.

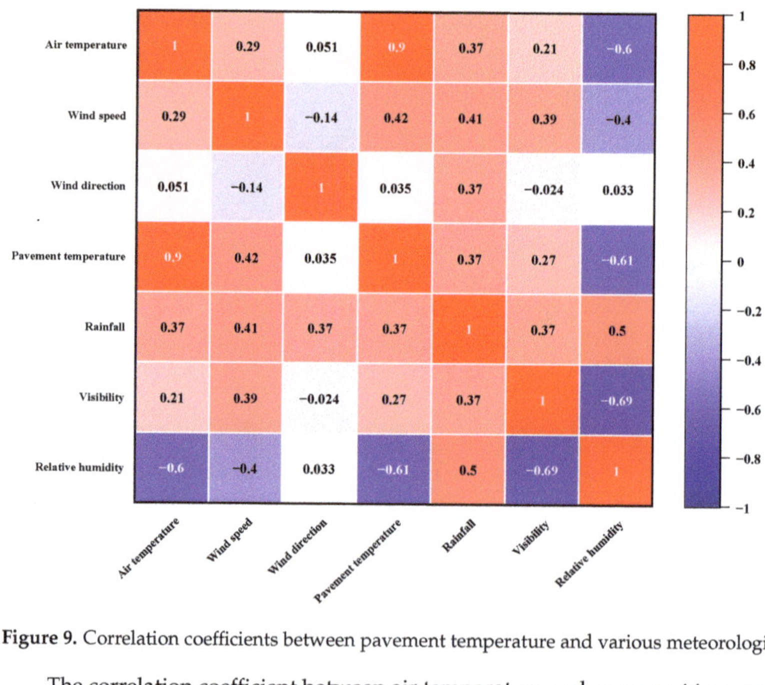

Figure 9. Correlation coefficients between pavement temperature and various meteorological factors.

The correlation coefficient between air temperature and pavement temperature was 0.9, indicating an extremely strong correlation between air temperature and road surface temperature. The correlation coefficients between wind speed, rainfall, visibility, and relative humidity were weakly correlated with pavement temperature, with correlation coefficients of 0.42, 0.37, 0.37, and −0.61, respectively. The correlation coefficient between wind direction and pavement temperature was 0.035, indicating a very weak correlation between wind direction and pavement temperature. Based on the Spearman correlation coefficient results, air temperature, wind speed, rainfall, relative humidity, and previous road surface temperature were selected as input features.

6.2. Optimal Parameters of the Att-BiLSTM Model

After building the model structure and determining the input features of the model, the next step was the training of the model. We divided the dataset into a ratio of 70% for the training set, 20% for the test set, and 10% for the validation set. The Keras application programming interface for TensorFlow was chosen to implement the model proposed in this paper. For the proposed model, there are several important parameters that have a significant impact on the prediction performance, including the size of the sliding window, the number of hidden layer neurons, the type of optimizer, and the number of training epochs. The grid search cross-validation method is used to find the optimal hyperparameters. The optimization process for these parameters is shown below.

6.2.1. The Size of Sliding Window

For time sequence data, the size of the sliding window is the most important parameter, as it directly affects the input features and the number of samples. Inputting sequences with high time correlation into the model can effectively improve the prediction accuracy of the model, while inputting sequences with low time correlation into the model will add irrelevant information. Considering the thermal inertia of the pavement temperature, the size of the sliding window was set to 1 h, 3 h, 5 h, 7 h, and 9 h, respectively. These values were tested, and the optimal value was selected by the evaluation metric.

The results of the calculations are shown in Table 3 and Figure 10, which indicate that when the size of the sliding window was 7 h, the prediction performance of the model proposed was the best.

Table 3. Performance of the models when the size of the sliding window is 1 h, 3 h, 5 h, 7 h, and 9 h.

Δt/h	*MAE*	*MSE*	*MAPE*
1 h	0.662	1.678	0.120
3 h	0.419	0.531	0.138
5 h	0.457	0.890	0.196
7 h	0.334	0.353	0.101
9 h	0.345	0.387	0.115

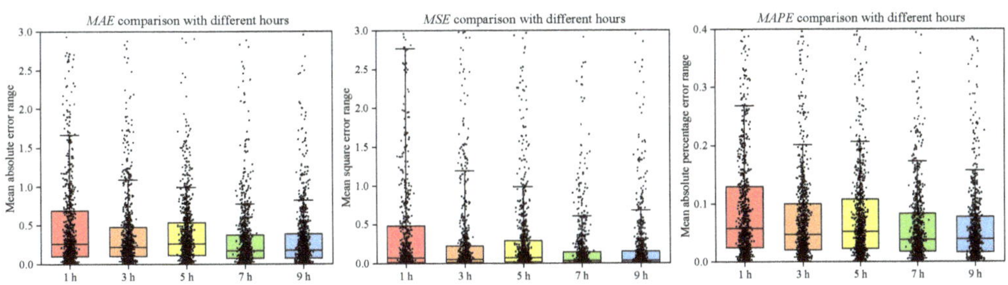

Figure 10. Errors comparison with different hours. *MAE* (**left**), *MSE* (**middle**), *MAPE* (**right**).

6.2.2. The Number of Hidden Layer Neurons

For neural networks, the number of hidden layer neurons also plays a significant role. Too few hidden layer neurons can lead to underfitting of the model and the inability to predict accurately, while too many can lead to overfitting and also increase the time complexity. The search space was set at 50 to 300.

As can be seen from Table 4, when the number of neurons was 150, the prediction performance of the model proposed was the best.

Table 4. Performance of the model with the different number of neurons.

The Number of Neurons	*MAE*	*MSE*	*MAPE*
50	0.391	0.389	0.239
100	0.368	0.392	0.178
150	0.334	0.353	0.101
200	0.393	0.430	0.164
250	0.398	0.434	0.189
300	0.430	0.454	0.239

6.2.3. The Optimizer

During the model training process, the model parameters were adjusted and changed to obtain the minimum loss function. The role of the optimizer is to guide the loss function to update in the right direction. In this paper, four commonly used optimizers are compared: Adaptive Moment Estimation (Adam), Stochastic gradient descent (SGD), Adaptive Gradients (Adagrad), and Root Mean Square Prop (RMSprop).

As can be seen from Table 5, when the optimizer was Adam, the prediction performance of the model proposed was the best.

Table 5. Performance of the models when the optimizer is Adam, SGD, Adagrad, or RMSprop.

Optimizer	*MAE*	*MSE*	*MAPE*
Adam	0.334	0.353	0.101
SGD	1.329	3.363	1.001
Adagrad	1.187	4.376	0.329
RMSprop	0.367	0.379	0.204

6.2.4. The Training Epochs

Figure 11 shows the prediction performance of the model in the training and validation sets, through which the performance of the model on the training and validation data can be evaluated to obtain the best epochs and prevent the model from overfitting or underfitting. It can find that in terms of training and validation data, the *MAE* gradually decreases as the epoch increases, which indicates that the model accuracy improves. The validation loss of the model is mostly lower than the training loss, and when the epoch is roughly 70, both the training loss and validation loss tend to be smooth, which suggests that the optimal epochs are around 70.

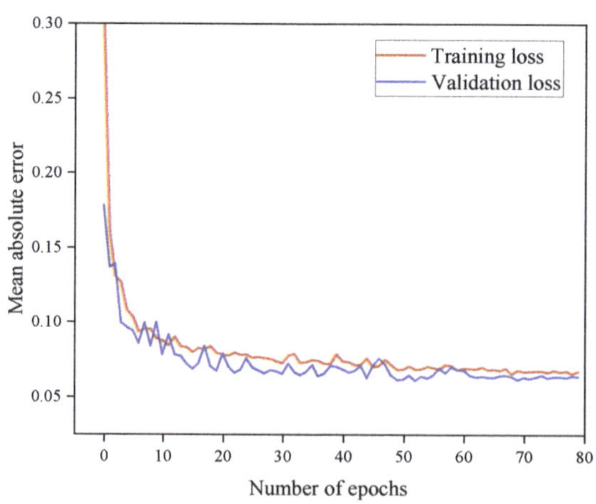

Figure 11. Performance of the model during training and validation error.

6.3. Performance Comparison

In this section, the predictive performance of the model proposed in the study is compared with that of other deep learning methods, including RNNs, GRU networks, LSTM networks and BiLSTM networks. The proposed Att-BiLSTM model and other baseline models were trained based on the same training data set. Table 6 shows the prediction performance comparison of the LSTM with other baseline models. The results show that RNNs have the largest prediction error among all the algorithms for all three metrics. This is because RNNs directly use the entire output as feedback and cannot forget and update the influence of meteorological factors, which leads to poor prediction. The GRU networks, as a variant of LSTM networks, have the ability to forget and update information. Compared with RNNs, the GRU networks achieved better prediction performance, where *MAE*, *MSE*, and *MAPE* were reduced by 3.2%, 9.7%, and 7.6% on average, respectively. However, the GRU network still falls short of the LSTM networks in terms of prediction performance due to its simplified cell states. Compared to the above two methods, the LSTM networks further improve the prediction performance, but the effective prediction of pavement temperature not only relies on past information but also considers the time sequence. The BiLSTM networks can integrate and process data from both front and back directions, which can solve the problem that LSTM only follows a one-way sequential order in information processing and can effectively capture the time sequence information of pavement temperature to achieve better prediction. The proposed model in this paper, by introducing the attention mechanism, adaptively calculates and adjusts the hidden layer state values corresponding to the original input features to highlight the important features and weaken the minor features to further explore the internal characteristics of the pavement temperature data. Therefore, the proposed model outperformed all models with the *MAE* of 0.334, *MSE* of 0.353, and *MAPE* of 10.1%. The comparison of pavement temperature truth and the predicted values of the proposed model on multiple days is visualized in Figure 12.

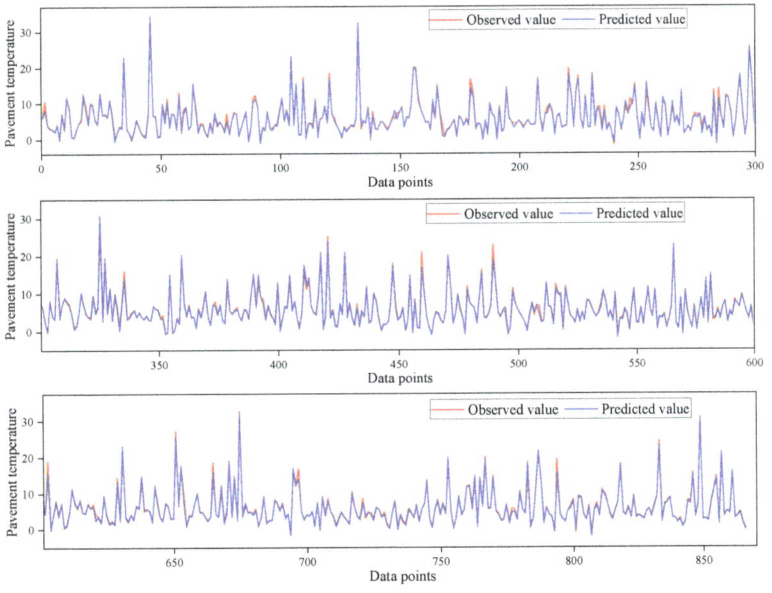

Figure 12. Comparison of pavement temperature truth and predicted values in the test set. Data points 1–300 in the test set (**first**), Data points 301–600 in the test set (**second**), Data points 601–866 in the test set (**third**).

Table 6. Predictive performance comparison of the proposed model with other models.

Algorithm	*MAE*	*MSE*	*MAPE*
RNN	0.460	0.544	0.184
GRU	0.445	0.491	0.170
LSTM	0.428	0.515	0.165
BiLSTM	0.386	0.437	0.140

6.4. Discussion

In this section, the prediction results and applications of the proposed model are analyzed and discussed.

Figure 13 shows the absolute errors of the predicted values of the proposed model and observed values. It can be seen that 93.4% of the absolute error is less than 1 °C, and 82.1% of the absolute error is less than 0.5 °C, which indicates that the model has a good predictive effect.

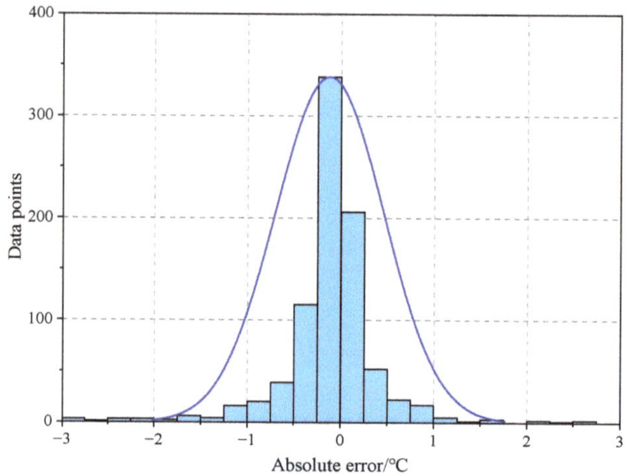

Figure 13. The error of the predicted and observed values.

From Figure 14, it can be found that the error values of the prediction model are more concentrated in low temperatures (−5 to 5 °C), which are prone to icing, and almost all of them are less than 1 °C. The prediction performance of the model in the high-temperature is weakened and is not as good as that in the low-temperature segment, with a more discrete distribution of error values. This may be due to the greater influence of meteorological elements such as solar radiation and total cloudiness on the road surface temperature in the high-temperature condition, which leads to fluctuations in the prediction errors. Overall, the Att-BiLSTM model has better performance in the low-temperature condition and has good prospects for engineering applications in winter low-temperature pavement temperature prediction.

The Att-BiLSTM pavement temperature prediction model proposed in this work can be combined with the pavement icing formation mechanism to determine future pavement icing and improve the accuracy and reliability of icing warning. Together with the facilities such as variable information boards or speed limit signs near the point (section), timely information on dangerous road conditions of bad driving conditions (or early warnings of road surface icing points) can be released to drivers, prompting them to control speed and drive carefully, thus reducing the occurrence of vicious traffic accidents such as vehicle skidding and rollover or rear-end collision.

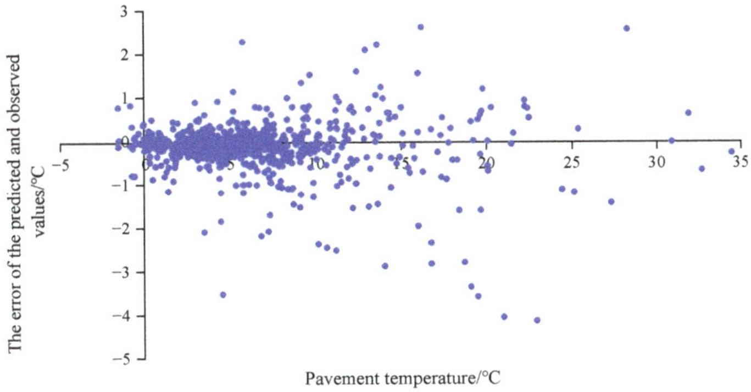

Figure 14. Distribution of errors.

7. Conclusions

The prediction of a pavement temperature at the microscopic scale has been a challenge to be solved. To address this problem, an Att-BiLSTM pavement temperature prediction model based on historical meteorological data and pavement temperature data was developed in this study. Pavement temperature data and meteorological data collected from road weather stations on route G85 from Maliuwan to Zhaotong in Yunnan, China, which covered 180 days. A feature vector was constructed to describe the influence of meteorological features on pavement temperature and the time series characteristics of pavement temperature by Spearman's correlation coefficient analysis. The Att-BiLSTM model predicted the future pavement temperature based on the feature vector. To demonstrate the validity of the model, RNNs, GRU, LSTM and BiLSTM networks were selected as benchmark models to compare their prediction performance with the prediction performance of the proposed model. The results show that the *MAE*, *MSE*, and *MAPE* of the proposed Att-BiLSTM model were 0.330, 0.339, and 10.1%, respectively, which were better than the other baseline models. It was shown that 93.4% of the predicted values had an error less than 1 °C, and 82.1% had an error less than 0.5 °C, indicating that the proposed model has a great prediction performance. The proposed prediction model has better performance at low temperatures (−5~5 °C). This shows that the method proposed in this paper has good prospects for engineering applications in low-temperature pavement temperature prediction in winter.

In future work, internal pavement or subgrade temperatures should be further considered to obtain better performance within the pavement temperature prediction model. In addition, the pavement temperature prediction model should be combined with the pavement temperature prediction model to further predict pavement conditions.

Author Contributions: Conceptualization, S.B. and W.Y.; methodology, S.B. and M.Z.; software, W.L. and W.Y.; data curation, D.L. and L.Z.; S.B. and M.Z. created the figures and table; and S.B. wrote the paper. All authors have read and agreed to the published version of the manuscript.

Funding: This research was funded by the General Program of Key Science and Technology in Transportation, the Ministry of Transport (2018-ms4-102 and zl-2018-04), the Science and Technology Demonstration Project of the Ministry of Transport (2017-09), the Science and Technology Innovation Program of the Department of Transportation, Yunnan Province, China (2019303 and 2021-90-2), Yunnan Fundamental Research Project (202101at070693), and Yunnan Key Laboratory of Digital Communications (grant NO. 202205AG070008).

Institutional Review Board Statement: Not applicable.

Informed Consent Statement: Not applicable.

Data Availability Statement: Some of the data, models, or code generated or used during the study are available from the corresponding author by request (code for ensemble model, data analysis method, and road weather station data).

Conflicts of Interest: The authors declare no conflict of interest.

References

1. Chen, S.; Saeed, T.U.; Labi, S. Impact of road-surface condition on rural highway safety: A multivariate random parameters negative binomial approach. *Anal. Methods Accid. Res.* **2017**, *16*, 75–89. [CrossRef]
2. Strong, C.K.; Ye, Z.; Shi, X. Safety effects of winter weather: The state of knowledge and remaining challenges. *Transp. Rev.* **2010**, *30*, 677–699. [CrossRef]
3. Chen, J.; Wang, H.; Xie, P. Pavement temperature prediction: Theoretical models and critical affecting factors. *Appl. Therm. Eng.* **2019**, *158*, 113755. [CrossRef]
4. Zapata, C.E.; Andrei, D.; Witczak, M.W.; Houston, W. Incorporation of environmental effects in pavement design. *Road Mater. Pavement Des.* **2007**, *8*, 667–693. [CrossRef]
5. Lei, D.; Liu, H.; Le, H.; Huang, J.; Yuan, J.; Li, L.; Wang, Y. Ionospheric TEC Prediction Base on Attentional BiGRU. *Atmosphere* **2022**, *13*, 1039. [CrossRef]
6. Sass, B.H. A numerical model for prediction of road temperature and ice. *J. Appl. Meteorol. Climatol.* **1992**, *31*, 1499–1506. [CrossRef]
7. Voldborg, H. On the prediction of road conditions by a combined road layer-atmospheric model in winter. *Transp. Res. Rec.* **1993**, *1387*, 231.
8. Meng, C. A numerical forecast model for road meteorology. *Meteorol. Atmos. Phys.* **2018**, *130*, 485–498. [CrossRef]
9. Chen, J.; Li, L.; Wang, H. Analytical prediction and field validation of transient temperature field in asphalt pavements. *J. Cent. South Univ.* **2015**, *22*, 4872–4881. [CrossRef]
10. Karsisto, V.; Nurmi, P. Using car observations in road weather forecasting. In Proceedings of the International Road Weather Conference, Fort Collins, CO, USA, 27–29 April 2016. Available online: https://www.researchgate.net/profile/Virve-Karsisto/publication/306099501_Using_car_observations_in_road_weather_forecasting/links/57b1524c08ae95f9d8f3bbb2/Using-car-observations-in-road-weather-forecasting.pdf (accessed on 1 August 2022).
11. Park, J.J.; Shin, E.C.; Yoon, B.J. Development of frost penetration depth prediction model using field temperature data of asphalt pavement. *Int. J. Offshore Polar Eng.* **2016**, *26*, 341–347. [CrossRef]
12. Asefzadeh, A.; Hashemian, L.; Bayat, A. Development of statistical temperature prediction models for a test road in Edmonton, Alberta, Canada. *Int. J. Pavement Res. Technol.* **2017**, *10*, 369–382. [CrossRef]
13. Kršmanc, R.; Slak, A.Š.; Demšar, J. Statistical approach for forecasting road surface temperature. *Meteorol. Appl.* **2013**, *20*, 439–446. [CrossRef]
14. Gedafa, D.S.; Hossain, M.; Romanoschi, S.A. Perpetual pavement temperature prediction model. *Road Mater. Pavement Des.* **2014**, *15*, 55–65. [CrossRef]
15. Yang, C.H.; Yun, D.G.; Kim, J.G.; Lee, G.; Kim, S.B. Machine learning approaches to estimate road surface temperature variation along road section in real-time for winter operation. *Int. J. Intell. Transp. Syst. Res.* **2020**, *18*, 343–355. [CrossRef]
16. Molavi Nojumi, M.; Huang, Y.; Hashemian, L.; Bayat, A. Application of machine learning for temperature prediction in a test road in Alberta. *Int. J. Pavement Res. Technol.* **2022**, *15*, 303–319. [CrossRef]
17. Milad, A.; Adwan, I.; Majeed, S.A.; Yusoff, N.I.M.; Al-Ansari, N.; Yaseen, Z.M. Emerging technologies of deep learning models development for pavement temperature prediction. *IEEE Access* **2021**, *9*, 23840–23849. [CrossRef]
18. Li, Y.; Chen, J.; Dan, H.; Wang, H. Probability prediction of pavement surface low temperature in winter based on bayesian structural time series and neural network. *Cold Reg. Sci. Technol.* **2022**, *194*, 103434. [CrossRef]
19. Hopfield, J.J. Neural networks and physical systems with emergent collective computational abilities. *Proc. Natl. Acad. Sci. USA* **1982**, *79*, 2554–2558. [CrossRef] [PubMed]
20. Wu, X.; Liu, Z.; Yin, L.; Zheng, W.; Song, L.; Tian, J.; Yang, B.; Liu, S. A haze prediction model in chengdu based on LSTM. *Atmosphere* **2021**, *12*, 1479. [CrossRef]
21. Graves, A. Long short-term memory. In *Supervised Sequence Labelling with Recurrent Neural Networks*; Springer: Berlin/Heidelberg, Germany, 2012; pp. 37–45. [CrossRef]
22. Huang, Z.; Xu, W.; Yu, K. Bidirectional LSTM-CRF models for sequence tagging. *arXiv* **2015**. [CrossRef]

www.ingramcontent.com/pod-product-compliance
Lightning Source LLC
Chambersburg PA
CBHW061808210926
51850CB00006B/643
* 9 7 8 3 0 3 6 5 8 4 6 1 4 *

MDPI
St. Alban-Anlage 66
4052 Basel
Switzerland
Tel. +41 61 683 77 34
Fax +41 61 302 89 18
www.mdpi.com

Atmosphere Editorial Office
E-mail: atmosphere@mdpi.com
www.mdpi.com/journal/atmosphere